Primate

Conservation

Biology

GUY COWLISHAW &
ROBIN DUNBAR

THE UNIVERSITY OF CHICAGO PRESS
CHICAGO AND LONDON

The University of Chicago Press, Chicago 60637
The University of Chicago Press, Ltd., London
© 2000 by The University of Chicago
All rights reserved. Published 2000
Printed in the United States of America
18 17 16 15 14 13 12 11 10 09   4 5 6 7 8
ISBN-13: 978-0-226-11636-5 (cloth)
ISBN-13: 978-0-226-11637-2 (paper)
ISBN-10: 0-226-11636-0 (cloth)
ISBN-10: 0-226-11637-9 (paper)

Library of Congress Cataloging-in-Publication Data

Cowlishaw, Guy.
    Primate conservation biology / Guy Cowlishaw & Robin Dunbar.
       p.   cm.
    Includes bibliographical references.
    ISBN 0-226-11636-0 (alk. paper)—ISBN 0-226-11637-9 (paper : alk. paper)
       1. Primates.   2. Wildlife conservation.   I. Dunbar, R. I. M. (Robin Ian MacDonald),
    1947–   II. Title.

    QL737.P9C69 2000
    333.95′9816—dc21

                                                                  00-023032

To our families

CONTENTS

Acknowledgments   xi

1 INTRODUCTION   1

2 DIVERSITY   8
2.1. The Primate Order   8
2.2. Patterns of Diversity   15
2.3. Origins of Diversity   22
2.4. Summary   27

3 BEHAVIORAL ECOLOGY   28
3.1. Life History   28
3.2. Ecology   33
3.3. Behavior   42
3.4. Summary   54

4 COMMUNITY ECOLOGY   56
4.1. Communty Species Richness   56
4.2. Community Structure   67
4.3. Competition in Communities   75
4.4. Primates in Plant Communities   84
4.5. Summary   91

5 DISTRIBUTION, ABUNDANCE, AND RARITY   93
5.1. Geographic Distribution   95
5.2. Population Abundance   105
5.3. Distribution-Abundance Relationships   114
5.4. Summary   117

6 POPULATION BIOLOGY   119
6.1. Demographic Variables   119
6.2. Population Dynamics   126
6.3. Metapopulation Dynamics   140
6.4. Population Genetics   143
6.5. Summary   156

7   EXTINCTION PROCESSES   158

7.1. Extinction Rates   158
7.2. Causes of Extinction   161
7.3. Species Differences in Extinction Risk   174
7.4. Case Studies in Primate Extinctions   181
7.5. Summary   189

8   HABITAT DISTURBANCE   191

8.1. Patterns of Habitat Disturbance   191
8.2. Effects of Habitat Loss   204
8.3. Effects of Habitat Fragmentation   208
8.4. Effects of Habitat Modification   215
8.5. Species Vulnerability Patterns   229
8.6. Summary   240

9   HUNTING   242

9.1. Optimal Foraging Theory   244
9.2. Hunting Patterns   247
9.3. Trade in Primates   257
9.4. Effects of Hunting   266
9.5. Species Vulnerability Patterns   280
9.6. Hunting with Habitat Disturbance   282
9.7. Summary   287

10   CONSERVATION STRATEGIES   289

10.1. Strategy Design Principles   289
10.2. Setting Taxon Priorities   291
10.3. Setting Area Priorities   313
10.4. Practical Considerations   321
10.5. Summary   328

11   CONSERVATION TACTICS   330

11.1. Protected Area Systems   331
11.2. Sustainable Utilization   343
11.3. Captive Breeding   359
11.4. Restocking and Reintroduction   365
11.5. Summary   377

12   CONCLUSIONS   380

12.1. The Past and Future of Primate Diversity   382
12.2. Diagnosing Populations in Trouble   388
12.3. Effective Conservation Action   393
12.4. Finding Unique Solutions   399

Appendix 1. Primate Species and Conservation Status   404
Appendix 2. Leslie Matrices   415
Appendix 3. Primate and Conservation Organizations   417
References   421
Index   479

ACKNOWLEDGMENTS

This book has evolved over several years and would not have been possible without the help and enthusiasm of a large number of people. We started the book following an invitation from Barrie Goldsmith and Bob Carling. Carel van Schaik and Liz Rogers gave us useful feedback on our initial proposal, and four anonymous referees made extremely detailed and helpful comments on the final manuscript. Jon Bridle, Mike Bruford, Emmanuel de Merode, Jan de Ruiter, Karen Strier, and Rob Wallace also read and commented on individual chapters or sections. Tom Butynski, Harriet Eeley, John Fa, Sandy Harcourt, Andrew Grieser Johns, Mike Lawes, Georgina Mace, John Oates, and Paul Williams supplied us with copies of unpublished manuscripts or reports. John Caldwell provided CITES trade data with the permission of the CITES Secretariat, and Dan Sellen furnished data from the World Ethnographic Sample (compiled by Pat Gray). Nicola Koyama and Susy Paisley helped us with the diagrams, and Jared Dunbar worked on the reference list. Andy Purvis, Bob Sussman, and the publishers acknowledged in the text gave us permission to use their figures. Rosalind Heywood drew the elegant chapter heads.

During the writing of the book, GC has been based at the Institute of Zoology (Zoological Society of London) and the Departments of Anthropology and Biology (University College London), and has been hosted by Phoebe Barnard and Rob Simmons in Namibia. The Economic and Social Research Council funded him during part of this period. RD has been based at the School of Biological Sciences at the University of Liverpool. Our editor, Christie Henry, and her team at the University of Chicago Press have guided us smoothly and efficiently through production. Finally, Jocelyn Hacker and Patsy Dunbar have provided logistical and moral support.

To all these people and organizations we are very grateful. Thank you again for all your help!

1 Introduction

 Not since the demise of the dinosaurs 65 million years ago has this planet witnessed changes to the structure and dynamics of its biological communities as dramatic as those that have occurred over recent millennia, and especially in the past four hundred years. The root cause of these changes can be attributed to the direct and indirect effects of human activity since the end of the Pleistocene some 11,500 years ago. These effects have been associated with the spread, growth, and development of human populations around the planet, which in turn have been strongly promoted by both agricultural and industrial revolutions.

Ironically, perhaps, human population processes are now destroying the very natural resources that have fueled them. Habitats have been devastated, and an unknown number of plant and animal species have already been hunted or harvested to extinction. Colonization of new regions and the introduction of exotic species has led to widespread extinctions of native taxa. And as one species disappears, so too have others, thanks to the way the effects of one loss can cascade through an ecosystem.

There is no doubt that we humans are dramatically changing the nature of this planet. What makes matters worse is that as the human populations continue to grow and their resource bases continue to diminish, the severity and rate of change are accelerating. We are entering a phase of mass extinction that is likely to be unique in the planet's history: never before has one species been responsible for the extinction of so many others.

Our closest relatives, the other primates, have not been spared from this impending catastrophe. Their populations are coming under increasing pressure from encroaching humans, and there is no doubt that several primate species are on the brink of extinction. The snub-nosed monkeys (genus *Rhinopithecus*) of Vietnam and China are a case in point. This group comprises four species with highly restricted distributions that are threatened by both habitat loss and hunting (Kirkpatrick 1995).

The Yunnan snub-nosed monkey (*R. bieti*) now numbers fewer than nineteen social groups in the wild (at most 1,200 individuals). Only four of these groups range in formally gazetted reserves (Long et al. 1994). The Tonkin snub-nosed monkey (*R. avunculus*) is even more severely endangered.

Not even the most common primate species, those that are well adapted to coexisting with people, are safe. Uncontrolled live-trapping, exacerbated by deforestation and the use of hunting as a pest control measure, resulted in the rhesus macaque *Macaca mulatta* population in India collapsing to about 10% of its original size in less than two decades (Southwick, Siddiqi, and Oppenheimer 1983).

With these statistics in mind, the future of the primates seems bleak. Nonetheless, all is not yet lost. If we act now, we may still be able to avert the worst of the impending catastrophe in biological diversity. Conservation biology, the "crisis-discipline," as Soulé (1991) termed it, provides the means to achieve this. Conservation biology is a scientific discipline that aims to provide the sound knowledge and guidance necessary to implement the effective conservation action that will be necessary to maintain in perpetuity the natural diversity of living organisms. Conservation biology thus underlies practical conservation action to preserve both the natural state and the biological processes that underpin living systems (Caughley and Sinclair 1994). Not only are the animals and the communities in which they exist important to conservation, but so too are the ecological-evolutionary processes that gave rise to the communities as we now find them and that continue to drive them.

Primates certainly justify such efforts. They have a diverse range of values that we cannot afford to lose. In terms of ecological value, primates play an unexpectedly important role in pollination and seed dispersal in many tropical forests (see sec. 4.4). If primates disappear, the viability of at least some forest communities may be threatened; if the forests disappear, then we lose numerous and indispensable benefits that range from climate regulation and water catchment to extraction of timber and other forest products (including natural medicinal compounds, many of which are almost certainly still unknown to biomedical science). Primates also have a more direct economic value. They are an important source of food in many tropical countries: in West Africa alone the bushmeat trade is worth millions of dollars (sec. 9.3). But if hunted to extinction, primates can no longer be a source of food or income. Similarly, primates can generate significant revenues through tourism (sec. 11.2.3).

In addition, the closeness of primates to ourselves, both in physical appearance and in social and cognitive skills, means that in many cultures they already have significant cultural value: it is not uncommon, for example, for primates to be considered sacred (e.g., the Hanuman langurs

Semnopithecus entellus of India and the guenons *Cercopithecus* spp. of the sacred groves in Nigeria). In fact, there are many who would argue that we have an ethical obligation to treat animal species that have such highly developed intelligence as primates with more than the average respect. There is also an intellectual value attached to primates, since the more we understand about comparative primate biology, the more we understand about ourselves.

Finally, primates are large, charismatic mammals, and as such they make powerful flagship species for conservation projects of all sorts (Dietz, Dietz, and Nagagata 1994). This means that primates make a good focus for the conservation of habitats and ecosystems, but it also means that if we lose these animals, our ability to raise support for future conservation action may suffer.

Our purpose in writing this book is to make a contribution to the conservation process. In doing so, our focus is on the primates, but our aim is to draw on all the diverse aspects of conservation and evolutionary biology that have emerged during the past few decades and apply them to a single taxon. In doing this we hope that we will be able to advance the broad field of primate conservation biology by synthesizing in a single coherent volume all these themes and, at the same time, to feed back into the discipline of conservation biology the lessons we learn.

Our approach is as follows. In the first part, we review the essential background of primate biology. The patterns and processes of primate diversity are examined in chapter 2, and species life history, ecology, and behavior are surveyed in chapter 3. The dynamics of primate communities, and the distribution and abundance of species, are the subject of chapters 4 and 5. Since conservation must ultimately be based on ensuring the survival of populations of species, we then review primate population biology (chapter 6). The next five chapters build on this knowledge base to explore key issues in primate conservation. Population viability is determined by both intrinsic and extrinsic processes: in chapter 7, we review the intrinsic processes that influence extinction risk, while the next two chapters introduce the two most important extrinsic threats to primates (habitat disturbance in chapter 8, hunting in chapter 9). The final group of chapters are concerned with the conservation strategies that can be assembled to tackle these processes most effectively (chapter 10) and the tactics these strategies might employ in order to achieve this end (chapter 11). In closing, we draw together some key issues that we feel need to be addressed for the successful conservation of primates in the future (chapter 12).

Only the last part of the book is, perhaps, typical of a traditional text on practical conservation: a description of the problems (habitat disturbance and hunting) and of the solutions (conservation strategies and tactics). Yet all the preceding material we cover is pivotal, since a clear understanding of problems and solutions is not possible without first understanding the biology of the systems we are trying to conserve. For example, only once the biological implications of large body size have been explained (chapter 3) does it become possible to start understanding differences between species in global population size (chapter 5), population dynamics and genetics (chapter 6), and the ability to cope with habitat disturbance (chapter 8), all of which contribute to the severity of extinction risk (chapter 7). In other words, we deal not only with patterns in conservation but also with the processes that underlie them. Without such an understanding, effective conservation is all the more difficult—if not impossible. This, then, is perhaps the greatest contribution that the science of conservation biology can make to the practical conservation of species and ecosystems.

We have attempted to review the existing literature as comprehensively as we can in the limited space available to us. Where possible, we present new analyses that provide further insights into the patterns and processes of the current primate conservation crisis. In discussing general principles in conservation biology, we have always striven to illustrate those principles with primate examples. Where this has not been possible, we have turned to the best available alternatives, giving preference to other tropical mammals wherever we can. We have sought out the most up-to-date material for our examples, although, given the escalating speed with which conservation situations change, it is inevitable that some of the material on the threats and conservation status of some primate taxa may already be out of date.

Two methodological points also need to be made. First, we assume that readers have a basic knowledge of statistics. The era when qualitative assessments provided adequate data on conservation issues is long since past: we can no longer afford the luxury of getting our recommendations wrong (Caughley and Gunn 1996). Statistics provides an essential tool for hypothesis testing in any science that deals with phenomena as complex and multidimensional as those of the biological world.

Second, we shall frequently attempt to look for functional (i.e., causal) relationships between variables across species. Traditionally, we might have used standard statistical regression techniques to show that there is an association between, for example, population density and body mass, and we would then have concluded that these variables are related through some underlying mechanism (e.g., large-bodied species need larger areas to provide sufficient food resources to sustain them and thus occur at lower densities). But species cannot be treated as statistically independent data points because closely related species are more likely to show similar patterns due to their shared evolutionary history (Harvey and Pagel 1991). Comparative methods based on phylogenies have been developed to deal with this problem, and where possible we have used these techniques to conduct such analyses. Where this has not been possible, we emphasize that a correlation involving species traits, such as body size versus extinction risk, cannot be fully verified without such controls. The same caution should be borne in mind in our discussions of previous analyses that have not employed these methods.

Finally, on a terminological note, throughout this book we use "primates" to refer to nonhuman primates, and we use "natural" to refer to a phenomenon that is nonanthropogenic. Although one can reasonably argue that human action is as natural as the behavior of any other animal, whether or not it is achieved through technology, it remains useful to discriminate between primate population and habitat phenomena that are the result of human forces on the one hand and nonanthropogenic forces on the other.

APPROACHES TO CONSERVATION

Conservation has a long history, although most of the early efforts at conservation were based on preserving hunting grounds or economic resources for the exclusive use of ruling elites. The establishment of national parks during the middle decades of the twentieth century reflects this tradition. After the establishment in 1872 of the world's first national park (Yellowstone National Park in the United States), preserving pristine habitats for the exclusive use of indigenous species became widely regarded as essential to the success of conservation.

Nevertheless, it was not until after the beginning of the 1960s that national parks and reserves of various kinds were established in large numbers throughout many parts of the world. Even as this process got under way, however, the seeds of its breakdown had already been sown. The pressure of human populations, and the associated demand for land (often from disfranchised communities), around the edges of many reserves and parks became so serious that effective protection of preserved areas came to be all but impossible in some cases. This demand for agricultural land at the local level was paralleled at the same time by increasing pressure from large-scale private or state-owned industries for access to the resources in preserved areas (principally for timber concessions, but large-scale ranching, mining, and hydroelectric schemes also played an important role).

This led, during the 1970s, to a shift in perspective among conservationists: the need for integrated conservation schemes (rather than preservation per se) came to be seen as the only viable way forward. Such schemes would integrate the interests of the local human population with the need to preserve biodiversity in ways that, it was hoped, would harness the power of local politics for the benefit of conservation. Local communities were viewed as essential allies in furthering conservation. In part this shift in emphasis reflected a growing recognition that the battle for conservation by exclusion had already been lost: impoverished Third World economies simply could not afford to set aside huge tracts of land for conservation unless they were prepared to risk civil unrest. At the same time, an important impetus was provided by the gathering view that, in the aftermath of the colonial era, it was inappropriate for the developed world to impose its own parochial demands and standards on the developing economies. The insistence on conservation-by-preservation was seen as imperialism by another name.

The resulting polarization of opinions as to how conservation should be furthered has continued to arouse heated debate. Inevitably, both sides probably have right on their side. Much of the pressure for a more integrated approach to conservation has come from social scientists concerned to protect the rights of indigenous tribal peoples whose traditional hunting or grazing lands have often been casualties of the preservationist approach. Social scientists' knowledge of ecology commonly varies between the lamentable and the nonexistent, yet they have sought to define good conservation practice—often based on secondhand opinions that lack empirical support. In contrast, the preservationist lobby has often failed to recognize the economic and social costs of conservation policies. Caught between encroaching state-backed big business and the conservation lobby's demands for habitat protection, the real losers have often been biodiversity and those tribal societies that depended on the habitat's resources for their traditional ways of life.

More recently, a new contestant has begun to make sallies onto the battlefield. Economists have begun to see conservation as an important field in which to exercise their muscle (Pearce and Turner 1990; Swanson 1995). The inevitable tenor of their argument has been that conservation is practicable only when it has obvious and direct economic benefits. If conservation strategies do not provide benefits to those who bear the costs of implementing them, the external and internal pressures for the exploitation of natural resources will simply overwhelm the good intentions.

The problem will not be easily resolved. The realpolitik of the human population explosion and the economic and social demands of development leave conservation caught in an impossible bind. The preservationist lobby may be overwhelmed by circumstances beyond its political control, while the participatory approach may be unwittingly presiding over species' extinctions and a tragic decline in biodiversity.

It is not our task here to engage in the debate over the aims and mechanisms of conservation, important as they are. Our purpose in raising these issues at the outset is simply to draw attention to the hard political and economic facts of conservation. Inevitably, we have to work with the realities of the situation as we find it. Our ability to do so will be strongly influenced by how well we understand the dynamics of living systems and the forces that threaten them: only then can we plan conservation strategies that will ensure their survival within the constraints imposed by the social, economic, and political realities of the habitat countries.

Note added in production: The rapid growth of research in conservation biology has meant that several relevant publications have appeared since this book went into production. Unfortunately it is too late to incorporate their findings into the present text, but we can take this opportunity to briefly recommend those that are likely to be most helpful to interested readers: Fleagle, Janson, and Reed (1999); Oates (1999); Robinson and Bennett (1999); and *Primate Conservation*, vol. 17 (1996–97 [published 1999]), containing the proceedings of the symposium "Primate Conservation: A Retrospective and a Look into the Twenty-first Century" (held during the Sixteenth Congress of the International Primatological Society, August 1996, Madison Wisconsin).

2 *Diversity*

The next two chapters provide an introduction to the order Primates. In this chapter we outline the diversity and evolutionary history of the primates and then consider the processes that lead to speciation. In the next chapter we review the main features of the life history, ecology, and behavior of primate species. The information in these two chapters provides background material essential to the rest of this book, but those familiar with these aspects of primate behavioral biology may prefer to proceed straight to chapter 4.

In this chapter we identify those traits that define the Primate order and the main groups within the order. We then consider the spatial and temporal patterns of primate diversity and finally the speciation processes that underlie these patterns and ultimately generate primate diversity. Understanding the processes of speciation and their end products (species as we see them today) is important for conservation because we need to know just what it is we are conserving and because we may wish to give some taxa higher conservation priority if they are of unusual taxonomic status (sec. 10.2.1).

2.1. The Primate Order

The living primates consist of some 200 to 230 or so species sharing a number of characters that together define a suite of biological traits marking primates out as different from other mammals. These are:

1. a shortened snout (with corresponding reduction in sense of smell);
2. an unspecialized skeleton, with hands and feet that retain the primitive five-digit pattern;
3. an opposable thumb permitting a precision grip;
4. nails rather than claws on the fingers and toes;
5. a large brain relative to body size;

6. an extended period of development, both before and after birth, that involves heavy energy investment by the mother as well as a long period of socialization;

7. a placenta that invades (or burrows into) the wall of the uterus (to extract nutrients more efficiently from the mother's bloodstream);

8. an increased dependence on vision, with forward-facing eyes that allow binocular vision.

Although not every primate species exhibits all these characteristics (marmosets, for example, have claws rather than nails, except on the big toe, and African colobus monkeys have only a vestigial thumb), these traits nonetheless define a set of closely related species that share a long evolutionary history.

Based on anatomical traits, the living primates were traditionally divided into two major groups. These represented two distinct grades (crudely speaking, stages) of primate evolution: the more primitive prosimians and the more advanced anthropoids. The evolutionary relationships between these two groups, and the taxa that compose them, are shown by the phylogeny in figure 2.1.

The prosimians differ from the anthropoids in several important respects. They are characterized by a bare muzzle, associated with greater dependence on olfaction; three premolars rather than two in each quadrant of the tooth row; and a reproductive biology that is more characteristic of mammals (a bicornuate uterus and a well-defined estrous period, with mating that is under close hormonal control; in anthropoid primates, sexual behavior is under less direct hormonal control, and females exhibit a true menstrual cycle rather than an estrous cycle: Martin 1990; Fleagle 1999). The prosimians include four major subsets: the lemurs of Madagascar, the galagos of the African mainland, the loris-potto group (whose four principal members are divided between the African mainland and Southeast Asia), and finally the tarsiers of Southeast Asia. Prosimians do not occur in the New World.

The anthropoid primates are in turn divided into two major groups: the New World monkeys (Platyrrhini) and the Old World monkeys and apes (the Catarrhini, the group humans belong to). The New World monkeys are found only in South and Central America, while the Old World monkeys and apes are broadly distributed throughout sub-Saharan Africa (excluding Madagascar) and Southeast Asia eastward from Pakistan to the Japanese archipelago. Although both Europe and southwestern Asia were widely populated by primates during Plio-Pleistocene times, native primate populations are now absent from Europe through to Afghanistan (except for a small hamadryas baboon population in Saudi Arabia).

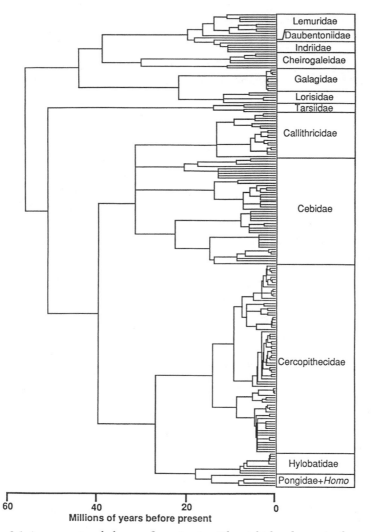

Figure 2.1 A composite phylogeny for primates. The right-hand margin shows the twelve families in the primate order. The Lemuridae, Daubentoniidae, Indriidae, Cheirogaleidae, Galagidae, Lorisidae, and Tarsiidae constitute the prosimians: the rest are anthropoids. (From Purvis 1995 with the permission of the Royal Society.)

Within the catarrhine primates, the apes and monkeys are classified into distinct groups (the Hominoidea and the Cercopithecoidea, respectively) on strictly anatomical grounds. The extant ape species share a number of traits that are not found in the monkeys. These include the absence of an external tail, a broad rather than deep (or doglike) chest, and a characteristic arrangement of the tubercles on the molar teeth (the so-called Y5 pattern).

The tarsiers have always remained anomalous within this classification, since they exhibit traits reminiscent of both prosimians and anthropoids. Although tarsiers are classed as prosimians (e.g., Smuts et al. 1987), it turns out that their resemblances to the prosimians may have arisen largely by convergence through a nocturnal lifestyle (Martin 1990). Modern phylogenies therefore place the tarsiers firmly in the same grouping as the anthropoids (e.g., Purvis 1995; see fig. 2.1), and the primates are now commonly divided into the Strepsirhini (prosimians without tarsiers) and Haplorhini (anthropoids plus tarsiers). However, the tarsier-anthropoid relationship remains debated. In light of this uncertainty, we prefer to follow Fleagle (1999) and use the less cumbersome terms prosimian and anthropoid, while treating tarsiers as a case apart where appropriate.

The basic primate groups, together with their characteristic patterns of life history, ecology, and behavior, are summarized in table 2.1 (further details are provided in chapter 3). A full list of species is given in appendix 1. Note that the number and identity of species listed in a taxonomy is highly variable depending on the source but that the number of primate species generally continues to increase. This increase is partially a result of new discoveries (e.g., Ferrari and Queiroz 1994) but mainly a result of partitioning taxa previously classified as single species, such as the formerly monospecific night monkey *Aotus* (see appendix 1). These increases have been particularly notable among the nocturnal species, largely as a result of the recognition of cryptic species (see below): Bearder (1999) notes that the number of recognized nocturnal primate species has increased by 285% since 1967. Current taxonomies record primate species richness at between 201 and 233 species (Corbet and Hill 1991 versus Groves 1993). These species are illustrated in Rowe (1996). To avoid confusion in the analyses that follow, we will generally use those taxonomies adopted in the original sources of the works we are citing.

SPECIES DEFINITION

The way species are defined has a variety of implications for conservation biology. A basic understanding of species definition, and the principles of taxonomy and phylogeny, will therefore be useful when we examine these implications in later chapters.

Taxonomies (or biological classifications) exist to allow us to reduce the natural complexity of the real world to a point where we can talk about generalities (i.e., types of individuals). These types are usually termed *species*. Traditionally, individuals were classified into species (and

Table 2.1 Diversity of living primates

Infraorder	Family Subfamily	Living species	Common names of some member species	Mass (kg)	Niche[1]	Group size[2]	Group type[3]
LORISIFORMES	Galagonidae	11	Galago (bushbaby)	0.1–1.1	A,N	1	S
	Loridae	6	Loris, potto	0.2–1.2	A,N	1	S
LEMURIFORMES	Cheirogaleidae						
	Cheirogaleinae	6	Dwarf and mouse lemur	0.1–0.4	A,N	1	S
	Phaerinae	1	Fork-marked lemur	0.5	A,N	2	Mo
	Daubentoniidae	1	Aye-aye	2.5	A,N	1	S
	Indriidae	5	Indri, sifaka	0.8–6.8	A,N/D	2–9	Mo,G
	Lemuridae	10	Mongoose lemur, ruffed lemur	0.7–3.5	A,D/C	2–20	Mo,G
	Megaladapidae	7	Sportive lemur	0.5–0.9	A,N	1	S
TARSIIFORMES	Tarsiidae	5	Tarsier	0.1	A,N	1–2	S,Mo
PLATYRRHINI	Callitrichidae	26	Marmoset, tamarin	0.1–0.8	A,D	2–8	Mo,G
	Cebidae						
	Alouattinae	8	Howler monkey	4.0–6.6	A,D	6–30	G
	Aotinae	10	Owl monkey	0.7–1.2	A,N/C	2–4	Mo
	Atelinae	9	Spider and woolly monkey	7.0–9.3	A,D	12–50	G
	Callicebinae	13	Titi monkey	0.8–1.4	A,D	2–4	Mo
	Cebinae	9	Capuchins, squirrel monkey	0.7–2.5	A,D	8–50	G
	Pitheciinae	9	Saki, Uakari	1.6–2.9	A,D	2–30	Mo,G
CATARRHINI	Cercopithecidae						
	Cercopithecinae	47	Baboon, guenon, macaque	1.1–15	A/T,D	4–150	G
	Colobinae	34	Colobus, langur, leaf monkey	4.2–15	A/T,D	4–150	G
	Hylobatidae	11	Gibbon, siamang	5.3–11	A,D	2–5	Mo
	Hominidae/Pongidae	5	Chimpanzee, orangutan, gorilla	33–30	A/T,D	1–120	S,G

Note: Infraorder from Smuts et al. 1987; family, subfamily, and species classification from Groves 1993; body mass data, describing the range of female values, from Smith and Jungers 1997; typical niche, group size, and group type from Fleagle 1999. Table after Dunbar 1988.

[1] Niche types: A, arboreal; T, terrestrial; N, nocturnal; D, diurnal; C, cathemeral.

[2] Group size: number of members.

[3] Group type: S, solitary (although individuals might share night nests); Mo, monogamous pair (often accompanied by young); G, group-living (multiple breeding females and/or breeding males).

species into higher-order taxa such as genera and families) based on physical resemblance. Since the 1940s, however, the consensus among biologists has been the *biological species definition* proposed by Mayr (1942, 1963): a species is a population of individuals capable of inter-breeding (i.e., producing fertile offspring). When two populations cease to be able to produce fertile hybrids they are then said to belong to different species. Although this definition works well enough most of the time, it does encounter some difficulties.

One such problem is that different species can sometimes interbreed and produce fertile hybrids. For example, hitherto five species of *Papio* baboon have been recognized, based on differences in physical appearance. However, they all interbreed in captivity, and those whose geographical distributions abut also do so in the wild. Should these be classified as different species, or should they be subspecies of a single species (Jolly 1993)? The current consensus appears to favor the latter view. Geographical clines of this kind are not uncommon: another well-known example (that of the superspecies *Cercopithecus aethiops*) is illustrated in figure 2.2. In the case of the *C. aethiops* group, geographically adjacent populations resemble each other, but those at opposite ends of the taxon's pan-African distribution are quite dissimilar in appearance. Opinions still differ on whether these constitute a single species or as many as six. A more disturbing problem is provided by instances where species of different genera can produce fertile hybrids. Examples include the gelada baboon (*Theropithecus*) with savanna baboons (*Papio*) and macaques (*Macaca*) with guenons (*Cercopithecus*) (Gray 1972). In contrast, other species may look very similar but be unable to interbreed: an example is provided by the several bushbaby species of the genus *Galago* (Masters 1993; Bearder, Honess, and Ambrose 1995). Although such problems are more commonly the exception than the rule, they do pose serious difficulties for taxonomy.

Partly because of dissatisfaction with the biological species concept, Patterson (1985) introduced the idea of the *specific mate-recognition system* as the defining basis for species. For Patterson, a species' identity is maintained by selection for compatible reproductive biology at the behavioral, physiological, and anatomical levels (in contrast, the classic biological species definition sees selection against less viable intermediates as the main factor reinforcing species identity). A specific example of this is offered by the African bushbaby group. Bearder and colleagues (Bearder, Honess, and Ambrose 1995; Bearder 1999) have suggested, based on detailed analyses of vocalizations, that there may be as many as six unrecognized species of the Galagonidae in addition to the eleven currently recognized by taxonomists. They explicitly point to the role of species-specific vocalizations in determining breeding patterns within a

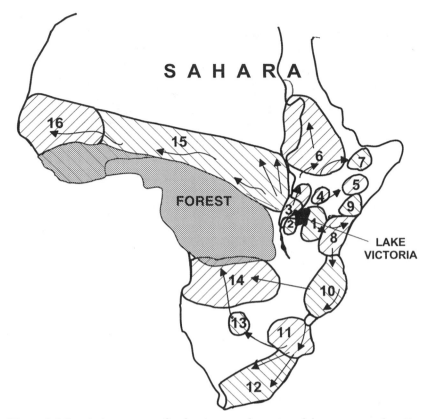

Figure 2.2 Speciation patterns for the sixteen subspecies of the vervet monkey *Cercopithecus aethiops* superspecies. As populations dispersed from the species' putative ancestral home in the vicinity of Lake Victoria, genetic drift (and perhaps local adaptation) has resulted in increasing divergence in pelage color and pattern. (After Hill 1966.)

natural population, arguing that some anatomically indistinguishable populations that live sympatrically may in fact be reproductively isolated by their respective vocalizations. Such *cryptic species* are likely to be particularly common among the nocturnal primates.

Contemporary taxonomies may therefore extend the traits used beyond the purely anatomical ones that form the basis of traditional classification. This is particularly true of the use of genetic data in taxonomic analyses. In addition, contemporary taxonomies place a much greater emphasis on incorporating *phylogeny* (the evolutionary relationships between species) in their structure. The use of molecular techniques to establish the degree of relatedness between different species has led to some unexpected findings that are in marked contrast to traditional taxonomy. Among these have been the observations that the mangabeys, all originally classified as belonging to the genus *Cercocebus*, derive from two distinct evolutionary roots and should belong to different genera (Cronin and Sarich 1976); that the drill-mandrill group is less closely related to the *Papio* baboon group than is the gelada (Disotell, Honeycutt, and Ruvulo 1992); and that the western and eastern subspecies of lowland gorilla are less closely related to each other than are the two recognized species of chimpanzees (genus *Pan*) (Ruvulo et al. 1994).

It is also important to note that emphasizing different traits can produce different taxonomies. Chimpanzees and gorillas share a quadrupedal gait that humans do not. On anatomical grounds of this kind, chimpanzees and gorillas were classified together (along with orangutans) as great apes (the Pongidae) and were clearly distinguished from humans (the Hominidae). The genetic evidence suggests that humans actually are more closely related to chimpanzees than either is to the gorilla, and that these three taxa should be classed together in the African great ape clade, with the orangutan as a distant cousin (Fleagle 1999). Unfortunately, similar problems can arise even with molecular data: comparisons of gene sequences from different segments of the genome may yield different classifications.

Finally, an important lesson from the molecular data has been that differences in anatomical characters may not always be a good guide to how long two taxa have been separated. An instructive example is provided by the gelada (genus *Theropithecus*) and the common baboons (genus *Papio*): physically and ecologically they are very different, but genetically they are no more different than chimpanzees and humans (Cronin and Meikle 1982). The converse may also apply: the Cheirogaleidae (the dwarf lemurs of Madagascar) share many anatomical similarities with the Galagonidae (the bushbabies of mainland Africa) and have, in at least one taxonomy, been classified together, but the molecular evidence indicates that the Cheirogaleidae share a common ancestry with the other Lemuriformes about 40–45 million years ago, whose collective common ancestry with the Galagonidae is at least 20 million years older (Yoder 1997). Anatomical differences may thus be a less reliable guide to common origins because they reflect characters (or genes) that interact directly with the environment: selection may force much more rapid changes in some characters than is the case for alleles that are under neutral selection, whereas other characters may remain stable for long periods because the selection pressures do not change.

2.2. Patterns of Diversity

Biodiversity is not a static phenomenon. It is highly dynamic, exhibiting complex patterns of variation over space and time.

Figure 2.3 Distribution of taxon richness in African primates for all lower-rank taxa (number of species or subspecies) at the 1° latitude-longitude grid cell scale. The grid cell with the maximum value is shown in black, whereas the other nonzero scores are grouped into five classes (corresponding to the gray scale on the right), containing approximately equal numbers of grid cells. The map has been smoothed by taking each cell's score as the mean score of the surrounding cells. The five cells containing the Barbary macaque in North Africa are not shown. (Redrawn with the permission of Elsevier Science from Hacker, Cowlishaw, and Williams 1998.)

2.2.1. Geographic Distribution

Primates are tropical animals. The vast majority of species occur in tropical and subtropical regions, where they exist in a variety of ecosystems including woodlands, savannas, and deserts. Nevertheless, it is in areas of equatorial tropical rain forest that the greatest number of taxa are found (e.g., Africa; Hacker, Cowlishaw, and Williams 1998: fig. 2.3). Although primates are also found in colder biomes (notably montane ecosystems and temperate forests), only five species have a geographic distribution entirely outside the tropics: two macaques (*Macaca fuscata, M. sylvanus*) and three snub-nosed monkeys (*Rhinopithecus bieti, R. brelichi, R. roxellana*). Primates are also primarily continental in distribution: they are largely restricted to the landmasses of Africa, Asia, and the Americas and are completely unknown in Australasia and the Pacific (major exceptions are Madagascar and the larger islands of Southeast Asia). The patterns of distribution of basic primate taxa have already been noted (sec. 2.1): prosimians occur only in the Old World and are the only primates found on Madagascar; the monkeys of the Old World and the New World are entirely distinct; and the apes are found only in the Old World.

The pattern of primate diversity differs between both continental regions and countries. Across regions, Asia has the most families of primates, Africa has the most genera, and the Americas have the most species (table 2.2). Among the fifteen countries scoring highest for primate species richness (table 2.3), most are African and few are Asian, although

Table 2.2 Distribution of primate taxa across major continental regions

Region	Families	Genera	Species
Africa	3	19	59
Madagascar	4	14	24
Asia	5	10	53
The Americas	2	16	64

Note: Taxonomic patterns based on Corbet and Hill 1991.

Table 2.3 The fifteen highest-scoring countries for primate diversity

Region and country	Number of primate species	Number of primate genera	Number of primate families	Percentage of endemic species
Africa				
Democratic Republic of Congo	31[1]	17	3	7
Cameroon	29	17	3	0
Nigeria	23	15	3	4
People's Republic of Congo	22	16	3	0
Equatorial Guinea	22	14	3	0
Central African Republic	20	14	3	0
Angola	19	11	3	0
Uganda	19	12	3	0
Gabon	19	13	3	0
Madagascar	28[1]	14	5	93
Asia				
Indonesia	34[1]	9	5	50
The Americas				
Brazil	52[1]	16	2	35
Colombia	27	12	2	11
Peru	27[1]	12	2	7
Bolivia	18	12	2	0

Sources: Data on species richness and endemism from Ayres, Bodmer, and Mittermeier 1991. Those on the number of genera and families are from Eudey 1987 (Asia); Oates 1996a (Africa); Mittermeier et al. 1992 (Madagascar); and Mittermeier 1987b (Latin America). Note that the taxonomies of these sources differ from that in table 2.2.

[1]Classed as "megadiversity countries" by Mittermeier et al. 1994.

Indonesia is clearly of great importance. However, the rank order of countries differs depending on which measure of diversity is used. Brazil, for example, has more species than any other country, but Indonesia and Madagascar have more families, while Madagascar has almost twice as many endemic species (i.e., those found only in the specified geographical area) as Indonesia. For conservation purposes, there is no single ideal measure; rather, the choice of measure must depend on the conservation goal (see sec. 10.1). Nonetheless, within this sample of high-scoring countries, it is possible to distinguish a subset of five that show unusually high richness and endemism in living organisms: these are the *mega-diversity countries* of Brazil, Indonesia, Madagascar, Peru, and Democratic Republic of Congo (Mittermeier 1988; Mittermeier et al. 1994).

Within these countries, the distribution of primate diversity is mostly contingent on the distribution of tropical forest and the biogeographic communities represented therein. The composition of these communities may largely reflect the location of Pleistocene forest refuges (e.g., Oates 1996a; see below), but a variety of other physical and ecological factors can play a role in determining primate species richness. Further discussion of these factors is postponed to chapter 4. The evolutionary history of the primates, which underpins this contemporary distribution, is the subject of the rest of this section.

2.2.2. Evolutionary History

Primates are one of the most ancient and anatomically least specialized lineages of mammals. Their origins probably date back to the time of the dinosaurs some 65 million years ago, although the earliest recognizable primates in the fossil record appear about 10 million years later at the start of the Eocene era. They soon became widely distributed throughout the (then tropical) Northern Hemisphere (North America and Eurasia). These early primates were very different from modern monkeys and apes, though they exhibit some affinities with today's prosimians. The Eocene witnessed a major radiation of these early primates, which exhibited very considerable diversity in both body size and ecological specialization (Fleagle 1999).

From about 35 million years ago, we enter a period when the fossil record is very poor (the Oligocene "fossil gap"). When we emerge from it toward the middle of the Oligocene period some 5 million years later, we find a dramatic change. Primate fossils are now no longer found in the higher northern latitudes but instead are found in the equatorial regions of the Old World. A dramatic cooling of the global climate occurred during the early Oligocene (mean sea surface temperatures dropped by an astonishing 30°C between 50 and 30 million years ago), which in turn resulted in a shift equatorward in the tropical forest belts to which the early arboreal primates were confined. The earliest known fossil sites of the mid-Oligocene come from North Africa and already reveal a wide variety of species. These species are very different from the earlier Eocene primates and resemble modern anthropoid primates much more closely (although in many ways they resemble South American primates more than they do contemporary Old World ones). Since the genetic data suggest that the modern prosimians owe their origins to a radiation that began between 50 and 60 million years ago (Yoder 1997), it is clear that a separate prosimian lineage was also in existence at this stage even though their fossil record is all but nonexistent.

Before the Oligocene, there appear to have been no primates in South America (though the fossil record on this continent is too poor to be certain of this). The earliest fossils (from a single site in Bolivia) are dated to about 26 million years ago, so this provides a latest date by which the invasion of South America must have occurred. It is currently assumed that at some time during the early Oligocene (some 30 million years ago, according to the genetic data: Purvis, Nee, and Harvey 1995), ancestral anthropoid primates crossed from Africa to South America, though exactly how this happened remains a mystery. However, it may be noted that the Atlantic Ocean began to open up only 65 million years ago, and its width at the end of the Eocene may have been as little as 500 km (Conroy 1990). For much of this period there appears to have been a series of island chains (now submerged) along the mid-Atlantic oceanic ridges; with the exposure of the continental shelves, the distance migrants would have had to travel across open water would have been relatively short by present-day standards. The most likely explanation for the invasion of South America thus seems to be island hopping, combined with rafting on clumps of vegetation (Conroy 1990; Fleagle 1999). In contrast, no prosimians successfully colonized the New World.

Once separated by the Atlantic, the New and Old World primate stocks underwent very different evolutionary radiations (with the Old World primates showing the most radical differentiation). The New World monkeys experienced an extensive radiation that gave rise to some fifteen extant genera (following Groves 1993), comprising the Callitrichidae and the six subfamilies of the Cebidae: the Alouattinae (howler monkeys), Atelinae (spider monkeys), Aotinae (owl monkeys), Cebinae (capuchins and squirrel monkeys), Callicebinae (titis and their allies), and the Pitheciinae (sakis and their allies) (table 2.1). The fossil record for South America is so poor that it is all but impossible to reconstruct the evolutionary history of the platyrrhines. Nonetheless, we know enough to be aware that during the course of their evolutionary history they occupied the whole of South and Central America as well as the Caribbean islands (in the latter case, probably as a single colonization event followed by speciation as individual islands were occupied). With subsequent climate change the southernmost populations (in the savanna regions of Patagonia and the Gran Chaco) became extinct. The extinction of the Caribbean populations was more likely the result of human activities (see sec. 7.4.2).

Meanwhile, in the Old World, the apes and monkeys split from each other relatively early. Throughout most of the Miocene the apes dominated the primate biomass with a remarkable radiation of species that filled most of the ecological niches now occupied by the catarrhines as a whole (the only major difference between the Miocene apes and modern cercopithecoid monkeys seems to be that there is no evidence for any of the fast quadrupedal running and jumping abilities that are so characteristic of the monkeys). This early period also witnessed some major radiations of the monkey lineages, giving rise to the ancestors of the modern colobine monkeys. Both colobine and ape radiations were associated with the first of many subsequent invasions of the Asian continent.

A further cooling of the global climate during the later part of the Miocene resulted in the gradual breakup of the great forest blocks of the tropical regions. The end of the Miocene thus witnessed the emergence of the savanna grassland communities that we associate in particular with Africa. This seems to have stimulated an increasing terrestrialization in the African primate fauna, and is allied with the emergence of several new monkey lineages—collectively termed the cercopithecines—that evolved from colobine-like ancestral stock but exhibited greater terrestrial adaptation. The cercopithecines could successfully handle the secondary compounds found in unripe fruit and seeds (evolved by plants to reduce levels of seed predation). This capacity gave them a significant selective edge over the apes: the ape lineages—including humans—cannot eat unripe fruits because they cannot detoxify the condensed tannins that give such fruits their astringent taste (Andrews and Aiello 1984). Perhaps because of the greater ecological competitiveness of these new monkey lineages, the ape lineages went into terminal decline during the Plio-Pleistocene (fig. 2.4), while the cercopithecine lineage underwent a series of explosive radiations. Only one ape lineage (the one that gave rise to our own species) emerged from this winnowing process with any success, and even its survival seems to have been associated with increased terrestrialization, the exploitation of poorer-quality food resources, and some major demographic bottlenecks.

The first of the cercopithecine radiations was associated with a new wave of migration into Eurasia from Africa about 10 million years ago: these were the ancestors of the macaque lineage. This lineage subsequently moved eastward as far as the Japanese archipelago, speciating repeatedly as it went. Within Africa itself, a series of radiations produced

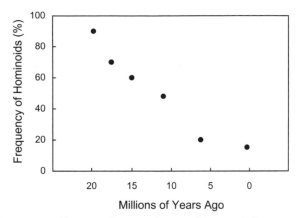

Figure 2.4 Frequency of hominoid species (apes and humans) shown as a percentage of all hominoid and cercopithecoid species over the past 10 million years. (After Fleagle 1999.)

first the baboons and their allies (geladas, mangabeys, drills) and then the guenons, with the latter radiation taking place within the past 2–3 million years. Purvis, Nee, and Harvey (1995) note that the Cercopithecinae as a whole exhibit rates of cladogenesis (lineage splitting) approximately double those seen in other primate lineages, with still higher rates observed within some taxa of the lineage (notably in the genus *Cercopithecus*). The reasons remain unclear, but the species of this family stand out from other primates in sharing a characteristic suite of sociodemographic characteristics associated with female philopatry and cohesive matrilineal coalitions (Di Fiore and Randall 1994): these features are especially likely to lead to greater demic substructuring, which in turn may hasten speciation (see sec. 6.4). In addition, they are characterized by large brains relative to body size compared with most other primate groups, and it may be that the behavioral flexibility associated with large brains has allowed them to invade or cope with a broader range of habitats, thus giving natural speciation processes greater opportunity to take hold.

Meanwhile, the Eocene prosimian faunas of the Old World went into eclipse, with those species that managed to hold their own ecologically against the anthropoid lineages apparently doing so only by exploiting a nocturnal niche. The one exception was on the faunistically depauperate island of Madagascar (where both anthropoid competitors and large cursorial predators were absent). The prosimians of Madagascar (which are now believed to have arrived on the island in a single colonization event: Yoder et al. 1996) speciated rapidly and produced a large number of descendant genera in a remarkable adaptive radiation. Many of the Madagascan primates occupied diurnal niches and came to fill the terrestrial niches occupied on the mainland by monkeys and apes and even antelopes (Fleagle 1999). In this respect they contrast strikingly with the exclusively arboreal and nocturnal mainland African and Asian prosimians.

Just how Madagascar was invaded by mainland species is uncertain, since the Mozambique channel opened up 100–200 million years ago and long predates the appearance of even the earliest primates (60 million years ago). It is possible that, as with the South American invasions, small numbers of animals rafted across the intervening seaways on trees or clumps of vegetation washed down from major rivers. However, McCall (1997) has suggested that islandlike land bridges may in fact have maintained contact between Madagascar and mainland Africa until as late as the Miocene. Either way, it remains unclear why prosimians but not anthropoids colonized Madagascar and why the reverse occurred in the New World.

For more details on primate evolutionary history, see Richard (1985), Conroy (1990), Martin (1990), and Fleagle (1998).

2.3. Origins of Diversity

Speciation, a fundamental mechanism in generating the diversity of extant primates, is the outcome of a complex hierarchy of processes operating at both macroecological and microecological scales.

Present evidence suggests that the predominant macroecological force may have been climate change. The number of climatic cycles within a given 0.5 million year period correlates strongly with the number of species in at least three primate groups: the theropiths, papionids, and hominids (Foley 1993). The causal mechanism that connects climate change and speciation is less clear, although it is notable that extinction events, like speciation events, tend to occur in relatively discrete phases separated by periods of quiescence; moreover, the phases of the two processes seem to coincide, with bouts of extinction preceding bouts of speciation (e.g., African primates; fig. 2.5). Further analysis (R. Foley 1994) indicates that, at least among the papionids and hominids, it is primarily extinction rates rather than speciation rates that are driven by climate change. This suggests that speciation in these groups has been dependent on ecological release following the extinction of ecologically dominant species. In other words, extinctions free up ecological niche space that is then occupied by new species emerging out of an ancestral stock that had previously been restricted by the activities of the species that went extinct.

Several major radiations of primates seem to have been triggered either by the loss of ecological competitors or by the invasion of habitats

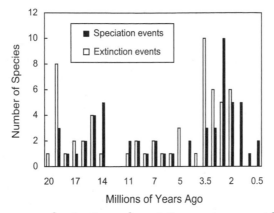

Figure 2.5 Frequency of extinction and speciation events among African primates during the past 20 million years. (After Delson 1985.)

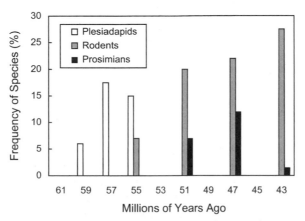

Figure 2.6 The relative abundance of plesiadapids, rodents, and prosimian primates during the Paleocene and Eocene of North America (as a percentage of total mammal fossils). The time scale shown on the x-axis is approximate. (After Fleagle 1999.)

that lacked ecological competitors. The rapid radiation of prosimian species during the earliest stages of primate evolution in the Eocene may have been promoted by the demise of the plesiadapids, whose extinction seems to have been hastened by the rise of the rodents, thereby freeing sufficient aboreal niche space to allow the early prosimians to diversify ecologically (fig. 2.6). Similarly, as we have already noted, the invasion of Madagascar by the ancestral lemurs provided the opportunity for a remarkable radiation because there were neither ecological competitors nor predators on the island.

2.3.1. Mechanisms of Speciation

At the microecological scale, any factor that promotes genetic isolation will promote speciation. This is because the biological species definition is based on the assumption that populations will remain part of the same genetic deme, or species, so long as gene exchange is possible between them: genetic isolation is therefore a prerequisite for speciation to occur. Genetic isolation can arise for three reasons (Ridley 1996): because a geographical barrier, such as a river or desert, intrudes between the two populations (*allopatric* speciation); because gene flow is reduced by socio-demographic structure between different parts of a species' range (*parapatric* speciation); or because a species undergoes sufficient genetic differentiation within its geographical range that two distinct species eventually emerge (*sympatric* speciation).

The relative importance of these three mechanisms of speciation is difficult to establish because each might have contributed at some point to the evolution of a given species that we recognize today: for example, the different traits that characterize a species might have different paths of divergence (with some traits diverging in allopatry and others in sympatry). In addition, contemporary patterns do not correlate in a simple way with historical processes: for example, hybrid zones may indicate parapatric speciation but can equally well result from allopatric speciation in two isolated populations that subsequently expand and make secondary contact.

At present we know surprisingly little about speciation mechanisms in primates. Most authors appear to assume that allopatric speciation has been the predominant force generating contemporary patterns of primate diversity, but there are problems with many models of allopatric speciation (e.g., Endler 1991), and we are unaware of any systematic attempt to quantitatively test either this hypothesis or the alternatives. In the absence of further information, we will spend the rest of this section outlining the present evidence for allopatric speciation and the forces that might contribute to it. However, we will postpone discussing the genetic aspects of speciation until after our review of primate population genetics in chapter 6 (see sec. 6.4.2).

ALLOPATRIC SPECIATION IN PRIMATES

Allopatric speciation may result most commonly from climate change and the associated loss of habitat. Within Africa, for example, during the past 2 million years climate change has driven complex patterns of expansion and contraction of the major forest blocks that stretch between eastern and western Africa. During the cool, dry periods corresponding to European glacial events, the forests became divided into a number of small refugia (isolated patches left after wide-scale habitat loss) (fig. 2.7),

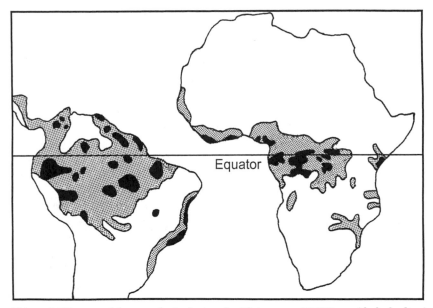

Figure 2.7 Proposed forest refugia during the Pleistocene 18,000 B.P. (dark shading) in relation to the present distribution of lowland tropical rain forest cover (light shading). (After Archibold 1995.)

which became joined again during subsequent warm periods (Haffer 1982). In the intervening periods, it is possible that the forest pockets harbored primate populations that went on to evolve in isolation. Adaptation to local conditions, genetic drift (see sec. 6.4), or both might have been responsible for the initial evolution of reproductive isolation during these periods of physical separation, but once the forest blocks joined up again the descendant populations in each block were no longer capable of interbreeding (or willing to interbreed because of behavioral or pelage differences) and were thus launched on separate evolutionary trajectories.

The effect of refugia on speciation in Africa may be particularly strong in the case of the guenons, genus *Cercopithecus* (Kingdon 1990), in which several species commonly coexist in modern forest blocks. The relatively rapid rate of speciation in this genus (nineteen species after a mere 2 million years of existence: see Purvis et al. 1995) has been attributed to the repeated occupancy of forest refugia during the glacial periods of the past million years or so by these small-bodied species that do not easily travel between forest blocks (Hamilton 1988; Kingdon 1990). In contrast, the larger-bodied and often more terrestrial mangabeys show much lower speciation rates despite facing the same conditions.

Similar conclusions have been drawn for South American primates.

Kinzey (1982) examined the present-day distribution of the species and subspecies of cebids and callitrichids in relation to putative forest refugia identified by herpetological and avifaunal studies (fig. 2.7). He found a strong tendency for the modern species distributions to overlap in the areas thought to have been Pleistocene forest refuges in southeastern Brazil, with their distributions often radiating outward in a nonoverlapping manner outside these refugia. He also found that there is a single forest refuge corresponding to the emergence of each of the species of *Callicebus* in the Amazon basin.

Brandon-Jones (1996) has also emphasized the role of climate-driven deforestation in shaping the present-day diversity and distribution of Asian colobines: it took only one such event about 190,000 B.P. to eliminate *Nasalis/Simias* in Sumatra, exterminate *Presbytis* throughout much of its former range, and fragment the geographic distribution of *Pygathrix/Rhinopithecus* (although several primate species survived this event on the Mentawai Islands, where the maritime climate afforded the forests some protection against the desiccating conditions). Climate-driven isolation of forest blocks in Southeast Asia has also arisen through changes in sea level. Chivers (1977) proposed that a sequence of changes in sea level since 800,000 B.P. (the early Pleistocene) have led to the repeated isolation and reconnection of ancient gibbon (genus *Hylobates*) populations on the islands of the Sunda Shelf, resulting in the nine species of gibbons extant today.

Allopatric speciation may be promoted not only by extrinsic forces such as climate change but also by certain intrinsic species traits. Species that are better able to cross geographic barriers, because of large body size, for example, may less commonly experience allopatric speciation. The existence of such a body-size effect receives mixed support. On the one hand, Ayres and Clutton-Brock (1992) have shown that larger Amazonian primates have fewer subspecies relative to their geographical ranges than smaller ones do, presumably because they are better able to cross large rivers like the major Amazon tributaries. On the other hand, on a broad taxonomic scale, no strong relationship has been found between body size and species richness within individual primate clades or across the primate order (Gittleman and Purvis 1998).

Similarly, when both sexes migrate between social groups the rate of genetic differentiation and hence speciation is likely to be slower. This can be illustrated by a comparison of speciation rates in two genera of arboreal African monkeys, *Cercopithecus* and *Colobus*. Speciation rates have been much higher in *Cercopithecus* (currently nineteen species, many with a number of subspecies) than in *Colobus* (currently just four species) (e.g., Groves 1993), even though *Colobus* is a much older genus than *Cercopithecus;* the extinct species *Colobus flandrini* is recognized from deposits dating from the late Miocene, about 10 million years ago, whereas the earliest members of the genus *Cercopithecus* date to only about 2 million years ago. Both taxa live in predominantly one-male social systems, but *Cercopithecus* has female philopatry (females remain in their natal territories), while *Colobus* does not.

Finally, a variety of other factors that might also constrain species' geographic ranges could equally well lead to isolation and allopatric speciation. These factors are discussed in the context of species' geographic distributions in section 5.1.2.

2.4. Summary

1. Primates are a diverse group of some 230 species, split into two major groups: the prosimians (split about equally between the nocturnal Lorisiformes and Tarsiiformes, and the Lemuriformes of Madagascar) and the anthropoids (largely diurnal herbivores). The anthropoids show two distinct lineages: the New World primates (the cebids and callitrichids) and the Old World primates (all other taxa).

2. Except for our own species, living primates are confined to the tropics or their immediate environs; the vast majority of species are forest dwelling, with an arboreal lifestyle.

3. Primate origins lie in the early mammal radiations after the extinction of the dinosaurs. Their evolutionary history occupies two major phases: an early phase characterized by an extraordinary radiation of prosimians (mainly confined to the Northern Hemisphere) and a later phase corresponding to the rise and radiation of the anthropoid primates (mainly confined to the tropics).

4. The primary cause of speciation and extinction events in recent primate history appears to have been climate change, mediated through the mechanisms of habitat loss, habitat fragmentation, and competitive release.

3 *Behavioral Ecology*

In the preceding chapter we introduced the primates and reviewed their evolutionary history. In this chapter we continue this theme by providing an overview of the key aspects of life history, behavior, and ecology of the order that are important for conservation biology. In the first section, we review primate life histories and their implications. We then review the broad principles of primate ecology, and finally we address a number of important issues relating to primate social behavior and grouping patterns.

3.1. Life History

In general the life histories of primates—the species-specific schedules of growth, reproduction, and mortality—are characterized by long life spans and low reproductive rates. Although these parameters are strongly influenced by body size, so that among larger taxa these patterns become more extreme, primates tend to have birthrates, death rates, and growth rates that are between one-quarter and one-half those of other mammals of similar size (Charnov and Berrigan 1993). In this section we discuss body size variation among primate taxa and the possible reasons why primates have such slow life histories.

3.1.1. Body Size Variation

Primate species show considerable variation in body size, ranging from the pygmy mouse lemur *Microcebus myoxinus* (female mass 30 g) to the gorilla (female mass 90 kg), with a median female body mass of 3 kg ($N = 206$ species: Smith and Jungers 1997). However, the distribution of body sizes among primate taxa is heavily skewed: there are many more small species than large species (fig. 3.1). In general the prosimians are small (with the exception of some diurnal prosimians on Madagascar) and the great apes are large. The processes that underlie interspecific variation in body mass in primates are still poorly understood (Harvey, Martin, and

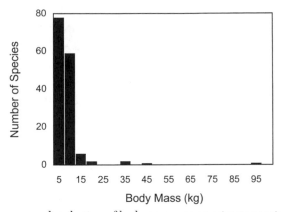

Figure 3.1 Frequency distribution of body mass among primate species. (Data from Happel, Noss, and Marsh 1987.)

Clutton-Brock 1987), although one pattern that does stand out is that terrestrial species tend to be larger than arboreal species. There are several possible reasons for this: arboreal species cannot grow too large because tree branches can support animals of only a relatively low mass; high predation rates on terrestrial taxa might select for large body size; and trees tend to be less abundant in drier, more seasonal environments where larger animals are at an advantage (owing to the scaling patterns of metabolic rate and fat reserves; see below), leading to a positive correlation between degree of terrestriality and body size.

There can also be considerable intraspecific variation in body mass. Marked differences can exist between males and females: in general males are larger than females, although in certain taxa the differences between the sexes are negligible (in particular, sexual size dimorphism is conspicuously absent among lemurs: e.g., Kappeler 1997). Sexual dimorphism tends to be a function of sexual selection: male-male competition selects for large body size in males because it makes males more successful competitors (Plavcan and van Schaik 1997), though this seems to be true only for anthropoids (Lindenfors and Tullberg 1998). In addition, the advantages of early reproduction may also drive female body size down in species that occupy unpredictable habitats (Demment 1983; Willner 1989). In addition, interpopulation differences in body mass can be marked within sexes. In baboons, both male and female body mass appears to be a complex function of mean annual rainfall (a measure of habitat quality) and mean ambient temperature (Dunbar 1990a; Barrett and Henzi 1997).

Finally, body size can also show cyclic fluctuations owing to seasonal and life history patterns of fat deposition. For example, male squirrel

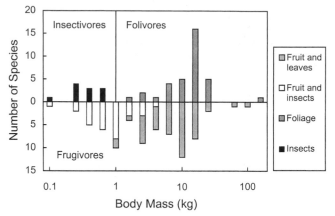

Figure 3.2 Frequency distribution of body mass among primate species in different dietetic categories. (After Kay and Simons 1980.)

monkeys become markedly heavier during the annual mating season, apparently through selection to maximize competitive ability in contests with other males (Boinski 1987a). In contrast, during lean seasons individuals in many populations can lose a considerable fraction of body mass (e.g., tamarins: Goldizen et al. 1988).

IMPLICATIONS OF BODY SIZE

Body size variation is important, since it has a host of effects on ecology and behavior. First, body size imposes a variety of constraints on diet because an animal's energy requirements are largely determined by its body size. Baseline energy consumption is a simple function of body mass. According to Kleiber's law (Kleiber 1969), $BMR = 70\ W^{0.75}$, where BMR is basal metabolic rate (the energy consumed per day while at complete rest, measured in kilocalories) and W is body mass (in kilograms). This relationship implies that there are savings of scale: larger species use less energy per kilogram of body mass than smaller animals. Nonetheless, it remains the case that large animals require *absolutely* more energy in total than smaller species (even though they might require *proportionately* less than might be expected based on their body mass).

Kay and Simons (1980) plotted dietary specialization against body size in living primates (fig. 3.2) and found that small species are typically insectivorous; very large species are typically folivorous; and frugivores can be divided into two groups depending on whether they obtain their proteins from insects (the smaller-bodied species) or from leaves (the larger-bodied ones). (Specific details about these different dietary types are given below: sec. 3.2.2.) These data suggest that there is a threshold body size above which species rarely adopt an insectivorous diet. This is mainly because insects typically come in very small packets, and the time cost of

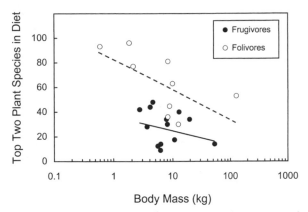

Figure 3.3 Percentage of diet accounted for by top two plant species plotted against body mass for primate species. The plotted lines are least-squares regression lines. (After Clutton-Brock and Harvey 1977.)

harvesting enough to provide the nutrients required by a large body soon becomes excessive. Hence, although insects provide a rich source of accessible nutrients, only animals that can survive on small absolute quantities of food can afford to be obligate insectivores. In contrast, larger species tend to be less dietetically selective, in part because large body size allows effective use of poorer-quality food sources such as foliage and in part because larger species have to use the resources available to them more intensively (and hence less selectively) in order to survive at all. Clutton-Brock and Harvey (1977) showed that the proportion of the diet made up from the top two plant species is negatively related to body size in primates; in addition, folivores tend to lie on a higher plane than frugivores, indicating that folivores are more selective (fig. 3.3).

The second implication of body size is its influence on individuals' ability to survive in variable environments. Because the amount of stored energy increases with body size at a faster rate than an animal's energy requirements do (Lindstedt and Boyce 1985), bigger animals tend to have relatively larger energy reserves and are therefore likely to survive periods of food scarcity (or inclement climatic conditions) more effectively than smaller species. Peters (1980) used experimental data on mass loss under starvation to estimate mean survival times under total starvation for birds of different body mass and concluded that survival time was positively related to body mass. That orangutans are extremely efficient at storing fat during fruit-rich seasons and subsequently utilizing these stores during leaner periods (Knott 1998) may provide a primate example of this, since theory predicts that similar capabilities are unlikely to be found in smaller primates.

The third and final implication of body size is that it imposes constraints on life history variables. It has been repeatedly shown for many different taxa that birthrates and death rates, as well as litter size, decline with increasing body size, whereas life expectancy and generation time increase (Peters 1980; Harvey, Martin, and Clutton-Brock 1987). Humans and other great apes, for example, weigh 35–100 kg and typically produce one offspring at a time at intervals of four years in a breeding life span that starts at about fifteen years of age and lasts until about age forty-five. Mouse lemurs, which typically weigh only 30–60 g, produce up to three infants at a time once (occasionally twice) a year, begin breeding at about one year of age, and rarely live beyond about fifteen years. The reasons for this relate to individual life history strategies and are the subject of the next section.

3.1.2. Life History Strategies

Charnov (1991, 1993) has suggested that the variations in life history characteristics across mammals might be explained in terms of the energy costs of rearing. All female mammals require energy for growth and reproduction (in the case of reproduction, energy for fetal and infant growth during gestation and lactation). This *production energy* must be derived from the mother's own metabolic energy (except, of course, while the female is herself an infant, when she derives all her production energy from her own mother through lactation). At maturity, production energy is divided between maintenance and reproduction. Since a simple growth law can be used to derive adult body mass from age at maturity, Charnov argued that fecundity and other life history variables can be explained as an attempt to maximize fitness (or, at a proximate level, lifetime reproductive output) through a trade-off between the energy available for reproduction and the risk of dying. Purvis and Harvey (1995) tested Charnov's model using life history data from sixty-four mammalian species (including primates) and were able to confirm five key assumptions and five novel predictions from the model.

Primates are both slow to reach maturity and slow to reproduce once adult and thus appear to be poor at deriving production energy from metabolic energy (performing at only about 40% of the standards of other mammals) (Charnov and Berrigan 1993). If Charnov's theory is correct, the key to understanding primate life histories will lie in determining why primates are not better at deriving production energy from metabolic energy. Charnov and Berrigan (1993) suggest that the answer may lie in the high energy demands of growing and supporting the big brains that are so characteristic of primates.

Brain tissue is one of the most energetically expensive tissues in the body (in humans it consumes eight to ten times as much energy as would be expected based on its mass alone: Aiello and Wheeler 1995). Sacher and Staffeldt (1974) argued that neural tissue grows slowly and imposes a limit on the rate at which adult brain size can be achieved, thus slowing down all the associated life history variables. In contrast, Martin (1981) and Hofman (1983) argued that neonatal brain growth was rate-limited by maternal metabolic turnover. Subsequent analyses have shown that, once phylogeny and body size are controlled, neonatal brain size does not correlate especially closely with maternal metabolic rate, although it does correlate with gestation length (Pagel and Harvey 1989; Harvey and Krebs 1990). This finding suggests that one of the reasons large-brained species have slow growth rates is that neural tissue is laid down at a fixed rate: developing a large brain simply takes longer. However, it is also clear that at least part of the extended developmental period is related to the need to program the brain during socialization: Joffe (1997) has shown that relative neocortex size (but not total brain size) in primates correlates with the relative length of the juvenile period, but not with any other period of the life cycle.

There is some evidence that strepsirhines may be different in these respects, however. Kappeler (1996) showed that among strepsirhines neither adult brain size nor basal metabolic rate (the factor most likely to limit maternal investment) correlates with the principal life history variables (fetal growth rate, relative litter mass, and postnatal growth rate) once body size and phylogeny are controlled. Kappeler argued that because the three life history variables are strongly correlated with each other but only weakly correlated with body size, they constitute a co-adapted suite responding to selection pressures different from those that influence brain size, body size, and metabolic rate. What this selection pressure might be remains unclear.

3.2. Ecology

Primates occupy a diverse range of niches and guilds, which can be defined by various combinations of habit (arboreal vs. terrestrial, nocturnal vs. diurnal) and diet (frugivore vs. folivore vs. insectivore) (table 3.1, following Bourlière 1985). Within these guilds, the ecology of primates is dominated by the need to avoid predators and find food. In this section we describe the distribution of primates through these guilds and review these two primary aspects of primate ecology.

Primate species are not evenly distributed around the eight guilds described. Diurnal, arboreal frugivores account for over half of all the extant primate species, with diurnal arboreal folivores accounting for a further 20%. Though widely assumed to have been characteristic of the ancestral primate stock, insectivory is now found only in a small group of relatively specialized primates, all of which are nocturnal. More impor-

Table 3.1 Ecological guilds occupied by primates

Guild	Contemporary examples	Number of species
Diurnal arboreal frugivores	Guenons, atelines, gibbons	101
Diurnal arboreal folivores	Colobines, howlers	35
Diurnal arboreal insectivores	[None]	0
Diurnal terrestrial frugivores	Baboons, macaques, gorilla[1]	15
Diurnal terrestrial folivores	Gelada	1
Diurnal terrestrial insectivores	[None]	0
Nocturnal arboreal frugivores	Dwarf lemurs, owl monkeys	15
Nocturnal arboreal folivores	*Lepilemur, Hapalemur*	3[2]
Nocturnal arboreal insectivores	Tarsiers, lorises, galagos	11[2]
Nocturnal terrestrial frugivores	[None]	0
Nocturnal terrestrial folivores	[None]	0
Nocturnal terrestrial insectivores	[None]	0

Note: Guilds after Bourlière (1985); taxonomic specifications and dietary categorizations after Smuts et al. (1987).

[1] Gorillas would traditionally have been considered folivores, but recent evidence suggests that the degree of folivory seen in the Virunga Mountain gorilla population reflects their habitat rather than their dietetic specialization.
[2] Prosimians only.

tant, a number of these guilds (specifically the diurnal insectivore and nocturnal terrestrial guilds) are not occupied by any primates (at least among extant species).

The absence of any species filling the three nocturnal terrestrial guilds probably reflects a combination of selection pressures including predation risk and the costs of searching for food. Frugivory is not a promising option for nocturnal species because fruits are largely designed to be located by sight: frugivory is closely linked to color vision and is thus strongly associated with a diurnal habit (Barton 1996; Barton and Dunbar 1997). The relatively high frequency of nocturnal frugivores is misleading in this respect: all but one of the fifteen nocturnal frugivores listed in table 3.1 are prosimians, and many of these place a greater emphasis on flowers and nectar (both strongly adapted to olfactory searching) than on fruits. In fact insectivory is a major dietary category for nocturnal species (40% of all nocturnal species are classed as insectivores) because insects are particularly active at night in the tropics.

Body size may, however, also be important in this respect: large species tend toward folivory-frugivory, whereas smaller ones tend toward insectivory (mainly because insects seldom come in large enough packets to support a large body size: sec. 3.1). In addition, competition tends to favor solitary foraging in insectivores, but this would be feasible only if these animals could exploit an antipredator strategy that does not depend on safety in numbers, for example, camouflage or behavioral crypsis.

Thus insectivory may represent a suite of traits involving small body size, solitary foraging, and nocturnal habits. However, while darkness can enhance crypsis, it can equally limit sociality, particularly among animals that are already heavily committed to vision. Given the additional problem of finding fruit in the dark (see above), nocturnal terrestrial primates may simply face too many ecological incompatibilities for this niche ever to be filled, other than in predator-free environments.

Two final points are worth stressing. One is the virtual absence of the nocturnal or terrestrial guilds from South American habitats according to Bourlière's scheme. Only one Neotropical taxon is nocturnal (*Aotus*, which until recently was also considered a monospecific genus), and none are terrestrial (although evidence suggests that large-bodied terrestrial taxa did exist in the past: Heymann 1998). Just why there should be no nocturnal or terrestrial platyrrhine primates remains unclear. The second is that research since Bourlière's study has revealed a considerable number of cathemeral primate species (species that are active during both the day and the night). Except for *Aotus* (which shows only limited cathemerality in certain populations: Wright 1989), they are all lemurs, primarily species of the genera *Eulemur* and *Hapalemur* (Kappeler 1997). Why cathermerality should be almost entirely restricted to lemurs is still debated, although the most convincing suggestion is that the recent extinction of diurnal predatory eagles from Madagascar has allowed previously nocturnal species to become diurnal (van Schaik and Kappeler 1996). It is certainly true that *Aotus* is active in the daytime only where diurnal predators (and large diurnal primate competitors) are absent (Wright 1989).

3.2.1. Predation Pressures

Primates are preyed on by a wide range of species, predominantly raptors and felids. In Kibale Forest (Uganda), Struhsaker and Leakey (1990) found that monkeys constituted 84% of the prey taken by crowned hawk eagles (the African monkey specialist). Gautier-Hion, Quris, and Gautier (1983) estimated that a single pair of crowned hawk eagles at Makokou (Gabon) took at least eight individuals a year from the local community of about one hundred guenons (representing three species). Peres (1993a) recorded an average of one attack by birds of prey every 8.8 days on tamarins (*Saguinus* spp.) on the upper Urucu River in Brazil, and similar rates of attack were reported by Goldizen (1987a) for tamarin groups in Peru. In fact, tamarins were so fearful of raptor attacks that they would risk serious injury by dropping in free fall as much as 35 m from the upper canopy, even in cases of false alarm. Aerial predators can also be a serious source of mortality for terrestrial primates, such as vervet monkeys (from eagles: Cheney and Seyfarth 1990), and nocturnal primates, such as mouse lemurs (from owls: Goodman, O'Connor, and Langrand 1993).

Among terrestrial primates, however, felids generally pose the greatest threat. In Africa, leopards appear to be the most dangerous predators of baboon populations and can attack groups by day or night with success rates over 75% (Cowlishaw 1994). In Asia, Schaller (1967) reported that langur hair appears in 6% of tiger feces and 27% of panther feces at Kanha (India). Similarly, in South America, jaguars and other large felids may be a source of predation risk (Emmons 1987; Peetz, Norconck, and Kinzey 1992). Although smaller primate taxa are generally more susceptible to predators, the hunting power of the big cats means that even the largest primates are often at risk (e.g., gorillas and chimpanzees from leopards: Fay et al. 1995; Boesch 1991; orangutans from tigers: Rijksen 1978). In Madagascar, where native felids are absent, this niche is occupied by the large viverrid *Cryptoprocta ferox,* or fossa, which can specialize on lemurs (Wright 1998).

In some cases primates can also be predators. Chimpanzees are important major predators for other primates that exist sympatrically with them: this is especially true for red colobus monkeys (*Procolobus badius*). Prey selectivity is often quite specific: sympatric diana monkeys (*Cercopithecus diana*) are much less frequently targeted than red colobus in the Taï Forest (Boesch and Boesch 1989). However, chimpanzee populations do vary in the species they hunt most frequently. The Taï and Gombe populations (a continent apart) concentrate on red colobus, but the Mahali population (located barely 100 km south of Gombe) exhibits a greater preference for *Cercopithecus* monkeys and forest ungulates (Boesch and Boesch 1989). In part these differences may reflect local availability in prey, but they may also reflect different cultural traditions in spatially separated populations or within populations over time (Boesch 1997).

Finally, many other predators can also take primates, including crocodiles and hyenas (e.g., baboons: Cowlishaw 1994). The most important of these remaining predators, however, are almost certainly snakes, which pose a significant risk to most small- and medium-bodied primates, whether terrestrial open-habitat species (e.g., vervets: Cheney and Seyfarth 1990) or arboreal forest species (e.g., tamarins: Heymann 1987; marmosets: Corrêa and Coutinho 1997; lemurs: Goodman, O'Connor, and Langrand 1993).

In response to the risk of predation, primates have evolved a variety of antipredator measures (Dunbar 1988; Isbell 1994). Some of these are morphological, and include camouflage (such as the drab coloration of the olive colobus monkey: Oates 1994b), large body size (terrestrial species tend to be larger than arboreal species: Plavcan and van Schaik

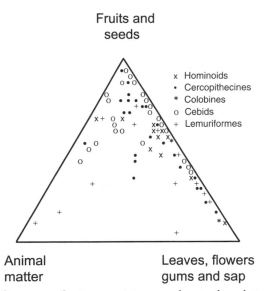

Figure 3.4 The diet space of primate species according to the relative consumption of three principal food types. Note that most primates are folivorous or frugivorous. (After Chivers 1994.)

1997), and large canine size (particularly among males, which often retaliate against predators: Plavcan and van Schaik 1992). Other antipredator measures are behavioral. These include vigilance (primates will often scan more actively for predators when they are distant from neighbors or in exposed positions: e.g., van Schaik and van Noordwijk 1989; Cowlishaw 1998; Treves 1998), alarm calls (many primates have a range of calls specific to different predators that allow them to rapidly adopt the most effective evasive response; for example, vervet monkeys: Cheney and Seyfarth 1990), cryptic behavior (e.g., red colobus monkeys will become silent and hide high in the canopy in positions that are hidden from the ground when predatory chimpanzees are nearby: Bshary and Noë 1997), the selective use of habitats (e.g., baboons will prefer to forage in poorer quality habitats if the risk of predation there is lower: Cowlishaw 1997a), the use of refuges during both the day and the night (e.g., tall trees and cliff faces in the case of baboons: Cowlishaw 1997b and Hamilton 1982, respectively), birth synchrony (e.g., squirrel monkeys: Boinski 1987b), group mobbing of predators (particularly by male coalitions, for example, by baboons against leopards and by red colobus monkeys against chimpanzees: Cowlishaw 1994 and Stanford 1995, respectively) and group living, both with conspecifics (sec. 3.3) and in polyspecific associations (sec 4.3).

3.2.2. Dietary Strategies

A great deal more is known about primate diets and foraging behavior than about predation and antipredator behavior. The predominant dietary preferences of primate species are illustrated in figure 3.4, which plots diet in relation to the three dimensions of faunivory, frugivory, and folivory. Although most primates fall toward one corner or another of the triangle, most species are dietetically quite flexible.

PLANT FOODS

Folivores can subsist on low-quality foods and have the advantage that leaves are widely available. However, leaves provide their own problems in that the nutrients are locked up within the cell walls, and the celluloses that compose the walls are often difficult to digest (van Soest 1982). Accessing the nutrients within plant leaves generally requires specialized adaptations (e.g., fermentation) or long gut passage times to permit the absorption of nutrients. Folivory is also often associated with dental specializations (e.g., hypsodont molars) for shredding or grinding leaves to promote microbial action (Dunbar and Bose 1991). Gut passage time is a direct function of the length of the digestive tract, and this in turn is a function of body size for the purely mechanical reason that a large body is needed to house a large intestine (Milton 1984). The latter is reflected in the fact that folivores are typically the largest of the primates (fig. 3.2).

In addition, leaves are often protected through chemical defenses (Waterman at al. 1988). Waterman (1984) analyzed the leaves of tree species eaten by three colobine species and found that the quantity of digestible protein is inversely related to the quantity of digestion inhibitors (principally fiber and condensed tannins): preferred items tend to have high digestibility (i.e., lower levels of digestion inhibitors and more accessible protein). Similar results have been reported for other Southeast Asian colobines (Davies, Bennett, and Waterman 1988) and lemurs (Ganzhorn 1992).

Many of the folivores (e.g., howler monkeys and African colobines) exhibit specialized adaptations to folivory in their digestive tract. These species rely on microbial fermentation to digest the plant cell walls so as to make the nutrients accessible (in many cases by subsequently digesting the bacteria in the hindgut after these have extracted the cellular nutrients). In effect, they are fermenters without the rumination seen in ungulates like cattle and antelopes (Bauchop and Martucci 1968; Kay, Hoppe, and Maloiy 1976; Kay and Davies 1994). Experimental studies indicate that howler monkeys may acquire as much as 35% of their daily energy requirement from the by-products of fermentation (Nagy and

Milton 1979; Milton, van Soest, and Robertson 1980). Ruminant-like strategies essentially require animals to have a large fermentation chamber where leaves can be processed. Two alternatives can be seen among primates: foregut fermenters (colobines), which have an enlarged stomach, and hindgut fermenters (howlers, several lemurs) which have an enlarged colon or cecum (the pocket at the junction of the small and large intestines). Hindgut fermentation is less efficient than foregut fermentation because nutrients are extracted mainly in the small intestine before the food reaches the cecum (Janis 1976).

Among primary consumers, the main alternative to folivory is frugivory. Frugivores are not well adapted to coping with leaves; although many nonfolivorous primates do eat leaves as a regular part of their diet (and sometimes rely on them as subsistence foods), their ability to wrest nutrients from this food source is relatively poor. This has been demonstrated experimentally by Mori (1979a), who showed that Japanese macaques (*Macaca fuscata*) fed on a diet of leaves had a calorie intake below that required to maintain body mass (and thus lost weight over time), whereas the same animals were able to maintain body condition quite satisfactorily on a predominantly fruit-based diet.

Fruits provide animals with more readily accessible nutrients than leaves do. Energy is likely to be especially important (Altmann 1998), although other nutrients may be important in particular cases (e.g., calcium in figs: O'Brien et al. 1998). But fruits suffer from their own intrinsic disadvantages. First, they tend to be more patchily distributed than leaves and are often highly seasonal in their availability. Second, although many plants may want primates to swallow their seeds whole (since primates are often a vehicle for seed dispersal: sec. 4.4.3), they do not want primates to consume seeds before they are mature or to destroy seeds by chewing them into pieces. Hence many plants have evolved defenses to protect seeds from premature dispersal or predation.

Fruits typically come in two varieties: those that consist of just the seed, often in some kind of casing or pod (e.g., the seeds of many palms), and those whose seed is encased in a soft, fleshy outer covering (e.g., figs, plums). Plants that produce the first kind of fruit tend to rely on physical defenses (e.g., shells that may require considerable strength to break open) to minimize seed predation; those that produce the second kind tend to rely on chemical defenses (including a whole range of toxins such as tannins and phenolics that inhibit digestion: for useful reviews, see Waterman 1984; Waterman and Kool 1994) to avoid both seed predation or premature dispersal. Coping with physical defenses often requires considerable strength or specialized features (or both): examples include the unusually strong molars that pithecines and *Cebus* monkeys use for cracking open palm nuts (Kinzey 1992) and the nut-cracking skills of chimpanzees (Boesch and Boesch 1981). In most cases the toxins used to prevent premature dispersal are designed to deteriorate naturally as the fruit ripens, so that the fruit becomes chemically accessible once the seeds are capable of independent survival. However, some species (notably the Old World cercopithecines) have evolved the ability to detoxify unripe fruits (Andrews and Aiello 1984) and can thus exploit these food sources before other frugivores (e.g., apes, frugivorous birds).

Another dietary specialization of some importance in primates is gummivory (feeding on plant gums and other exudates). Two groups of primates exhibit specializations in this respect: the marmosets (*Callithrix* spp.) and some of the prosimians (e.g., the lemur *Phaner furcifer* and the galago *Euoticus elegantulus*). Among the prosimians, the procumbent (forward pointing) lower incisors form a *tooth comb* that is sometimes used to gouge the bark of trees to stimulate a flow of exudate. The marmosets similarly have incisiform canines and elongated (though less procumbent) incisors that appear to be used in a similar way (Sussman and Kinzey 1984). Both groups also tend to have an enlarged cecum to enhance the digestion of gums. Exudates are an important source of energy for many of these species, especially during the dry season when fruits and other preferred foods are in short supply. This seems especially clear in the marmosets, which are thus able to maintain larger groups on smaller territories than the closely related tamarins (*Saguinus* spp.), which do not eat exudates (Ferrari and Lopes Ferrari 1989).

Finally, it should be stressed that leaves, fruits, seeds, and gums are not the only plant foods primates consume. Flowers and nectar are commonly eaten, particularly by frugivores (e.g., *Miopithecus talapoin:* Gautier-Hion 1970), and bark may sometimes be eaten in lean seasons, particularly by folivores (e.g., gorillas: Tutin, Fernandez, et al. 1991; White et al. 1995). In addition, some primate species have specialized on particular plant taxa. For example, *Rhinopithecus* spp. spend a great deal of time foraging on lichens (e.g., Kirkpatrick 1995), and *Hapalemur* spp. appear to specialize on bamboo. In the latter case, the various species seem to prefer different parts of the bamboo plant (Fleagle 1999), which might help explain how they can occur sympatrically. One example of an unusual dietary specialization is provided by the gelada baboon (*Theropithecus gelada*): a true grazer, it is the only living primate that feeds exclusively on grasses (Dunbar 1977).

ANIMAL FOODS

Even though most primates are vegetarians, most also eat small amounts of animal matter, since it is highly nutritious and contains vitamin B_{12} (which primates cannot synthesize or obtain from nonanimal sources). In most cases, carnivory involves predation on insects and other invertebrates (e.g., worms), small birds and their eggs or nestlings, or other small vertebrates (e.g., lizards, frogs). Only one primate taxon is exclusively faunivorous:the genus *Tarsius*. With a diet that consists of 90% arthropods and 10% vertebrates, the tarsiers may fulfill the same ecological role as the owl in dense undergrowth where aerial predators cannot penetrate (Bearder 1987).

In contrast to feeding on small animals, the active hunting of animals as large as small ungulates or even medium-sized primates is confined to chimpanzees (e.g., Boesch and Boesch 1989; Stanford at al. 1994; for a review, see Uehara 1997) and to a lesser extent *Papio* baboons (Hausfater 1976; Strum 1981; Davies and Cowlishaw 1996) and perhaps mandrills (*Mandrillus sphinx:* Kudo and Mitani 1985).

Accounts of predation in a number of Old World monkeys (including baboons, vervets, and green monkeys) suggest that hunting is more characteristic of dry (i.e., poor-quality) habitats than wet ones and, within habitats, of the dry season rather than the wet season (Dunbar 1988). Hausfater (1976), for example, reported a significant increase in predation during the dry season in the Amboseli baboons. This seems to reflect a broadening of diet in response to a seasonally impoverished habitat. Similarly, chimpanzees' hunting appears to be highly seasonal, being more common during the dry season at all three sites where they have been intensively studied (Gombe and Mahali in Tanzania, and Taï in Ivory Coast: Uehara 1997).

DIETARY FLEXIBILITY

Finally, it is worth emphasizing once again that categorizing primates dietetically in the way we have done here is a gross oversimplification. The emphasis on one or other dietary category can vary from one habitat to another, even within a species. African colobines are more folivorous in East Africa, where forests on nutrient-rich volcanic soils are characterized by a high diversity of palatable species (Dunbar 1987a), but they are more typically seed eaters in West Africa, where the nutrient-poor soils tend to produce forests whose vegetative parts are heavily defended chemically (Dasilva 1992; Oates 1994a). Similarly, gorillas in Rwanda are typically terrestrial folivores, whereas those in Gabon are more typically arboreal frugivores (Tutin, Fernandez, et al. 1991b). In contrast, although chimpanzees have always been considered the archetypal frugivores, in Kibale they eat a great deal of folivorous material (pith and terrestrial herbs), especially in those seasons of the year when fruits are in short supply (Wrangham et al. 1991). Most extraordinarily, *Papio* baboons routinely exhibit a full range of diet types across their geographic range, with fruit, foliage, or underground items (such as roots and bulbs) predominant in their diet depending on the population in question (fig.

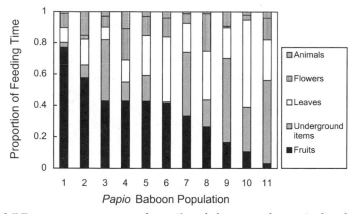

Figure 3.5 Dietary variation across eleven *Papio* baboon populations (indexed as proportion of time spent feeding): 1, Mount Assirik, Senegal (*Papio papio*); 2, Gombe, Tanzania (*P. anubis*); 3. Suikerbosrand, South Africa (*P. ursinus*); 4, Mikumi, Tanzania (*P. cynocephalus*); 5, Cape Reserve, South Africa (*P. ursinus*); 6, Bole, Ethiopia (*P. anubis*); 7, Amboseli, Kenya (*P. cynocephalus*); 8, Laikipia, Kenya (*P. anubis*); 9, Ruaha, Tanzania (*P. cynocephalus*); 10, Gilgil, Kenya (*P. anubis*); 11, Drakensberg, South Africa (*P. ursinus*). According to some taxonomies, all these populations would be considered to belong to a single species: e.g., *Papio hamadryas* (see appendix 1). Note that where these data did not add up to one they are given as a proportion of the available total. (Data from Whiten et al. 1991.)

3.5). Thus, even though species can be characterized as favoring one type of diet (and indeed may exhibit morphological specializations to that effect), we do well to remember that dietary flexibility is an important feature of the ecology of all primates.

3.3. Behavior

This account briefly explains the evolutionary forces that drive primate behavioral patterns. We begin with an examination of the differences in male and female behavioral strategies. We then describe the implications of these strategies for food and mate acquisition and, ultimately, social systems. Finally, we focus on the determinants of size and composition of primate groups. More detailed reviews of the evolution of primate behavior can be found in Smuts et al. (1987), Dunbar (1988), and Kappeler (1997). For further details on mammalian mating systems and mating systems in general see, for example, Clutton-Brock (1989) and Davies (1991), respectively.

As in all living organisms, the behavior of primates ultimately revolves around the fact that individuals seek to maximize their personal genetic contributions to the next generation. Since female primates invest much more in each reproductive event than males (compare the costs of egg production, gestation, and lactation with the costs of sperm production), the behavioral strategies adopted by each sex are markedly different. Female reproductive success is primarily constrained by the number of healthy offspring produced, which in turn is a function of females' access to key ecological resources (mainly food and safety from predators). In contrast, male reproductive success is mainly a function of the number of females fertilized, which in turn is a function of males' access to fertile females. Because males have the potential to produce many more offspring than females (in principle, a single male can fertilize all the females in a population), the intense competition between males can lead to some males' failing to reproduce at all. As a result, males (the lower investor) experience a much higher variance in reproductive success than do females (the higher investor).

Access to key resources such as food and safety (among females) and fertile females (among males) depends at least partially on the number of conspecific competitors that are also seeking those resources. Such resource competition can take one of two forms (fig. 3.6). *Scramble* competition occurs where a resource is limited but no single individual can monopolize it, so that all competitors suffer equally from the effects of competition, which intensifies with the number of competitors. Scramble competition is likely to be most common among females foraging on evenly distributed and dispersed foods like leaves (e.g., folivorous species such as colobines, gorillas, and howler monkeys: van Schaik 1989) and among males where females are seasonal breeders (e.g., many macaque species: Cowlishaw and Dunbar 1991; Dunbar and Cowlishaw 1992). In contrast, *contest* competition occurs where a resource is limited and monopolizable, so that competitors of high dominance rank (see below) can monopolize the resource at the expense of low-ranking competitors. This is likely to be typical of females foraging on patchy foods like fruits (e.g., frugivorous species such as many cebids, guenons, and mangabeys: van Schaik 1989) and males where females are not seasonal breeders (e.g., baboon species: Cowlishaw and Dunbar 1991; Dunbar and Cowlishaw 1992). The competitive regime that characterizes the limiting resource in a given environment has a strong influence on the social relationships of primates in that environment (females: van Schaik 1989; males: van Hooff and van Schaik 1994).

Dominance rank among contest competitors is generally determined by those factors that combine to determine fighting ability (resource holding potential, or RHP: Parker 1974). These can include individual size, condition, and experience. Among the more social primates, individuals often attempt to boost their intrinsic RHP by forming a coalition

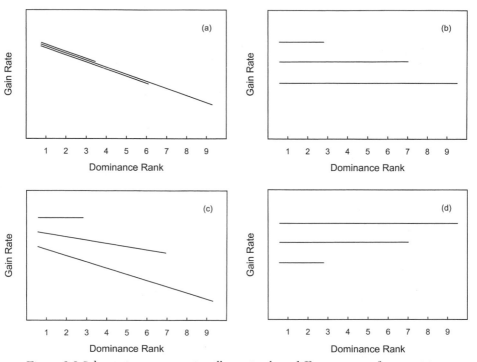

Figure 3.6 Schematic representation illustrating how different types of competition can differentially affect the gain rate of individuals of different dominance rank or group membership. The different competitive regimes shown are (a) pure within-group contest competition, (b) pure within-group scramble competition, (c) a mixture of within-group scramble and contest, and (d) pure between-group contest competition. In each case, gain rate might be measured in a variety of units including food consumed, nutrients obtained, or, in the case of males, copulations or fertilizations achieved. (After Janson and van Schaik 1988.)

with one or more conspecifics (e.g., Harcourt 1989). Since, under contest conditions, individuals of high rank have better access to those resources that limit their reproductive success, they also tend to experience higher reproductive success than subordinates. This is true both for females (Harcourt 1987) and for males (Cowlishaw and Dunbar 1991), although the effects are often stronger in males, since they are the sex characterized by high variance in reproductive success.

Scramble and contest competition can also occur between groups (fig. 3.6). In between-group contest, for example, dominant groups (or coalitions) can outcompete subordinate groups for access to resources. This can be seen when two social groups arrive at a food patch at the same time and fight for the right of access (e.g., capuchin monkeys: Robinson

1988). In contest competition between groups, the dominant group is typically the larger group, because in most cases RHP is determined by the number of individuals in the coalition (Harcourt 1992). One common outcome of between-group contest is territoriality: in this case it is the space that is defended rather than the resources themselves. There is evidence for both food defense and female defense territorial systems in primate populations (e.g., food defense in female gibbons: Cowlishaw 1992, 1996; female defense in male langurs: van Schaik, Assink, and Salafsky 1992). Whether territoriality does emerge from between-group contest appears to depend largely on whether the area of a group's home range is economically defensible in terms of its size and the mobility of the resident group (Mitani and Rodman 1979; Lowen and Dunbar 1994).

3.3.1. Behavioral Strategies

Since female behavior is a key determinant of male behavior in primates, we take females as our starting point. The fundamental question for females is whether access to food and safety can be maximized by foraging alone or with others. In the first case, food acquisition generally becomes more difficult in the presence of coforagers owing to feeding competition. In the second case, by contrast, the avoidance of predation becomes easier in the presence of coforagers owing to the antipredator benefits associated with group living (e.g., Alexander 1974; van Schaik 1983, 1989; Dunbar 1988). In some circumstances coforaging females may also gain a foraging benefit owing to greater competitive ability in between-group feeding competition (Wrangham 1980, 1987; van Schaik 1989).

The magnitude of foraging costs that females are willing to bear in order to maintain safety from predators by living in groups can be substantial. Females may suffer reduced reproductive success through lower nutritional status, direct suppression of the reproductive system as a result of social stress (for physiological details, see Abbott at al. 1986), or both. In the former case, correlations between female body condition and birthrate have been reported for wild long-tailed macaques in Indonesia (van Schaik and van Noordwijk 1985) and wild vervet monkeys in Kenya (Whitten 1983). The effects of the latter have been demonstrated experimentally in captive primates, including both callitrichids and talapoin monkeys (Abbott 1984; Abbott et al. 1986), as well as from observational data in enclosure-living macaques (Silk 1988) and from free-ranging gelada (Dunbar 1980) and *Papio* baboons (Smuts and Nicholson 1989; Wasser and Starling 1988). The effects of reduced food intake and direct stress-induced reproductive suppression can be particularly costly among subordinates. For example, Dunbar (1984) found that each unit drop in social rank incurred by a female gelada baboon resulted in the loss of about half an offspring over a lifetime. Since females produce only about five offspring in a lifetime, this is a very considerable cost: the lowest-ranking female in a group of ten reproductive females stands to have a lifetime reproductive output close to zero.

Given that differences in the costs and benefits of grouping lead females to forage either without other females (e.g., most lorisiformes, hylobatids), in small female groups (e.g., many cebids, colobines), or in large female groups (e.g., many cercopithecines), what are the options for males? Since males wish to maximize the number of females they fertilize, they must distribute themselves around females (and other males) in such a way that they maximize their access to the females.

EMERGENT MATING SYSTEMS

If females forage alone, males face a choice between roving male polygyny and monogamy. Since monogamy constrains male mating opportunities, this strategy should be encountered only when males are forced into it because either females are widely dispersed or males perform a service for females that is essential for the successful production of offspring. On balance, there is little evidence that male primates are forced into monogamy because females occupy ranges so large that a male cannot defend more than one female's territory. Instead, van Schaik and Dunbar (1990) suggested that monogamy evolved in large primates such as gibbons because males provide protection against infanticidal males (see below). In the smaller monogamous primates, this service may relate to reducing predation risk (e.g., the callitrichids: Goldizen 1987b). Direct paternal care is rare in all primates, including monogamous taxa (e.g., hylobatids: van Schaik and Dunbar 1990; monogamous lemurs: Kappeler 1997). Even where such care is observed, it appears to be a consequence of monogamy rather than its cause (the callitrichids: Dunbar 1995a).

Where male services are not essential for successful reproduction, males are expected to adopt roving male polygyny, whereby dominant males achieve higher reproductive success by defending access to several females at once. This form of mating system is common among the nocturnal prosimians (e.g., genus *Galago:* Bearder 1987) and perhaps also the orangutan (Rodman and Mitani 1987). The extent to which a single dominant male can obtain a disproportionate share of paternity by monopolizing breeding access to a number of females is termed the *polygyny skew.* In monogamous mating systems, such skew is minimal; roving male polygyny is likely to be associated with relatively high skew.

If females opt to live in small groups, males will prefer to monopolize the whole group where possible, thereby forming a one-male group (e.g., most guenons and colobines). However, if female group size is large

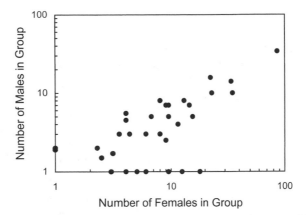

Figure 3.7 The number of males in primate groups is a linear function of the number of females. The plotted data are the mean number of adult males and females in social groups for individual primate species. This result holds up when phylogeny is controlled by independent contrasts analysis. (Data from Dunbar 2000)

(beyond about six to ten females: Andelman 1986) or if the females' reproductive cycles are highly synchronized (Ridley 1986), the male may be unable to prevent influxes by rival males, and the result will be a multimale group in which males compete for access to individual females as they come into estrus. In species that normally live in multimale groups, the number of males is a function of the number of females likely to be in estrus on the same day (Dunbar 1988), which in turn is a function of the number of females in the group (fig. 3.7). Where breeding is seasonal, one-male groups may become multimale during the breeding season, but this pattern appears to be largely restricted to the guenons (Cords 1987).

Polygyny skew tends to be greater in one-male than in multimale groups, owing to the difficulty that dominant males experience in preventing subordinate males from interacting with the females when both exist in the same group. These difficulties become increasingly severe as the number of male competitors in the group increases (Cowlishaw and Dunbar 1991). In multimale groups, dominant males can lose paternity even where they have priority of access to estrous females simply because subordinate males may later mate with the same female (the dilution effect of matings by multiple males); sperm competition within the female's reproductive tract may exacerbate this effect (see Dunbar and Cowlishaw 1992).

These principles may sometimes act to create more complex social systems. Most notable among these are those taxa that live in one-male groups but can also form larger congregations when food is clumped or predation risk is high (e.g., gelada and hamadryas baboons: Barton, Byrne, and Whiten 1996; snub-nosed monkeys: Kirkpatrick 1995); those taxa in which females form groups around males because males provide protection from infanticidal males (e.g., gorillas: Doran and McNeilage 1998; see below); and those taxa in which females are members of large stable groups called *communities,* whose territorial boundaries are defended by coalitions of males but where the group itself is dispersed and association and subgrouping patterns among individuals are highly variable and dynamic (e.g., chimpanzees, spider monkeys: Goodall 1986; Chapman, Wrangham, and Chapman 1995, respectively). Owing to the fluidity in grouping patterns that characterizes the last situation, these social systems have been termed *fission-fusion* societies.

The complexity of primate social systems has recently been underscored by the recognition that behaviors characteristic of fission-fusion systems may be observed in a range of species with quite disparate social patterns. Such complexity may be found, for example, in the social behavior of "solitary" taxa such as nocturnal prosimians (Bearder 1999) and orangutans (van Schaik 1999) (although in the latter the stable community that distinguishes typical fission-fusion societies may be absent). The dynamic aggregation-disaggregation patterns of one-male groups among hamadryas and gelada baboons also have their parallels in the fragmentation of social groups in some lemurs (e.g., *Eulemur* and *Varecia* species: Kappeler 1997), Old World monkeys (e.g., savanna baboons: Dunbar 1992b; long-tailed macaques: van Schaik and van Noordwijk 1988), and New World monkeys (primarily cebids: Kinzey and Cunningham 1994). Encompassed in this range of species it is possible to find aggregations of monogamous and multimale groups in addition to the one-male groups described above. Van Schaik (1999) has suggested that such systems be described as "individual-based fission-fusion" for solitary foragers and those living in communities and "group-based fission-fusion" for group-based foragers. The benefits to individuals of ephemeral grouping are likely to be social in the former (increased mating opportunities and protection from harassment) but ecological in the latter (reduced predation risk).

Finally, an important consideration in many of these systems is the infanticidal strategies of males. Infanticide by males is a reproductive tactic designed to remove the contraceptive effects of lactation and bring females rapidly back into estrus so they can then be fertilized (e.g., Hausfater and Hrdy 1984; Borries et al. 1999). Males typically commit infanticide only against infants they are unlikely to have fathered; consequently, an infanticidal act can both reduce a competitor's reproductive success and increase the infanticidal male's personal fitness. Such behavior is common in one-male groups when a new male takes over the group or in multimale groups when a new male becomes dominant (either from within or from outside the group). Infanticide is particularly common in one-male social systems, perhaps because new males are almost certain not to be the fathers of current infants and because the tenure of the new male in the group may be short, so that males that wait for females to return to estrus naturally may not have much time to reproduce (e.g., Hanuman langurs: Sommer 1987; Sommer and Rajpurohit 1989).

Primates are particularly at risk of infanticide because of females' slow reproductive rates (van Schaik and Dunbar 1990; see sec. 3.1). Given the high risk of infanticide, it has been argued that this selective pressure may have led to the evolution of male-female bonds across the Primate order in those cases where females carry infants rather than park them in nests (van Schaik and Kappeler 1997; Palombit 1999). Such bonds are seen in monogamous, one-male, and multimale social systems. Close bonds between a male and the mother(s) of his offspring may allow the male to adopt defensive behavior to minimize the risk of infanticide.

3.3.2. Grouping Patterns

From basic principles, we can understand why females form groups and why males join these groups. In this section we consider the factors that determine the finer points of group size and composition.

GROUP COMPOSITION AND DISPERSAL PATTERNS

Many primates leave their natal groups and disperse to new areas. Three general patterns may be discerned in this respect(Pusey and Packer 1987; Kappeler 1997): both sexes disperse, though one may travel farther than the other (most colobines and almost all Neotropical primates, as well as many lemurs); males disperse but females remain within the same social group for their entire lives—that is, females are philopatric (most cercopithecines); and females disperse and males are philopatric (chimpanzees, and possibly gorillas and spider monkeys). For the migrating sex, most individuals will migrate at least once during their lives, and some may migrate several times. At Amboseli, for example, Cheney and Seyfarth (1983) found that twenty-six of twenty-eight vervet males transferred from their natal groups at some point in their lives, and Samuels and Altmann (1991) recorded a total of forty-nine migrations involving thirty-seven males over a five-year period in two baboon groups (with some males migrating several times).

Although female philopatry used to be considered the norm for primates, there is growing evidence that bisexual dispersal is probably more typical. This leaves the striking (and near universal) female philopatry of the cercopithecines in need of explanation. There are several factors that might underlie this pattern (Moore and Ali 1984; Pusey and Packer

1987). First, females' knowledge of local food resources may enhance their foraging efficiency and therefore increase reproductive success; consequently, females benefit from philopatry. Second, if females benefit by forming coalitions with other females to obtain access to resources, there are genetic advantages to forming such coalitions with relatives. Third, if a mature male stays in an area where he was born when females are philopatric, he is likely to breed with close relatives, thereby risking increased infant mortality through inbreeding (see sec. 7.2.2). By dispersing at maturity, males can avoid such costs. In addition, males may maximize their reproductive success by ranging over large areas to increase the number of estrous females they come into contact with (Alberts and Altmann 1995). These points may also explain why males often disperse several times during their lives; such behavior ensures that they do not breed with their own female offspring and also allows them to move on to new areas where more fertilizable females may be available.

Where females gain little by philopatry in terms of resource acquisition, feeding competition or the risk of infanticide may make dispersal advantageous for them (see Sterck 1997). In such cases males may benefit from forming groups to defend territories that provide them with exclusive mating access to the females that range within it. Kin selection may then make coalitions between related males ("brotherhoods") advantageous. Common chimpanzees may provide one example of this, although there is some evidence that, while kinship may explain male chimp associations at the community level (Morin et al. 1994), it is less effective at explaining individual patterns of coalition formation within communities (Goldberg and Wrangham 1997). Further details on primate dispersal and intergroup movement are provided in section 6.4.1.

In addition, there are two further points concerning the relative numbers of adult males and females in primate social groups. First, although the number of males in a social group generally correlates with the number of females, there is evidence that the relative number of males in groups increases disproportionately under conditions of high predation risk (van Schaik and Hörstermann 1994; Hill and Lee 1998). This may be for one of two reasons: males may be reluctant to disperse if they are at high risk of predation while transferring between groups (van Schaik and Hörstermann 1994), and males play an important role in reducing predation risk for females and offspring (e.g., red colobus: Busse 1977; capuchins: van Schaik and van Noordwijk 1989; baboons: Cowlishaw 1994; see Hill and Lee 1998). Second, although females generally outnumber males at all group sizes, lemur social groups are exceptional in that the numbers of males and females are usually about equal and one-male groups are absent. The reasons for this divergence are currently uncertain (see van Schaik and Kappeler 1993; Kappeler 1997).

GROUP SIZE

Although we know that female group size ranges from one to many depending mainly on the trade-off between feeding competition and predation risk, we have thus far skirted the question of what determines the precise size of group. We can view this as an optimization problem where a number of constraints act to limit the range of possible group sizes that a species can occupy in a particular habitat (see Dunbar 1996). The range of permissible group sizes is limited by a set of variables that can be divided into benefits, costs, and structural constraints.

The benefits (which relate to predation risk and between-group contest: see above) set the minimum group size in a particular environment. As group size increases, the benefits also increase (e.g., predation risk gets lower—and therefore an individual's fitness increases—as groups get bigger). In contrast, the costs (which relate to within-group scramble and contest competition) place an upper limit on the size of group the animals can cope with. These costs affect time budgets by imposing additional feeding, moving, or social (grooming) time requirements on the animals. Since the amount of time available each day is limited, these time budget elements ultimately place an upper limit on the size of the group. If the cohesion of social groups depends on their members' maintaining social bonds with one another, groups will begin to fragment and may ultimately fission permanently into two or more daughter groups when there is not enough time to devote to servicing these bonds.

The precise values for maximum potential group size in any given habitat can be estimated using systems models that quantify the explicit connections between environmental variables and the behavior of individuals in populations (based on empirical data from those populations). Such models have been developed by Dunbar (1992b,c) using data for gelada and *Papio* baboons and have been extended to chimpanzees by Williamson (1997). These models assume that climatic variables influence time budget and other behavioral ecology variables (e.g., day journey length) either directly (through metabolic costs) or indirectly (through their influence on plant biomass and nutritional content). The flow diagram for the *Papio* model is given in figure 3.8: arrows indicate putative causal influences given by the best fit multivariate regression equations for individual time budget or behavioral ecology variables (for details, see Williamson and Dunbar 1999).

In simplified terms, feeding time is determined by a combination of plant nutritional quality (more feeding time is required as plant quality deteriorates at extremes of temperature and rainfall) and ambient temperature (more feeding time is required to fuel the costs of thermoregulation in cooler climates). Travel time is influenced by the length

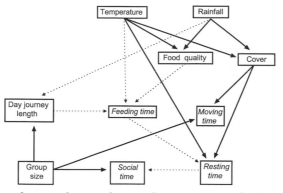

Figure 3.8 Flow diagram showing the causal connections in the *Papio* baboon systems model. Solid lines indicate positive relations, dashed lines negative relations: activity time budget components are italicized. In some cases positive and negative relationships may switch signs where a quadratic function is involved. (After Dunbar 1992d.)

of the day journey (itself a function of group size–dependent feeding demands and habitat patchiness), while social time is a direct function of group size (being the principal mechanism used to maintain group cohesion and integrity). Resting time may itself be subject to climatic influence (resting time increases in high-temperature habitats as animals are forced to rest in order to avoid heat overload), but otherwise it acts as a reserve of uncommitted free time from which additional feeding, travel, and social time can be drawn as changing circumstances demand. Once the model has been built, it can be used to estimate the maximum size of group the species can maintain in a given habitat when all spare resting time has been allocated to feeding, travel, or social time. This value sets the upper limit on group size for that habitat, and this in turn limits species viability, since a species that cannot maintain groups of the minimum size required by a habitat's predation risk obviously cannot survive there.

By combining the species-specific cost, benefit, and structural constraint components in a conventional linear programming model, we can predict the range of habitat-specific realizable group sizes for a given species. Figure 3.9 shows this for the *Papio* model. For graphical convenience, group size is plotted against just one climatic variable (total annual rainfall), although in fact three other climatic variables (mean temperature and two measures of rainfall seasonality) also act as driving variables in the model (see Dunbar 1996). The range of realizable group sizes within the state space is shown by shading. Baboons cannot survive outside the limits of this shaded range either because the ecologically

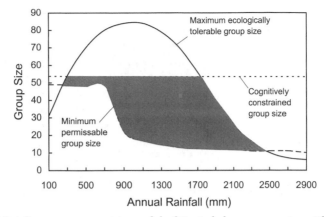

Figure 3.9 A linear programming model of *Papio* baboon group sizes. The plotted value is the permissible upper or lower limits of group size for populations living under different annual rainfall regimes at a mean annual temperature of 20°C (approximately the center of the genus's temperature range). The upper solid line is the maximum ecologically tolerable group size permitted by time budget constraints; the middle dotted line is the approximate limit on group size imposed by cognitive constraints (the exact position of this line is not known at present but may well be somewhat higher than shown); and the lower dashed line is the minimum permissible group size set by predation risk. Baboons can survive in those habitats only where the minimum group size is less than both the maximum group size and the cognitive group size, indicated by the shaded area. (After Dunbar 1996.)

maximum tolerable group size (imposed by time budget constraints from feeding competition) is exceeded or because maximum group sizes are less than the minimum permissible (for the habitat's predation risk), or because the species lacks the neocortex-based cognitive abilities to manage the number of relationships involved (see below).

The final size of group adopted in a particular habitat will depend on whether the animals are more interested in maximizing birthrates or their own (or their offspring's) survival rates. Since birthrates are reduced by grouping, animals that are mainly concerned to maximize birthrates will prefer the minimum group sizes commensurate with minimizing predation risk (i.e., those that lie along the lower bound in fig. 3.9). Those that prefer to maximize survival will opt for the largest possible group sizes commensurate with the constraints imposed by time budgets (i.e., they will lie along the upper bound). Note, however, that a population's options may be limited by the fact that primate groups can undergo fission only if they are at least twice the minimum permissible group size; otherwise the smaller of the two groups will be at too much risk from predation to survive on its own. Some groups may thus overshoot the maximum tolerable group size for short periods despite the ecological (and perhaps reproductive) costs that the animals incur by doing so.

Finally, there may be cognitive constraints on the size of groups that individual species can maintain through time as coherent social groups: this constraint is a function of the size of the species' neocortex and seems to be related to the number of individuals with which an animal can maintain a coherent and stable relationship (Dunbar 1992a, 1998a; Joffe and Dunbar 1997). Although we do not yet know how to integrate these constraints formally into the systems model of figure 3.9, we do at least know that there is a linear relation between neocortex size and mean social group size for individual species. When a group drifts above this limit, it will start to become socially unstable, and sooner or later fission will occur.

These models have been found to be robust under detailed testing, including the prediction of time budgets for individual baboon populations (Dunbar 1992c; Williamson 1997); the prediction of both altitudinal distributions of *Papio* baboons and geladas (Dunbar 1996) and the geographical distribution of chimpanzees (Williamson 1997); and the prediction of patterns of troop fission in baboons (Henzi, Lycett, and Weingrill 1997). These models, however, are likely to apply only to broad interpopulation comparisons, and caution needs to be exercised when applying them to individual populations through time (see Bronikowski and Altmann 1996), since other climatic and environmental variables may become important on the scale of within-habitat comparisons (Williamson and Dunbar 1999).

3.4. Summary

1. Compared with other mammals, primates have unusually slow rates of growth and reproduction, with long interbirth intervals, very small litter sizes, slow development, and extended life spans. It is likely that these features are related to the costs of growing the relatively large brains characteristic of primates as a group.

2. Body size varies markedly across the Primate order. This variation has important biological implications: large primates tend to have relatively slow life histories (leading to low reproductive rates), relatively low basal metabolic rates (allowing them to subsist on poorer-quality foods), and relatively large fat reserves (enhancing their ability to fast for long periods).

3. Predation is an important factor influencing the ecological and behavioral patterns of all primates. Two behavioral strategies adopted to reduce predation risk but with important ecological implications are sociality and selective habitat use.

4. Primates are typically omnivorous, although most species tend to be mainly either folivores (leaf eaters) or frugivores (fruit or seed eaters), with some of the smaller species being insectivorous. Individual primate species can exhibit considerable dietetic flexibility, however; folivores can survive on fruit or seeds (and frugivores on leaves) when circumstances require.

5. The grouping and foraging patterns of female primates tend to be dictated by the availability and defensibility of the resources required for successful reproduction, primarily food and safety from predators. In contrast, male grouping and foraging patterns tend to be dictated by the distribution and reproductive characteristics of the females. Primate social systems are further influenced by the risk of infanticide by males.

6. Primate group sizes are a trade-off between maximizing the benefits of group living (reducing predation risk or increasing group defense of resources) and minimizing the costs of grouping (ecological competition and reproductive suppression), subject to a species-specific cognitive constraint on the upper size limit for coherent social groups.

4 Community Ecology

An understanding of the position of primates in ecosystems is essential for predicting how ecosystem changes will affect primate populations and, equally, how changes in those populations will affect their ecosystems. In this chapter we first describe the processes that might determine species richness in primate communities and the structure and biomass of those communities. We then investigate the fundamental driving force behind the patterns of community diversity and structure, namely ecological competition. Finally, we consider the ecological services that primates might provide to plant communities.

4.1. Community Species Richness

A range of factors can ultimately determine how many species exist in a given community. Our discussion of such factors initially looks at those physical factors that may be most important according to island biogeography theory (MacArthur and Wilson 1967; Rosenzweig 1995), namely the area of a region and the isolation of that region. We then examine the more biologically explicit factors of climatic stress and habitat heterogeneity.

4.1.1. Area and Isolation

The effects of land area and isolation on species richness have been key components of island biogeography theory. Although developed with islands in mind, much of the theory can be applied equally well to other kinds of ecological isolates (e.g., areas of forest isolated by agriculture), as well as to contiguous areas of habitat.

EFFECTS OF AREA

In general, the number of species in a region increases with the area of that region. However, this increase is not linear but logarithmic (fig. 4.1).

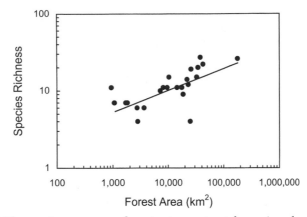

Figure 4.1 The species-area curve for primate species richness (number of species) and original closed forest area in African countries with high primate species richness. The plotted line is a least-squares regression line. (After Cowlishaw 1999.)

This relation is known as the *species-area curve*. On a log-log plot, it produces a straight line with a regression equation of the form

(4.1) $S = cA^z$,

where S is the number of species, A is the area surveyed, and c is a constant. The crucial parameter is z, which describes the slope of the relationship between S and A (i.e., the rate at which the number of species increases with area). There is a lot of scatter in this relationship, mainly because a number of other processes influence species richness (see below).

In table 4.1 we compute z-values from a variety of different databases and compare these with similar estimates from other studies. These values can be used to illustrate several points relating to the species-area curve. First, primate species richness increases with area as predicted, but the relationship is stronger when forest area is used rather than land area (compare r^2 values in set 1). This is probably because total land area includes a variety of nonforest habitats that are unsuitable for primates.

Second, comparing land mammal richness between insular and mainland countries shows that true isolates tend to have larger z-values than samples of equal area taken from continuous habitat (compare z-values in set 2). This is a common pattern across most taxa: z-values for islands tend to range between 0.25 and 0.33, while area samples taken from continental landmasses usually range from 0.13 to 0.18 (Rosenzweig 1995). This means that a change in the area of an isolate will have a greater impact on species richness than a change in the same area of continuous

Table 4.1 The relationship between species richness (number of species) and area in primates and land mammals

Set	Region	Area variable	Taxon	N	r^2	Z-value	Source
1	Global countries	Land area	Primates	15	0.38	0.14	This study[1]
	Global countries	Forest area	Primates	15	0.51	0.12	This study[1]
2	Continental tropical countries	Land area	Land mammals	53	0.56	0.19	Ceballos and Brown (1995)
	Insular countries	Land area	Land mammals	43	0.55	0.36	Ceballos and Brown (1995)
3	Brazilian sites	Land area	Primates	30	n.s.	n.s.	This study[2]

Note: Z-values calculated using linear regression on log-transformed data. Results are only given where $p < .05$ two-tailed (n.s. indicates relationship is non-significant). Note that different studies often use dissimilar sampling methodologies, and different statistical techniques produce very different slope estimates (e.g., line-fitting by reduced major axis produces substantially steeper gradients: Vasárhelyi and Martin 1994). In this table, all tests are conducted using linear regression (the standard technique in analyses of species-area relationships).

[1] Data from Ayres, Bodmer, and Mittermeier 1991; World Resources Institute 1994.
[2] Data from Rylands and Bernardes 1989.

habitat. In addition, for any given area, isolates tend to have fewer species per unit area than nonisolates because the latter will inevitably contain both species that occur locally but make use of resources outside the sampled area (e.g., MacArthur and Wilson 1967) and species that are extinction prone but that are maintained locally by high levels of immigration (the *rescue effect:* see sec. 6.3). Crucially, the difference between area samples and isolates becomes greater the smaller the area under comparison, hence producing larger z-values for isolates.

Third, in the only example of a microscale analysis of the species-area relationship (set 3), the predicted patterns hold up less well. A plot of primate species richness against size of protected area among thirty national parks, biological reserves, and ecological stations in Brazil fails to produce the predicted relationship. This may be because additional processes that influence species richness become increasingly important at a higher resolution (see below). Rylands and Bernardes (1989), for example, point out that one national park and one biological reserve in this data set have particularly high species richness because they extend across rivers: rivers are important biogeographic boundaries and often have different primate faunas on opposite banks.

Finally, why does the number of species increase with the size of the area? At the level of the landmass, such a relationship might reflect an increase in the number of habitats surveyed, but within habitat types this explanation is obviously inadequate. The most probable explanation in this case is that habitat area is a surrogate for population size. Larger areas hold larger populations, and because extinction rates are lower in larger populations (chapter 7), they contain more species. It is difficult to test this explanation directly, although Lawes (1992) has shown that the probability that a forest will be occupied by samango monkeys increases with forest area (as this explanation would predict): very small forests do not contain any monkeys, but as forest size increases so too does the proportion of those forests that do contain monkeys (fig. 4.2).

EFFECTS OF ISOLATION

Regions that are distant from a colonizing source of species contain fewer species than those closer to the source (in marine island systems, for example, this source may be the mainland). This is because shorter isolation distances are associated with higher rates of immigration, and high immigration rates are associated with an increased incidence of both rescue effects, which reduce extinction risk in small populations (sec. 6.3), and recolonization, which reintroduces a species to an area after its local extinction. In primates this is illustrated by the strong relationship between diurnal primate species richness in East African forests and the distance of these forests from a Pleistocene refuge (fig. 4.3).

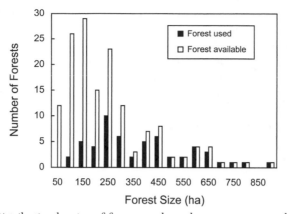

Figure 4.2 Distribution by size of forest patches where samango monkeys *Cercopithecus mitis* are found in South Africa. Open bars indicate the frequency of all forest patches within the range of the samango monkey; closed bars indicate those that are occupied. (After Lawes 1992.)

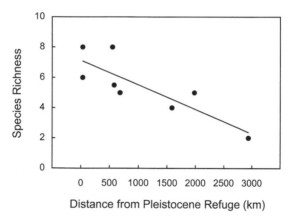

Figure 4.3 Primate species richness (number of species) in East African forests plotted against the isolation distance from the nearest Pleistocene refuge. The plotted line is a least-squares regression line. (After Struhsaker 1981.)

The effects of isolation are particularly important in island communities, whether these communities are on *oceanic islands* (which have always been isolated from mainland areas) or on *continental* or *land-bridge islands* (which have only recently become isolated by a rise in sea level, typically within the past 2 million years). The islands of the Sunda Shelf (including Java, Sumatra, and Borneo), for example, are all land-bridge islands, while the nearby island of Sulawesi is an oceanic island. However, there are also some important differences between these island types. In particular, oceanic islands may never have contained the same number of terrestrial mammal species as a corresponding area of mainland (i.e., they may be ecologically incomplete, or *undersaturated:* Lawlor 1986). This is because the number of such species on oceanic islands is limited by low colonization rates (since mammals are poor over-water colonists: Lawlor 1986; Patterson and Atmar 1986). In contrast, land-bridge islands are likely to have had the same number of species as an equivalent area of mainland at the time of isolation. That these isolates now typically contain fewer species than mainland areas of the same size suggests they have subsequently lost a proportion of their species (for the reasons outlined above). In other words, land-bridge islands are initially *supersaturated* (Lawlor 1986) but subsequently undergo *faunal collapse* (Wilcox 1980). Hence it is argued that the richness of terrestrial mammal species on land-bridge islands is limited not by colonization rate but by extinction rate (Lawlor 1986).

Within island communities, Lomolino (1986) further emphasizes that the interactions between colonization rate and extinction rate are impor-tant determinants of species richness (and indeed the types of species that make up the island communities). In particular, many species with poor colonization ability may nonetheless be found on islands if they are resistant to extinction, because their low colonization rates are balanced by their low extinction rates (see also Peltonen and Hanski 1991). In this context, Patterson and Atmar (1986) found that the nested subsets of mammal species that remain on land-bridge islands after isolation are not a random subset of the nearby mainland mammalian fauna, a pattern they attributed to differential extinction risk (rather than colonization ability) among the mammalian taxa (see chapter 7). However, it is diffi-cult to gauge the relative colonization abilities of different primate taxa. Although island colonizations have occurred (e.g., Madagascar: Mitter-meier et al. 1994; West Indies: Horovitz and MacPhee 1999), they appear to be very rare (on the order of hundreds of thousands of years for a single event: Heaney 1986; Morgan and Woods 1986).

Given the importance of extinctions following isolation on land-bridge islands, it would be useful to know how rapidly they might occur. Heaney (1986) plotted species-area curves for terrestrial mammals in peninsular Malaysia (not isolated), the Sunda Shelf islands (isolated approximately 12,000 years ago from peninsular Malaysia), and the Palawan Islands of the Philippines (isolated from Borneo approximately 160,000 years ago). These three curves, which represent a time series of isolation events, are shown in figure 4.4. As we would expect, the isolates lie below the main-land areas and have higher z-coefficients. Note that the difference be-tween the two isolate groups is minimal compared with the difference between these two isolate groups and the mainland, suggesting that there

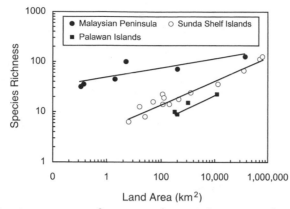

Figure 4.4 Species-area curves for terrestrial mammals in peninsular Malaysia, the Sunda Shelf islands, and the Palawan Islands of the Philippines. The plotted lines are least-squares regression lines. (After Heaney 1986.)

may be a high rate of extinction soon after isolation and a much lower rate subsequently. For an island of 10,000 km² this equates to a 61% loss of species within the first 12,000 years, but only a further 17% loss during the next 150,000 years.

Heaney (1986) also emphasized the importance of speciation events in determining species richness on older isolates: in some parts of the Philippines the speciation rate appeared to be double the colonization rate (again, this is perhaps not surprising given the poor overwater colonization ability of mammals). For long-isolated primate communities such as those on Madagascar, speciation has been a particularly important determinant of current species richness.

4.1.2. Environmental Processes

A variety of environmental factors might also influence primate species richness. We focus here on two potential key forces: climatic stress and habitat heterogeneity.

EFFECTS OF CLIMATIC STRESS

There is evidence that climatic stress might reduce primate species richness. In general, cold, seasonal environments tend to have fewer primates than warmer, less variable environments. Hence species richness tends to decline at higher altitudes and higher latitudes. High-altitude grassland habitats, for example, tend to support far fewer primate species than low-lying savanna habitats: compare the Simen Mountains of Ethiopia (3,300 m above sea level [asl]) and the Drakensberg Mountains of South Africa (2,200 m asl), each with their single resident primate species, with the lower-lying Amboseli National Park in Kenya (1,100 m asl, three species) and the Awash National Park in Ethiopia (950 m asl, four species). Similarly, primate species richness declines with distance from the equator in most geographical regions, including Africa (fig. 2.3: see Cowlishaw and Hacker 1997) and South America (Ruggerio 1994; see also Freese et al. 1982).

This pattern may exist because primates find it difficult to cope with low temperatures and variable environments. This explanation is supported by the fact that among Neotropical primates the small-bodied taxa, which are less likely to be able to sustain high energetic costs of thermoregulation or to survive long periods of food scarcity (see sec. 3.1.1), are less likely to be found at higher latitudes (Pastor-Nieto and Williamson 1998). However, whether temperature or seasonality plays the key role remains difficult to ascertain. Current evidence suggests the latter may be more important, since Cowlishaw and Hacker (1997) found that the latitudinal extent of species geographic ranges among African primates was not correlated with mean temperature but was strongly correlated with the seasonality of rainfall.

EFFECTS OF HABITAT HETEROGENEITY

Across primate communities, the most important determinant of fine-grained patterns of species richness may be the diversity of the habitat. In undisturbed forest (western Madagascar: Ganzhorn 1994), logged forest (Kibale Forest, Uganda: Skorupa 1986) and fragmented forest (north of Manaus, Brazil: Schwarzkopf and Rylands 1989), there is a positive association between primate species richness and habitat heterogeneity. Ganzhorn (1994), for example, reported a positive correlation between lemur species richness and tree species richness. Skorupa (1986) described a similar correlation between primate species richness and both tree species richness and a tree species diversity index. Finally, Schwarzkopf and Rylands (1989) found that the number of primate species was higher when the number of trees increased, the proportion of large trees declined, and the number of lianas increased (although the independent effect of each of these variables was not gauged). The simplest explanation for these patterns is that diverse habitats have more ecological niches, permitting more species to exist there (see also Rosenzweig 1995).

Regional diversity at a broader scale may similarly be influenced by habitat heterogeneity, given that forested habitats typically harbor more primate species than woodland or savanna and that these habitats are floristically more complex. (This pattern may also help to explain why primate diversity declines with distance from the equator, since the distribution of tropical forests is centered on the equator and gives way to woodland and savanna at higher latitudes.) Primate communities in

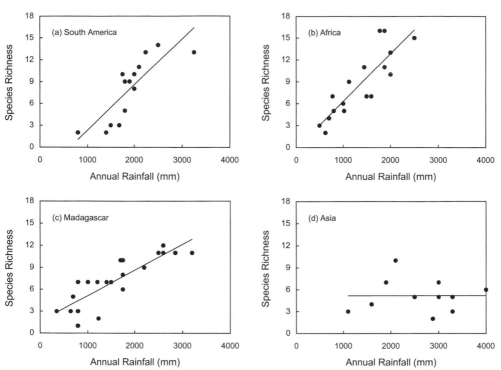

Figure 4.5 Primate species richness (number of species) plotted against annual rainfall across sites in (a) South America, (b) Africa, (c) Madagascar, and (d) Asia. The plotted lines are least-squares regression lines. (After Reed and Fleagle 1995.)

forested habitats typically contain as many as ten to fifteen species, while savanna habitats typically contain only one to five (see Bourlière 1985). Since the degree of forest development is partly dependent on the amount of rainfall, it is perhaps not surprising that there is frequently a positive relationship between the number of primate species and both minimum rainfall (for lemurs in western Madagascar: Smith, Horning, and Morre 1997) and mean annual rainfall (for primates in South America, Africa, Madagascar, and Asia: Reed and Fleagle 1995; see fig. 4.5).

Examining the regression slope parameters for the distributions in figure 4.5 reveals that Africa and South America do not differ significantly but that Madagascar has a slope that is about half the value for these two continental masses ($b = 0.0035$ vs. $b \approx 0.007$). Reed and Fleagle (1995) suggest that this may reflect the recent extinction of many Madagascan prosimian species: approximately a third of the forty-seven known extant and subfossil lemur species went extinct in the 1,500 years following the arrival of humans on the island (see sec. 7.4.1). Increasing all community sizes by a third in the Madagascan sample produces a regression equation with the same slope parameter ($b \approx 0.007$) as the mainland African and South American samples, thus supporting Reed and Fleagle's suggestion.

The reason the Asian habitats should be so different is less clear, however. Reed and Fleagle (1995) propose that this might be either because Asian habitats are broken up into a large number of small islands (a factor that is likely to limit species number: see above) or because Asian monsoon climates are very seasonal. Alternatively, Terborgh and van Schaik (1987) argue that the tropical forests of Southeast Asia are much less productive than those of Africa; in addition to low productivity (as measured by fruit and litter production), many Southeast Asian forests are dominated by dipterocarps whose leaves and fruits are heavily defended chemically (Proctor 1995). As a result, we might suppose that Asian forests will not become progressively richer for primates as rainfall increases.

Kay et al. (1997) have argued that these studies underestimate the complexity of primate community structure because they are based on a small sample of sites with a very limited range of rainfall values. In a sample of more than one hundred South American sites with rainfall volumes ranging up to about 7,000 mm per year, they claim that the distribution of primate species numbers is in fact ∩-shaped and that this is because, although wetter sites are more forested, their productivity progressively declines when annual rainfall exceeds about 3,000 mm. However, their evidence for the downturn in species number is based on just six sites at which rainfall exceeds 4,000 mm: if these sites are excluded as outliers, it is very difficult to justify anything other than a linear relation between rainfall and species numbers. Similarly, their claim that forest productivity declines when rainfall exceeds 3,000 mm a year is based on just five sites: if these are excluded, the graph has an asymptotic shape that mirrors the distribution of tree species number and the number of wet months (months with >100 mm rainfall).

THE AFRICAN EVIDENCE

To explore the ecological determinants of species richness in primate communities in more detail, we determined the number of species at all study sites where baboons (*Papio* and *Theropithecus*) have been studied in Africa. Baboons make a useful sampling marker because they occur in almost all habitat types. There are twenty-nine sites in the sample, drawn from across sub-Saharan Africa and representing a wide range of habitat types and altitudes (table 4.2). Area and isolation effects were not incorporated in this analysis, but potential effects were examined for three key geophysical variables: altitude, latitude, and rainfall. A multiple regres-

Table 4.2 Numbers of primate species at sites where baboons have been studied

Site	Country	Latitude (°lat)	Altitude (m)	Annual rainfall (mm)	Number of primate species
Niokolo-Koba NP	Senegal	13°00'	150	940	3
Simen Mountains	Ethiopia	13°15'	3,300	1,385	1
Debra Libanos	Ethiopia	9°40'	2,000	1,100	2
Mulu	Ethiopia	9°25'	2,300	1,100	2
Bole Valley	Ethiopia	9°25'	1,700	1,100	4
Erer Gota	Ethiopia	9°20'	1,200	665	1
Awash Station	Ethiopia	9°00'	915	665	2
Metahara	Ethiopia	8°50'	950	665	2
Awash NP	Ethiopia	8°50'	950	640	2
Murchison NP	Uganda	2°12'	1,100	1,140	5
Budongo Forest	Uganda	1°45'	1,015	1,570	5
Ishasha	Uganda	0°40'	950	970	4
Kibale Forest	Uganda	0°15'	1,590	1,475	9
Chololo	Kenya	0°20'	1,660	475	2
Gilgil	Kenya	0°30'	1,770	690	1
Masalani	Kenya	2°40'	920	620	1
Amboseli NP	Kenya	2°40'	1,130	775	3
Diani	Kenya	4°11'	25	1,077	6
Ruaha NP	Tanzania	7°40'	1,230	580	3
Mikumi NP	Tanzania	7°15'	550	850	5
Okavango Swamps	Botswana	19°00'	300	457	2
Tsaobis Park	Namibia	22°23'	1,000	85	1
Kuiseb	Namibia	24°00'	1,000	18	1
Honnet Reserve	South Africa	22°36'	310	305	2
Mkuzi	South Africa	23°00'	100	630	2
Drakensberg	South Africa	29°00'	2,000	1,140	1
Mount Zebra NP	South Africa	32°10'	1,500	364	1
Cape Point	South Africa	34°15'	50	635	1
de Hoop Reserve	South Africa	30°27'	40	380	1

Source: Data from Dunbar (1992b and unpublished).

NP = National Park.

sion analysis with number of species as the dependent variable yielded the following equation:

$$N_{species} = 2.36 - 0.0013\ (Alt) - 0.56\ \ln(Lat) + 0.0035\ (Rain)$$

($r^2 = .70$, $F_{3,25} = 19.00$, $p < .0001$), where $N_{species}$ is the number of species present, Alt is the altitude of the site (in meters), Lat is absolute latitude (in degrees), and $Rain$ is the mean annual rainfall (in millimeters) (fig. 4.6). Each of the slope parameters for the three independent variables is significant (Alt: $t = -4.16$, $p < .001$; Lat: $t = -3.41$, $p = .002$; $Rain$: $t = 5.76$, $p < .001$). Altitude and latitude are the major influences on both climatic variability and ambient temperature, while rainfall is a correlate of the degree of forestation (and hence habitat diversity). In Africa

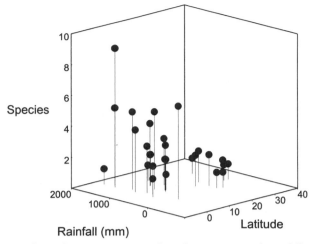

Figure 4.6 Number of primate species plotted against annual rainfall and latitude across sites in sub-Saharan Africa. The sampled sites are all locations where baboons have been studied. (Data from table 4.2.)

at least, the number of species in a primate community increases in habitats that are warmer, more stable, and more heterogeneous.

4.2. Community Structure

The mammalian herbivore/frugivore biomass in tropical communities can vary considerably between sites, even within similar habitat types (table 4.3). Across tropical forest sites, Oates et al. (1990) and Tutin et al. (1997) argue that the variance in biomass may be due to differences in resource availability during ecological "crunch" seasons (usually the dry season when most tree species are reproductively quiescent). Many African tropical forests, for example, experience annual bottlenecks of this kind that can last up to three months. These are likely, in turn, to reflect regional and local differences in climate and soil fertility. We return to these issues below in the specific context of primate communities.

The primate contribution to total biomass also shows marked variation between habitat types. In the open savanna woodland habitat of the Amboseli basin (Kenya), primates are both absolutely and relatively rare, whereas at a rain forest site like Manu National Park (Peru) they make up nearly half the total mammalian biomass. But vegetational complexity does not explain all the variance in primate biomass: some otherwise ecologically poor communities can support very high primate biomasses, although these tend to involve ecological specialists that can exploit a particular kind of habitat. Geladas, for example, can maintain extremely high densities on the species-poor Afro-alpine grasslands of the Ethiopian

Table 4.3 Composition of the mammalian herbivore/frugivore biomass in a number of tropical habitats occupied by primates

Region and site	Habitat type	Biomass (kg/km²)[1]					Primates as percentage of total	Source
		Primates	Terrestrial mammals	Arboreal mammals	Bats	Total		
Africa								
Amboseli, Kenya (ca. 1970)	Wood- and grassland	11 (2)	4,848	0	0	4,859	0.2	1,2
Simen Mountains, Ethiopia	Afroalpine moorland	648 (1)	1,890[2]	0	0	2,530	26	3
Karisoke, Rwanda	Montane forest	84 (1)	3,017	0	0	3,101	3	4
Southwest Gabon	Lowland rain forest	255 (5+)	758	?	?	1,013+	~25	5
Lopé, Gabon	Lowland rain forest	319 (7)	2,776	5	?	3,095	~10	6
Asia								
Kuala Lompat, Malaysia	Lowland rain forest	945 (6)	?	?	?	?	?	7
Ketambe, Sumatra	Lowland rain forest	837 (6)	?	?	?	?	?	7
Kutai, Borneo	Lowland rain forest	324 (5)	?	?	?	?	?	7
The Americas								
Los Tuxtlas, Mexico	Relict rain forest	171 (2)	345	?	?	516	33	8
Manu NP, Peru	Riverine forest	650 (10)	512	170	(50–100)	ca. 1,400[3]	46	9
BCI, Panama	Lowland rain forest	421 (5)	230	793	?	1,444	29	10
Guatopo, Venezuela	Lowland rain forest	167 (3)	773	208	?	946	18	11

Sources: (1) Coe, Cummings, and Phillipson 1976; (2) Altmann, Hausfater, and Altmann 1985; (3) Dunbar 1978; (4) Plumptre and Harris 1995; (5) Prins and Reitsma 1989; (6) White 1994; (7) Terborgh and van Schaik 1987; (8) Estrada and Coates-Estrada 1985; (9) Terborgh 1983; (10) Eisenberg and Thorington, R. 1973; (11) Eisenberg, O'Connell, and August 1979.

[1]Numbers of species are given in parentheses. Terrestrial and arboreal mammals exclude primates.

[2]Includes domestic stock (cattle and horses).

[3]Birds account for an additional ca. 200 kg/km².

NP = National Park; BCI = Barro Colorado Island.

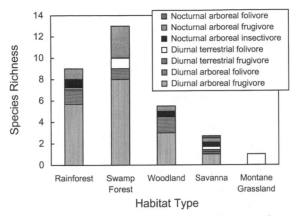

Figure 4.7 Mean number of sympatric primate species (species richness) in different feeding guilds at a selection of sites. Guilds are classified by activity period, arboreality, and diet type. The diurnal, arboreal frugivore guild is by far the commonest. (Data from Bourlière 1985.)

plateau only because they are grazers; in these habitats they may constitute as much as 58% of the wild herbivore biomass (Dunbar 1978).

In the following sections we examine in detail patterns of primate community structure and the dependence of primates on keystone resources. The role of primates themselves as keystone species we return to subsequently.

4.2.1. Primate Community Structure

In this section we address two issues. First, we describe how the distribution of primate species across guilds varies between communities. Second, we examine how the distribution of primate species biomass across niches varies between communities and investigate the ecological-evolutionary processes that underpin this variation.

PRIMATE GUILDS

In section 3.2 we described Bourlière's (1985) categorization of primate ecological guilds (table 3.1). Although a total of twelve guilds can be differentiated in terms of the three variables Bourlière considered (activity period, diet, and arboreality), only seven are in fact occupied by living primates, and two of those are occupied by fewer than four species each. The distribution of these seven guilds across ecological community types is also uneven (fig. 4.7). Savanna habitats in particular tend to be more uniform in their species composition (Shannon's diversity index $h = 1.75$ in savanna habitats, compared with $h = 0.90$ for woodland habitats and $h = 1.13$ in rain forest habitats: data from Bourlière 1985). The main reason is that savanna habitats tend to have a single representative species in each guild (or group of guilds), whereas forested habitats tend to have a large number of species in the diurnal arboreal frugivore guild (approximately 60% of all species present in a given habitat may belong to this one guild, hence its overriding dominance in table 3.1). The possible reasons for the preponderance of species in this guild in forest habitats are discussed below.

Other important patterns to notice in figure 4.7 are that forested habitats invariably contain absolutely more species than open savanna or woodland habitats; that swamp forests seem to be especially rich even by forest standards; and that two guilds (nocturnal and diurnal arboreal frugivores) are particularly associated with forest habitats and are largely responsible for their characteristic community structure.

PRIMATE BIOMASS

Breaking down the primate biomass into different trophic niches provides further insight into community structure (table 4.4). The most important features can be summarized as follows. First, insectivorous primates account for a relatively small proportion of total primate biomass (often as little as 1%). Second, terrestrial primates also typically contribute a very small proportion of the total biomass. Third, the bulk of primate biomass is made up of arboreal frugivores. Marsh and Wilson (1981) found that in Malaysia two folivores (both *Presbytis* spp.) averaged 82% of the primate biomass at five lowland rain forest sites and about 75% at three montane forest sites, but the pattern does vary across habitats: the equivalent figure was only about 26% in six swamp forest sites. Fourth, even though folivores are a significant proportion of the total community biomass (and a mean of 32.6% of the primate biomass), they tend to be represented by a single species (whereas frugivores, which account for only twice as much biomass, are typically represented by four to nine species). The number of frugivore species exceeds the number of folivores in all eleven of the sites listed in table 4.4, and in only two of the sites is there more than one species of folivorous primate.

This result seems to run counter to the suggestion that communities that have an imbalanced structure across trophic categories are intrinsically unstable. Fox (1987) has argued that new additions to communities will normally come from a different "functional group" (i.e., trophic category) than those already represented until they are evenly balanced, after which the cycle will repeat itself. Fox's hypothesis predicts that the distribution of species among the main trophic categories will normally be in approximate balance, and that when this is not so either a new species will be added by invasion or some of those present will become locally extinct through competition. Ganzhorn (1997) tested Fox's rule on

Table 4.4 Biomass of different primates at various rain forest sites

Biomass (kg/km²)[1]

Region and site	Insectivores	Frugivores		Folivores		Source
		Arboreal	Terrestrial	Arboreal	Terrestrial	
Africa						
Lopé, Gabon	?	137 (4)	89 (2)	91 (1)	0	White 1994
Douala-Edéa, Cameroon	?	211 (6)	? (1)	198 (1)	0	Oates et al. 1990
Campo, Cameroon	?	273 (5)	c. 10 (1)	0	0	Mitani 1991
Tiwai, Sierra Leone	?	997 (7)	0	382 (1)	0	Oates et al. 1990
Madagascar						
Andasibé	3–5	233 (6)	0	225 (4)	0	Ganzhorn 1988
Ampijoroa	3	774 (4)	0	51 (1)	0	Ganzhorn 1988
Asia						
Kuala Lompat, Malaysia	25 (1)	(583 [4])		337 (2)	0	Terborgh and van Schaik 1987
Ketambe, Sumatra	0	(702 [5])		>135 (1)	0	Terborgh and van Schaik 1987
Kutai, Borneo	0	(242 [4])		82 (1)	0	Terborgh and van Schaik 1987
The Americas						
BCI, Panama	0	29 (4)	0	416 (1)	0	Terborgh and van Schaik 1987
Manu NP, Peru	1	467 (9)	0	180 (1)	0	Terborgh and van Schaik 1987

[1]Number of species given in parentheses; NP = National Park; BCI = Barro Colorado Island.

Madagascan lemur communities using three trophic categories (folivores, frugivores, and omnivore-faunivores). He found that, in eleven out of fourteen communities, the species composition (including subfossil species) was significantly more evenly distributed between the three categories than would be expected if species had been drawn at random from the thirty-two species available in the total sample. In at least one of the three exceptions (Berenty), the imbalance in community structure results from the recent introduction of a new species (*Eulemur fulvus*); Ganzhorn (1997) predicts that this will eventually result in the local extinction of one of the three frugivore species now present at this site.

The data presented in table 4.4 suggest that Fox's rule may not apply in continental tropical forests. This is puzzling given the evidence for its validity elsewhere. One possible explanation is that the imbalance in favor of frugivores is redressed by an equivalent number of nonprimate folivore and omnivore species in these habitats. In other words, one reason Fox's rule might hold for Madagascan lemur communities is that prosimians constitute most of the species in these communities. The data to test this explanation are not readily available. A second possibility might be that the imbalanced communities are of relatively recent origin and we may yet see a significant winnowing of frugivore species until the communities are once more in trophic balance. If so, then the unusually speciose and commonly sympatric *Cercopithecus* genus is likely to bear the brunt of these competitive exclusions.

Why should folivores contribute such a large biomass but such a small number of species in many primate communities? One explanation might be the recency of evolutionary divergence among catarrhine frugivores. However, the contrast still holds up when genera are considered and is as true of New World as of Old World communities. Alternatively, this contrast might reflect a need for greater ecological separation among frugivores in forest habitats. Fruits are, broadly speaking, less abundant than leaves, and their availability is often highly seasonal. Janson and Emmons (1990) point out that fruiting cycles are strongly synchronized at Manu National Park (Peru), with the result that frugivores face a serious bottleneck during the summer months. Such bottlenecks are likely to require considerable ecological diversification to allow several sympatric frugivorous populations to persist through lean seasons (sec. 4.3.1). At Lopé (Gabon), the biomass of individual primate species correlates negatively with their degree of frugivory (Tutin et al. 1997). In contrast, the biomass of folivores may be less severely affected by seasonality because such taxa can survive on poorer-quality resources (sec. 3.2.2; see also sec. 5.2).

Similarly, Terborgh and van Schaik (1987: see also Fleagle and Read 1996; Kappeler and Heymann 1996) describe important regional differences in the biomass of primate communities, with South American habitats often supporting a lower biomass than Old World sites even though their communities contain a similar number of species. This seems to be associated with an underrepresentation of specialized folivores among the Neotropical primate communities. In fact, such communities are consistently less ecologically diverse than those elsewhere (Fleagle and Reed 1996; Kappeler and Heymann 1996; Jernvall and Wright 1998): for example, they also exhibit a relatively narrow range of body sizes. Terborgh and van Schaik (1987) have argued that most of these differences can be explained by the very different climatically driven phenological patterns characteristic of Neotropical forests. Because leafing and fruiting cycles are both more seasonal and more tightly synchronized in South American forests, there are seasons when neither leaves nor fruits are widely available. This severely limits the opportunities both for folivorous species (which largely depend on leaves throughout the year) and for other large-bodied species (which in Old World communities typically fall back on leaves in lean seasons): in contrast, nonfolivorous small-bodied species are less severely affected, since they can survive on the sparse resources remaining (e.g., nectar, insects).

In a further analysis of regional patterns, Fleagle and Reed (1996) compared the ecological profiles of primate species from two representative forest habitats in each of the four main biogeographic areas occupied by primates (South America, Southeast Asia, Africa, and Madagascar). Although overall there were broad similarities in the patterns of primate communities from all biogeographical regions, some crucial differences emerge. The most important feature of the clustering patterns was phylogenetic relatedness: closely related species tended to have similar ecological characteristics. Because the biogeographic regions have different histories, the species that make up their communities tend to be different, and this in turn creates marked differences between the regions in the details of their community profiles. Communities within the same biogeographic region therefore tend to be more similar to each other than they are to communities in other regions (and they often share many of the same species or genera).

Finally, it is important to remember that a variety of other factors, in addition to phenology and phylogeny, might underlie patterns of convergence and nonconvergence across primate communities, including differences in soil fertility, forest structure, climatic history, and predator guilds (see Oates at al. 1990; van Schaik, Terborgh, and Wright 1993; Kappeler and Heymann 1996; Tutin et al. 1997).

4.2.2. Keystone Species

The term *keystone species* has come to occupy an important place in community ecology. A keystone species is a taxon on whose presence the structure or processes of the community depend (e.g., Terborgh 1986).

Figs (*Ficus* spp.) are keystone species for many frugivorous forest primates (e.g., Southeast Asia: Leighton and Leighton 1983; South America: Terborgh 1986). *Saimiri,* for example, are dependent on access to fig fruits in many parts of their distribution (Terborgh 1983). Because *Ficus* species tend to fruit asynchronously, it will usually be possible to find at least one fig tree in fruit somewhere, although the area required to ensure this will be inversely proportional to the density of fig trees (Dunbar 1988). Hence, when fig trees are at low densities, *Saimiri* require large ranges. In contrast, marmosets (*Callithrix* spp.) resort to gums scraped from tree trunks during the dry season (Sussman and Kinzey 1984; Ferrari and Lopez Ferrari 1989) and are thus able to survive in much smaller ranges.

Keystone species may be crucial to the survival of primates at some locations. Terborgh (1983), for example, found that figs and the nuts of *Scheelea* palms accounted for 73% of the diet of *Cebus apella* and 97% of that of *C. albifrons* at Manu National Park during the five-month dry season. Without access to these species, *Cebus* might be hard pressed to survive at this site. In addition, it is probably access to these foods that allows *Cebus* to live in considerably smaller home ranges than the sympatric tamarins (*Saguinus* spp.), even though the tamarins are physically much smaller (Terborgh 1983). *Scheelea* nuts are also a secondary keystone resource for *Saimiri* in the same area, though in this case the monkeys depend on *Cebus* monkeys to open the kernels.

Keystone resources will thus tend to be species-specific and to reflect the dietary adaptations of the species concerned. However, it does not necessarily follow that the same food species will be a keystone resource for all populations of a given species. Wrangham et al. (1993) emphasized the importance of figs as a keystone resource for chimpanzees in East African forests, but Tutin et al. (1997) argued that figs are not sufficiently dependable at Lopé (or elsewhere in Gabon) to be a keystone resource for any primate species (see also Gautier-Hion and Michaloud 1989). Instead, White (1994) pointed to herbaceous plants of the Marantaceae and Zingiberaceae as being particularly important in maintaining high population densities of gorillas and elephants at Lopé, precisely because (unlike figs) they are available year round. He found that the biomass of both elephants and gorillas was highest in the two habitat types at Lopé where these plants were most abundant (essentially second-stage colonizing forest in the succession from savanna to full forest). Indeed, Oates (1996b) has argued that terrestrial herbaceous vegetation (THV) may be a keystone resource for gorillas throughout Africa.

The continued survival of an entire primate community at a particular site may depend on there being sufficient numbers of different keystone resources to reduce ecological competition between species during the dry season when preferred foods like fruits are not widely available. This has been illustrated for primate communities in Africa (Lopé: Tutin et al. 1997), Asia (Kutai, Borneo: Leighton and Leighton 1983), and the Americas (Urucu River, Brazil: Peres 1994). For example, at Lopé, mangabeys (*Lophocebus albigena*), colobus (*Colobus satanas*), and mandrills (*Mandrillus sphinx*) rely on seeds that they can exploit because of their greater strength and their ability to detoxify them. Gorillas exploit pith and bark, while the chimpanzees rely on leaves; only the three small-bodied guenons (*Cercopithecus nictitans, C. pogonias,* and *C. cephus*) continue to eke out a fruit-dominated diet (Tutin et al. 1997).

4.3. Competition in Communities

Ecological competition between members of the same community has long been viewed as an important factor driving the evolution of community structure. One of the fundamental principles underlying this view has been the assumption that ecologically similar species cannot coexist: one or the other will change its ecological niche or become locally extinct (Begon, Harper, and Townsend 1986). This is often referred to as the *competitive exclusion principle.* Although the issue of competition and coexistence is considerably more complex, the difficulty species face when competing for the same resources provides a useful starting point for considering the ecology of primate communities. Indeed, that at least some communities contain large numbers of primate species raises pressing questions about how ecological competition is avoided in these cases. In this section we examine how primates compete with other primate species and with nonprimates.

4.3.1. Competition between Primate Taxa

The most important determinant of niche partitioning is likely to be diet (see Cords 1986; Struhsaker 1978). Tutin et al. (1997) found that, of the four hundred species of plants eaten by the eight species of primates in the Lopé Reserve (Gabon), almost half (46%) were eaten by a single species and just four (1%) were eaten by all eight species. Dietary differentiation was most apparent during the dry season when fruit was scarce. This was particularly true for gorillas and chimpanzees, whose wet season diets overlapped considerably: during the dry season, gorillas often resorted to feeding intensively on foliage and bark (food sources for chimpanzees only during famine years), and the two apes had mutually exclusive keystone resources (*Duboscia macrocarpa* for gorillas and *Elaeis guineensis* for the chimpanzees) (Tutin, Fernandez, et al. 1991).

Yet despite different dietary strategies, evidence still suggests that different species compete for the same food resources in many primate

communities. Rodman (1973) showed that there were significant negative relationships between the frequencies with which pairs of species
were sighted in 200 × 200 m quadrats among the five species of primates
present in a primate community in Borneo. He interpreted this as implying that the competition between species was sufficient to generate
competitive exclusion. The only species whose distribution did not seem
to affect (or be affected by) the presence or absence of other species
was *Macaca nemestrina:* this species had least ecological overlap with the
other four (*Pongo pygmaeus, Hylobates muelleri, Presbytis aygula,* and
Macaca fascicularis), perhaps because it mainly used the ground-level
vegetation whereas the others were largely arboreal.

We noted earlier (sec. 4.2) that the frugivore guild tends to contain
many species in most tropical forest communities, whereas the folivore
guild rarely contains more than one. That there is rarely more than one
folivorous monkey species in a community might be interpreted as implying that ecological competition is important. But the same data should
also prompt us to ask how it is that the guild of frugivores does not seem
to suffer the same fate. Analyses of the ecological overlap between frugivore species at several African rain forest sites suggest that species may
separate out in physical space even when they appear to be competing
for the same resources (fruits).

Forest guenons belong to the same genus and share a similar diet, and
many species are frequently sympatric. In the South Bakundu Forest
(Cameroon), Gartlan and Struhsaker (1972) found that the various species present tended to forage at different heights in the canopy: *Cercopithecus erythrotis* and *C. mona* were found in the lower strata (at about
10 m), whereas *C. pogonias* preferred the upper layers (about 25 m);
C. nictitans (the largest of the four species) ranged widely throughout
the various canopy layers, though with a tendency to prefer the intermediate levels between 15 and 25 m (fig. 4.8). The two smallest species,
C. erythrotis and *C. pogonias,* were found at the lowest and highest levels, respectively, suggesting that their use of the canopy may be affected
by the preferences of the three larger species. In the Kibale Forest
(Uganda), Struhsaker (1978) found that *Cercopithecus ascanius* (the
smallest of the five monkeys present) exhibited the least marked preferences in feeding height, apparently being obliged to feed at whatever
levels the other four species were not using. Similar patterns have been
reported from other West African guenon communities (Gautier-Hion,
Quris, and Gautier 1983; Mitani 1991), as well as from Amazonian forests
(Terborgh 1983; Peres 1993). Peres (1993b), for example, found that
67% of the variance in foraging height among five species of cebids was
explained by body mass.

This pattern is not limited to closely related frugivores. In the Krau

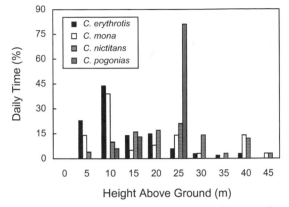

Figure 4.8 Foraging heights for four sympatric species of *Cercopithecus* monkeys at South Bakundu, Cameroon. (Data from Gartlan and Struhsaker 1972.)

Reserve (Malaysia), Mackinnon and Mackinnon (1978) found that frugivorous species belonging to different genera (e.g., *Hylobates* vs. *Macaca*) overlapped less in diet than folivorous species of the same genus (e.g., *Presbytis melalophus* vs. *P. obscura*). These folivores minimized competition by using different layers of the canopy (see also Fleagle 1978).

Little is known about the dynamics of any of these systems, and we cannot say for certain whether these foraging heights reflect real preferences or enforced patterns. Nonetheless, some studies do provide evidence that body size may be a determinant of relative dominance among the species in a community. Terborgh (1983) reports that the larger *Cebus* monkeys would occasionally lose patience with the *Saimiri* that foraged with them and expel them from the tree in which they were all feeding. Similarly, in competition for access to the small feeding sites (e.g., vines and small trees), the larger *Saguinus imperator* were always able to displace the smaller *S. fuscicollis* (Terborgh 1983). Peres (1996) reported that the physically larger *S. mystax*, which also benefited from having larger social groups, were invariably able to monopolize small food patches at the expense of the smaller *S. fuscicollis avilapiresi*. Vervet monkeys are also nervous around, and give way to, the very much larger baboons in most African habitats where they coexist (e.g., Dunbar and Dunbar 1974a). But sheer size need not always decide dominance: Kummer (1968) and Dunbar and Dunbar (1974b) noted that gelada baboons (*Theropithecus gelada*) usually give way to the smaller but considerably more aggressive hamadryas baboons (*Papio hamadryas*).

Ganzhorn (1989) conducted the only study that provides a detailed analysis of the relative importance of diet and foraging height in niche partitioning. He used a principal components analysis to assess niche overlap in seven species of lemurs at Analamazoatra (a rain forest site in eastern Madagascar). He found that the species could be separated ecologically along two key dimensions (foraging height and diet composition). In general, species that were similar in diet differed in preferred feeding heights, and those that made similar use of the habitat tended to have different food preferences. Only *Cheirogaleus major* (the smallest of the set aside from *Microcebus*) seemed to show marked overlap with other species. But ecological competition was probably reduced in its case by the sheer breadth of its niche space, which extended into sectors of the state space not occupied by any other species. It is possible that, once again, small body size has forced this species to broaden its dietary niche to avoid competition. *Microcebus*, in contrast, probably reduces competition with the other species by being more insectivorous.

In addition to dietary and feeding height preferences, ecological competition was reduced still further in these species by differences in activity pattern. Of the seven taxa at the site, four (*Microcebus, Cheirogaleus, Avahi,* and *Lepilemur*) were nocturnal and thereby avoided direct confrontation with the three larger diurnal taxa (*Eulemur fulvus, Indri,* and *Hapalemur*). Ganzhorn (1988) found that activity period may be an important factor permitting coexistence because the species partition naturally into pairs with similar diets but different activity periods: the nocturnal *Avahi* had a dietary profile similar to that of the diurnal *Indri*, the nocturnal *Lepilemur* partnered *Eulemur fulvus*, and the nocturnal *Cheirogaleus* partnered *Hapalemur* (fig. 4.9).

Contrasts in dietary preference, feeding height, and activity pattern may not explain all cases, however. It may be that greater ecological overlap between species can be tolerated in some (e.g., richer) habitats than in others. It is conspicuous that the number of members of the frugivore guild is much smaller in poorer-quality open-country habitats than in highly productive rain forest (see fig. 4.7). Moreover, in these habitats spatial differentiation between species may extend to physical separation of ranging areas rather than just vertical separation. Hall (1965), for example, plotted home range distributions for sympatric baboons (*Papio* sp.), patas (*Erythrocebus patas*), and vervet monkeys (*Cercopithecus aethiops*) in Kabalega National Park (Uganda). The vervets were confined to the gallery forest along the edge of the Nile watercourse, whereas the highly terrestrial patas ranged widely over the savanna grasslands well beyond the confines of the forest. Although the baboons used the forest as a base for sleeping, they ranged out into the bordering woodlands to forage during the day, with only limited overlap with the vervets or the patas. A comparable spatial separation was noted between *Papio* baboons, geladas (*Theropithecus* sp.), and vervets in the Bole Valley (Ethiopia) by Dunbar and Dunbar (1974a).

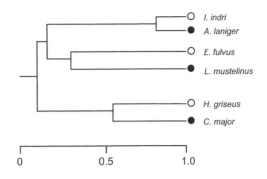

Dietary Similarity

Figure 4.9 Dietary similarity, indexed by the similarity of microhabitats used while feeding, in six species of sympatric Lemuriformes: *Indri indri, Avahi laniger, Eulemur fulvus, Lepilemur mustelinus, Hapalemur griseus,* and *Cheirogaleus major.* Solid circles are nocturnal species, open circles are diurnal species. (After Ganzhorn 1988.)

Spatial separation consequent on differences in habitat preference has also been noted in some forest species. Mittermeier and van Roosmalen (1981) found that microhabitat preferences effectively differentiated many of the eight species of primates in the Voltzberg Reserve (Suriname). *Cebus nigrivittatus* and *Ateles* were found only in high forest, whereas *Chiropotes* and *Pithecia* occurred only in high forest and montane forest; in contrast, *Cebus apella* and *Saguinus* occurred in all five forest types present in the reserve, and *Saimiri* occurred in all except high forest. Similarly, Peres (1993b) found that *Saimiri* and *Cebus albifrons* were only seasonal visitors to *terra firme* (unflooded) forest in the Amazon basin in Brazil. These two species used ranges of well over 1 km² and ranged widely outside the small 900 ha study plot; their movements appeared to be dictated by the distribution of fruiting trees in different types of flooded and unflooded forests in the vicinity. Drills and mandrills (*Mandrillus* spp.) are also erratic visitors in some West African forest communities because they range over very large distances (Gartlan 1970; Lahm 1986). Ranging patterns may thus also reduce interspecific competition.

However, not all studies of niche overlap have reported such clearcut results. Chapman (1987a) reported very significant dietary overlap between three species of monkeys occupying Santa Rosa National Park (Costa Rica) and concluded that interspecific competition was unlikely to have been a significant selection pressure influencing their diets. He did not, however, examine the ways the three species used the habitat. Mitani (1991) did examine both diet and habitat use among five sympatric monkey species in the Campo Reserve (Cameroon) but could find no striking differences between the species in either respect. His sample sizes on diet were probably too small in most cases to be meaningful, however. One of the problems these particular studies highlight is the difficulty of proving similarity. If striking differences emerge when comparing species, then it is clear that niche separation exists even when sample sizes are small. But if species appear to behave similarly, this does not necessarily mean they are not ecologically distinct on other (as yet unknown) dimensions. Apparent similarity can often simply be an index for ignorance.

POLYSPECIFIC ASSOCIATIONS

A number of studies have reported that primates form foraging groups with members of other species. These include associations between closely related species (e.g., forest guenons, East and West Africa: Gartlan and Struhsaker 1972; Gautier-Hion, Quris, and Gautier 1983; Cords 1990; tamarins, South America: Terborgh 1983; Peres 1993a); between species of different genera (*Cebus* and *Saimiri* in Central and South America: Mittermeier and van Roosmalen 1981; Terborgh 1983; *Procolobus* and *Cercopithecus* in West Africa: Oates and Whitesides 1990; Noë and Bshary 1997; *Theropithecus* and *Papio* in Ethiopia: Dunbar and Dunbar 1974a); and between primates and ungulates (baboons and gazelle in Africa: DeVore and Hall 1965; langurs and deer in India: Newton 1989).

The functional significance of such behavior has been the subject of considerable discussion. Three main alternative explanations have been suggested: polyspecific associations are a by-product of random foraging patterns (mangabeys and guenons: Waser 1982; *Papio* and *Theropithecus*: Dunbar and Dunbar 1974a); they are a consequence of one species' exploiting another to gain a feeding advantage (*Saguinus fuscicollis* from *S. mystax*: Peres 1993; *Saimiri* from *Cebus*: Terbogh 1983; Mittermeier and van Roosmalen 1981; *Cercopithecus cephus* from other guenons: Gautier-Hion, Quris, and Gautier 1983); and they allow animals to minimize the ecological costs of forming large groups as an antipredator strategy (forest guenons: Cords 1990; *Procolobus badius* with *Cercopithecus diana*: Noë and Bshary 1997).

Even though polyspecific associations may be chance events in some instances, in others they occur far too frequently for all cases to be explained in this way. Terborgh (1983) found that *Saimiri-Cebus* associations in Manu National Park (Peru) could last from a few hours to as long as ten days, and Peres (1992) reported that specific groups of different *Saguinus* species formed mixed-species associations for as much as 97% of the time at his study site on the Urucu River (Brazil). Moreover, it is clear that the species involved often incur costs by remaining in poly-specific associations. Gautier-Hion, Quris, and Gautier (1983) found that polyspecific guenon groups traveled farther during the day than did monospecific groups. Similarly, Terborgh (1983) found that the distance traveled per day increased by as much as 40% when a *Cebus* group was joined by *Saimiri* and by about 30% when *Saguinus imperator* was in a mixed-species association with *S. fuscicollis*. He also reported that *S. fuscicollis* was likely to be displaced from fruit trees by *S. imperator* and that *Saimiri* was likely to be displaced from food sources by the larger-bodied *Cebus* monkeys with which they habitually associated.

On balance, there is little evidence that would allow us to discriminate unequivocally between the three alternative hypotheses, and that in itself may suggest that there is no single explanation for all cases. The most convincing evidence, however, is that provided by Noë and Bshary's (1997) experimental demonstration that *Procolobus badius* and *Cercopithecus diana* in the Taï Forest (Ivory Coast) were significantly more likely to form mixed species associations (or, if already in one, to remain together longer) after playback of chimpanzee loud calls than after control noises. In this case the evidence suggests that polyspecific associations between these two species form mainly because the colobus are anxious to use the diana monkeys as protection against predatory chimpanzees. This inference is strengthened by the fact that polyspecific associations are more common during the dry season when the chimpanzees hunt most often.

In all these cases, the existence of polyspecific associations probably depends crucially on compatibility of the two species' diets and ranging patterns. When species have incompatible diets, or when their ranging patterns differ markedly, the groups will inevitably drift apart too quickly to form stable associations (e.g., *Papio* and *Theropithecus* in Ethiopia: Dunbar and Dunbar 1974a). In contrast, species with similar ranging patterns and complementary diets (e.g., leaf eating colobines and fruit eating diana monkeys: Noë and Bshary 1997) can effectively partition the local niche space. It is not so obvious how closely related species can form mixed-species associations without ecological conflict, unless vertical separation allows them to minimize competition (see above).

Mixed-species associations appear to be surprisingly common among forest primates, though rare among woodland and savanna species. This probably reflects a combination of availability (high primate population densities) and opportunity (greater complementarity of foraging niches in forest primates). The large range sizes and low population densities of most open-country primates almost certainly militate against polyspecific associations. Although it is difficult to draw general conclusions, current evidence suggests that predation may most commonly be responsible for mixed-species groups. In some cases there may be additional benefits in terms of access to food sources (e.g., *Saimiri* probably benefit from access to partially eaten *Scheelea* palm nuts that only *Cebus* are strong enough to open: Terborgh 1983) or the defense of joint territories (e.g., associations between groups of different *Saguinus* species: Terborgh 1983), but on balance complementary food preferences seem to be more important as a facilitating factor than as a selection pressure favoring mixed-species groups.

4.3.2. Competition with Other Taxa

Finally, we need to consider briefly the question of competition between primates and other species of animals. By and large, primates do not compete to any significant extent with most of the major groups of mammals, particularly the ungulates, rodents, and carnivores, although in some areas there may be some competition with frugivorous bats and squirrels and with arboreal folivores (e.g., sloths in the New World). Unfortunately, there are relatively few data with which to investigate this issue.

In the case of folivorous primates, Dunbar (1978) studied the ecology of the mammalian herbivore community in the Simen Mountains (Ethiopia), and concluded that there was relatively little competition between geladas and the other herbivores (duikers, klipspringers, bushbuck, ibex, horses, and cattle). Since population growth was positive for all wildlife species over a three-year period, it seems unlikely that any of them was being ecologically suppressed through competition. The seven species could be distinguished dietetically into two groups (graminivores and herbivores) between which there was little overt competition. Competition between species in the same dietetic category was reduced by differences in spatial habitat use and activity cycles. At least as far as the only primate in the guild (the gelada) was concerned, ecological overlap during the dry season was reduced by the gelada's inability to digest dry grasses: at this time of year, the gelada exploited a resource (grass roots) that was not available to any of the other species because they lacked the specialized ability to dig. Dietary overlap between geladas and horses (their principal ecological competitor) fell from 94% in the wet season (when green graze was plentiful) to 24% in the dry season (when horses continued to feed on desiccated grass blades while the geladas shifted to grass roots).

Similarly, Plumptre's (1995) analyses of the synecology of the mammalian herbivore community in Parc National des Volcans (Rwanda) suggest that the gorillas are not under any significant pressure from the other species (duikers, bushbuck, buffalo, and elephants). Plumptre (1995) modeled plant-herbivore dynamics to assess the likely impact of increases in the population size of each species and concluded that only an increase in the number of elephants (currently irregular visitors to the Park) would have a serious effect on gorilla numbers. Plumptre attributed this lack of impact to the fact that there is relatively little dietary overlap between the four resident herbivore species.

Across a range of dietary types, Emmons, Gautier-Hion, and Dubost (1983) examined the ecological characteristics of some sixty-six species of mammals that occur at Makokou (northeastern Gabon) and found a lack of direct competition between species comparable to that described for the folivorous species in the case studies above. They noted that 79% of the 491 pairs of species that share similar habitat, feeding height, activity pattern, and diet characteristics are separated by body mass ratios greater than 1:2.2, in close conformity with Hutchinson's (1959) rule of thumb that ecological competition is negligible if species differ in length by a ratio greater than 1:1.3 (mass being a cubic function of linear body measurements). They take this to imply that ecological competition between species is limited. Of the rest, no fewer than forty-eight pairs (10%) consist of closely related sister species.

Nonetheless, there is evidence that under certain conditions primates can suffer from ecological competition with other mammals. Strum and Western (1982) showed that the birthrates of savanna baboons (*Papio anubis*) at Gilgil (Kenya) were not significantly affected by annual rainfall, estimated primary production, or the growth rate of the baboon population, but they were negatively correlated with a measure of the level of competition with other herbivores (mainly ungulates and domestic stock). Their index of feeding competition (total herbivore biomass divided by total primary production) explained most of the year-to-year variance in fecundity rate.

Estrada and Coates-Estrada (1985) studied the synecology of eight mammalian species inhabiting the Los Tuxtlas forest reserve (Mexico), a community that divided naturally into three trophic groups: a folivore (howler monkeys), six frugivores (spider monkeys, kinkajou, arboreal porcupine, two squirrel species, and an opossum) and two frugivore-faunivores (ring-tailed "cat" and an opossum). Spider monkeys thus face considerable competition from other frugivores. Although howlers seem to be ecologically unconstrained by competition from other mammals at this site, leaf eating ants might be a significant ecological competitor. Leigh and Windsor (1982), for example, estimated that insects (including leaf cutter ants) consumed about 12% of the 6.5 tons dry weight per hectare of leaf fall on Barro Colorado Island (Panama).

Similarly, frugivorous primates may suffer more ecological competition from nonmammals in some cases, e.g., frugivorous birds. Terborgh (1983) found that birds and primates between them accounted for about half the total biomass of Manu National Park (Peru), with birds representing about one-quarter of their combined biomass (although only a portion of these would be frugivorous species). Although contest competition between birds and primates may occur only rarely (e.g., Davies and Cowlishaw 1996), the frequency and severity of scramble competition remain to be assessed.

On balance, although it is clear that competition from other mammalian and nonmammalian herbivores may have an impact, this appears to be less severe in terms of ecological competition than that due to other primate species (sec. 4.3.1). However, the paucity of studies probably dictates caution in drawing conclusions at this stage.

4.4. Primates in Plant Communities

Animals are important to plants in two respects: as predators that eat them or their reproductive parts and, inadvertently, as collaborators in reproduction. The latter, which can involve primates as both pollinators and seed dispersers, can have critical implications for the long-term viability of forest habitats, in many of which primates seem to play a keystone role.

4.4.1. Plant Predation

Animals that destroy the vegetation of their habitat do not survive. However, in some cases plant species may be restricted in their growth form or population dynamics by grazing pressure exerted by animals, including primates. Struhsaker (1978) noted, for example, that heavy consumption of the flowers of *Markhamia platycalyx* by red colobus (*Procolobus badius*) in the Kibale Forest (Uganda) suppressed fruit production in this species. A colobus group could strip a *Markhamia* tree of its flowers in an hour of concentrated feeding. The only *Markhamia* specimens at Kibale that managed to fruit during a twenty-four-month period were outside the forest or in areas of forest with low primate densities. When flowering of *Markhamia* trees subsequently became highly synchronized, however, the colobus were in effect swamped, and a significant fruit crop resulted. Struhsaker (1978) argued that synchronized flowering is a mechanism evolved by some species of forest trees to minimize the impact of predation by arboreal herbivores like primates.

Not all tree species adopt this strategy. The common forest species *Celtis africana* exhibits no tendency toward reproductive synchrony. Struhsaker suggested that this species has instead opted for being cryptic in an attempt to outwit its predators. Flowering asynchronously but briefly, as *C. africana* does, may be an effective strategy when individual trees are widely scattered (and thus difficult to find).

Primates may also damage plants while feeding on their structural parts. Glander (1975), for example, claimed that the deaths of two of the four remaining specimens of *Cecropia pelatata* in a relict patch of Costa Rican riparian forest were due to feeding pressure by howler monkeys. However, in this case the relict status of the forest may have been crucial: where a primate population is squeezed into a small area of habitat, it is quite possible that destructive foraging becomes unavoidable.

Indeed, there is little direct evidence to support the claim that primate foraging pressure ever normally reaches a sufficient level to prevent recovery. Gelada baboons habitually dig for grass roots and rhizomes during the dry season and can leave areas of up to 500 m² heavily dug over. Yet grass cover does not appear to be especially diminished by their impact (recovery appears to be complete with the following rains), probably because the geladas do no more than dig over the top few centimeters of soil. Iwamoto (1979) estimated that geladas consume only about 5.5% of the net primary production at Gich in the Simen Mountains National Park (Ethiopia) and concluded that despite their high population density (sixty-five animals per km²) the geladas had minimal impact on the local habitat. Similar observations have been made for *Papio* baboons and rhesus macaques (*Macaca mulatta*), which dig up the taproots and tubers of a variety of plants (Richard 1985).

In at least some cases, browsing pressure can provoke compensatory strategies by the plant. Oppenheimer and Lang (1969) examined the impact of feeding by *Cebus* monkeys on *Gustavia superba* trees on Barro Colorado Island (Panama) and found that although there were no differences in tree height and diameter between a fifty-tree site where the monkeys fed and a similar site where they did not, the trees in the first site had significantly more branches. It seems that the monkeys' removing terminal buds during feeding released the lateral buds from dominance by the apical buds and encouraged increased branching.

While the evidence for an adverse effect on whole plants is at best equivocal, the situation with respect to seeds is perhaps less ambiguous. Mangabeys (*Lophocebus albigena*), for example, destroy the seeds of *Diospyros abyssinica* by breaking them open and chewing them (Waser 1977), and chimpanzees and *Cebus* monkeys destroy the seeds of palms. Mittermeier and van Roosmalen (1981) found that the seeds of some plant families (notably Lecythidaceae) are ground into a pulp by *Chiropotes, Pithecia,* and *Cebus nigrivittatus* (but not by *Cebus apella, Saimiri, Saguinus, Alouatta,* or *Ateles*). Rowell and Mitchell (1991) found that the feces of two East African guenons (*Cercopithecus mitis* and *C. ascanius*) contained many fragments of seeds and seed cases but no intact seeds larger than 2 mm in diameter: only the seeds of *Solanum giganteum* passed through the animals undamaged. Gautier-Hion (1984) similarly reported that seed damage by forest guenons in Gabon (West Africa) increases with seed size. However, little is known of the real impact that this level of seed predation has on the population dynamics of the plant species concerned. Serious damage probably occurs only in relict habitats where the animals themselves are living under marginal survival conditions. In any case, the level of damage primates impose has to be balanced against the positive benefits they provide as seed dispersers (see sec. 4.4.3).

4.4.2. Pollination

The traditional assumption has been that most vectored plant pollination is done by insects, with birds and volant mammals (primarily bats) as secondary sources. But evidence has accumulated in recent years to suggest that nonvolant mammals may also be involved (Proctor, Yeo, and Lack 1996; Carthew and Goldingay 1997). Although much of the evidence is circumstantial, a number of marsupial, rodent, carnivore, and primate species have been identified as putative pollinators.

Sussman and Raven (1978) argued that lemurs on Madagascar had coevolved with plants in a pollinator role (perhaps because of the paucity of alternative vertebrate pollinators). At least five species of lemurs (*Eulemur mongoz, Eulemur macaco, Microcebus murinus, Phaner furcifer,* and *Cheirogaleus medius*) have been observed licking nectar from tree flowers (Sussman 1978; Petter, Schilling, and Pariente 1975; Charles-Dominique et al. 1980; Birkinshaw and Colquhoun 1998). Indeed, *E. mongoz* was reported to spend up to 84% of its time in nectar-related feeding, mostly on a single species, the kapok (*Ceiba pentandra*). Coe and Isaac (1965) also report that *Otolemur crassicaudatus* feeds extensively on the nectar of baobab trees (*Adansonia digitata*) in East Africa. In South America, Janson, Terborgh, and Emmons (1981) reported seven species of primates (including two species of *Saguinus, Saimiri sciureus,* and two species of *Cebus*) lapping at the nectar in the flowers of several tree and liana species at Manu National Park (Peru). Although they did some damage to flower petals, the animals' faces were often covered with pollen, and the flowers' ovaries invariably remained intact. Ferrari and Strier (1992) reported that *Brachyteles arachnoides* and *Callithrix flaviceps* spent as much as 25–30% of their feeding time exploiting the nectar of *Mabea fistulifera* when this species' flowering season was at its height. They note that sixteen of the eighty species of platyrrhine primates have been reported to feed on the nectar of at least eleven different plant species.

However, the evidence that primates (or other nonvolant mammals) are active pollinators of plants is at best circumstantial: no experimental trials to test the hypothesis have been run for any species. The only study to provide any evidence of a possible coevolutionary origin for primate-based pollination is that by Nilsson et al. (1993) in the case of the indigenous Madagascan liana *Strongylodon craveniae* (although see also Birkinshaw and Colquhoun 1998). They argued that the heavy mechanical spring action of the flower keel, which has to be opened to gain access to the anthers, is specifically adapted to a relatively strong animal like a lemur rather than to a bird or an insect. They recorded a total of four lemur species feeding on the flowers (*Cheirogaleus major, Eulemur rubriventer, E. fulvus,* and *Microcebus rufus*), but only *Cheirogaleus* (and possibly *E. rubriventer*) did so nondestructively, and only *Cheirogaleus* did so in a way that caused the tip of the flower keel to hit the animal's forehead so as to transfer pollen. Even so, Nilsson et al. (1993) doubt whether this is a case of true coevolution. For one thing, *Cheirogaleus* is nocturnal, yet the heavily scented flowers open in the morning rather than at night, suggesting an original adaptation to a diurnal pollinator (though this may still have been a lemur). Rather, they suggest that this may be a case of the plant's adapting itself to exploit the particular characteristics offered by a local guild of primates.

Although Chapman (1995) doubts whether there has generally been sufficient opportunity for plants and primates to coevolve to this degree, plant species that are habitually used by primates as nectar sources share a number of characteristics, suggesting some degree of coevolution between plants and primates. Carthew and Goldingay (1997) note that species exploited by primates typically have brightly colored flowers (whereas those used by other vertebrates tend to have dull colored or cryptic flowers), nonpungent nectar, and simultaneous flower opening. If nothing else, these observations suggest that more research is needed to clarify the role that (some) primate species might serve as pollinators. We know too little about most plants' pollination strategies to be able to predict such matters with any confidence.

4.4.3. Seed Dispersal

A great deal more is known about the role of primates as seed dispersers. So far, a wide range of species have been identified as contributing to the active dispersal of seeds, including gorillas (Tutin, Williamson, et al. 1991, Voysey et al. 1999a,b), chimpanzees (Idani 1986; Wrangham, Chapman, and Chapman 1994), orangutans (Payne 1995), baboons (Lieberman et al. 1979), guenons (Jackson and Gartlan 1965; Gautier-Hion 1984; Gautier-Hion et al. 1985; Rowell and Mitchell 1991), howler monkeys (Estrada and Coates-Estrada 1986; Chapman 1989; Julliot 1997; Andresen 1999), spider monkeys (Chapman 1989; Andresen 1999), capuchin monkeys (Chapman 1989; Rowell and Mitchell 1991), and tamarins (Hladik and Hladik 1967; Garber 1986). Seed dispersal after ingestion by animals is known as endozoochory. A useful review of

Table 4.5 Seed dispersal by chimpanzees and arboreal forest monkeys in
the Kibale Forest, Uganda

	Chimpanzees	Monkeys
Mean number of seeds per defecation	22.0	0.4
Mean number of defecations per day	6.7	6.7
Seeds defecated per km^2 per day per individual	147	2.5
Seeds defecated per km^2 per day by population	369	446

Source: After Wrangham, Chapman, and Chapman 1994.

this topic is provided by Chapman (1995).

Five points of importance emerge from this literature. First, species differ markedly both in the intensity of their use of seeds (and hence their potential to act as dispersers relative to other sympatric vertebrate species) and in how badly they damage seeds during ingestion. Rowell and Mitchell (1991) point out that *Cercopithecus* species tend to destroy seeds by grinding them very finely during chewing, whereas *Cebus capucinus* pass the seeds they ingest more or less intact. Wrangham, Chapman, and Chapman (1994) found that although chimpanzees accounted for under 2% of the primate frugivore population in the Kibale Forest (Uganda), they accounted for 45% of all the seeds defecated by frugivorous primates (table 4.5). Their importance as seed dispersers is thus quite disproportionate to their numbers in the population. Some seeds may also be too large to be eaten (or eaten intact) by any but the largest animals. Tutin, Williamson, et al. (1991) concluded that the very large seeds of *Cola lizae* can be handled only by animals as large as a gorilla, so these apes are crucial seed dispersers for *C. lizae*. Mittermeier and van Roosmalen (1981) found that all eight species of primates in their Suriname study site acted as seed dispersers at some time or another but that they differed markedly in both the plant species whose seeds they ate and in how often they destroyed seeds rather than dispersed them (fig. 4.10).

Second, seed dispersal need not always involve gut passage. In some cases the fleshy outer layers of fruits may be eaten while the seeds are spat out. Corlett and Lucas (1990), for example, found that *Macaca fascicularis* could swallow only seeds smaller than 3–4 mm in diameter and that the seeds of 69% of all fruit species eaten are spat out (often cleaned of all flesh).

Third, most studies report that germination rates are significantly improved after seeds have passed through the gut. Several have reported that seeds germinate faster or more prolifically after passing through a primate gut than do fresh seeds "off the branch" (baboons: Lieberman et al. 1979; *Ateles, Alouatta,* and *Cebus:* Chapman 1989; Julliot 1996; chimpanzees: Wrangham, Chapman, and Chapman 1994; Idani 1986).

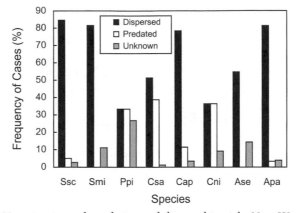

Figure 4.10 Variation in seed predation and dispersal in eight New World monkeys: Ssc, *Saimiri sciureus;* Smi, *Saguinus midas;* Ppi, *Pithecia pithecia;* Csa, *Chiropotes satanas;* Cap, *Cebus apella;* Cni, *Cebus nigrivittatus;* Ase, *Alouatta seniculus;* and Apa, *Ateles paniscus.* (Data from Mittermeier and van Roosmalen 1981.)

The role of the digestive system in stripping flesh from seeds and in weakening the hard outer casing of the seed may promote rapid germination, but the additional fertilization provided by associated fecal material may also contribute. Tutin, Williamson, et al. (1991), for example, found that germination rates for *Cola lizae* were significantly higher for seeds deposited in dung at gorilla nest sites than for seeds deposited elsewhere.

However, it does not follow that ingestion by primates is always beneficial for seeds. Hladik and Hladik (1967) found that although germination rates for *Ficus insipida* were higher after ingestion by spider monkeys or capuchins, they were very poor after passage through the gut of howler monkeys. One reason may be that the howler monkey's digestive system is designed to break down cellular material by microbial fermentation, and this may be highly damaging for seeds (though perhaps not for all plant species: see Julliot 1997). Indeed, it has been suggested that leaf fermentation in Old World colobines may provide the basis for detoxifying seeds in many of these species (Andrews and Aiello 1984; Kay and Davies 1994). In a series of experimental studies, Happel (1988) found that five species of West African monkeys all destroyed seeds by chewing them. *Cercopithecus aethiops* was found to destroy over 80% of the seeds ingested. Happel argued that the bilophodont molars characteristic of the cercopithecine monkeys may have evolved to crush and grind seeds so as to extract nutrients more effectively.

Even if a seed survives this process, though, germination is still not guaranteed. Postdefecation consumption by secondary predators can take a very heavy toll on both seeds and seedlings. Chapman (1989) reported that spiny mice destroyed up to 52% of all seeds and seedlings in artificial dung piles in Costa Rica within five days of deposition (with up to 100% loss in some cases after 180 days). Estrada and Coates-Estrada (1991) found that 90% of the seeds dispersed by howler monkeys at Los Tuxtlas (Mexico) were ultimately destroyed by rodents, but seeds that were inadvertently buried by dung beetles exploiting the howlers' dung had a very high chance of germinating successfully. However, the protection afforded by such burial is strongly contingent on depth (Andresen 1999).

The impact a primate species has on plant reproduction may also depend on the kinds of food sources available to it. Gautier-Hion, Gautier, and Maisels (1993) found that *Cercopithecus pogonias* in Gabon mainly eats fruit pulp, whereas the closely related *C. wolfi* in the Democratic Republic of Congo is a seed eater: fleshy fruits were significantly more abundant at the Gabon site in terms of both species number and fruit density. Because fleshy fruits are seasonally in very short supply for *C. wolfi*, the monkeys are forced to feed much more heavily on the seeds produced by fruit species lacking a fleshy fruit (mainly because the alternative—leaves—is even more unpalatable and highly defended chemically). The monkeys' impact on the seeds differed in consequence: in Gabon they were seed dispersers for 82% of the species they ate and seed predators for only 4%, whereas in the Democratic Republic of Congo they were dispersers for 58% of species and predators for 40%.

Fourth, there are important differences in seed dispersal owing to differences in behavioral ecology between primate species. Because chimpanzees have larger home ranges and travel farther each day than arboreal forest monkeys, they are able to disperse seeds over a much wider area (Wrangham et al. 1991). Voysey et al. (1999a,b) describe gorillas as high-quality dispersers because they eat large quantities of seeds and fortuitously deposit them in areas that enhance the survival and growth of seedings. Rowell and Mitchell (1991) point out that Howe's (1980) conclusion that primates may be poor seed dispersers because they deposit seeds in clumps (thus increasing both secondary predation rates and seedling competition) may be biased by the fact that most of his data derive from howler monkeys. Whereas howlers tend to travel short distances and pass firmer, more cohesive stools, capuchins travel farther and pass loose stools that become widely scattered as they fall through the forest canopy. Rowell and Mitchell estimated that capuchins can carry seeds as much as 1 km away from the parent tree.

New World species are more likely to drop seeds picked out of fleshy fruits below the parent trees because they tend to eat the fruits in situ, whereas Old World monkeys are more likely to store the fruits in their cheek pouches (a feature unique to Old World monkeys) so they can travel away from the parent tree to eat the fruits in safety (often in a tree of a different species). This can be crucial for seed dispersal, since many forest trees actively suppress germination of their own seeds when these fall beneath their canopy (Howe, Schupp, and Westley 1985). Schupp (1988) found that being deposited as little as 5 m away from the parent tree increased a seed's chances of germinating successfully by 340%. Even within the New World monkeys, some species are more effective dispersers than others. Mittermeier and van Roosmalen (1981), for example, point out that spider monkeys (*Ateles* sp.) were more effective seed dispersers in their Suriname forest study area than howler monkeys (*Alouatta* sp.) because they traveled much farther during the day, thereby depositing seeds over a wider area. Even though Julliot (1996) estimated that howler monkeys can disperse seeds up to 550 m from the source (though the mean distance was closer to 260 m), the long gut throughput time for howlers (about 20.6 hours) resulted in a bimodal defecation rhythm that caused 60% of defecations to be deposited under sleeping trees.

Finally, since seeds may be transported some distance before being deposited, they may be deposited outside the immediate forest environs. Jackson and Gartlan (1965) noted that vervets (*Cercopithecus aethiops*) on Lolui Island (Lake Victoria) transported seeds to termite mounds in grassland beyond the forest edge, where germination benefits from the nutrient-rich soils of the termitaria. This dispersion had the effect of spreading the forest species into the surrounding grassland and so contributing to the expansion of the local forest. Similarly, Lieberman et al. (1979) suggested that baboons in the Shai Hills (Ghana) may be seeding new plots of woodland by depositing seeds on rocky outcrops adjacent to the forested parts of the hills. Hladik and Hladik (1967) attributed the overdispersed nature of the forest tree species on Barro Colorado Island (Panama) to endozoochory by the monkeys.

Chapman (1995) has argued that frugivorous primates may be among the most important seed dispersers in tropical forests, simply by virtue of their numbers (see sec. 4.2). The consequences of losing primate communities may thus be potentially disastrous. This conclusion is substantiated by a recent comparison of tree recruitment patterns between forest containing an intact primate population and forest fragments where primate populations had been severely reduced: Chapman and Onderdonk (1998) found that the fragments had lower seedling density and fewer species of seedlings than the forest.

4.5. Summary

1. Primate species richness increases with an increase in area (probably because across habitats more niches are sampled, and within habitats population sizes are bigger); a decline in isolation (which increases immi-

gration rates and therefore also rescue effects and recolonization events); a decline in climatic stress (which reduces the ecological demands on species); and an increase in habitat heterogeneity (which increases the number of ecological niches available).

2. Primate communities are richest and most complexly structured in forested habitats, mainly because frugivorous taxa are more speciose in these habitats. Primates can often represent a very significant proportion (up to 50%) of the total mammalian herbivore/frugivore biomass. Although each folivorous primate species tends to occur at higher biomass than each frugivorous primate species, the frugivorous taxa tend to make the greatest contribution to total primate biomass by virtue of their greater number of species.

3. Primates often depend on keystone plant resources to see themselves through the lean period of the year. The identity of these keystone resources varies between different geographic areas and different primate species.

4. Competition is an inevitable feature of ecological communities. Although there is some evidence for niche partitioning among the primate species in a community, through patterns of dietary preference, foraging height in canopy, and activity period, the intensity of competition appears to be highly variable and to depend on the density of competitor species. Moreover, some guilds (notably diurnal arboreal frugivores) appear to be able to cope with several species in any given forest. Perhaps as a result, polyspecific associations are common (in some cases involving nonprimate species); predation risk seems to be the most likely explanation for such associations.

5. Primates play an important role in the dynamics of plant communities. Although some primates may damage or destroy the plants and seeds they feed on, there is increasing evidence that they play an important (and in some cases crucial) role both as seed dispersers and, less certainly, as pollinators.

5 Distribution, Abundance, and Rarity

In chapters 3 and 4 we examined the fundamental features of primate behavioral ecology and community ecology. We now move on to investigate the importance of these characteristics in determining the size of primate populations (this chapter), their population dynamics (chapter 6), and thus ultimately their intrinsic vulnerability to extinction (chapter 7).

The global population of a species can be described in terms of its geographic distribution (range) and its population abundance (density). Species with restricted distributions or low abundances are often considered rare, a characteristic commonly equated with a high risk of extinction. Given the importance of distribution, abundance, and rarity, this chapter reviews the patterns and processes that might underlie these characteristics in the Primate order. Here we confine our focus to the natural (non-anthropogenic) determinants of abundance and distribution. Although humans can have an enormous influence, it is important to understand what underlies natural variation in these species characteristics before addressing the additional effects of human action (chapters 8 and 9). After a brief consideration of what constitutes rarity, we examine in detail geographic range size and population density. In the final section we investigate the relationship between distribution and abundance.

DEFINING RARITY

Rabinowitz (1981) defined rarity in terms of three key parameters: population size, geographic range, and habitat specificity. Seven categories of rarity were generated using different combinations of these parameters, with an eighth defining common species (see table 5.1). However, this system may be unnecessarily complex. Here we define rarity simply based on low population density (an abundance measure) or small geographic range (a distribution measure). Although both abundance and

93

Table 5.1 A classification of rarity

Range size	Habitat specificity	Local population size	Possible primate candidates	Range size (× 1,000 km²)	Number of habitat types	Average population density (n/km²)
Large	Wide	Large	*Otolemur crassicaudatus*	3,810	4	100
		Small	*Pan troglodytes*	2,500	5	1
	Narrow	Large	*Alouatta caraya*	2,570	2	130
		Small	*Hylobates lar*	1,930	2	4
Small	Wide	Large	*Phaner furcifer*	60	3	250
		Small	*Hapalemur griseus*	70	4	54
	Narrow	Large	*Trachypithecus vetulus*	40	2	154
		Small	*Simias concolor*	8	1	9

Note: Classification follows Rabinowitz 1981. Possible primate candidates are suggested for each category, based on geographic range size, the number of habitat types they are known to occupy and their average population density (data from Happel, Noss, and Marsh 1987).

distribution can be measured on a continuous scale, rarity is probably most useful as a functional concept when defined in terms of a threshold criterion, since this allows rare species to be distinguished from common ones. We therefore treat a taxon as rare within its group (which may, for example, be defined in terms of phylogeny, ecosystem, or geopolitical area) whenever its abundance or range size falls into the lower quartile of the frequency distribution for that group (following Gaston 1994).

Note that rarity, when defined in this way, is a relative property (Dobson, Yu, and Smith 1995): it is impossible to say what is rare without reference to other members of the group. Since group identity can change, it is not unusual to find that a species may be rare at one spatial scale but not at another. Similar problems may arise when comparing two groups of taxa that have different population size distributions: a species in one group may be classified as rare whereas one from another group is not, even though both species are of the same population size. These inconsistencies can be instructive, however, since they may reflect different processes operating at different spatial scales (May 1994). Moreover, they have important implications for the development of conservation strategies (see chapter 10).

Rare species are often also *endemics*. By definition, endemic taxa are limited to a single area and are commonly a focus of conservation attention. But endemism can be a confusing term. In some cases it refers to a species distribution limited to a single country (e.g., the Ethiopian gelada baboon or the Philippine tarsier); in other cases it may refer to a region that is independent of political boundaries (e.g., Müller's gibbon is endemic to Borneo, which encompasses Brunei, Kalimantan, Sabah, and Sarawak: Marshall and Sugardjito 1986). Moreover, endemic species are not always rare: several primate species that occur in multiple countries have substantially smaller ranges than those occurring in only one country. Thus the pygmy marmoset *Callithrix pygmaea* occurs in five countries (Bolivia, Brazil, Colombia, Ecuador, and Peru) and has a geographic distribution on the order of 1.47×10^6 km^2, whereas the common marmoset *Callithrix jacchus* has a distribution of almost twice that area (2.59×10^6 km^2) but occurs in only one country (Brazil) (data from Wolfheim 1983). For these reasons we do not treat rarity and endemism as corresponding characteristics, and we specify species as country endemics or regional endemics where appropriate.

5.1. Geographic Distribution

Primates are unusually well known in terms of their global distribution, with geographic range maps available for all species (Wolfheim 1983) as well as many subspecies (e.g., guenons: Lernould 1988). These range

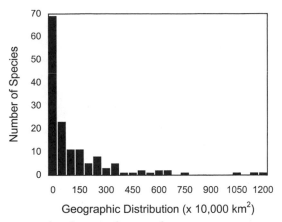

Figure 5.1 Frequency distribution of geographic range size among primate species. (Data from Happel, Noss, and Marsh 1987.)

maps are compiled from a variety of sources but usually rely on the sites where museum specimens were collected or on the oral and written reports of firsthand observers. Distributions are described for species at a variety of scales, including local sites (e.g., Rodman 1978), regions (e.g., Butynski and Koster 1994), and countries (e.g., Tutin and Fernandez 1984; Blom et al. 1992; Marchesi et al. 1995). However, since comprehensive cross-species data on distribution exist only at the global scale (e.g., Wolfheim 1983), much of the following discussion will focus on this level of analysis. We deal with primate distributions in two stages. First we describe the patterns of distribution and rarity among primates. Then we explore the primary processes that may be responsible for these patterns (independent of anthropogenic forces).

5.1.1. Patterns of Distribution

Among the primates, there are many more species with small ranges than with large ranges: the frequency distribution of the geographic ranges of primate species is strongly right skewed, with most possessing ranges of less than 1,000,000 km² (fig. 5.1). According to our criterion for rarity (the lower quartile of the area distribution), primates become rare when their ranges are smaller than 170,000 km².

Rarity is often a characteristic of island species. This pattern is also evident among primate taxa. In only two of nine major primate groups are most species rare (according to our 170,000 km² criterion), and these are the only major groups restricted to islands: the lemurs of Madagascar and the tarsiers of Southeast Asia (table 5.2). Hacker, Cowlishaw, and Williams (1998) have also recently confirmed that Madagascar is a center

Table 5.2 Geographic distribution and population abundance for primate taxa

Taxon	Geographic distribution (per 1,000 km^2)		Population abundance (per km^2)	
	Median value	*Species in sample*	*Median value*	*Species in sample*
Lorisiformes	960	11	33	8
Lemuriformes	70	18	233	13
Tarsiiformes	60	3	98	1
Callitrichidae	230	16	15	13
Cebidae	600	27	13	21
Cercopithecinae	890	41	26	27
Colobinae	525	24	31	20
Hylobatidae	420	6	4	9
Pongidae	545	4	2	4

Source: Data from Happel, Noss, and Marsh 1987.

of rarity for African primates as a whole (fig. 5.2). The reason for this pattern is that species whose ranges are bounded by largely impassable barriers, such as oceans, have little opportunity to extend their geographic ranges. This point is also visible in Hacker, Cowlishaw, and Williams's (1998) analysis, which found rare species accumulating along continental coastal fringes and in the Ethiopian highlands (fig. 5.2).

Rarity is also often associated with equatorial species, and once again this pattern is illustrated in African primates (fig. 5.2). The tendency for species to have smaller ranges in equatorial regions is probably related to *Rapoport's rule:* species tend to exhibit progressively narrower latitudinal limits nearer the equator (Stevens 1989; Rapoport 1982). Rapoport's rule seems to be a common biogeographic pattern for terrestrial taxa, including Palearctic mammals (Letcher and Harvey 1994), North American mammals (Pagel, May, and Collie 1991), and South American primates (Ruggerio 1994; Pastor-Nieto and Williamson 1998), although African primates obey this rule consistently only south of the equator (Cowlishaw and Hacker 1997; see below). Rapoport's rule appears to emerge from another important constraint on species distribution in addition to geographic barriers: niche specialization. Both of these factors, together with a third constraint (interspecific competition), are the subject of the following section.

5.1.2. Processes of Distribution

The constraints on a species' geographic range are likely to be manifold. Whatever might be responsible for limiting one species might have a negligible effect on another; similarly, within a single species it is likely that a restriction in one part of the range may be irrelevant in another.

Figure 5.2 Distribution of rarity-weighted taxon richness in Africa and Madagascar for all lower-rank taxa (species or subspecies). Rarity-weighted richness is calculated by taking the reciprocal of the number of 1° latitude-longitude cells in which the species occurs and then summing these scores across all those species present: the grid cells that score highest are those that have the largest number of taxa with limited distributions. The grid cell with the maximum value is shown in black, whereas the other nonzero scores are grouped into five classes (corresponding to the gray scale on the right), containing approximately equal numbers of grid cells. The map has been smoothed by taking each cell's score as the mean score of the surrounding cells. The five cells containing the Barbary macaque in North Africa are not shown. (Redrawn with the permission of Elsevier Science from Hacker, Cowlishaw, and Williams 1998.)

Unfortunately, most previous studies have adopted a univariate approach to this problem, making it difficult to clarify the relative importance of different causal factors and their interrelationships (but see Taylor and Gotelli 1994). Here we review three key factors: the ability to disperse to another area, the ability to find food there on arrival, and the ability to compete successfully with other species for that food. These factors incorporate both intrinsic constraints (such as body size) and extrinsic constraints (such as dispersal barriers). These constraints can operate in current ecological time, although modern-day distributions inevitably are also strongly influenced by historical events.

DISPERSAL ABILITY

Geographic ranges are commonly smaller among organisms that have limited powers of dispersal and establishment. Unfortunately, little is known about either the distances dispersing primates travel or their ability to establish new populations. A recent study of the genetics of the Mikumi baboon population suggests that about two-thirds of infants are sired by males that have traveled less than 22 km from their natal group (Rogers and Kidd 1996). However, some males may clearly travel farther than this (e.g., the 60 km traveled by two Japanese macaques: Pusey and Packer 1987). On the other hand, individual dispersal may be less important than the movement of new groups after group fission (if dispersal is strongly sex-biased: see sec. 3.3.2).

The most important extrinsic constraint on the expansion of distributions is geographical barriers such as high mountain ranges, rivers, and oceans. Less abrupt changes in habitat type, such as the Dahomey Gap (a forest-savanna mosaic separating the rain forests of Ghana and Nigeria) can similarly constitute major barriers (Oates 1981, 1988). Such barriers undoubtedly play an important role in determining primate species distributions; island taxa cannot extend their distributions owing to the presence of seas and oceans, and even a cursory glance at the ranges of the Callitrichidae shows that these taxa are largely limited in their distributions by major river systems (e.g., Rylands, Coimbra-Filho, and Mittermeier 1993; cf. guenons: Colyn 1988).

Ayres and Clutton-Brock (1992) show that Amazonian primate communities on facing riverbanks are more similar across rivers that are both narrower and of lower discharge (each variable having an independent effect). In addition, comparing slow-flowing white-water rivers with fast-flowing black- or clear-water rivers showed that similarity is reduced across faster flowing rivers (fig. 5.3). This, they argued, might reflect an increased risk of drowning for monkeys swimming across faster rivers; an increased occurrence of the formation of oxbow lakes on slower rivers, leading to more opportunities for connections between opposite banks; or an increased likelihood of arboreal bridges over slower rivers (because the meanders of these rivers increase their length and hence the probability that such a bridge will form by chance).

Although sharp boundaries of this kind restrict the movements of primates, they are not always impenetrable. Indeed, single male baboons have been seen crossing the Kalahari Desert 80 km from the nearest baboon group or free water source (Hamilton and Tilson 1982). Colonizing oceanic islands would also have required passage across large stretches of inhospitable terrain. In contrast, this is not an issue for primates currently found on land-bridge islands isolated after climate-driven changes in sea level (sec. 4.1.1). Similarly, Lawes (1990) shows how climate-induced changes in forest distribution explain the patchy distributions of *Cercopithecus mitis* in southern Africa today. Lawes (1990) argues that *C. mitis* do not currently inhabit a number of suitable

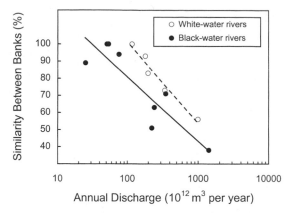

Figure 5.3 Degree of taxonomic similarity between platyrrhine species occupying opposite banks of major rivers in South America as a function of the size of the river (indexed as the volume of annual discharge). Taxonomic similarity is indexed as the percentage of species that occur on both riverbanks. Fast-flowing nutrient-poor black-water rivers and slow-flowing nutrient-rich white-water rivers are distinguished. The plotted lines are least-squares regression lines. (After Ayres and Clutton-Brock 1992.)

forests in South Africa because corridors did not exist between these patches and those the monkeys colonized during their last radiation. Time itself may also be a factor in this context (Willis 1922): colonizing new areas takes time even after climate change has made such expansion possible.

As a rule of thumb, body size may be an important correlate of intrinsic dispersal ability, since it may allow individuals to cross barriers more easily and travel farther. In their analysis of Amazonian primate distribution, Ayres and Clutton-Brock (1992) found positive correlations between river size (in terms of annual discharge, flow rate, and width) and the body size of the largest primate whose range bordered that river; smaller primates did not appear to be able to cross larger rivers, but larger primates were not so constrained. In addition, along the Tana River, species with high colonization rates between forest patches tend to have either larger source populations (which presumably have more group fissions per unit time and therefore produce more migrant daughter groups: e.g., Syke's monkeys) or mixed-sex dispersal (which presumably promotes the formation of new groups: e.g., red colobus) (Cowlishaw, n.d.).

ECOLOGICAL SPECIALIZATION

The geographical availability of their ecological requirements inevitably plays a large part in determining species' distributions. As climatic conditions and soils change over space, so too do the biomes they underpin and the species that exist within them. Thus, in tropical ecosystems, forest gives way to woodland, then savanna, and finally desert as rainfall declines. Populations of woodland primate species, such as the lesser bushbaby, may occur at high densities in woodland areas, but they become sparser in both forest and savanna areas and disappear altogether in deserts (Bearder and Doyle 1974). An unusual example in this case is the *weed macaques,* which thrive when in contact with people (Richard, Goldstein, and Dewar 1989). Richard and her colleagues suggest that the spread of these species through Asia may have been at least partially dependent on the spread of cultivation and pastoralism over recent millennia, particularly in high-latitude and high-altitude habitats. But such a pattern is highly exceptional.

If the modification of a habitat from suitable to nonsuitable is responsible for limiting the geographic range of a species, there is likely to be a gradual decline in density toward the borders of the distribution (sec. 5.3). Moreover, the borders themselves are likely to be quite fuzzy, since many of the low-density populations in these regions are likely to be sink populations with a relatively high turnover through colonization and extinction (see sec. 6.3). The gaps in the distribution of *Callicebus moloch* in central and southern Bolivia (Freese et al. 1982) may be an example of this. In some cases small "fingers" of suitable habitat might allow species to extend their distributions into the harshest of environments. An example is provided by the baboon populations that exist in the groundwater woodlands along the Kuiseb and Swakop River courses in the Namib Desert (Hamilton, Buskirk, and Buskirk 1976; Cowlishaw 1997a, respectively).

As a general rule, species that show a high degree of ecological specialization tend to have more restricted ranges. Nevertheless, even generalists can be limited by the resources they require, and the identity of this resource can vary along different borders of the range. Biquand et al. (1992) suggest that the distribution of hamadryas baboons in Saudi Arabia is limited by rainfall along one boundary of its range but by the availability of sleeping cliffs along another. In most cases the key limiting resource is likely to be food; species will be limited to those geographic regions where their dietary requirements can be met. Among African catarrhine primates, Eeley and Foley (1999) have shown that the mean size of geographic range increases with the mean number of habitats utilized and the mean number of food types eaten (where the means are calculated across all species present in grid cells 500 km square). Similarly, Rodman (1978) found that the local distributions of primates at a site in east Kalimantan was strongly correlated with diet: each species was more likely to be found in habitats where the availability of fruit and leaves most closely resembled the mixture eaten by that species. Despite the potential circularity of this finding, it is consistent with the suggestion that food resources can influence distribution.

The importance of niche breadth in determining species' distribution patterns is well illustrated by the case of *Papio* and gelada baboons. The geladas are grazers whose diet is restricted to the grasses of the cool, wet highlands of Ethiopia: they cannot cope with the more fibrous and silicated grasses that grow below about 1,500 m altitude in tropical Africa. Their distribution over the past 2 million years has thus largely been dictated by the altitudinal movement of the montane grassland zone in response to climate change. As a result, the species' geographical range has contracted from a pan-African one during the Pleistocene to that of a very restricted Ethiopian country endemic today. In contrast, the *Papio* baboons are omnivores (albeit with a preference for fruits), and they have proved ecologically more successful during these climate changes; indeed, if anything they have been able to expand their geographical range (Jolly 1966).

The impact of past climate change on the altitudinal distributions of these two taxa has been modeled by Dunbar (1992a, 1993) using a systems model of the way climatic variables influence the taxons' respective time budgets (sec. 3.3.2). The altitudinal distributions of the two genera are shown in figure 5.4, both under present climatic conditions and when global temperatures are just 2°C cooler or warmer than now. The model suggests that the two species move up and down the altitudinal gradient in tandem, with geladas becoming commoner when climates are cooler and rarer during warmer episodes.

Patterns of niche breadth may also underlie Rapoport's rule: species at higher latitudes may show a greater latitudinal range because they have the broader ecological tolerance necessary to exist at higher latitudes where seasonality is more marked (the climatic variability hypothesis: e.g., Lawton et al. 1994). Ruggerio's (1994) study supports this hypothesis for South American primates, since the mean latitudinal extension of species at each latitude correlates marginally more strongly with seasonality than with latitude itself. Cowlishaw and Hacker (1997) found no consistent latitudinal gradient in geographic range for African primates, but they did find a strong effect of climate, with primates having larger latitudinal ranges where rainfall was more seasonal. That this effect persisted where the correlation with latitude was absent provides good support for the niche breadth hypothesis.

Increased dietary breadth among larger animals may also help explain the common pattern that the area of species' distribution tends to increase with body size (e.g., Neotropical forest mammals: Arita, Robinson, and Redford 1990; North American land mammals: Brown and Maurer 1989). Although the body size distribution pattern is not seen

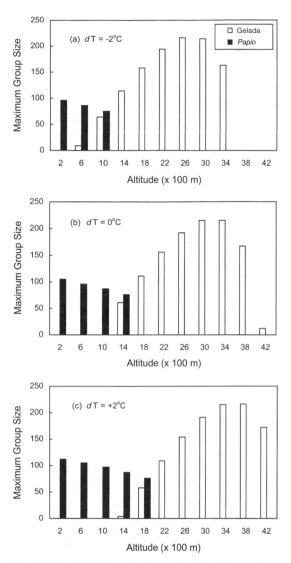

Figure 5.4 Predicted altitudinal distributions of gelada *Theropithecus* and savanna *Papio* baboons when the mean global temperature (a) declines by 2°C, (b) remains the same, or (c) increases by 2°C. The plotted values are the maximum ecologically tolerable group sizes generated by the systems model shown in figs. 3.8 and 3.9. (After Dunbar 1993.)

for primates on a global scale (Happel, Noss, and Marsh 1987; Gaston 1994), there is a good correlation both for Amazonian primates (Ayres and Clutton-Brock 1992) and for South American primates as a whole (Pastor-Nieto and Williamson 1998). Primates with bigger bodies may have larger distributions because they tend to be ecological generalists (see sec. 3.1.1). In support of this hypothesis, Emmons (1984) noted that small primate species that were known dietary specialists were especially patchy in their distribution across several sites in Amazonia. Another important factor may be that larger animals are better able cope with fluctuations in food availability (sec. 3.1.1). This is corroborated by Pastor-Nieto and Williamson's (1998) finding that the broader distribution of large Neotropical primates correlates with the variability of rainfall.

Two other explanations besides niche breadth might be considered for the correlation between body size and distribution (although there are others: see below). One is that body size is simply an indirect measure of another life history trait that is the real causal influence: reproductive rate might be one possibility. Happel, Noss, and Marsh (1987) found no correlation between primate body size and distribution, but they did find that species with longer gestation lengths had larger geographic ranges. Why a slow reproductive rate should lead to a large distribution remains far from clear, although it is possible that this might reflect increased ecological flexibility: since longer gestation lengths are associated with longer periods of development to adulthood, this might allow greater opportunity to develop a broad behavioral repertoire.

An alternative possibility is that bigger taxa have larger distributions because animals with higher absolute energy requirements need larger areas of feeding habitat per capita; as a result they have populations with lower densities, which require wider distributions to remain at a viable size (see reviews in Gaston 1994 and Brown 1995). However, this hypothesis fails to explain the observed covariation with climatic variability. In fact, the consistency with which studies show that species have larger distributions where rainfall is more seasonal (Cowlishaw and Hacker 1997; Pastor-Nieto and Williamson 1998) provides a powerful argument in support of the hypothesis that ecological flexibility is an important determinant of the association between body mass and distribution.

INTERSPECIFIC INTERACTIONS

Since species richness progressively declines with increasing latitude (sec. 4.1.2), the corresponding increase in species latitudinal range with latitude (Rapoport's rule) means that species richness and species range size in an area are often correlated: where species richness is high, range sizes are smaller. Several authors have argued that there may be a causal connection between the two. Rapoport (1982) has argued that high species richness is maintained at low latitudes by ephemeral populations of high-latitude species. Conversely, Brown (1995) has argued that the high-latitude boundaries of species' ranges might be determined by physical conditions but that the lower-latitude boundaries might be restricted by interspecific interactions. However, Cowlishaw and Hacker (1997) have shown that for African primates a latitudinal gradient in species richness can occur independent of a latitudinal gradient in range size. Indeed, in regions where these two patterns do occur together, their association may simply arise from a third variable; for example, high species richness and small range size might coincide in equatorial areas because a high density of rivers can both limit population distribution and lead to allopatric speciation.

In general, the evidence that interspecific competition might restrict species distributions is limited (e.g., Pagel, May, and Collie 1991; Eeley and Foley 1999), although three lines of evidence suggest it might have a significant effect in some cases. First, Emmons (1984) notes that it is the smaller primates (those typically excluded from food sources by larger primates where they coexist) that show the patchiest distributions across several sites in Amazonia. Second, patterns of allopatry between closely related taxa with very similar ecological requirements suggest that they might compete for the same resources (e.g., gibbons: compare maps in Wolfheim 1983). Third, the geographic distributions of fossil theropithecines may have been limited by the distributions of other grazing herbivores, notably warthogs (Pickford 1993), although it remains possible that these two taxa's distributions were independently affected by global patterns of climate change.

Interspecific competition is not the only interaction between species that can limit distributions. Predators and parasites can also have an effect, although little is known about either (Pagel, May, and Collie 1991). In the case of predation, a possible effect can be surmised from the finding that the geographic range size of primate species increases with their ability to live in open habitats (Happel, Noss, and Marsh 1987). If it is valid to assume that primates in open habitats are at higher risk of predation, this finding suggests that species that are more resistant to predators might have wider distributions because their populations find it easier to survive in high-risk areas.

5.2. Population Abundance

An important starting point for the study of primate abundance is the question, Do species that occur at low density in one community also occur at low density in others? That is, are some species consistently rare or consistently common? Data on primate abundance from five sites at

Table 5.3 Abundance of forest primates at five sites in the Lopé Reserve, Gabon

Taxon		Individuals per km² at each site				
		1	2	3	4	5
Cercopithecus nictitans	Spot-nosed guenon	24ᶜ	7	18ᶜ	27ᶜ	19ᶜ
Cercopithecus pogonias	Crowned guenon	5	2	6	4	6
Cercopithecus cephus	Moustached guenon	6	2	4	10	4
Lophocebus albigena	Gray-cheeked mangabey	9	3	7	10	12
Colobus satanas	Black colobus	14	13ᶜ	4	11	12
Mandrillus sphinx	Mandrill	2	5	8	5	0ᴿ
Gorilla gorilla	Gorilla	1ᴿ	1	0ᴿ	0ᴿ	1
Pan troglodytes	Common chimpanzee	1ᴿ	0ᴿ	1	1	0ᴿ

Note: Each site varies in vegetation type and logging history, although all logging is low intensity. Zero values indicate that the species was present but at a density below 0.5 km⁻². The rarest and most common species are denoted by R and C respectively. Data from White 1994.

Lopé (Gabon) (table 5.3) can be used to address this question and to demonstrate two important points.

First, these data show that the ordering of species according to their abundance in a community is relatively consistent, with the same species tending to be either common (*Cercopithecus nictitans*) or rare (*Gorilla gorilla, Pan troglodytes*). The determinants of such interspecific variation in abundance will be the subject of the second part of this section (5.2.2). Second, although these patterns clearly exist, their consistency is not always strong: correlations between sites of the species abundance rank values reveal that in only five of ten pairwise correlations are the rank values significantly similar (site 1 with sites 2, 4, and 5, and site 4 with sites 2 and 5: Spearman rank correlations $p < .05$). Moreover, most species show a fourfold variation in population abundance even within this small geographic area. These points demonstrate the importance of intraspecific variation in primate abundance patterns, the subject of the first part of this section (5.2.1).

5.2.1. Abundance Patterns within Taxa

The key factors influencing population size in primates are food availability, predation risk, and disease. We examine their mechanisms of action in section 6.2. Here we review the specific evidence that food availability influences interpopulation variation in abundance. (Unfortunately, the comparable effects of predation and disease are difficult to assess at present, owing to the absence of good comparative data.)

FOOD CONSTRAINTS ON ABUNDANCE

Although food availability may be among the most important constraints on abundance, several authors have suggested that food may not always be limiting; in fact it has even been suggested that some primate populations live under conditions of food surplus (e.g., Coelho et al. 1976; Coelho, Bramblett, and Quick 1977). Although Coelho et al.'s analysis of howler and spider monkeys in Guatemala has been criticized on methodological grounds (Cant 1980), other authors using more rigorous techniques have found food surpluses (e.g., geladas: Iwamoto 1979; blue monkeys: Butynski 1989). This result is not entirely counterintuitive given that populations naturally fluctuate in size and can undergo crashes because of processes that operate independent of food resources (e.g., disease epidemics). Precisely this explanation was proposed by Butynski (1989) for the differences in abundance observed over a six-year period between two nearby populations of blue monkeys in Kibale Forest (Uganda).

It is important to distinguish between those populations that are genuinely not food-limited and those that are food-limited during particular seasons (Cant 1980; Emmons 1984). "Ecological crunches" or "nutritional bottlenecks" might be particularly important for frugivorous primate populations (sec. 4.2). For example, Peres (1994) has argued that the low abundance of primate frugivores in the Urucu Forest of the Brazilian Amazon is due precisely to the strong seasonality of fruit production there. The effects of nutritional bottlenecks are by no means confined to this single dietary group, however; indeed, one of the most detailed case studies comes from the omnivorous chacma baboon population of the Drakensberg Mountains of South Africa (Byrne et al. 1993).

During the lean periods that characterize bottlenecks, populations often have to rely on keystone food resources (sec. 4.2.2). Consequently the availability of these resources may be the critical determinant of population density. McFarland Symington (1988) suggests that the availability of keystone foods provides a better explanation of intersite variation in spider monkey abundance (in four sites across the Neotropics) than either predation, interspecific competition, climate and seasonality, or the availability of fruiting trees. But one problem with testing the hypothesis that a single dietary component (or a small subset of such components) might determine population abundance is identifying just which food resource plays the crucial role. This is particularly difficult for primate populations given the great dietary flexibility observed in most species (sec. 3.2.2).

One useful model system might be that of the colobine monkeys, since their dietary options are constrained by their specialized digestive tract.

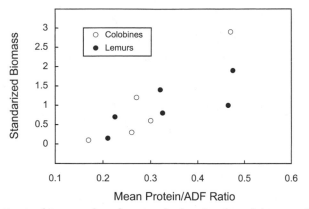

Figure 5.5 Species biomass plotted against leaf quality for colobines and folivorous lemurs. Biomass is standardized to control for variation in species body mass, and leaf quality is indexed by the ratio of protein to acid detergent fiber (which is difficult to digest). (After Ganzhorn 1992.)

Yet even among these species there are regional differences. Although both Marsh and Wilson (1981) and Davies (1994) have shown that the biomass of Asian *Presbytis* species in different populations is strongly correlated with the proportion of trees that are legumes, the same correlation is not seen for African colobines (Davies 1994). A second, more robust, pattern that seems to hold both across colobine genera and across continents is that colobine biomass is determined by mature leaf chemistry, being higher at sites where leaf quality is high (Waterman et al. 1988; Oates et al. 1990; Davies 1994). Ganzhorn (1992) found a similar pattern for folivorous lemurs. The relation between the relative increase in species biomass and leaf quality is similar for both primate taxa (fig. 5.5), although the absolute biomass of colobines is always higher than that of lemurs (Ganzhorn 1992). Thus the abundance of folivorous primates may be limited by the same key dietary components in markedly different geographic regions and taxonomic groups.

The importance of nutrients has also been highlighted for frugivorous taxa by Caldecott (1980). He suggested that energy limitation is responsible for the decline in gibbon abundance at higher altitudes. He argued that the higher energy costs of both thermoregulation (owing to reduced ambient temperatures) and locomotion (owing to a lower and more tangled canopy) at higher altitudes increases an animal's energy demands, while less energy is available to meet these demands owing to reduced plant productivity and floristic diversity in these areas. This hypothesis is supported by the observation that siamangs exist at greater biomass at higher altitudes than lar gibbons; their larger body size and ability to subsist on a folivorous diet would be expected to buffer them against such problems. However, the recent finding that frugivorous *Ateles* occur at higher altitudes than folivorous *Alouatta* (Estrada and Coates-Estrada 1996) suggests that species differences in altitudinal distribution cannot be explained by the degree of folivory or frugivory alone.

Given the importance of plant nutrients and secondary compounds, it is possible that the ultimate predictors of primate abundance may be the determinants of leaf quality. Thus far, two such determinants have been identified. The first is climate: Ganzhorn (1992) reports a consistent negative correlation between rainfall and leaf quality in each of three tropical forest regions (Madagascar, Africa, and Asia), a pattern that might arise if areas of high rainfall receive less sunshine, thus reducing photosynthesis. The second is soil quality: plants growing on nutrient-poor soils may find the recovery of nutrients lost from leaf predation (by herbivores) difficult and therefore invest more heavily in chemical defenses to prevent leaf predation (Janzen 1974). This second hypothesis has received considerable attention, but only mixed support. In the Old World there is little evidence for a correlation between soil quality and primate biomass across sites in either Africa (Oates et al. 1990) or Asia (West Malaysia: Marsh and Wilson 1981). In the New World comparisons have been conducted between floodplain forests (with nutrient-rich soils that result from seasonal flooding) and unflooded *terra firme* forests (with nutrient-poor soils). These comparisons suggest that the abundance of Amazonian primates is higher in floodplain forest than in *terra firme* forest, as predicted (Freese et al. 1982; Emmons 1984; Peres 1997a). In addition, within floodplain forests, howler monkey densities are high along nutrient-rich white-water rivers and decline with distance from the river but are low and uniform with distance around nutrient-poor black-water rivers (where flooding does not lead to such heavy deposition of nutrients) (Peres 1997a). However, although primate abundances are often lower along black-water rivers, this effect can be more marked among frugivores than folivores, bringing into question the validity of the proposed mechanism (seeds and vegetative parts should be defended more strongly than the fruits, which the plant wants to be eaten: Emmons 1984). These results suggest that soil geochemistry might be ultimately responsible for variation in primate abundance in some cases, but that other factors are clearly commonly involved and that these may overwhelm any effects of soil chemistry (if any are present).

In addition to the physical factors of soil quality and climate, tree species diversity may also influence the availability and quality of food resources and so primate population abundance (note that there is also evidence that tree species diversity can promote primate species diversity: sec. 4.1.2). The potential influence of tree species diversity on abundance is particularly significant because it has been reported from four independent studies. First, data from Caldecott (1980) show a strong correlation between the population density of both lar gibbons and dusky langurs with floristic diversity, although this pattern is not seen for sympatric populations of siamangs or banded langurs. (Reanalysis of Caldecott's data reveals no other significant predictor variables.) Second, Marsh and Wilson (1981) report a correlation between lar gibbon population density and plant species diversity across eleven sites in West Malaysia. Third, Thomas (1991) found that the abundance of *Cercopithecus* monkeys increased in areas of higher tree diversity in the Ituri Forest, Democratic Republic of Congo. Finally, Lawes (1992) describes the same relationship for *Cercopithecus mitis* across forest patches in South Africa (although in this case the trend was less consistent owing to an anomalously low population density in the most speciose forest). The mechanism that gives rise to this general pattern is unknown. One possibility is that areas that are more species rich are also likely to have a greater abundance of edible plants available in the lean season. More research into this pattern is clearly required, particularly in nondisturbed environments (human disturbance was present in three of these four studies and may have played a confounding role).

Finally, it remains to evaluate the influence of interspecific competition on abundance. The earlier studies that have evaluated the effects of food resources on primate abundance have tended to consider only standing crop rather than productivity. By definition, competitors play a pivotal role in determining that standing crop. Ultimately it is very difficult to separate the effects of interspecific competition from those of species' ecological requirements (see sec. 4.3). For example, the negative association across sites between the abundance of howler monkeys and that of more frugivorous primates, such as capuchins and squirrel monkeys, may be related simply to the habitat preferences of these species rather than to competition between them (Glanz 1990).

Nevertheless, there do appear to be cases where interspecific competition influences abundance, particularly where it is known that both species have similar habitat needs. For example, Davies (1994) cites competitive exclusion to explain why *Presbytis rubicunda* and *Presbytis hosei*, which are virtually indistinguishable in both their ecology and their morphology, can coexist throughout Sabah but never show codominance: one always predominates at any given site. Similarly, Tutin and Fernandez (1984) suggest that the relatively low densities of chimpanzees and gorillas in Gabon might result from competition between the two species. Waser (1987) summarizes some further possible cases of density compensation, and we reproduce his summary here (table 5.4). For the present, the relative importance of variation in niche overlap and competitor

Table 5.4 Possible cases of primate interspecific competition

Taxon	Population density[1] with competitors		Potential primate competitors	Source
	Absent	Present		
Hylobates spp.[2]	7	0.3–4	*Hylobates syndactylus*	1
Macaca fascicularis[2]	215	0–107	*Hylobates* spp.	2
Pan troglodytes[2]	2	1	*C. mitis*	3
Lophocebus albigena	80	9	*Procolobus badius, C. mitis*	4, 5
Lophocebus albigena[2]	18	10	*C. mitis*	3
Cercopithecus mitis	185	35	*L. albigena, P. badius*	6
Alouatta seniculus[2]	102	30	*Cebus* spp., *Callicebus* spp., *Ateles* spp.	7,8
Saimiri sciureus[2]	>316	41	*Cebus* spp., *Callicebus* spp., *Saguinus* spp.	9

Sources: (1) Mackinnon 1997; (2) Marsh and Wilson 1981; (3) Struhsaker 1981; (4) Struhsaker 1978; (5) Chalmers 1968; (6) Rudran 1978; (7) Terborgh 1983; (8) Rudran 1978; (9) Robinson and Janson 1987. After Waser 1987.

[1] Number of individuals per km[2].

[2] Given as biomass or group density per km[2] in Waser 1987 and converted to individual density using mean group sizes given by Smuts et al. 1987 and mean adult body weights given by Smith and Jungers 1997. For *P. troglodytes,* groups are taken to be foraging parties (mean size = 2.6; Ghiglieri 1984).

density, compared with plant productivity, remains an important but unanswered question.

5.2.2. Abundance Patterns between Taxa

Patterns of variation in abundance between species have been the subject of considerable research in recent years (Gaston 1994; Brown 1995). However, most of this research has focused on taxa other than primates (e.g., Blackburn et al. 1993) or included primates within the broader heading of mammals (e.g., Robinson and Redford 1986; Fa and Purvis 1997). Across primates alone, a frequency distribution plot of species abundances exhibits the familiar skewed pattern (fig. 5.6) common to many taxa (Gaston 1994; Brown 1995). At the global scale, these data suggest that rare primates can be defined as those with a population density below seven individuals per km[2], according to the quartile criterion for rarity. Among taxonomic groups, it is clear that the prosimians (particularly the lemurs) tend to achieve the highest densities while the pongids exhibit the lowest; indeed, all the apes can be recorded as rare according to our criterion (table 5.2). Both body size and ecological specialization might be responsible for these patterns, as we describe below.

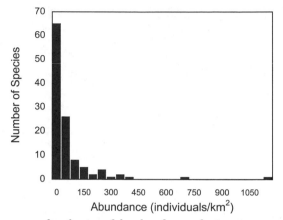

Figure 5.6 Frequency distribution of the abundance of primates among primate species. Abundance is indexed as population density: the mean number of individuals per km². (Data from Happel, Noss, and Marsh 1987.)

BODY SIZE AND ABUNDANCE

In most taxa, body size and abundance show a strong association, with population density scaling with body mass to the power of -0.75 (e.g., mammals: Damuth 1981; Peters and Raelson 1984; Silva and Downing 1995). This relationship is conventionally explained by the "energetic equivalence rule," which states that equal amounts of energy are available to each species in a community (Damuth 1981, 1991). The finding that, for mammals in general, abundance scales with body mass to the power of -0.75 while metabolic rate scales to the power of 0.75 is consistent with this hypothesis. However, it has recently been suggested that this pattern might hold only across samples drawn from many different communities, often at a global scale. Within a variety of assemblages, the body size–abundance relationship does not always obey the -0.75 rule, nor is it always linear (Blackburn et al. 1993). For a recent review, see Blackburn and Gaston (1997).

In primates, at a global scale, the log–log negative relation between body size and abundance is linear but has a regression coefficient of -0.41 (adjusted $r^2 = .12$, $F_{1,109} = 16.6$, $p = .0001$: fig. 5.7) (cf. Happel, Noss, and Marsh 1987). One explanation for this lower slope is that it might indicate that larger species obtain proportionately more energy than smaller species (Pagel, Harvey, and Godfray 1991). But there are several problems here. First, these allometric relationships do not control for phylogeny. Second, Blackburn et al. (1993) show that the steepness of the slope is strongly dependent on the regression technique used (reduced major axis and least-squares regression models can produce

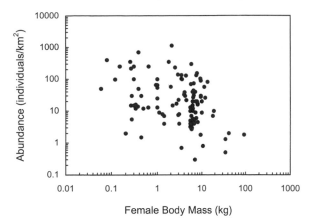

Figure 5.7 Mean abundance plotted against body mass across primate species. Abundance is indexed as population density: the mean number of individuals per km². (Data from Happel, Noss, and Marsh 1987.)

slopes that are above or below −0.75 for the same data set: see Rayner 1987). Finally, Blackburn and coworkers (Blackburn et al. 1993; Blackburn, Lawton, and Pimm 1993) note that there is little ecological evidence that energy should limit the abundance of all species in a community, although it might limit the most common ones, and that the negative relationship between body size and abundance might simply be the result of concatenating the frequency distributions of two variables that just happen to have opposite skews.

Finally, one alternative interpretation of the body size–abundance association is that body size is not directly related to abundance at all but rather is correlated with another life history trait that influences abundance. In a study of abundance patterns in British birds, Blackburn, Lawton, and Gregory (1996) found that traits associated with fast reproductive rates correlated more strongly with abundance than did body size and that this association remained when the effects of body size were removed. Happel, Noss, and Marsh (1987) also found that reproductive rate variables provided stronger correlations with abundance than did body size in primates, where abundance was greater in those species that had shorter gestation lengths and earlier ages of female sexual maturation. Both studies suggest that the mechanism underlying this pattern might be that species with higher reproductive rates are generalists that can utilize a wide range of resources and therefore achieve both high rates of reproduction and high population densities. This brings us directly to the second covariate of rarity: niche specialization.

NICHE SPECIALIZATION AND ABUNDANCE

Species that are less specialized might be able to achieve higher densities because they have a wider range of food resources available to them. Caldecott (1980) has argued that banded langurs occur at higher altitudes than dusky langurs because they can exploit more food resources in such depauperate areas: in lowland forest, the banded langur is more of a generalist, exploiting 137 food species in comparison with the 87 used by the dusky langur over the same one-year period. Similarly, a capacity to eat seeds may allow colobines to achieve higher abundances than if they were more strictly folivorous (Oates et al. 1990). In addition, Oates (1996b) has suggested that folivores generally achieve higher abundances than frugivores in tropical forests in Africa because their ability to eat mature leaves allows them to sustain large populations during periods of seasonal food scarcity.

Diet type may be as important as dietary breadth. Hladik (1975) has argued that since the availability of food energy increases at lower trophic levels, the abundance of folivores should be greater than that of frugivores, which should be greater than that of insectivores. Robinson and Ramirez (1982) tested this hypothesis for Neotropical primates but also incorporated the effects of body size. Their analysis confirmed two key predictions: within each dietary type, abundance tended to decline with increasing body size; and for any given body size, folivores exceeded frugivores in abundance and frugivores exceeded insectivores. Similar results have been found for Neotropical mammals as a whole (Robinson and Redford 1986) and Afrotropical mammals as a whole (Fa and Purvis 1997). Similarly, Clutton-Brock and Harvey (1977) have shown that home range size increases per unit biomass across species, but that for any given group biomass frugivorous species use larger ranging areas than folivorous species (fig. 5.8). Consequently, in addition to body size, diet type and dietary breadth may contribute significantly to interspecific differences in primate abundance.

5.3. Distribution-Abundance Relationships

In this section we discuss how abundance and distribution may be related and how distribution-abundance patterns might be used to index vulnerability to extinction.

5.3.1. Covariation between Measures

The relation between distribution and abundance can be examined at two scales: the texture of abundance across the geographic range of a single species, and the association between geographic distribution and average abundance across species.

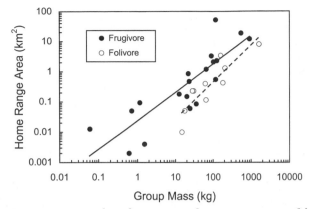

Figure 5.8 Home range area plotted against social group mass across folivorous and frugivorous species of primates. The lines are least-squares regression lines. (After Oates 1987.)

We have already explored how abundance can vary markedly within a species' geographical distribution (sec. 5.2). But does the population density of a species vary in a consistent way with position in its geographical range? It is commonly argued that species abundances are highest in the center of the distribution and progressively decline toward the edge (e.g., Brown 1984; Brown, Mehlman, and Stevens 1995). This argument assumes that the availability of key ecological resources is greatest at the center of the range but declines toward the periphery. Other authors, in contrast, have stressed the role of population dynamics (e.g., Lawton 1993, 1994). In this case it is hypothesized that the crucial differences between the core and the edge of the distribution lie not in population abundance but in population growth rate, with core and edge populations connected in a series of source-sink relationships (although variation in the availability of ecological resources may still provide the underlying cause of the demographic differences).

Although this may be true for distributions that are limited by gradual changes in habitat from the core to the periphery, it may be less true where they are limited by "hard" boundaries such as rivers and oceans. Currently there is little evidence for this one way or the other among primate taxa. Nonetheless, one may gain insight into this issue by studying the patterns of collapse of species' geographic ranges as a result of anthropogenic activities. Lomolino and Channell (1995) found that the shrinking of the golden lion tamarin's range resulting from habitat loss occurred not as a contraction toward its center (as predicted by the core-edge patterns proposed above), but as a contraction toward its eastern edge. This eastern bias was a consistent trend among other mammalian taxa in other regions. This suggests that populations on the periphery of species geographical ranges might be as important in terms of their abundance and growth rates as those in the center.

The second pattern to consider here is the association between geographic distribution and population abundance across species. Species with a relatively large geographical distribution frequently also have relatively high population abundances within that distribution (e.g., Brown 1984; Lawton 1993, 1994). A variety of explanations have been proposed for this pattern (e.g., Gaston, Blackburn, and Lawton 1997). Ultimately, it is the degree of ecological specialization that may be most important: generalist species are more likely to achieve both a high density (owing to an ability to utilize a variety of resources in a habitat) and an extensive distribution (owing to an ability to exist in a variety of habitats).

However, the same pattern is not seen across primates globally. Analyses of the available data (from Happel, Noss, and Marsh 1987) finds no correlation between species' geographic range area and either mean or maximum recorded population density ($n = 127, p = .27; n = 107, p = .14$), but a strong negative association between range area and minimum density ($n = 107$, Spearman $r_s = -.27, p = .008$): that is, species that have a large geographic range also tend to have low minimum abundance. One explanation might be that some species find it difficult to maintain populations at low density and therefore exist only in high-quality habitats (where they live at relatively high density), thereby restricting their distribution (reasons some species may find it difficult to sustain populations at low density are explored in sec. 7.3).

5.3.2. Compound Rarity

Since there is generally a strong positive correlation between distribution and abundance, there are many species that will have both restricted ranges and low population densities. Identifying these species may prove useful in recognizing species that are especially rare (e.g., Arita, Robinson, and Redford 1990; Lawton 1994, 1995). Arita, Robinson, and Redford (1990) undertook such an analysis for a sample of one hundred Neotropical mammals. Among the thirty-eight primates in the sample, most had a restricted range but were split equally between the high and low abundance categories. Nevertheless, 39% of the Neotropical primate species fell into the category of greatest rarity in this sample (i.e., in the lower half of the distribution for both geographical distribution and abundance across all one hundred species). In fact, the primates were the rarest order of mammals, contributing 71% of all species in this category.

Dobson and Yu (1993) developed this method further by arguing that restricted-range/low-abundance species should be identified only after removing the effects of body mass, on the grounds that body mass is correlated with a variety of potentially confounding biological variables (see chapter 3). Hence, although we might expect smaller species to occur at higher population densities than larger species, the key issue for defining rarity may be not whether a species occurs at low density but whether it occurs at low density for a given body mass. Repeating Arita, Robinson, and Redford's (1990) analysis but using residual rarity rather than actual rarity, they found a similar number of primate taxa with restricted ranges and low population densities. They also ranked the relative rarity of each mammal by the deviation of its position (on a residual-distribution/residual-abundance plot) from a straight line of slope 1.0 through the origin (the position and gradient of this line assume that distribution and abundance play an equally important role in determining rarity). The primates were again the rarest order, with fifteen species among the twenty-five highest ranked mammals and a mean rank of thirty-eight out of one hundred across all species in the sample.

Although this ranking procedure gave a measure of compound rarity for the entire sample of animals in their analysis, there are a variety of complicating factors. These include assumptions about the relations between body mass, population abundance, and geographical distribution that are at best poorly understood. In addition, the effects of phylogeny were not considered. Overall, then, the value of this approach in identifying the rarest species remains open to interpretation (see Gaston and Blackburn 1995; Dobson, Yu, and Smith 1995; Gaston and Blackburn 1996a,b, and Dobson, Smith, and Yu 1997).

5.4. Summary

1. Geographic distribution and population abundance are key determinants of global population size and therefore global extinction risk. Primate taxa might be considered rare if they fall in the lower quartile of the range of values for either distribution or abundance.

2. The geographic ranges of different primate species show a log-normal distribution (there are more species with small ranges than with large ranges). Rare primates may be defined as those with ranges smaller than 170,000 km^2. Primates with small ranges tend to be found in equatorial regions and on islands.

3. Primate species ranges may be limited in size by a variety of factors, including the intrinsic constraints of small body size and of low ecological flexibility and the extrinsic constraints of geographic barriers to dispersal and of interspecific interactions (both competition and predation).

4. The population abundances of different primate species show a log-normal distribution (there are more species with low abundances than high abundances). Rare primates may be defined as those with abundances below seven individuals per km^2.

5. Intraspecific variation in abundance may be strongly influenced by

annual food availability or, in seasonal environments, food availability during the lean season (which may equate with the availability of keystone resources, such as fig trees). Among folivorous primates, mature leaf quality appears to be a fundamental correlate of interpopulation variation in abundance. There is also evidence of higher primate abundances in habitats of greater floristic diversity and in areas where interspecific primate competitors are absent.

6. Interspecific variation in primate abundance is negatively correlated with body mass, but not in the way predicted by the energy equivalence rule. Primate abundances appear to be further influenced by diet: species at high trophic level (e.g., insectivores) are less abundant than species at low trophic level (e.g., folivores) for any given body mass.

6 *Population Biology*

The survival of a species under threat of extinction depends in large measure on the behavior of its constituent populations in the face of that threat. An important element in this equation is the relationship between population dynamics, population genetics, and extinction risk. But before we can discuss the precise links involved (see chapter 7), it is necessary to consider the biology of populations under normal conditions.

Population dynamics and genetics are a product of the frequencies with which individual animals give birth, migrate, and die. These processes are in turn subject to the influence not only of a species' evolutionary history (its life history strategy, reviewed in sec. 3.1), but also of the environmental and social milieu within which individual animals live. These links between local conditions, individual behavior. and population biology are the focus of this chapter. First we introduce life tables, which describe the characteristic demographic schedule of a population. We then examine in more detail those forces that drive the dynamics of primate populations and metapopulations. In the final section we explore the genetic consequences of these processes.

6.1. Demographic Variables

In this section we review life table theory, together with a fundamental life table variable, the birthrate (and the environmental factors that influence it). For a general introduction to population demography, we recommend Caughley (1977) and Wilson and Bossert (1977). For summaries that specifically relate to primates, see Dunbar (1987b, 1988) or Lyles and Dobson (1988).

6.1.1. Life Tables

The rates at which individuals give birth and die vary with age. The most convenient way of showing this is a life table that sets out the age-specific

119

Table 6.1 Life table for female rhesus macaques on Cayo Santiago

Age (yrs)	q_x	b_x	m_x	p_x	l_x	e_x
0–1	0.196	0.000	0.000	0.804	1.000	7.5
2–3	0.165	0.128	0.064	0.835	0.804	10.5
4–5	0.146	1.268	0.634	0.854	0.671	12.6
6–7	0.129	1.600	0.800	0.871	0.573	13.5
8–9	0.188	1.292	0.646	0.812	0.499	15.9
10–11	0.153	1.424	0.712	0.847	0.405	16.4
12–13	0.166	1.174	0.587	0.834	0.343	17.9
14–15	0.336	0.000	0.000	0.664	0.286	18.5
16–17	0.247	1.000	0.500	0.753	0.190	19.2
18–19	0.503	0.000	0.000	0.497	0.143	19.5
20–21	0.000	1.000	0.500	1.000	0.071	22.0
22–23	1.000	0.000	0.000	0.000	0.071	22.5

Source: After Sade et al. 1976.

Note: Variables are defined as follows, using an age interval (x) of two years: q_x: death rate (probability of dying between age x and age $x + 1$); b_x: birthrate (births per interval for individuals aged x); m_x: fecundity rate (number of same-sex offspring born per interval to individuals of age x: $m_x \approx 0.5\,b_x$); p_x: survival rate (probability of surviving from age x to age $x + 1$: $p_x = 1 - q_x$); l_x: survivorship (probability of surviving from birth to age x: $l_x = 1 - \Pi p_x$); e_x: life expectancy (mean age at death for individuals who survive to age x). Because of small-sample effects, estimates of variables tend to be unreliable in the later age classes.

rates at which these occur. Life tables also provide the fundamental basis for studying population dynamics. Life tables are population- and sex-specific: table 6.1 shows the life table for female rhesus macaques from Cayo Santiago in the Caribbean, using a two-year age interval. The two left-hand columns show the rates at which females of a given age-class died (mortality rate, q_x) and gave birth (birthrate, b_x) during each age interval. The choice of age interval is largely a matter of convenience, usually being longer with longer-lived species.

These two variables provide the basis for calculating a number of important derivative indices. The birthrate gives us the fecundity rate (m_x), defined as the rate at which an individual produces same-sex offspring during a given age interval. For obvious reasons, this is usually half the birthrate; however, it need not be if there are selective advantages to be gained from investing more heavily in one sex.

There are good theoretical reasons to expect females to vary the sex ratio of their offspring in relation to their ability to influence the future reproductive success of the two sexes. This may in turn be influenced either by their rank or their physical condition (the *Trivers-Willard effect:* Trivers and Willard 1973; Dittus 1998) or by the ecological state of the population (the *local resource competition model:* Clark 1978; Silk 1983; Johnson 1988). The Trivers-Willard effect argues that when the variances in lifetime reproductive success of the two sexes differ, females in good condition (usually high-ranking individuals) will prefer to invest in the sex with the higher variance (usually sons) if investing in this sex allows the offspring concerned to achieve greater numbers of grandchildren (perhaps because bigger sons achieve higher average lifetime dominance rank). Females in poor condition (usually low-ranking individuals) will prefer the safer option of investing more heavily in the sex with the lower variance in reproductive success (daughters) if investment in the high-variance sex (sons) runs the risk that the offspring may be poor competitors and thus have low (or even zero) lifetime reproductive success. By contrast, the local resource competition model predicts that when levels of within-group resource competition are high, females in female-philopatric species will prefer sons in order to minimize the competition their daughters will face (because sons are more likely to disperse as adults). An extension to this idea proposes that when cooperative alliances are important it may pay high-ranking females in such circumstances to prefer daughters, with which they can build large coalitions, whereas low-ranking females will prefer sons, which can avoid competition by dispersing (van Schaik and Hrdy 1991).

The two models should be seen as complementary rather than mutually exclusive: both can be expected to operate. For example, van Schaik and Hrdy (1991) were able to show that in Old World monkeys ecological competition promotes a preference for daughters among high-ranking females in populations that are at carrying capacity (when ecological competition will be at its fiercest: see sec. 6.2.2) but a preference for sons in populations undergoing expansion. They interpreted these results as implying that the resource competition model effects are more intrusive (i.e., take precedence) when ecological competition is significant (e.g., in stationary populations), but the Trivers-Willard effect is the stronger when competition is weak. In sum, current findings suggest that the two models interact and that individual species (or even populations) may exhibit the two effects in different measure.

The mortality rate (measured as the proportion of animals that enter age-class x that die before entering the next class, $x + 1$) can be used to calculate the survival rate p_x (the proportion of animals entering age-class x that survive to join age-class $x + 1$), and this in turn can be used to calculate survivorship (l_x, defined as the proportion of all animals born that survive to age x) from the expression

$$l_x = \prod_{i=0}^{i=x} p_i.$$

Other important indexes are life expectancy e_x (defined as the mean age at which an animal that survives to age x can expect to die) and the generation time T (the mean age at which females produce their first offspring). Note that e_x is sometimes calculated as the mean number of future years of life (following standard practice in human demography).

A final index of some importance in evolutionary biology is reproductive value v_x (the number of future same-sex offspring an individual of age x can expect to produce during the remainder of its life, relative to the average at birth for all individuals of that sex). Reproductive value is a ratio that is standardized by setting the average lifetime average reproductive output for all females to $v_0 = 1$ at birth. The formula for calculating v_x is complex, since it depends on the demographic state of the population:

(6.1)
$$v_x = l_x^{-1} e^{rx} \sum_{x=y} l_y m_y,$$

where r is the intrinsic growth rate of the population (see sec. 6.2.1). In stationary populations (where $r = 0$) this reduces to

$$v_x = l_x^{-1} \sum_{x=y} l_y m_y.$$

Reproductive value is of particular importance in calculating the genetic consequences of different reproductive strategies. An important implication of equation 6.1 is that it pays to reproduce early in expanding populations (doing so maximizes the number of descendants), but it pays to reproduce later in declining populations (when your descendants will constitute a larger fraction of the population alive at any given future time). Although rarely used in conservation biology, we include this index here because its importance in relation to population genetics makes it possible that it will become a more prominent part of conservation biology's mathematical tool kit in the future.

A life table is habitat and time specific, as well as sex specific. Both fecundity and mortality rates can vary across time (or between populations) so as to raise or lower the baseline levels for all members of the population. We therefore cannot assume that a life table determined for one population will automatically apply to another population elsewhere or even to the same population at a later time.

6.1.2. Birthrates

Overall birthrates vary considerably both between populations of the same species and, within populations, over time (fig. 6.1: see also Dunbar 1988). Perhaps the most important factors influencing birthrates are food availability and quality. Birthrates increase in well-fed populations both because females start reproducing at an earlier age and because interbirth intervals are shorter (Dunbar 1988; Lyles and Dobson 1988).

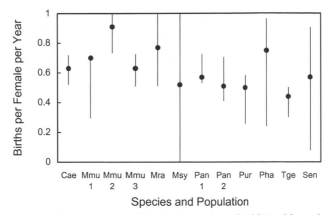

Figure 6.1 Mean and range in annual birthrates in several Old World monkey species: Cae, *Cercopithecus aethiops* (Amboseli, Kenya, six group-years: Lee and Hauser 1998); Mmu, *Macaca mulatta* (1 is Cayo Santiago, Puerto Rico, 10: Drickamer 1974) (2 is Chattari, India, 15: Southwick and Siddiqi 1977) (3 is Kathmandu, Nepal, 8: Teas et al. 1981); Mra, *Macaca radiata* (Davis, California, 16: Smith 1982); Msy, *Macaca sylvanus* (Gibraltar, 82: Fa 1986); Pan, *Papio anubis* (1 is Gilgil, Kenya, 9: Strum and Western 1982) (2 is Tana River, Kenya, 5: Bentley-Condit and Smith 1997); Pur, *Papio ursinus* (Kuiseb, Namibia, 5: Hamilton 1986); Pha, *Papio hamadryas* (Erer Gota, Ethiopia, 5: Sigg et al. 1982); Tge, *Theropithecus gelada* (Simen, Ethiopia, 4: Ohsawa and Dunbar 1984); Sen, *Semnopithecus entellus* (Jodhpur, India, 12: Winkler, Loch, and Vogel 1984).

The effects of provisioning on primate populations illustrate this particularly clearly.

In the semiprovisioned free-ranging Cayo Santiago rhesus population, females achieve puberty at about four years of age, whereas those from the wholly provisioned captive colony at the California Primate Research Center (CPRC) do so at about three years. Loy (1988) reported similar results for Japanese macaques (*Macaca fuscata*): mean female age at first reproduction was significantly lower (5.2 versus 6.0 years) in provisioned populations. In rich environments animals also grow, and therefore mature, faster: Phillips Conroy and Jolly (1988), for example, found that dental eruption patterns were delayed by about 1.5 years in wild hamadryas and yellow baboons compared with animals living in well-fed captive populations in Texas. Fecundity exhibits a similar pattern. On Cayo Santiago, rhesus females reproduce at intervals of 1.60 years, whereas those from the wholly provisioned CPRC colony reproduce at intervals of 1.49 years. Figure 6.2 shows how birthrate (measured as the proportion of females giving birth each year) increases with level of provisioning in Japanese macaque populations. The relationship is highly significant (Spearman $r_s = .74, p = .002$).

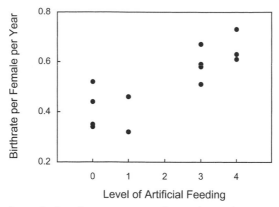

Figure 6.2 Birthrate (indexed as proportion of reproductively mature females giving birth each year) plotted against the level of artificial provisioning for Japanese macaque *Macaca fuscata* populations. Intensity of provisioning is classified as 0, none; 1, sporadic light provisioning; 2, provisioning as a minor proportion of total dietary intake; 3, provisioning as a major proportion of total intake; 4, provisioning as main or only source of food. (Data from Lyles and Dobson 1988.)

Wild primates with access to garbage dumps or provisioning by humans are also significantly heavier, with a higher body fat content (baboons: Altmann et al. 1993; see also review in Dittus 1998), and thus have significantly higher birthrates than other nearby groups (Japanese macaques: Mori 1979b, Sugiyama and Ohsawa 1982).

In a comparison of two populations of wild Barbary macaques, Ménard and Vallet (1996) found that the population inhabiting the richer high-altitude evergreen forest had an annual birthrate significantly higher than that of the population inhabiting a low-altitude deciduous forest (annual birthrates of 0.64 vs. 0.57 per female). Similarly, Ohsawa and Dunbar (1984) found that high-altitude gelada baboon populations had lower birthrates than lower-altitude populations, either because forage quality declines with altitude or because temperatures decline (thereby forcing females to divert energy from reproduction to thermoregulation) or both. Strum and Western (1982) found that baboon birthrates correlated inversely with the standing crop available to the baboons (indexed as habitat productivity discounted by the offtake removed by other herbivores).

There is, in addition, some evidence that the timing of births in seasonally reproducing species may be related to the availability of food resources in seasonal habitats (lemurs: Sauther 1998; squirrel monkeys: Boinski 1987b; guenons: Chism, Rowell, and Olson 1984; Butynski 1988). Indeed, under more benign foraging conditions, seasonally reproducing species may become less seasonal: Srivastava and Dunbar (1997), for example, found that provisioned populations of langurs (*Semnopithecus entellus*) were significantly less seasonal in their reproduction than nonprovisioned populations at similar latitudes.

However, multivariate analyses of birthrate data for two taxa (Hanuman langurs and *Papio* baboons) suggest that a number of factors besides forage quality may be implicated. Srivastava and Dunbar (1997) carried out a stepwise regression analysis of data from twenty *Semnopithecus entellus* populations and obtained the following regression equation for annual birthrates:

$$\ln(B) = -0.29 + 0.59\ \ln(D) - 0.99\ \ln(OMG) + 0.34\ \ln(F)$$

($r^2 = .50$, $F_{3,16} = 5.39$, $p < .05$), where B is the birthrate per female per year, D is the number of dry months (months in the year with less than 50 mm of rainfall, a measure of habitat quality), OMG is the percentage of social groups that have a single breeding male, and F is the mean number of females per group. Other environmental variables not selected by the analysis included mean annual rainfall, mean annual temperature (although the number of dry months may well be a surrogate for mean annual temperature), and latitude. (Note that, since birthrate is a probability, its natural log is a negative number, so the signs should be reversed when interpreting the directional influence of the independent variables.) Birthrates fall as the habitat gets drier (and perhaps warmer?) and more seasonal (ecologically stressing). In addition, the birthrates fall as the number of females in the group increases (social competition and stress) but rise with the proportion of one-male groups (possibly reflecting high rates of infanticide following male takeovers).

A comparable analysis of data from fourteen *Papio* populations yields the following best-fit equation:

$$\ln(IBI) = 20.74 - 2.91\ \ln(F) + 0.55\ (\ln(F))^2 - 8.95\ \ln(T) + 1.45\ (\ln(T))^2$$

($r^2 = .84$, $F_{4,9} = 11.68$, $p = .001$), where IBI is the interbirth interval (in months), F is the number of females in the group, and T is mean annual temperature (in °C) (Hill, Lycett, and Dunbar 2000). In this case total annual rainfall, habitat seasonality, altitude, latitude, group size, and the number of males in the group were not selected as predictor variables. Birthrates are a ∩-shaped function of both the number of adult females in the group (once again, presumably reflecting competition effects: see also Smuts and Nicholson 1989) and mean annual temperature. Differentiation indicates that maximum birthrates are achieved in groups with fourteen females living in habitats characterized by a mean annual temperature of 22°C. The low birthrates (long interbirth intervals) in low- and high-temperature habitats appear to reflect the costs of thermoregulation (habitat seasonality appears not to be the issue here, since rainfall does not contribute). The ∩-shaped relationship to number of females probably reflects the interaction of two separate effects: that group size increases with habitat quality (i.e., groups with few females tend to live in poor-quality habitats) and that intragroup competition becomes more intense as group size becomes larger.

The importance of temperature in the baboon case is particularly striking, since we might not expect temperature to be important to tropical species. Temperature has also been found to influence birthrates positively in geladas (Ohsawa and Dunbar 1984), callitrichid primates (Dunbar 1995b), and at least some small African antelope species (e.g., klipspringer: Dunbar 1990b).

6.2. Population Dynamics

The demographic structure of a population (or group) is the outcome over time of the interaction between the fecundity and mortality schedules in the population-specific male and female life tables. These interact to produce a sex-specific age structure for the population. In demographically stationary populations (see below), this has a pyramidal form with the cohort of newborns at the base; attritional mortality gradually reduces the number of animals in each successive age cohort. Growing populations have a disproportionately large base (number of younger animals), while declining populations have age pyramids that are often lozenge-shaped (because each successive cohort of newborns is smaller than its predecessor). In this section we describe two basic indices used to study population growth, and evaluate the influence of food, predation, and disease on population growth and decline.

6.2.1. Indices of Population Growth

Populations grow at a rate determined by the balance of births (plus immigrations) over deaths (plus emigrations) (sometimes referred to as the *BIDE* model). The growth rate of a population can be characterized in one of two ways: by the net reproductive rate (R_0) or by the intrinsic rate of increase (r_m).

NET REPRODUCTIVE RATE

The net reproductive rate is essentially the number of same-sex offspring an individual can, on average, expect to produce over the course of its lifetime, given the local sex- and population-specific survivorship and fecundity characteristics. It is calculated directly from a life table as

$$R_0 = \sum_{x=0}^{x=\infty} l_x m_x,$$

where l_x is the age-specific survivorship and m_x is the age-specific fecundity (see table 6.1). This equation can also be given in continuous form by integrating across time, but here we prefer to consider only the mathematically simpler discrete time form.

Because R_0 is defined as the lifetime number of same-sex offspring produced by the average individual, it provides a useful index of a population's demographic state. When $R_0 = 1$, each individual can expect to replace itself when it dies. As a result, the population will be in a *stationary* state: births will exactly balance deaths, and the composition of the population in terms of age and sex will remain stable across time. When $R_0 < 1$, the population will be in decline and will be able to maintain its numerical size only through in-migration. Finally, when $R_0 > 1$, each individual will contribute proportionally more offspring to the next generation, and the population will increase. If the rate of change under either of the last two cases is constant, the population is said to be *demographically stable*. The technical difference between being stationary and being stable is important.

Because it offers an instantaneous index of the demographic health of the population, R_0 is sometimes known as the *replacement rate*. Note, however, that this variable does not take migration into account: it looks simply at the intrinsic demographic health of the population, given its current birthrate and death rate.

R_0 is influenced by four key parameters: litter size, age at first reproduction, interbirth interval, and life span (Dunbar 1988). Unfortunately, the number of primate species for which it is possible to calculate complete life tables is small, so variances in R_0 cannot be easily illustrated. However, the net reproductive rate for the Cayo Santiago rhesus population shown in table 6.1 is $R_0 = 1.878$. Since each female is producing nearly two daughters in a lifetime, this population is almost doubling in size every generation. Juvenile mortality will, of course, slow this down, but since survivorship to puberty (at 3.5 years) is $l_x = 0.788$, the impact of preadult mortality is small. Lyles and Dobson (1988) modeled population growth rates for medium-sized monkeys with the life history characteristics of vervets (fig. 6.3). They showed that birthrates below about 0.5 per year and infant survival rates below about 50% both resulted in $R_0 < 1$, and thus population decline.

Leslie matrices (Leslie 1948) use the same life-table variables to project future population size and composition. This approach offers a valuable alternative to R_0 as a means of assessing a population's demographic viability, and a number of population viability models are based on it. A brief guide to Leslie matrices is given in appendix 2.

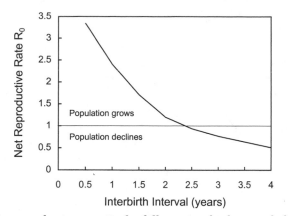

Figure 6.3 Net reproductive rate, R_0, for different interbirth intervals for a medium-sized primate with the life history characteristics of the vervet monkey. Populations will grow so long as interbirth intervals are less than about twenty-four months. (After Lyles and Dobson 1988.)

INTRINSIC RATE OF INCREASE

The alternative way of describing a population's growth is in terms of its rate of increase, r. This is defined by the equation

$$N_t = N_0 e^{rt},$$

where N_t is the size of the population at time t, N_0 is the size of the population at time 0, and e is the base of the natural logarithm. Inverting, we obtain

(6.2)
$$r = (\ln(N_0) - \ln(N_t)) / t.$$

In effect, r is the difference in the size of the population at two censuses divided by the time interval between the censuses. In this case $r = 0$ if the population is stable, $r < 0$ if it is declining, and $r > 0$ if it is increasing. It is obvious that r represents the actual rate of growth of the population, and for this reason it is sometimes called the *Malthusian parameter*.

The quantity r_m (or $r_{\max}$) is sometimes used to identify the maximum possible rate of increase that a species is capable of under the best possible conditions, and it is termed the *intrinsic* (or *maximum*) *rate of increase*. Technically this is calculated from Cole's (1954) classic equation:

$$1 = e^{-r_m} + be^{-r_m(a)} - be^{-r_m(w+1)},$$

where r_m is the intrinsic rate of increase, b is the mean fecundity rate (female offspring per female per year), a is the mean age at first reproduction for females (equivalent to the generation time, T), and w is the mean age at last reproduction. But following Caughley (1977), it can also be calculated from the life table, using the fact that

(6.3) $$\sum l_x e^{-rx} m_x = 1.$$

Using equation 6.3 and the data shown in table 6.1, Sade et al. (1976) calculated a value of $r_m = 0.071$ for the Cayo Santiago rhesus population. In effect, this population was capable of growing at a maximum rate of approximately 7% a year. It is difficult to know how close the population was to achieving this growth because animals were periodically removed. However, Ohsawa and Dunbar (1984) used equation 6.2 to calculate actual values of r for four bands of gelada baboons in the Simen Mountains (Ethiopia) and obtained values that ranged between $r = -0.003$ and $r = 0.162$ (with population growth rates being inversely related to altitude).

A variety of factors might determine interspecific variation in r_m. Robinson and Redford (1986) tested the hypothesis that diet correlated with r_m using data for Neotropical forest mammals but found no evidence of such a relationship. In contrast, Rowell and Richards (1979) suggested that reproductive rates are higher in primate species that occupy more seasonal open-country habitats (e.g., baboons) than in those that occupy stable forested environments (e.g., forest guenons). They argued that species living in unstable environments need to be able to respond faster if good conditions for breeding happen to come along, whereas those in stable habitats can rely on the constancy of conditions and can thus afford to delay reproduction in the interests of achieving larger body size or investing more heavily in current offspring.

Ross (1992) tested this hypothesis by comparing the intrinsic rate of population increase, r_m (calculated using Cole's equation as given above), for seventy-two species of primates. Since r_m also correlates with body mass (larger species breed more slowly: Ross 1988), she considered both absolute and relative r_m (i.e., the residual of r_m plotted against body mass). She found that relative (but not absolute) r_m correlated negatively with the amount of forest cover characteristic of a species' typical habitat fig. 6.4), as well as with eight measures of climatic variability (including the coefficient of variation of mean temperature, annual precipitation, and temperature and rainfall in the most extreme months). The more variable the climate, the higher was relative r_m. Latitude (measured either at the center of the species' geographical range or at the greatest absolute limit of its range) did not, however, correlate with r_m. Ross suggested that the relationship with climate was found only with relative r_m (i.e., when body size is discounted) because large-bodied species are less likely to be affected by climate than small-bodied ones, and this may confound the underlying relationship between r_m and climate.

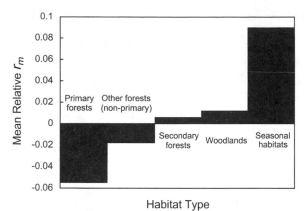

Figure 6.4 Mean relative intrinsic rate of population increase (r_m) for primate species that occupy different types of habitat. The mean relative r_m is calculated as the observed $\log_{10} r_m$ minus the expected $\log_{10} r_m$, where the latter is obtained from the allometric relationship between r_m and body mass. The mean relative r_m thus controls for confounding variation due to body mass (reproductive rates progressively decline in larger species: sec. 3.1). (After Ross 1992.)

6.2.2. Population Fluctuations

Populations commonly fluctuate in size as a result of changes in individual birth and death rates. Such changes are primarily driven by three environmental forces: the availability of food (and conspecific competition for that food: sec. 3.3), the intensity of predation (sec. 3.2.1), and the prevalance of disease. Under extreme conditions, these environmental sources of demographic fluctuation can cause population extinction (see sec. 7.2.2). In this section we examine what is known about the usual dynamics of each of these three forces in primate populations.

EFFECTS OF FOOD AVAILABILITY

The maximum size a population can reach is most commonly limited by the foods it consumes and the availability of these food resources (for a review of such effects on population density, see sec. 5.2). Consequently, food availability places an upper limit on the size of the population, and the maximum size is referred to as the *carrying capacity* of the habitat. This concept also emphasizes the importance that competition for food has on population growth: as the number of competitors increases, per capita food availability declines, leading to reduced reproductive rates and therefore slower population growth. This means that population growth is usually *density dependent*. Such processes can be modeled by the logistic equation

(6.4) $dN / dt = r_m N (1 - N/K)$,

where dN/dt is the instantaneous growth rate of the population and K is the carrying capacity. The growth in population size over time described by equation 6.4 is S-shaped, with the population showing a slow increase in size both early and late in its history (fig. 6.5).

How far reproduction is density-dependent in a population can vary, and this may have substantial effects on the recovery rates of populations after they have been perturbed away from carrying capacity by events such as droughts or disease outbreaks (Dobson and Lyles 1989). This is illustrated in figure 6.6. When recruitment is only loosely related to population density effects on resource availability (that is, b is low), populations are likely to return asymptotically to equilibrium when perturbed. As the relationship between recruitment and density becomes stronger, perturbations are more likely to give rise to damped oscillations that die out as the population approaches carrying capacity. When recruitment is strongly dependent on density, stable limit cycles will be common. But when adult survival is below about 70% per birth interval, extinction is inevitable.

Empirical evidence for the impact of density-dependent effects on primate population growth rates is provided by two populations of howler monkeys (*Alouatta* sp.). In each case a plot of the mean annual growth rate (dN/dt) against the detrended population size (how far actual population size deviates from that predicted by the long-term linear trend) produces a negative relationship (fig. 6.7: Spearman $r_s = -.53$, $p = .04$, and $r_s = -.9, p = .04$, respectively), indicating a classic density-dependent effect. The population grows when its size is below the predicted level and declines when it exceeds the predicted level.

The logistic equation emphasizes the importance of food availability on population growth, but it assumes a constant carrying capacity. In reality, things are more complicated. A habitat's carrying capacity can increase in good years (e.g., mast years in forest habitats) and decline in bad years (especially as a result of events such as hurricanes and forest fires). When carrying capacity drops below the current size of the population, famine and starvation become a serious prospect.

Under normal conditions, food resources probably influence population growth mainly through their effects on birthrate, although during famine periods a severe drop in food availability may also increase mortality through starvation. When provisioning was withdrawn from two populations of Japanese macaques, body mass declined, age at first birth increased, and birth intervals lengthened dramatically (table 6.2). The main demographic cause of the decline in this case was a 50% decline in fecundity, although infant mortality was also high in one of the

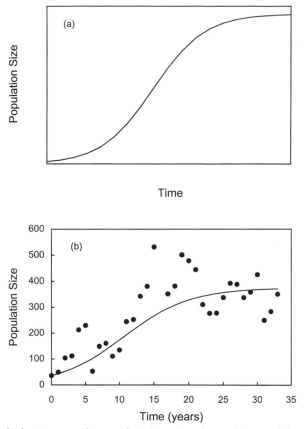

Figure 6.5 The logistic growth curve for (a) a notional population and (b) a population of Barbary macaques *Macaca sylvanus*. In the case of the Barbary macaques, the points are original data describing population size over time for a closed population on Gibraltar (Fa 1986), while the line is a theoretical projection based on an estimated carrying capacity of $k = 375$, a known initial population size $N_{t=0} = 36$, and a species-typical intrinsic rate of population increase $r_m = 0.21$ (the last from Ross 1992). Time $t = 0$ is 1937 (data follow through to 1970). One data point ($t = 16$, $N = 626$) is not shown. The projected line generally shows a reasonably good correspondence to the observed data, indicating density-dependent growth in this population. However, the substantial scatter also indicates how populations fluctuate around the baseline pattern owing to the complexities of the real world such as variation in food availability (k) and disease. These data suggest that the population initially exceeded k ($t = 15$, 16) but then crashed back down to carrying capacity as food shortages took effect. Note, however, that these patterns must be interpreted with care because of limited population management throughout this period (primarily the provisioning of food to supply two-thirds of the animals' nutritional requirements and the importation of a small number of monkeys in 1944 to replenish dwindling numbers).

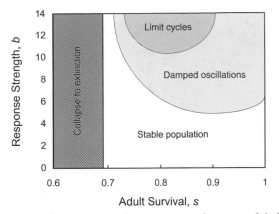

Figure 6.6 Responses of primate populations to perturbation modeled as a function of the response strength (the way the birthrate is affected by resource availability), b, and adult survival, s. When the recruitment rate s is less than about $s = 0.7$, populations will invariably collapse; when $s > 0.7$, populations will be stable providing the response strength is less than about $b = 6$ but otherwise will oscillate or slip into limit cycles. (Redrawn with the permission of Blackwell Science from Dobson and Lyles 1989.)

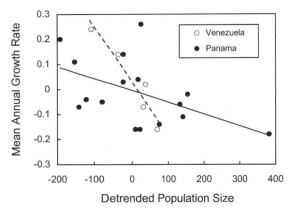

Figure 6.7 Mean annual population growth rate (r) plotted against detrended population size for two populations of howler monkeys: Barro Colorado Island (*Alouatta palliata*: Panama) for 1977–1994 (data from Milton 1996) and open woodland habitat at Hato Masaguaral (*Alouatta seniculus*: Venezuela) for 1969–1984 (data from Crockett and Eisenberg 1987). The detrended population size is the residual of actual population size from the regression line for population size plotted against time: this removes the effect of the secular trends (i.e., natural increases or decreases) in population size with time. The plotted lines are least-squares regression lines.

Table 6.2 Effects of withdrawal of provisioning on life history variables in two populations of Japanese macaques

| | Koshima Island | | Ryozenyama | |
	Before	After	Before	After
Age at first birth (yrs)	6.2	6.8	5.2	6.7
Birth interval (yrs)	1.49	3.13	1.69	2.94
Survivorship to 2 yrs	0.85	0.31	0.85	0.72
Population growth rate (%)	15.4	−4.0	13.4	4.9

Source: Data for Koshima from Mori 1979a; data for Ryozenyama from Sugiyama and Ohsawa 1982. After Dunbar 1988.

populations (table 6.2). Indeed, in some circumstances the role of famine-induced mortality may become paramount in precipitating population declines: the early decline of vervet monkeys at Amboseli appeared to be the result of high juvenile mortality from starvation (Struhsaker 1973).

Much of the evidence reviewed in section 6.1.2 suggested that birthrates are higher in richer habitats than in poorer ones. Comparisons of survival patterns in natural populations indicate that mortality rates also tend to be lower in richer habitats or in more productive years (e.g., macaques: Dittus 1977; howler monkeys: Milton 1982; baboons: Hamilton 1986; lemurs: Gould, Sussman, and Sauther 1999). Food-poor habitats may also indirectly lead to high mortality if malnourished animals become more susceptible to predation and disease (see below).

In many cases mortality may fall particularly heavily on subordinate and young animals. Low-ranking animals, for example, may experience higher mortality rates than high-ranking animals because their access to key resources is restricted by more dominant animals (e.g., vervets: Cheney, Lee, and Seyfarth 1981). Likewise, the survival prospects of young animals may be influenced by habitat quality: survival to age four years (i.e., sexual maturity) was 88% in geladas occupying optimal habitat in the Ethiopian mountains (Dunbar 1980b) but 78% in *Papio* baboons occupying gallery forest on the upper Tana River (Kenya) (Bentley-Condit and Smith 1997) and as low as 50% in *Papio* baboons living in poor-quality open woodland habitat at Amboseli (Kenya) (Altmann et al. 1977).

Similarly, when populations collapse during drought and famine, immature animals often bear the brunt of the mortality (e.g., vervets: Struhsaker 1973; howler monkeys: Milton 1982; baboons: Altmann et al. 1985). For example, Dittus (1977) reported high juvenile mortality (especially among females) in a population of Sri Lankan toque macaques (*Macaca sinica*) when these animals underwent a rapid 15% decline in population size following the worst drought in forty-four years. In one Namibian baboon population, 60% of infants alive at the start of one particular drought (or born during it) died (Hamilton 1986). Adult females (26% mortality) suffered slightly more heavily than juveniles (18% mortality) and a great deal more than adult males (8%), probably because of the added costs of reproduction (particularly lactation) under these stressful conditions. Gould, Sussman, and Sauther (1999) reported that juvenile *Lemur catta* experienced an 80% mortality rate during a drought year compared with only 18% the following year. They note that lemur milk is especially low in lipids and proteins and suggested that this may place infant lemurs at especially high risk when deteriorating feeding conditions prevent mothers from producing as much milk.

Perhaps the most striking natural example of the effects of environmental change is the precipitous decline of the baboon population of the Amboseli basin (Kenya) (Altmann, Hausfater, and Altmann 1985). During the early 1960s, the woodlands of the Amboseli basin supported a population of about 2,500 baboons in fifty-one troops. By 1969 this had dropped to 255 baboons in eight troops (representing an astonishing annual rate of decline of about 38%, or $r = -0.38$). The population declined further to 215 animals by 1971 and to 123 individuals by 1979. Thereafter it seemed to stabilize, with a slight increase in numbers to about 150 by 1983, with slow recovery continuing thereafter. A similar crash in the population of the sympatric vervet (*Cercopithecus aethiops*) was recorded at the same time, with an annual decline of 6.3% over the twenty-four years 1963 to 1986 (Lee and Hauser 1998). These declines were attributed principally to the die-off of the *Acacia* trees that had constituted the bulk of the woodlands, caused by a rise in the water table that both inundated root systems and forced a salt layer to rise nearer to the surface (Western and van Praet 1973). The rise in water table was in turn attributed to a combination of unusually heavy rainfall during the early 1960s and increased snowmelt from the ice cap of nearby Mount Kilimanjaro. Such changes may be cyclic over the long term: severe droughts seem to occur on a ten- to twelve-year cycle in East Africa and may be under the direct influence of the sunspot cycle (Wood and Lovett 1974).

EFFECTS OF PREDATION

Predation in prey populations acts primarily on mortality rates, often with dramatic effects. This is especially true where there is a sudden increase in predator specialization or density. Butynski, Werikhe, and Kalina (1990), for example, attributed a crash in one Ugandan population of mountain gorillas to the fact that a single leopard in the area had learned to specialize on the gorillas. Similarly, predation normally accounted for about 50% of the mortality experienced by vervet monkeys

Table 6.3 Causes of mortality in two primate populations

Cause of death	Number of cases	
	Vervets	*Chimpanzees*
Predation	54	0
Injury, cannibalism	0	17
Illness	19	23
Orphaning	5	6
Old age	0	3
Unknown	24	8
Total number of deaths	102	57

Source: Data for vervets (the Amboseli population) from Cheney et al. 1986; data for chimpanzees (the Gombe population) from Goodall 1983. After Dunbar 1988.

in an area at Amboseli (table 6.3), but in one year fourteen vervets disappeared in one thirty-day period (70% of all deaths that year) when a leopard took up residence in the area (Isbell 1990).

More commonly, predation rates are less variable over time and more attritional in their impact. Red colobus (*Procolobus badius*), for example, suffer consistently high levels of predation from chimpanzees in a number of East and West African populations. Stanford et al. (1994) estimated that annual offtake rates in Gombe National Park (Tanzania) are in the order of 17–33% of the total population (cf. Busse 1977; Wrangham and Bergman Riss 1990). In contrast, the hunting pressure exerted by the Taï (Ivory Coast) chimpanzees seems to be much lower (Boesch and Boesch 1989): Wrangham and Bergmann Riss (1990) estimated that about 4% of the red colobus population was killed by chimpanzees each year at this site (i.e., about an order of magnitude lower than they estimated for Gombe). Other primates are by no means immune from chimpanzee predation: Basabose and Yamagiwa (1997) estimated that chimpanzees in the Kahuzi-Biega National Park (Democratic Republic of Congo) kill about 11–18% of the local *Cercopithecus* monkey population.

Other more conventional predators may also vary considerably in their impact on prey populations. During a thirteen-month study of mouse lemurs (*Microcebus murinus*) in the Beza Mahafaly Reserve (Madagascar), a minimum of fifty-eight individuals (thirteen immatures and forty-five adults) were collected from the pellets sampled from barn owls and a further sixteen from the pellets of Madagascar long-eared owls (Goodman, O'Connor, and Langrand 1993). Taking into account the frequency of pellet sampling, Goodman, O'Connor, and Langrand (1993) estimated that some 522 to 580 *Microcebus* were taken each year by the nine or ten pairs of owls in the reserve. This represents about 25% of the total *Microcebus* population in the area. By comparison, Struhsaker and Leakey (1990) estimated that monkey-eating eagles removed only about 1% of the red colobus population at Kibale (Uganda), even though monkeys accounted for about 88% of all prey items. The difference in this case might reflect prey body size, with the smaller mouse lemurs being more vulnerable than larger-bodied species.

Finally, it is important to stress that mortality from predation does not necessarily fall equally on all members of the population. Monkeys constitute 84% of the prey taken by crowned hawk eagles in the Kibale Forest (Struhsaker and Leakey 1990); however, the eagles took disproportionately more immatures from the bigger species (e.g., red colobus) and more adults from the smaller species (e.g., blue monkeys). Struhsaker and Leakey concluded that the eagles had a significant impact on the adult population of most species, but particularly so on the adult males of blue monkeys and guereza colobus monkeys, thus leading to unusual biases in the adult sex ratios of these species. Similarly, chimpanzees generally prefer to take female adult red colobus rather than males, probably because females are less able to defend themselves than the larger-bodied males (Busse 1977). Much probably depends on the hunting style of the predators, however. Cowlishaw (1994) noted that leopards are more likely to take adult baboons than juveniles and, among adults, to take males rather than females. This may reflect both the profitability of prey (in terms of body mass) and their ability to escape in trees. Male baboons may also be more at risk because they are more often alone: not only are they more likely to sleep apart from others in the sleeping trees, but their patterns of dispersal (see sec. 3.3.2) may make them more vulnerable.

Little is known about how populations respond demographically to predation. Sudden acute levels of predation may distort natural population trajectories, especially if they involve differential mortality on breeding females or immature animals. But primates can also be expected to respond behaviorally to sudden increases in predation mortality by forming larger groups or altering their ranging patterns (see chapter 3). In some cases demographic responses may also be possible. That the *Microcebus* population referred to above can withstand a 25% harvesting rate by predators can almost certainly be attributed to its high natural birthrate. Hill and Dunbar (1998) showed that birthrates correlated positively with predation rates in a number of primate species and argued, following Lycett, Henzi, and Barrett (1998), that species (or populations) with high birthrates may be able to withstand higher predation rates. Which is cause and which is effect here remains to be determined. These findings, however, should caution us against assuming that high birthrates necessarily imply a growing population: a population may have high birthrates simply because it has high mortality rates.

EFFECTS OF DISEASE

Like predation, disease is more likely to affect mortality rates than birth-rates in primate populations. Although rather less is known about the impact of disease on primate populations because autopsies are rarely performed, it is clear that disease outbreaks can have devastating consequences. Examples include the polio epidemic that afflicted the Gombe chimpanzee population in the 1970s (see table 6.3) and the yellow fever epidemics that have repeatedly decimated howler monkey populations in South and Central America (Collias and Southwick 1952; James et al. 1997).

Epidemics can cause dramatic increases in mortality rates. Barrett and Henzi (1998), for example, reported a case in which, over a three-month period, forty-four out of eighty-five animals in two troops of baboons succumbed to a viral infection with symptoms involving hemorrhagic diarrhea. The outbreak began in one troop (85% of whose members succumbed) and spread to the second troop after the immigration of a sick adult male. After the onset of heavy rains, the disease disappeared without affecting either of the other two neighboring troops. This outbreak had important long-term demographic consequences: all six adult males in the two troops died, whereas only eleven of twenty-two adult females succumbed. Similarly, Pope (1998a) reported an 85% crash in the size of one red howler (*Alouatta seniculus*) population in Venezuela over four years owing to an unidentified pathogen, resulting in the complete extinction of some groups; a second population suffered a 40% decline over the same period (in this case with the brunt of the mortality being borne by the adult females).

The effects of disease can also be chronic. Milton (1996) found that botfly and screwfly infestations were a major problem for the howler monkeys on Barro Colorado island. Although both infestation and mortality rates varied considerably from one year to the next, as many as 16% of the total population died during one particularly bad year (though not all these deaths could necessarily be attributed to flies). Within years, both infestation rates and mortality were significantly higher during the wet season (July–December) than during the dry season (possibly because animals' resistance is lowered by the need to divert energy from fat reserves into maintenance during periods of food shortage). There appeared to be no obvious environmental correlates of year-to-year variability; but since rainfall correlated with monthly infestation and mortality rates, it is possible that wetter years (or years with longer rainy seasons) are associated with higher botfly densities or higher levels of infestation. Other studies have also reported links between heavy rainfall and higher incidence of disease (respiratory infections in mountain gorillas: Watts 1998; strongyle nematode infections in chimpanzees: Huffman et al. 1997). Among the gorillas, months with high rainfall were also associated with high mortality. Similarly, Stoner (1996) found that nematode loads were higher in more humid or wet conditions both within and between howler monkey populations.

Like other sources of mortality, disease can affect mortality rates in the various age-sex classes differently. Milton (1996) reported that juveniles suffered significantly higher botfly infestation rates than adults, whereas infants had significantly lower rates than either of these two classes. Brain and Bohrmann (1992) reported that tick infestations in one population of baboons inhabiting the Namib Desert in southwestern Africa were responsible for the deaths of ten infants when tick attachment or damage (to their mouths or the mother's teats) prevented them from sucking. Another five infants died indirectly when high-ranking females that had lost their infants kidnapped other females' infants, which then died from dehydration (Brain 1992). Ticks therefore led, directly or indirectly, to the deaths of 68% of all the infants born into the group over a four-year period, accounting for 83% of all infant deaths.

However, the impact of chronic disease need not always be severe. Although Eley et al. (1989) reported moderate loads of nematode, helminth, and ascarid parasites in Kenyan baboons, there was little evidence that these had a significant effect on mortality rates. Some gelada baboon populations suffer heavy infestations of a tapeworm that can lead to death when subcutaneous cysts burst (Dunbar 1980b). Approximately 12.4% of the subadults and adults in one gelada population were afflicted, although actual mortality caused by burst cysts was low (<1% of animals per year). Although populations may often be able to cope with moderate levels of chronic infection, it is worth emphasizing that a disease that is chronic and tolerable in a large population could become an agent of rapid population collapse in a small, fragmented population (May 1988).

In an extensive review of parasite loads in wild primates, Stuart and Strier (1995) found that high parasite infestation rates appeared to be correlated with humid conditions (because ova and larvae survive better in feces under these conditions, thus promoting infection of other individuals); with arid conditions where only standing water was available (especially where domestic stock contaminated the water); and with proximity to humans (either through contaminated human fecal matter or because habitat destruction may increase the habitat available to parasite hosts or vectors such as mosquitoes).

KEY FACTORS IN POPULATION FLUCTUATION

The material reviewed in the preceding sections shows that food availability, predation, and disease can all have substantial influences on population size, but it is difficult to assess the relative importance of these three pressures with the evidence available. In a recent review of population crashes, Young (1994) found that disease was the key factor in four of eight primate cases; in three other cases, habitat change was most important (presumably through its effects on food availability), and mortality from predation was responsible in the final case. Young noted that this pattern is strikingly different from that observed in other mammalian herbivores (where 80% of documented population collapses were due to starvation) but very similar to that observed in carnivores (where disease accounted for 55% of collapses). Primates may be resistant to population collapse from starvation owing to their unusual dietary flexibility (see sec. 3.2.2). However, individuals that are barely managing to survive in poor-quality habitats are likely to become more vulnerable to predation risk (e.g., when habitat quality prevents group size from increasing in response to higher predation levels) and disease (e.g., when in poor condition as a result of reliance on low-quality foods). This may help to explain Young's findings. In addition, it emphasizes that although famine may rarely be the proximate mechanism of population collapse, it might still be implicated as the ultimate cause.

6.3. Metapopulation Dynamics

Thus far we have discussed primate populations as independent entities. In many cases, however, individual populations are not isolated. Rather, they are interconnected, through the exchange of individuals, as part of a larger network. This system is termed a *metapopulation* (i.e., a population of populations). Metapopulations have been the focus of much discussion in population biology in recent years (e.g., Harrison and Hastings 1996; Hanski 1998, 1999), not least because the macroscale features of metapopulation structure can introduce unexpected and sometimes complex effects into the dynamics of their constituent populations, with important implications for processes of extinction. This is especially important in populations that are individually small and thus extinction prone, as is likely when a species occupies a fragmented habitat where individual patches vary considerably in size.

The essence of the metapopulation approach is that the presence of a given species in an area depends on the balance between the rates at which individual populations go extinct and the rates at which new ones are established by migrants from other populations elsewhere in the system. This dynamic process is a direct result of the fact that, although a number of habitat patches may be suitable for a population within a given geographical region, only some of these patches are occupied at any one time. For example, along the Tana River (Kenya), only a fraction of the riparian forest patches are currently inhabited by any one of the diurnal

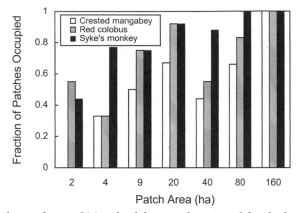

Figure 6.8 The incidence of (a) red colobus monkeys *Procolobus badius,* (b) crested mangabeys *Cercocebus galeritus,* and (c) Syke's monkeys *Cercopithecus mitis* in forest patches along the Tana River (Kenya), plotted against patch size. Incidence is indexed as the proportion of all available patches occupied by each species. The distribution of patch sizes in the Tana River forest system is uneven, and the four largest patches are excluded as outliers. From Cowlishaw (n.d.), based on data from Butynski and Mwangi (1994).

forest primate taxa that exist there (red colobus, crested mangabey, Syke's monkey: fig. 6.8); notably, species incidence usually increases in larger patches because they hold larger populations (which are less extinction prone: chapter 7) (see also fig. 4.2). As time passes, populations in some habitat patches disappear through natural extinction processes (see sec. 7.2) and new populations become established in previously unoccupied patches (through natural colonization, involving in-migration from other populations in the network). For metapopulations in equilibrium, the number of population extinctions approximates the number of new populations established. Communication between populations through migration is an essential component of these systems; in fact, according to some formulations a metapopulation is defined as a set of populations that would not be individually viable in isolation.

Little is known about primate metapopulation dynamics. Only two primate systems have been studied in detail, in both cases using a modeling approach based on empirical data. The first study deals with a samango monkey metapopulation (Swart and Lawes 1996) and is reviewed in section 10.2.3. The second deals with the red colobus, crested mangabey, and Syke's monkey metapopulations on the Tana River. Using an approach based on incidence functions (Hanski 1998), Cowlishaw (n.d.) estimated colonization and extinction rates for populations of each taxon in the Tana River system. The results indicate that the pattern of inci-

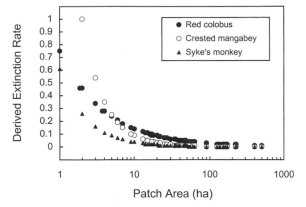

Figure 6.9 Covariation between derived extinction rates and forest patch area in three Tana River (Kenya) monkeys: red colobus monkeys *Procolobus badius,* crested mangabeys *Cercocebus galeritus,* and Syke's monkeys *Cercopithecus mitis.* Derived extinction rates are predicted based on the patterns of patch incidence observed in the Tana River forest system (fig. 6.8) using the simplest version of the incidence function model (in which the effects of interpatch distance are not incorporated: Hanski 1999). From Cowlishaw (n.d.), based on data from Butynski and Mwangi (1994).

dence varies between species (fig. 6.8). A preliminary analysis of the derived extinction rates (fig. 6.9) suggested that Syke's monkeys showed the lowest extinction rates; probably they are buffered by their relatively large populations. In contrast, in most patches red colobus had the highest extinction rate, probably owing to their low dietary flexibility. Crested mangabeys showed the highest extinction rates in small patches, however, perhaps because of the larger ranging requirements of frugivores or the existence of Allee effects (see sec. 7.2.2) in small populations composed of multimale groups. (For more detailed discussion of this case study, see secs. 7.2 and 7.3; for a conventional population model that treats the isolated patches as a continuous population, see sec. 10.2.3)

6.3.1. Source-Sink Systems

An important feature of metapopulation structure is that the constituent populations may differ in their dynamics. Those populations in better-quality microhabitats have positive growth rates ($r > 0$) and can act as *sources* from which migrants are drawn, whereas those in poorer microhabitats have negative growth rates ($r < 0$) and act as *sinks*—groups that absorb migrants without growing numerically. This can give rise to a *source-sink* relationship (Dias 1996). In the limiting case, emigrations from the source balance the source's natural fecundity rate and immigrations into the sink balance the sink's natural mortality rate, so that the metapopulation as a whole remains in demographic equilibrium.

The "rescuing" of populations otherwise fated for extinction by the immigration of individuals from neighboring populations is termed the *rescue effect* (Brown and Kodric-Brown 1977). Rescue effects have been particularly useful in explaining how island populations, which normally tend to suffer high extinction rates owing to small population size, can be maintained by immigration from nearby larger islands or the mainland itself (e.g., Pagel and Payne 1996; Lande, Engen, and Saether 1998). The prevalence of rescue effects is likely to differ between metapopulation systems and, in principle, between species in the same system depending on the characteristic colonization and extinction rates of the different species. This point is well illustrated in the distribution of rescue effects among the diurnal primates of the Tana River forest system (Cowlishaw, n.d.). Here rescue effects seem to be important in reducing extinction rates among red colobus and crested mangabey populations, but not in Syke's monkey populations. This appears to result from the intrinsic resilience of the local Syke's monkey populations (possibly owing to their relatively large population sizes).

Although rescue effects have received little attention in the primate literature, there is evidence that they might be important in some systems. Dunbar (1987a) reported that the marginal gully habitat in the Bole Valley (Ethiopia) acted as a demographic sink for the local *Colobus guereza* population: it absorbed excess animals from the subpopulation occupying the richer gallery forest, which seemed to be at carrying capacity. Crude population growth rates (estimated as birthrate minus death rate) were $r = 0.042$ per individual per year for the forest subpopulation, but only $r = 0.013$ for the gully subpopulation. Without the continued inward flow of migrants, the gully subpopulation would have had some difficulty maintaining itself. Similarly, gelada baboon population growth rates dropped below $r = 0$ at altitudes above 4,000 m, even though populations at lower altitudes were growing at rates as high as 13% a year (Ohsawa and Dunbar 1984). It seemed that the high-altitude populations were able to persist only because in-migration from below continually replenished them.

6.4. Population Genetics

Many primates live in highly structured populations, and this structuring has important implications for the genetics of populations, which may in turn have important consequences for their ability to survive when they become very small, particularly their capacity to resist the erosion of genetic diversity (see sec. 7.2). In this section we examine the determinants of genetic diversity within and between populations. Before doing so, however, it is necessary to introduce some of the conceptual tools of population genetics.

INDICES OF HETEROZYGOSITY

Heterozygosity (or genetic diversity) is measured as the proportion of alleles that are heterozygous in the average individual (that is, the proportion in which the two copies of the allele carried by a given individual are different). In the absence of localized inbreeding or genetic admixture, heterozygosity H is given as

$$H = \sum h_j / L,$$

where h_j is the proportion of individuals heterozygous at the jth locus ($h_j = 1 - p_{ij}^2$, where p_{ij} is the frequency of allele i at locus j), and L is the number of loci sampled. Note that the choice of loci to be sampled can affect estimates of H: if, by chance, a high proportion of naturally nonvariable loci are selected, then the estimated value of H will be lower than it ought to be. Furthermore, if the index of heterozygosity is intended as a measure of divergence by drift (or of time since divergence), it may be crucial to include only alleles subject to neutral selection: including any alleles that are under directed selection will imply misleadingly high levels of heterozygosity.

Heterozygosity can be lost from populations as the result of *inbreeding* and *genetic drift*. Genetic drift reflects chance variation in the frequency of certain alleles that results from sampling only a small part of the genetic diversity of a population during breeding; inbreeding (reproduction between close relatives) exacerbates this. Both are more common and thus problematic in small populations (see sec. 7.2.2). In addition, *founder effects* can occur during the colonization of new habitats: the smaller the founding population, the smaller will be the proportion of the parent population's genetic variation that it represents and the more intense the founder effect (and thus genetic drift) will be.

In mammals, the heterozygosity of allozyme markers typically has a range of $0.0 \leq H \leq 0.26$, with about 11% of species having $H = 0$ (Caughley and Gunn 1996). A range of mean heterozygosity values for a number of New and Old World primates is given in table 6.4. Although lack of direct comparability between studies means one must take care in interpreting these data (since different studies can use different loci in their analyses), three points are worth noting.

First, some species, such as callitrichids, have exceedingly low heterozygosities (close to $H = 0$). This may reflect a combination of factors characteristic of the Callitrichidae, including relatively small effective population sizes (low N_e/N ratios: see sec. 7.2.2) owing to a combination of polyandry, female reproductive suppression, and high variance in the

Table 6.4 Estimates of heterozygosity in primates

Taxon	Species	Country	Number of populations	h^1	H^2
Prosimian	*Lepilemur mustelinus*[3]	Madagascar	4	0.27	0.04
Callitrichidae	*Leontopithecus rosalia*	Brazil	1	0.03	0.01
		Captive	1	0.04	0.01
	L. chrysomelas	Captive	1	0.03	0.01
	L. chrysopygus	Brazil	1	0	0
		Captive	1	0.03	0.03
	Saguinus fuscicollis	Brazil	2	0.15	0.04
	S. midas	Brazil	2	0.23	0.04
	Callithrix humeralifer	Brazil	1	0.20	0.04
	C. emiliae	Brazil	2	0.15	0.02
	C. jacchus	Brazil	1	0.10	0.02
	C. penicillata	Brazil	1	0.20	0.05
	C. geoffroyi	Brazil	1	0.05	0.03
Cebidae	*Brachyteles* sp.	Brazil	2	0.20	0.11
	Alouatta belzebul	Brazil	4	0.46	0.07
	A. seniculus	Venezuela, Brazil	2	0.35	0.10
	A. pigra	Belize	1	0.05	0.02
	A. palliata	Costa Rica	1	0.10	0.01
Cercopithecinae	*Macaca mulatta*	Indian subcontinent	4	0.28	0.07
		China	1	0.30	0.08
	M. fascicularis	Southeast Asia	7	0.19	0.05
	M. sinica	Sri Lanka	1	0.20	0.07
	M. fuscata	Japan	9	0.07	0.01
	M. sylvanus	Algeria	1	0.20	0.07

Source: After Pope 1996.

[1]Proportion of individuals heterozygous at a given locus.
[2]Heterozygosity index.
[3]After Tomiuk et al. 1997.

reproductive success of both sexes; potentially high levels of inbreeding owing to frequent philopatry and low rates of migration in both sexes; and short generation times (Pope 1996).

Second, some species seem to show unusually high levels of variability (e.g., *Brachyteles*: $H = 0.11$ in both races). This suggests that the ancestral populations might have been very heterozygous. In contrast, the Central American *Alouatta* populations generally exhibit much lower heterozygosity ($H \approx 0.02$), almost certainly as a result of several major population crashes in the past century. James et al. (1997) attributed the low genetic variability of *Alouatta pigra* populations in Belize to four successive population bottlenecks (three due to hurricanes and one to a yellow fever epidemic) over the past sixty years.

Third, the macaques in general exhibit modest levels of within-population heterozygosity ($H \approx 0.05$–0.08) despite extensive habitat fragmentation leading to small, isolated populations in recent historical times. The exception is *Macaca fuscata* on the Japanese archipelago: in the small, isolated populations that now remain, values of $H \approx 0.00$–0.02 are typical. One reason might be that population fragmentation has a longer history in Japan, where human pressure in a region of intensive agriculture has been characteristic for much longer than in any other macaque habitat. Nonetheless, despite these low levels of within-population heterozygosity, heterozygosity across populations is very high ($H = 0.66$: Nozawa et al. 1991). In other words, much of the original heterozygosity of the species as a whole seems to have been preserved despite the fragmentation of its range.

Further discussion of the factors that influence heterozygosity is given in sections 6.4.1 and 6.4.2.

The genetic structure of a population can be described by a set of five indices determined from observed allele frequencies:

H_i = the observed heterozygosity of a randomly chosen individual in group i

H_I = the observed heterozygosity of an individual, averaged over all k groups

H_{si} = the expected heterozygosity of individuals in group i, based on the actual allele frequencies p_j in group i [$H_{si} = 1 - \Sigma p_{ji}^2$]

H_S = the expected heterozygosity of an individual based on the actual allele frequencies in group i, averaged over all k groups [$H_S = \Sigma H_{si}/k$]

H_T = the expected heterozygosity of an individual, based on the mean allele frequencies across all k groups [$H_T = 1 - p_j^2$].

From these, we can derive the following statistics, commonly known as the F-statistics (Nei 1977) or Wright's *fixation indices:*

$$F_{IS} = 1 - H_I / H_S$$
$$F_{ST} = 1 - H_S / H_T$$
$$F_{IT} = 1 - H_I / H_T,$$

each of which varies between -1 and $+1$. These indices describe the departure from Hardy-Weinberg equilibrium (defined as the expected frequencies of each possible genotype under conditions of random mating and in the absence of selection and migration). Since $F_{IS} = F_{IT} = 0$ under Hardy-Weinberg equilibrium, the values of these two indices measure the deviation from equilibrium within groups (in the case of F_{IS}) and (in the case of F_{IT}) for the population as a whole (i.e., between groups). F_{IS} and F_{IT} are both positive when the sample population is more homozygous than expected if individuals mate at random with each other, and it is negative when they are more heterozygous. In effect, they measure the level of inbreeding: positive values imply high levels of inbreeding, negative values imply outbreeding. F_{ST} is the difference between within-group and between-group levels of heterozygosity and thus provides a measure of substructuring within the population. F_{ST} is positive when matings between similar genotypes are favored (resulting in more divergence between population units) and negative when assortative matings between different genotypes are favored (resulting in more mixing across population units). In principle,

$$(1 - F_{IS})(1 - F_{ST}) = (1 - F_{IT})$$

(at least for individual loci). When groups exchange members freely, it follows that

$$F_{IS} = F_{ST} = F_{IT}.$$

6.4.1. Heterozygosity within Populatons

Several factors contribute to the genetic architecture of a population. These include group fission, intergroup migration, and social structure. We consider each of these in turn. In addition, as we noted above, levels of heterozygosity may also be affected by the size of N_e (the effective population size: sec. 7.2.2) and by population history (e.g., population bottlenecks): these we will return to in later chapters.

GROUP FISSION EFFECTS

Perhaps the most basic factor promoting genetic heterozygosity in a population is group fission. When social groups undergo fission, closely related animals frequently stay together, so that groups often split along lines of "least genetic resistance." This has been documented in a number of species, including rhesus macaques on Cayo Santiago (Chepko-Sade and Sade 1979), Japanese macaques (Koyama 1970), baboons (Nash 1976), and (if grooming indicates relatedness) forest guenons (Cords and Rowell 1986). When relatives remain together, daughter groups are not a random sample of the genetic diversity of the parent group; as a result, genetic relatedness is higher within daughter groups than in the parent group (Chepko-Sade and Olivier 1979; Scheffrahn et al. 1993).

One consequence of this is that when daughter groups remain near each other populations will exhibit a concentric ripple pattern of relatedness. Shotake and Nozawa (1984), for example, found that gelada reproductive units are genetically more homogeneous than the bands they belong to and that the bands in turn are more homogeneous than the local population as a whole. On a broader scale, Nozawa et al. (1982) found that Japanese macaque populations are structured into a series of "local concentrations" of social groups that are genetically more homogeneous than the population as a whole. Populations may thus maintain high population-level heterozygosity even when the groups themselves are individually homogeneous (providing the groups differ genetically from each other). In such cases the loss of one group from the population could dramatically reduce population heterozygosity if an entire allelic lineage is lost as a result.

Although close relatives often do seem to stay together after fission, this is by no means a universal rule. Melnick and Kidd (1983) found that wild rhesus macaque groups did not fission by matrilineal relatedness but seemed to do so in random pattern. Similarly, Ron, Henzi, and Motro (1994) found that chacma baboon groups at Mkuzi (South Africa) did not seem to fission along matrilineal lines; instead, individuals of alternate dominance ranks stayed together according to a rule of "abandon your immediate superior." Although kinship relationships were not sufficiently well known in this second case to determine how this pattern of fission mapped onto genetic relatedness, this process may explain why Melnick, Pearl, and Richard (1984) found that genetic profiles of individual groups of wild *Macaca mulatta* in northern India were a random sample of the population as a whole.

There are sound socioecological reasons why fission processes in populations might differ in this way, and these have important implications for population genetics. One reason the provisioned Cayo Santiago macaque groups fission along genetic lines whereas the unprovisioned Himalayan macaque groups fission at random may be that they differ in the size and integration of their respective matrilines (Dunbar 1988; Datta 1989). The provisioned Cayo Santiago population has a high growth rate, associated with short interbirth intervals (mean twelve months) and high offspring survival rates, leading to large matrilines of closely related females. In contrast, the wild Himalayan population lived under more challenging environmental conditions, with longer interbirth intervals (mean twenty-six months) and lower survival rates; such conditions inevitably lead to small matrilines, and hence to less interest in (or even opportunity for) staying with relatives when groups fission. Moreover, if groups living in poor environments grow more slowly and hence fission less often, they will tend to represent a greater proportion of the population's genetic variation (i.e., groups will be less genetically differentiated) (Olivier et al. 1981; Pope 1998a).

The complex interaction between demographic processes and genetic structure has been investigated by Pope (1998a) in a study of four Venezuelan red howler (*Alouatta seniculus*) populations. She found that between-group relatedness (estimated from six allozyme loci) was higher when the population was at or near carrying capacity than when it had undergone catastrophic collapse, presumably because in the latter case the few lineages that survive are reduced to just one or two individuals. However, she also found that mean within-group relatedness could increase during episodes of population collapse because the elimination of breeding adults made it unnecessary for maturing subadults to leave the group to find breeding opportunities, thereby giving rise to significant levels of inbreeding. This effect was reinforced later as the population recovered when new groups were created by fission or when sets of related males transferred together into other groups. In sum, opportunities for intergroup differentiation were reduced when populations were increasing (and within-group relatedness was low), but increased during periods when the population was stable or in decline.

MIGRATION EFFECTS

Groups are not isolated islands, and the size of the *deme* (the interbreeding population) is determined mainly by the migration patterns of the animals between groups: in contrast to the effects of group fission, migration can promote genetic similarity between groups (mainly by counteracting the effects of genetic drift). The key issues in this context are which sex disperses and where the dispersing animals go (see sec. 3.3.2).

In the first case, although both sexes disperse in most species of primates, there are a number of important exceptions: among the cercopithecine monkeys, it is invariably males that disperse, whereas female dispersal is the norm among the great apes. In the second case, dispersing animals may either establish new groups (e.g., howler monkeys: Pope 1992; leaf monkeys: Sterck 1997) or join existing groups. The identity of these groups may not be random. Among the Amboseli vervets, particular groups seem to engage in a reciprocal exchange of males (Cheney and Seyfarth 1983; see also Henzi and Lucas 1980). Similar findings have been reported for the Cayo Santiago rhesus macaques (Meikle and Vessey 1981; Colvin 1983) and some Japanese macaque populations (Koyama 1970). In contrast, Pope (1992) found that male transfers were random with respect to receiving group in at least one howler monkey population, perhaps because male transfer was the result of new males' taking over the breeding role in an existing group. Notably, groups that exchange members need not occupy adjacent territories. For example, Pope (1992) found that the migration distances of male howler monkeys were typically about 800 m in primary forest and 1,100 m in gallery forest (equivalent to about three home ranges).

One important consequence of sex-biased dispersal is the relative strength of the kinship structures among the philopatric sex. Female philopatry in the cercopithecoid primates seems to be associated with strong matrilineal bonds among the females (Di Fiore and Randall 1994), and social groups are thus more likely to be relatively homogeneous (at least compared with the broader population). The females of six groups of long-tailed macaques in Sumatra (where males disperse and females are philopatric) were much more closely related than were the males (de Ruiter and Geffen 1998). Similar results were reported for baboons (Altmann et al. 1996). In contrast, in chimpanzees (where males are the philopatric sex and females disperse), Morin et al. (1994) found that males are more closely related to each other than are the females within at least one well-studied community; moreover, maternally transmitted genotypes can be detected over distances of 600–900 km (indicating a surprisingly high level of geographical penetrance owing to dispersal). Note that the latter finding does not of itself imply that females are dispersing over these distances; it means that their genes are being transmitted into more distant communities over successive generations as daughters and then granddaughters disperse beyond their respective natal community ranges.

In addition, when only one sex disperses there may be striking differences in the distribution of nuclear and nonnuclear (e.g., mitochondrial) DNA. In five well-studied species of macaque (all of which are female philopatric), Melnick and Hoelzer (1992, 1994) found that maternally inherited mitochondrial DNA (mtDNA) was characterized by greater local homozygosity and greater interpopulation heterozygosity than was nuclear DNA (which is passed on by both parents, but where dispersing males are responsible for gene flow across populations). In contrast, mtDNA alleles were more homogeneously distributed throughout the regional populations of *Trachypithecus auratus*, a species in which both sexes habitually disperse (Rosenblum et al. 1997). However, curious anomalies can emerge in some cases. Gagneux, Boesch, and Woodruff (1999) found that male genotypes are more widely dispersed among the Taï chimpanzees than might be expected for a male-philopatric species, apparently because the females frequently crossed community boundaries to mate with males from adjacent communities. In this population, paternity determination indicated that about 50% of infants were sired by males from outside the community.

It is possible that in those species where either males migrate into the same groups as their male kin or a single male is responsible for most of the fertilizations in a social group over an extended period, group members may be more closely related through their paternal lineages even though females are philopatric (Altmann 1979). Paternity relationships are known from only two primate populations in sufficient detail to allow us to establish whether paternal kinship can act as a genetic "binding force." The molecular data from the Gombe chimpanzee population (Morin et al. 1994) confirm that, at least in this particular male-

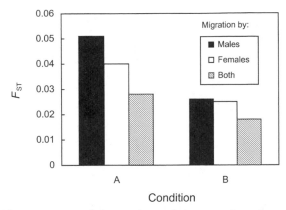

Figure 6.10 The proportion of the total population genetic variance found within groups, F_{ST}, depends both on which sex migrates and on how evenly sirings are distributed among the breeding males. In condition A only the top two ranking males gain any sirings, while in condition B all sirings are shared equally by all males. (After de Jong, de Ruiter, and Haring 1994.)

philopatric population, the males in the community may indeed be more closely related to each other than the females are. Similarly, the molecular data for one group of the Amboseli *Papio cynocephalus* population confirm Altmann's original hypothesis that when individual males dominate the mating opportunities over periods of several years, age cohorts will often be paternal sibships and groups may then be genetically substructured by age (Altmann et al. 1996).

It seems likely that the genetic homozygosity of a local population may depend primarily on a complex interaction between migration patterns and the rate of genetic drift. De Jong, de Ruiter, and Haring (1994) sampled seven social groups from one Sumatran population of *Macaca fascicularis* and found that genetic drift can be significant in populations of spatially close groups even when they exchange males. The estimated heterozygosities for seventeen blood proteins and enzymes revealed that both F_{ST} and F_{IS} values were slightly positive (0.045 and 0.016, respectively), indicating a slight preponderance of homozygotes within groups over what might be expected given random matings. De Ruiter and Geffen (1998) also found higher than expected levels of relatedness between groups in the same data set (especially between the high-ranking matrilines of neighboring groups). They concluded that this reflected patterns of gene flow through male dispersal within this population rather than group fission or the effect of geographical barriers (e.g., rivers or mountains) on drift.

MATING SYSTEM EFFECTS

Traditionally, population genetics has assumed that demes are panmictic populations: in other words, the opportunity for a free exchange of gametes between all members of the population is not constrained by any factors other than sex. This assumption is probably unrealistic for species, such as primates, whose populations exhibit spatial or social structuring. Chesser (1991) has pointed out that social structuring of this kind can have significant implications for population genetics and cannot be ignored.

De Jong, de Ruiter, and Haring (1994) simulated the genetic histories of a population with the breeding characteristics of *Macaca* and showed that the F-statistics are influenced in different ways by social structure. On the one hand, F_{ST}—which reflects relatedness—declines as the number of females in the group increases, is lower when all males contribute equally to sirings than when only some males are responsible, and is lower when only females migrate between groups than when only males migrate (fig. 6.10). In contrast, F_{IS} exhibits the reverse pattern. In effect, the opportunities for genetic drift between groups are greater when they have fewer females, when fewer males contribute to fertilizations, and when only the males transfer between groups. In addition, increasing the rate at which the dispersing sex transfers between groups is equivalent to increasing the number of breeding members of that sex in each generation, and this in turn will reduce the genetic differentiation between groups (Pope 1996).

This point has been made in more general form by Chesser (1991),

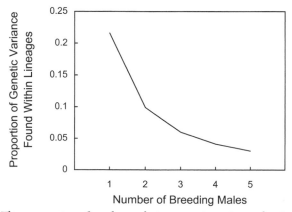

Figure 6.11 The proportion of total population genetic variance that is found within lineages (social groups) is inversely related to the number of males that sire offspring. These results derive from Chesser's (1991) model of the genetics of primate populations. (After Chesser 1991.)

who developed an analytical model of the genetic consequences of social structure in populations characterized by female philopatry. He showed that even in the absence of inbreeding, the number of males that sired the next generation is a crucial factor affecting the variance in heterozygosity within populations (cf. de Jong's study above). In particular, the proportion of the total population genetic variance that is found within female lineages (by which he means social groups) was inversely proportional to the number of males (range tested: 1–5) that contribute to the sirings within each lineage (fig. 6.11) and, to a lesser extent, lineage size (range tested: 5–20 females). The number of lineages within the population (range tested: 10–50) had only a marginal effect.

This finding can be extended to the more general case of monogamous versus polygamous mating systems by recognizing that monogamy is, in effect, a population of female lineages, each consisting of a single female with a single breeding male. Monogamy thus promotes genetic differentiation within populations, whereas between-group differentiation tends to be lower in polygynous mating systems: this is because proportionately more males are able to breed (and do so more evenly) in monogamous mating systems. This last prediction appears to be contradicted by the high levels of genetic homogeneity found in callitrichid populations: one possible reason is that callitrichids may not in fact be quite as monogamous as conventionally supposed (Dunbar 1995b).

If Chesser's model is correct, then we might expect species living in multimale groups to show lower rates of population (and hence species) differentiation than is typical of species living in one-male groups. *Cercopithecus aethiops* provides a possible example of this: this widely distributed "superspecies" runs to some sixteen recognized subspecies that border on specific status (fig. 2.2). In this respect it contrasts strikingly with the rest of its congeners (the forest guenons), which tend to be more taxonomically diverse and geographically more confined; *C. aethiops* is also the only member of the genus that habitually lives in multimale groups. Asian colobines may offer another example: there are fifteen species of the predominantly one-male-grouping genus *Presbytis* with only one of these species, *P. entellus*, being multimale (taxonomy follows Corbet and Hill 1991). Although other explanations might be offered, these examples clearly fit the pattern predicted by Chesser's (1991) model. The point is further reinforced by Dracopoli et al.'s (1983) finding that within-group genetic diversity was virtually the same as between-group diversity in Kenyan *Cercopithecus aethiops* populations (see below), a fact that is otherwise somewhat unusual and difficult to explain.

6.4.2. Population Genetics and Speciation

When gene flow between groups via migration of individuals is restricted, adaptation to local conditions combined with genetic drift can be expected to give rise to increasing genetic divergence between populations over time, and thus eventually to speciation (see sec. 2.3). Large river systems are one such barrier to gene flow. Peres, Patton, and da Silva (1996) found that gene flow in Brazilian populations of the tamarin *Saguinus fuscicollis* was limited to the headwaters of the major Amazon tributaries. Elsewhere the width of the rivers impeded migration, so that parallel divergent clines could be identified along opposite riverbanks. Only where the river beds were of a more meandering type was there any evidence for gene flow between banks. Body size, however, may be an important factor here, since de Ruiter and Geffen (1998) did not find significant genetic differentiation between opposite banks in a macaque population.

Pope (1996) has pointed out that most studies have failed to distinguish variance between social groups from variance between genetic demes. Where this distinction has been made, the results suggest that whether the genetic differences between groups within a population are greater or less than those between populations depends on a number of social and demographic factors. Dracopoli et al. (1983), for example, found that within- and between-locality heterozygosity was similar across groups in Kenyan vervet monkey populations ($F_{ST} = 0.08$ within populations and between population $F_{ST} = 0.09$, respectively), whereas Melnick, Jolly, and Kidd (1986) found that interpopulation heterozygosity in macaques was greater than that between groups within the same population ($F_{ST} = 0.08$ between populations compared with $0.03 < F_{ST} < 0.06$ within populations). In contrast, Pope (1992) found higher levels of heterozygosity within populations ($F_{IS} = 0.22$ and 0.14 for two populations) than between populations ($F_{ST} = 0.02$) for Venezuelan howler monkeys. Pope concluded that the high levels of genetic differentiation between neighboring groups and populations in the howlers arose because bisexual territoriality made it difficult for both sexes to transfer between groups (all females and many of the males that dispersed did so by forming new groups rather than joining existing groups). The more similar levels of within- and between-population heterozygosity in the vervets and macaques may be the result of high levels of male transfer between groups and populations.

Once populations become physically isolated, divergence in gene pools can be expected to arise through drift as well as selection. Scheffrahn et al. (1993) found significant differences in blood protein polymorphisms between two isolated populations of Algerian Barbary macaques. The Algerian metapopulation consists of seven isolated populations with a mean size of about 750 animals (range 50 to 1,750), most of them separated by more than 100 km. Scheffrahn et al. (1993) found high levels of genetic divergence between the two largest and most adjacent central populations (they share only 89% of the variation within the gene pool of the Algerian subpopulation), indicating that there had been significant divergence over a relatively short time. Since this is considerably more than is typical of most primate populations, these authors suggest that they are already beginning to move toward subspecific status. Both of these samples also differed markedly at the genetic level from the Moroccan subpopulation farther to the west (although genetic data for the latter were obtained only from derivative captive populations in Europe).

Where species' distributions are more continuous, genetic homogeneity across the species' range may be greater. Melnick and Hoelzer (1994) concluded that in species such as *Macaca mulatta, M. fascicularis,* and *M. sinica,* as much as 99% of the total genetic diversity of the species can be found in the local population and as much as 96% of the variation in the local regional population can be found in any one of its constituent groups. In such cases most of the genetic heterozygosity of a species may be represented by differences between individuals. By contrast, in species such as *M. nemestrina* and *M. fuscata,* whose ranges were highly fragmented (particularly by intervening water barriers), as little as 75% of the species' genetic diversity may be represented in any one population.

The Algerian Barbary macaque populations occupy what are in effect ecological islands. Similar results have been obtained from primate populations isolated on marine islands. Kondo et al. (1993) found that the long-tailed macaque (*M. fascicularis*) population on Mauritius exhibits lower genetic heterozygosity than the populations in peninsular Malaysia or on the islands of the Sunda Shelf (from which the Mauritian ancestral stock almost certainly derived: Lawler, Sussman, and Taylor 1995). Indeed, Lawler, Sussman, and Taylor (1995) found that mitrochondrial nucleotide variation in the Mauritius population was about one-tenth that in the native Indonesian and Philippine populations of this species, suggesting a strong founder effect. In fact, the Mauritius population has grown rapidly to about thirty thousand or so from a founder population of just a handful of animals released onto the island about four hundred years ago.

The effects of isolation on population differentiation depend not only on the size of the population and the degree of isolation (and hence the rate of immigration from neighboring populations), but also on the time since isolation. Scheffrahn, de Ruiter, and van Hooff (1994) sampled the variability in seven blood protein markers in thirteen populations of *M. fascicularis* in Sumatra and its adjacent islands. They found that the time since an island became separated from the mainland as a result of past sea level rises was an important factor influencing genetic drift. Islands that became separated more than 100,000 years ago were more highly differentiated (and had lower heterozygosity) than islands that became isolated only at the end of the last ice age some 10,000 years ago. As might be expected, these effects appeared to be influenced by island size as well: differentiation was greater on smaller islands (i.e., where the founding population would have been smaller). Although the differentiation of mainland populations was not necessarily as great as that of populations on long-isolated islands, the mainland populations did exhibit a classic pattern of increasing differentiation with increasing geographic distance from each other.

6.5. Summary

1. Populations can be described in terms of characteristic age-specific life history parameters (notably, fecundity and survivorship), which can be used to calculate several important indices of the population's demographic health (the net reproductive rate and the intrinsic rate of increase). Both fecundity (or birthrates) and mortality rates are influenced by environmental conditions, and both may also be influenced by body size, social competition, and other demographic variables.

2. Populations are ultimately limited by food, with maximum population size set by the habitat's carrying capacity. The impact of periodic population crashes (triggered in some cases by density-dependent effects, but in others by climatically induced famine or disease outbreaks), combined with demographic lag effects, means that populations are rarely stable through time. Some species may be especially prone to predation (in some cases by other primates), but predation will usually have a chronic rather than an acute impact on populations.

3. When populations crash, the brunt of mortality is often borne by subordinate animals and juveniles, although in other cases the decline in population may be driven primarily by reduced fecundity among breeding females.

4. Individual populations are often connected to one another in a larger network—a metapopulation—whose dynamics are often closely interlinked. Components of the system may be functionally related as a source-sink system in which the continued survival of subpopulations in marginal habitat is maintained through time by immigration from demographically more productive subpopulations (the rescue effect). The dynamics of metapopulations are poorly understood in primates, but they are likely to be of fundamental importance to understanding their population biology.

5. The genetic architecture of a population can be described by a number of indices of genetic heterozygosity. The dynamics of these indices depend on a number of key behavioral processes (including pat-

terns of group fission, migration, and mating). Because primates are highly social, the structure and dynamics of social groups may have an unusually strong effect on their population genetics. Even in the absence of inbreeding, populations will be more heterozygous when mating is more evenly distributed among males (e.g., monogamous or multimale mating systems), females are philopatric, migration rates between groups are low, and female group size is small.

6. When populations are highly differentiated genetically, genetic drift may exacerbate the level of differentiation and ultimately result in speciation. Speciation processes may be enhanced if physical barriers to migration and dispersal (e.g., habitat fragmentation, rivers, or isolation on islands) restrict gene flow.

Extinction Processes

Extinction is a natural process. The number of plant and animal species existing today represents only a small fraction of all the species that have existed since life began to diversify on this planet 3.5 billion years ago (May, Lawton, and Stork 1995). Whereas extant primate species number 200 to 230 (chapter 2), Martin (1993) estimates that as many as 6,500 primate species could have existed at one time or another, although the sample of these taxa known through fossils is tiny (ninety-seven extinct genera incorporating just 181 species: Martin 1990). This chapter reviews our limited knowledge of extinction processes in primates. First we tackle the question of how often primate taxa become extinct, then we address the processes that cause primate populations to become extinct. The influence of species' characteristics on population extinction are investigated next, and finally we examine the lessons we can learn about the extinction of primate populations through the prehistoric record.

7.1. Extinction Rates

Although cataclysmic "mass extinctions" may wipe out large numbers of taxa, the vast majority of species are probably lost as part of the natural and continuous turnover in species in what are termed "background extinctions" (e.g., Jablonski 1991, 1995). This is particularly true for primates, for which there is little evidence of mass extinctions (Martin 1990). Obtaining estimates of background extinction rates is problematic, but we are fortunate in that one particular group of primates, the hominids, has one of the best-studied fossil records.

R. Foley (1993, 1994) examined patterns of speciation and extinction in both the hominids and the fossil baboons. His analyses show that in the past five million years the number of known hominid and papionid

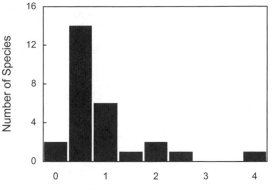

Figure 7.1 Frequency distribution of the "life span" (duration of taxon existence) of fossil primate species. (Data from R. Foley 1993.)

species present at any one time has fluctuated greatly. He also shows that the length of time these primate species existed is highly variable even within particular taxonomic groups, ranging from 400,000 years (in the case of *Paranthropus aethiopicus*) to 4 million years (*Theropithecus oswaldi*). However, the median "life span" across the Hominidae and Papionini is only half a million years, half the average duration for mammals as a whole (Martin 1993). Note that the frequency distribution of these duration scores indicates that most species are geologically short-lived, although some taxa persist much longer (fig. 7.1). This pattern is similar to that seen in most other animal taxa (Jablonski 1995).

Using the data collated by R. Foley (1994) and the method given by Caughley and Gunn (1996), it is possible to estimate the "normal" rates of extinction for the Hominidae. Calculating the proportion of hominid species that became extinct in each of ten half-million-year time bands, it is clear that the background extinction rate among the hominids has been highly variable over the past 5 million years (table 7.1). Nonetheless, these data suggest that the median proportion of species becoming extinct per million years is 0.22. This value translates to 0.02% of hominid species becoming extinct every thousand years. If this figure can be taken as a reasonable guide to background extinction rates in other primate groups, it gives us a baseline with which to view primate extinctions in recent history. In this respect it emphasizes how dramatic the recent loss of Malagasy primates has been (see sec. 7.4.1).

It is now widely accepted that we are currently witnessing a mass extinction event that is being driven by our own species, *Homo sapiens*. Since 1600 it is estimated that 1.3% of all mammal species have become extinct (Smith et al. 1993a,b). Yet, surprisingly, no primate species has

Table 7.1 Extinction rates in hominid species

Time period (million years)	Species present	Species lost	p_E	d	p_{Em}
0.0–0.5	4	3	0.75	2.78	0.94
0.5–1.0	3	2	0.67	2.22	0.89
1.0–1.5	3	0	0.00	0.00	0.00
1.5–2.0	4	1	0.25	0.58	0.44
2.0–2.5	5	2	0.40	1.02	0.63
2.5–3.0	3	1	0.33	0.80	0.55
3.0–3.5	2	0	0.00	0.00	0.00
3.5–4.0	1	0	0.00	0.00	0.00
4.0–4.5	1	0	0.00	0.00	0.00
4.5–5.0	1	0	0.00	0.00	0.00

Source: Data from R. Foley 1994.

Note: Variables are defined as follows: p_E: the proportion of species extinct in each 0.5 million year time band; d: the instantaneous rate of extinction per million years (calculated as $d = [-\ln(1 - p_E)]/t$); p_{Em}: the proportion of species lost per million years (calculated as $p_{Em} = 1 - e^{-d}$).

been reported extinct during the past four hundred years. Cole, Reeder, and Wilson (1994) list mammal species that have gone extinct in the past fifty years, and Groombridge (1992) lists animal species extinct since about 1600; neither of these inventories includes a single primate taxon. In fact, contrary to the general trend, the number of primate species extant today appears to be growing rather than diminishing! Not only are species of ambiguous status being rediscovered (the hairy-eared dwarf lemur: Meier and Albignac 1989), but new species are being discovered (the sun-tailed guenon: Harrison 1988). In the past decade in Brazil alone, five new primates have been discovered (Ferrari and Queiroz 1994). This reflects the increased sampling of tropical forest areas in recent times and should in no way be seen as evidence that the rate of speciation is greater than the rate of extinction among modern primates.

The probable reasons why no known primate species has become extinct within the past four hundred years, when up to 20% of known species have already disappeared in almost half of all other mammalian orders during this period (Cole, Reeder, and Wilson 1994), are discussed in section 12.1. On current evidence, the good fortunes of primate diversity in the recent past are unlikely to continue in the future. That we have not lost primate species such as the golden lion tamarin in recent years largely reflects the intensive efforts of conservationists. Moreover, the enviable record that exists for primate species does not apply to subspecies. The closing years of the past century may have already witnessed the extinction of Miss Waldron's red colobus (*Procolobus badius waldroni*) from Ghana and Ivory Coast (Primate Specialist Group 1996; McGraw 1998).

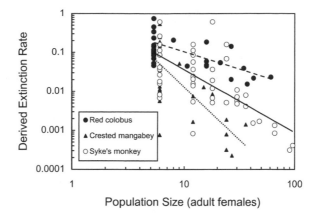

Figure 7.2 Covariation between derived extinction rate and population size across forest patches in the Tana River forest system (Kenya) for red colobus monkeys *Procolobus badius*, crested mangabeys *Cercocebus galeritus*, and Syke's monkeys *Cercopithecus mitis*. In each species, extinction rates increase in smaller populations. Derived extinction rates are predicted based on the patterns of patch incidence observed in the Tana River forest system (fig. 6.8) using the simplest version of the incidence function model (in which the effects of interpatch distance are not incorporated: Hanski 1999). Population size is based on census data. The plotted lines are least-squares regression lines. From Cowlishaw (n.d.), based on data from Butynski and Mwangi (1994).

7.2. Causes of Extinction

The key to understanding species extinctions is to study the extinction of populations, since the only distinction between the extinction of any one population of a given species and the loss of its last population is that the loss of the last results in the extinction of the species itself (Andrewartha and Birch 1954). The crucial issue, then, is Why should a population go extinct? The simple answer is that the population has become small. Extinction risk is a function of population size: risk is very low in large populations but rapidly increases in small ones. Among primate populations, the problem of small population size can be illustrated by the patterns of extinction risk in the Tana River (Kenya) metapopulations of red colobus, crested mangabey, and Syke's monkey (Cowlishaw, n.d.). Extinction risk in the local populations that make up the metapopulation becomes progressively greater as local population size declines (fig. 7.2).

The relationship between population size and extinction risk is universal. Nevertheless, it is important to distinguish between the reasons populations become small (the "declining population paradigm") and the reasons small populations are more vulnerable to extinction (the "small population paradigm") (Caughley 1994). The rest of this section deals with each of these processes in turn.

7.2.1. Declining Population Processes

Populations can decline in size for a variety of reasons. We have already discussed how population size can fluctuate dramatically through the effects of famine, predation, and disease (chapter 6). Of more immediate concern are those declines precipitated by human activities, either directly through habitat disturbance and hunting or indirectly though the introduction of exotic species or a chain of secondary extinctions. Diamond (1984, 1989) terms these four agents of population decline "the evil quartet." Since the evil quartet are external forces that operate on populations to cause their decline, they can be considered "extrinsic" factors in the extinction process.

For primates, habitat change and hunting are the most important threats to survival. In light of the importance of these two processes, we dedicate the two following chapters to them: chapter 8 deals with habitat disturbance and chapter 9 with hunting. In contrast, introduced species and secondary extinctions are not thought to pose such a serious threat for primates, although their effects may still be felt in some circumstances.

Introduced species affect native populations either through niche competition or through predation (Atkinson 1989). There are two probable reasons why primates are rarely at risk from this sort of threat. First, most primates occupy arboreal niches in tropical forest habitats (see sec. 3.2). Since most alien species are introduced from temperate Eurasia (e.g., rabbits, black rats, cats, dogs), few if any would be able to exploit this niche as either predator or competitor. Second, most primates are continental in their distribution (see sec. 2.2.1). Since introduced species have tended to have their most devastating effects on oceanic island species (such communities tend to be undersaturated, so oceanic island species have limited experience with predators and competitors: sec. 4.1.1), primates have not been common victims.

Nevertheless, there may be some exceptional cases. *Callithrix jacchus*, for example, may be threatened by predation from the recent immigration of burrowing owls into one region of the Atlantic Forest. However, in this case the marmosets are themselves new residents in this area (Stafford and Ferriera 1995). Although many primates are also threatened by exotic livestock (e.g., *Lemur catta:* Mittermeier et al. 1992), this is almost always through habitat disturbance rather than competition or predation. On the other hand, exotic plant species may adversely affect primate populations: the ability of *Macaca nigra* to make use of logged forest habitats, for example, may be compromised when these habitats are colonized by the introduced shrub *Piper aduncum* (Rosenbaum et al. 1998).

In the case of secondary extinctions, the loss of one species in an ecosystem can lead to changes that precipitate an extinction cascade (e.g., Rogers and Caro 1998). This is particularly true if the taxon that is lost plays a keystone role (sec. 4.2.2), as Jackson (1997) soberingly demonstrates in his account of how coral reef systems have been irreversibly transformed by the loss of keystone species. In tropical terrestrial ecosystems, the loss of large predators on Barro Colorado Island may have led to an increase in smaller terrestrial carnivores, resulting in higher rates of nest predation on forest-dwelling birds that subsequently led to their extinction (Terborgh and Winter 1980). Similarly, the loss of the subfossil lemurs might have been at least partially driven by a series of secondary extinctions after the introduction of domesticated animals to Madagascar (see sec. 7.4.1).

There are currently no accounts of contemporary primate populations being threatened by such processes, although this may reflect the difficulty of identifying these indirect effects. Nonetheless, Sterck (1998, 1999) has argued that the loss of predators from many primate habitats (owing primarily to human persecution and habitat disturbance) has reduced mortality so that more primate populations now exist close to carrying capacity. Sterck argues that in langurs (and perhaps some other primate species) the consequent "habitat saturation" reduces dispersal opportunities for females, leading to higher rates of male group takeover and thus infanticide (see sec. 3.3.1). Yet this does not appear to pose a significant threat to the viability of populations.

7.2.2. Small-Population Processes

In contrast to the factors leading to declining populations, those that lead to extinction once populations are small are largely "intrinsic" to populations of that size. These are demographic stochasticity, environmental stochasticity, and the loss of genetic diversity (Caughley 1994; Caughley and Gunn 1996). These forces all imperil small populations for fundamentally similar reasons: random events (such as the accidental death of particular individuals) have a much greater impact when the population consists of fewer individuals. The following subsections examine each of these three processes in turn, together with a fourth possible process, the loss of integrity of social structure. (For further background to these issues, see Lacy 1992; Körn 1994; and Caughley and Gunn 1996.)

DEMOGRAPHIC STOCHASTICITY

Although life history parameters will have a "typical" or mean value for a given population, this inevitably derives from a distribution of values whose extremes may vary considerably across individuals and, within individuals, over time. In large populations, stochastic variation in fecundity and mortality schedules from one year to the next will be absorbed by the large numbers of individuals in the population; but in small populations such variation can lead to population collapse and extinction.

In small populations, for example, a year with few births will have a serious effect on recruitment to the adult breeding cohort five or six years later (e.g., Dunbar 1988), and this can be sufficient to push the population over the threshold of viability. A low birthrate in one year might arise simply from the particular size and age structure of the cohort of breeding females: a population or group that contains mainly old or young females will have a lower mean birthrate than one made up of females in their reproductive prime. Strum and Western (1982), for example, reported a correlation between the overall mean birthrate and female age structure in a wild baboon troop.

The same effect can arise from purely stochastic variance in the natal sex ratio. Although the natal sex ratio typically averages about 50:50 males:females, this is normally true only of large samples. Most local primate populations are relatively small, and small-sample bias effects can lead to very considerable departures from this value in individual years. In addition, however, there is some evidence of consistently biased neonatal sex ratios in some primate populations (see sec. 6.1.1): Clark (1978), for example, reported a natal sex ratio of 57:43 in favor of males in one captive *Otolemur crassicaudatus* population, and Sigg et al. (1982) reported a natal sex ratio of 32:68 in favor of females over a seven-year period in one band of wild hamadryas baboons.

Natal sex ratio imbalances can have a dramatic effect on recruitment to the cohort of breeding males or females, especially if all the infants born in a series of years happen by chance to be male (Dunbar 1988). The effects of biased sex ratios on population viability have been modeled by Strier (1993–94): she found that even slight deviations from an equal neonatal sex ratio in favor of males could lead to dramatically smaller population sizes in muriquis (sec. 10.2.3). For this reason Liu et al. (1989) expressed concern at the large number of male offspring recently born in an isolated population of twenty-one concolor gibbons in Hainan, China. Eight males were born into the four family groups in the population, but there were only three female infants, and these were restricted to two groups. Similarly, Mace (1988) feared for the viability of captive gorilla populations in light of the finding that captive female gorillas are increasingly likely to give birth to male offspring as they become older.

ENVIRONMENTAL STOCHASTICITY

Environmental stochasticity comprises regular variation in environmental conditions, such as seasonal fluctuations, and less predictable variation such as El Niño events, together with catastrophes such as hurricanes, fires, and droughts. (Catastrophes are simply an extreme case of environmental stochasticity: Lande 1993; Caughley 1994.) Environmental stochasticity drives most natural fluctuations in population size, even in large populations, through its effects on food availability, predation, and disease (sec. 6.2.2): in small populations it can cause extinction. This is particularly true in the case of catastrophic events, which can occur surprisingly frequently (e.g., over 360 hurricanes hit Madagascar between 1920 and 1972) and over large areas (e.g., two earthquakes off the coast of Panama in 1976 affected over 450 km^2 of tropical forest and denuded about 54 km^2) (see Chapman et al. 1999 and references therein).

At Polunnaruwa (Sri Lanka), a single cyclone destroyed over 45% of all the emergent trees, 29% of all subcanopy trees, and 21% of all shrubs. This resulted in an overall loss of 50% of the foliage-bearing canopy area, followed by a drop in flower and fruit production. Local populations of *Semnopithecus entellus* and *Trachypithecus vetulus* modified their dietary and ranging patterns in an attempt to cope with these dramatic changes, yet their populations still showed a subsequent decline of 5–15% (Dittus 1985). Similarly, Boinski and Sirot (1997) describe how, in 1993, the only protected population of the critically endangered Costa Rican squirrel monkey *Saimiri oerstedi citronellus* was decimated by a hurricane, which knocked down 25% of the protected forest (including all the commonly used sleeping trees).

Environmental stochasticity can have complex effects on both birthrates and death rates. These effects are partially dependent on whether the main problem is famine, drought, predation, or disease (see sec. 6.2.2), but complex interactions can also occur. For example, the early decline of the vervet monkey population at Amboseli (Kenya) was caused by high mortality rates among juvenile animals, attributed to the mass die-off of fever trees (sec. 6.2.2; see Struhsaker 1973). However, as the decline progressed it was further associated with high predation rates of adult and immature animals (resulting from the territorial shifts driven by the food shortages), and it was finally predation that appeared to become the predominant force in the extinction of this local population (birthrates seemed constant through the period: see Cheney et al. 1988; Young and Isbell 1994; Lee and Hauser 1998).

The differential effects of environmental stochasticity on primate populations have recently been documented for the endemic Tana River primates by Cowlishaw (n.d.; see sec. 6.3). Here, changes in the course of the river through natural meanders and erosion has reduced the supply of groundwater to riparian forest patches on the old river course, leading to forest senescence (e.g., Butynski and Mwangi 1994). Populations in these patches therefore slowly dwindle as their food base gradually disappears beneath them. All species are vulnerable to this, although Syke's monkeys *Cercopithecus mitis* appear to be the most resilient, simply because they occur in larger populations in these patches. In contrast, red colobus monkeys *Procolobus badius* are the most sensitive to this environmental stochastic process, apparently as a result of their relatively narrow diets.

Even where immediate extinction does not occur, environmental forces may be sufficient to cause fluctuations in population size that become progressively more severe in successive cycles (May 1971; cf. Dobson and Lyles 1989), thereby ultimately driving population size below the threshold at which demographic or genetic spiral effects can take hold (e.g., P. Foley 1994). For example, Ginsberg, Mace, and Albon (1995) modeled the population dynamics of the now-extinct Serengeti population of wild dogs (*Lycaon pictus*) and were able to show that the population had a reasonable chance of withstanding up to three catastrophes during a thirty-year period but that a fourth catastrophe increased the chances of extinction in a given year from 15% to 56%.

LOSS OF GENETIC DIVERSITY

Smaller populations are likely to lose genetic diversity at higher rates as a result of genetic drift (see sec. 6.4) because stochastic variance in lifetime reproductive success has a greater effect on the proportion of all possible genes that are passed on the next generation. Reducing the genetic diversity of a population (i.e., reducing heterozygosity and concomitantly increasing homozygosity) is considered undesirable for two reasons. First, loss of heterozygosity means that populations have less genetic flexibility with which to respond to changes in environmental conditions, thus making extinction because of failure to cope more likely. Second, it increases the risk of inbreeding depression: the phenotypic expression of deleterious recessive genes. As population size gets smaller, the impact of such recessives may become disproportionately severe.

How large must a population be to retain its present levels of heterozygosity in the long term? The *minimum viable population size* (MVP) is the minimum number of individuals required to give 95% certainty that genetic variation will be maintained at its current level for the next t years (Franklin 1980). Franklin estimated that this was about 50 individuals for $t = 100$ years and about 500 individuals for $t = 1,000$ years (the so-called 50/500 rule). Two points need to be borne in mind. First, Franklin's estimates are at best very rough, and many researchers have questioned their value (Caughley 1994; Caughley and Gunn 1994; Lande 1995). Second, since not all members of a population contribute genes to the next generation, the actual population size does not give a realistic picture of how fast alleles are likely to be lost from the gene pool.

The *effective population size*, N_e, is the number of individuals that contribute offspring to the next generation (i.e., the number that actually reproduce). A simple estimate of N_e is given by $N_e = 4N/(2 + s^2)$, where s^2 is the variance between individuals in lifetime reproductive success. In monogamous species such as gibbons, N_e is more or less equivalent to the number of breeding adults. But in polygamously mating populations, where there is high polygyny skew, males do not contribute equally to the next generation (sec. 3.3.1). Where only a handful of males reproduce, only a small fraction of the male genes currently present in the population will be transmitted to the next generation, and this needs to be taken into account. In this case, a better estimate is given by $N_e = (4N_f N_m)/(N_f + N_m)$, where N_f is the number of females and N_m the number of males. A more complex, but more accurate, formula is given by Nozawa (1972). As a rule of thumb, however, the effective genetic population size can be estimated as approximately one-quarter of the census population. Hence, if Franklin's calculations are correct, his estimates for the minimum population size to maintain genetic diversity at its current level should, at the very least, be on the order of 100–200 and 1,000–2,000 animals of all ages and both sexes, respectively.

Other factors may also influence N_e in addition to the mating system. In particular, the size of N_e required to ensure 90% survival of the population for two hundred years decreases as a negative exponential with the length of the generation time (fig. 7.3). Long life span provides a buffering effect. In very long-lived species such as elephants or apes, values as low as $N_e \approx 40$ may be sufficient to ensure population survival over periods as long as two hundred years, whereas for short-lived species such as mouse lemurs an N_e of 400–500 may be necessary to achieve the same end.

Effective genetic population size has been estimated for several primate populations. Duggleby (1978) calculated N_e for the individual groups of Cayo Santiago rhesus macaques. Her estimates ranged from 20.5 to 32.9 (depending on the formula used to calculate N_e), for a mean census group size of 70.8. For the whole population (four groups), the population N_e was 81 to 129.6 (in a total population of 183 animals). Rogers and Kidd (1996) estimated the degree of genetic heterozygosity for four social groups of baboons (*Papio cynocephalus*) in the Mikumi National Park (Tanzania) and used this to estimate the effective population size for the whole area (i.e., the genetic deme size). The average heterozygosity for each of five autosomal alleles was 0.0033 (higher than that obtained for human populations), and this gave an estimate for N_e of 14,000.

The influence of different mating system and life history characteristics on N_e are well illustrated by Pope's (1996) estimations of N_e for two populations of forest primates in the Kibale Forest (Uganda). She found

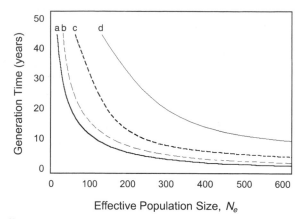

Figure 7.3 Effective population size, N_e, required to ensure 90% survival of the population over a two hundred year period in relation to the length of the generation time and the size of the founder population. The four populations assume (a) no founder effect, (b) a founding population of twenty individuals, (c) eight founders, and (d) six founders. (After Körn 1994.)

that for red colobus monkeys $N_e = 14,160$ for a total demographic population size of 40,520 animals, whereas for red-tailed guenons (*Cercopithecus ascanius*) the equivalent figures were 3,196 and 17,710 (table 7.2). Note that the effective population size is much smaller in the latter case than in the former ($N \approx 3N_e$ for the colobus monkeys but $N > 5N_e$ for the guenons). This difference can be attributed to the fact that male guenons were, individually, much less likely to contribute genetically to the next generation than male colobus monkeys, partly because they were about half as likely to survive to adulthood and partly because the variance in their lifetime reproductive success was much higher. In contrast, there were few or no differences in these characteristics between the females (see table 7.2).

An important criticism of Franklin's 50/500 rule is that it considers only the risks of genetic collapse and ignores demographic and environmental stochasticity. Moreover, Lande (1995) has pointed out that Franklin's formulation of the problem focuses solely on the loss of heterozygosity through genetic drift and ignores the fact that extinction risk may be influenced by other genetic factors, notably the appearance of deleterious mutations. Although highly deleterious mutations will be removed from the gene pool fairly rapidly (and thus have little effect on the genetic state of the population), mildly deleterious mutations are much more likely to reach fixation and thus to adversely affect the long-term viability of the population. Extinction induced by this process has been dubbed *mutational meltdown*. Based on observed mutation rates and

Table 7.2 Determinants of N_e and N_e/N for two species from Kibale Forest, Uganda

	Red-tailed guenon	Red colobus
Social system	Unimale	Multimale
Dispersing sex	Males	Females
Ecological niche	Small omnivore	Large folivore
Generation time (years)	9.6	9.3
Survival to adulthood (%)		
Males	19	34
Females	60	47
Mean LRS		
Males	3.67	2.29
Females	1.37	1.71
Variance in LRS		
Males	18.13	8.03
Females	1.58	1.78
Population size	17,710	40,520
N_e	3,196	14,160
N_e/N	0.18	0.35

Note: Parameters are calculated using the procedure described in Lande and Barrowclough 1987. After Pope 1996.

their effects, Lande (1995) estimated that the population size required to minimize the risk of genetic collapse over a thousand years was closer to 5,000 than 500 (i.e., an order of magnitude greater than Franklin's rule suggests). However, more recently an experimental study with *Drosophila* suggested that the accumulation of mutations may be much less serious than Lande proposes, at least for populations with N_e of 25–500 over forty-five to fifty generations (Gilligan et al. 1997). Clearly, the question of how many individuals constitutes the minimum necessary to ensure the long-term genetic viability of a population remains open (see Franklin and Frankham 1998; Lynch and Lande 1998; Frankham and Franklin 1998).

Inbreeding depression can give rise to both direct and indirect costs. The direct costs lie in the fact that the chances of an individual's inheriting two copies of a deleterious recessive gene are much higher if its parents are close relatives than when outbreeding (breeding between distantly related or unrelated individuals) is the norm. Inbred individuals commonly have depressed viability, reduced fertility, or both. The indirect cost is that inbred populations lose heterozygosity at an exponential rate: the proportion of heterozygotes in an inbred population is equivalent to $(0.5)^t$, where t is the number of generations of inbreeding (Maynard Smith 1989). (For useful reviews of inbreeding in the context of conservation, see Lande 1988; Lacy 1992; and Körn 1994.)

Ralls and Ballou (1982) found that infant mortality was significantly higher in inbred populations than in those that were not inbred in fifteen of sixteen primate colonies. Ralls, Ballou, and Templeton (1988) subsequently estimated the effects of inbreeding depression in forty populations of captive primates, rodents, and ungulates and found that, although they were highly variable, the mean survival rates for inbred populations were 33% lower than in outbred populations. Such high levels of reduced survival are likely to impose a considerable strain on the viability of populations that are already on the margins of demographic collapse. Although Pray et al. (1994) caution against extrapolating inbreeding effects from captive to wild populations (based on experiments with beetles), there is some evidence for inbreeding depression in wild primates: Packer (1979) and Alberts and Altmann (1995) found that mating between relatives led to high infant mortality rates in two separate *Papio* baboon populations. In both cases, however, the sample sizes were extremely small.

Inbreeding need not necessarily imply a high genetic load, however: some natural populations may have low heterozygosity yet show relatively high levels of inbreeding without the associated infant mortality. For example, the cheetah exhibits virtually no heterozygosity at any of the alleles that have been assayed (O'Brien et al. 1983) (although Caughley [1994] has cast considerable doubt on the validity of these results), yet there is little evidence so far of any disadvantage arising from this (Caro and Laurenson 1994). Many strains of domestic animals also exhibit normal fertility despite deliberate and intense inbreeding, and many human populations for which uncle-niece and first cousin marriages have been the preferred norm for many hundreds of generations exhibit few effects attributable to inbreeding depression (e.g., the Dravidians and other populations of southern India: Rao and Inbaraj 1980; Bittles, Radha Rama Devi, and Rao 1990). Similarly, Lacy (1992) reports that, as expected, heterozygosity was significantly lower in island populations of Florida deer mice than in mainland populations that were subject to much higher levels of gene flow through migration, yet there were no detectable effects on fertility. However, in a later experimental study Jiménez et al. (1994) found that deer mice taken from the wild, inbred in captivity, and then released back into the wild survived less well (and weighed less when they did survive) than individuals that had been outbred in captivity.

One reason for these apparently conflicting results may be that some populations have previously passed through natural bottlenecks that led to high levels of inbreeding, during which deleterious recessive alleles were purged from the population. These populations are subsequently likely to cope with inbreeding more successfully than those that have not experienced such bottlenecks, provided such naturally inbred animals still exist in the environment where the bottleneck occurred. Otherwise, if these animals are forced to cope with a modified or entirely different environment, their low heterozygosity may not give them sufficient flexibility to reproduce successfully (see Pray et al. 1994).

On balance, then, the message seems to be that inbreeding depression may well be a problem for species with high heterozygosity that currently live in small populations, such as *Brachyteles* and *Macaca fuscata* (sec. 6.4), because these have had the opportunity to accumulate relatively large numbers of deleterious recessive alleles. In contrast, species with low heterozygosity, or species that have recently been through a population bottleneck, such as callitrichids and howler monkeys (sec. 6.4), are likely to be at lower risk from the deleterious genetic effects of small population size.

DISRUPTION OF SOCIAL STRUCTURE

Once population size drops below a certain level, a process known as the *Allee effect* can become important. The Allee effect alludes to the fact that, below a certain population density, population growth rates may decline disproportionately with population density (Allee 1931). Consequently, density-dependent population dynamics may be associated with reduced growth rates not only at high densities (as the population approaches carrying capacity: see sec. 6.2.2) but also at low densities (termed *negative depensation*). Several mechanisms might underlie Allee effects (Reed 1999). In primates, these could include the inability of individuals in small populations to maintain sufficiently large social groups to provide either adequate protection from predators (e.g., baboons: Dunbar 1992d, 1993; sec. 7.4.2) or adequate success in group foraging (although this may have limited importance to primates: Janson and Di Bitetti 1997). They could also include a high rate of infanticide in low-density populations (e.g., samango monkeys: Swart, Lawes, and Perrin 1993; sec. 10.2.3).

Dobson and Lyles (1989) evaluated the role of a third Allee effect mechanism in primate population dynamics: the inability of individuals in low-density populations to find mates. Using a Leslie matrix simulation, they found that the minimum population density necessary to avoid population collapse was significantly lower in populations characterized by male-biased sex ratios and high polygyny skew (fig. 7.4). They concluded that solitary and monogamous populations may thus be more at risk of demographic collapse from these effects than populations where polygynous mating is typical. Such differences might even exist across polygynous systems: hamadryas baboon populations, for example, may be more at risk of extinction from Allee effects than gelada populations because the mean number of females a male mates with is only 1.9 com-

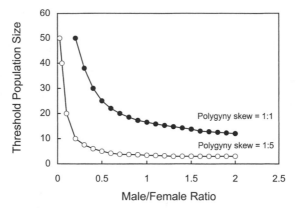

Figure 7.4 Effect of adult sex ratio and mating system on population persistence time in primates. The graphs show how the threshold population density below which population collapse occurs is influenced by the sex ratio of the population, for alternative mating systems: a population with a polygyny skew of $p = 1$ (monogamy) and a population with a polygyny skew of $p = 5$ (polygamy). (After Dobson and Lyles 1989.)

pared with 3.7 in the gelada (Stammbach 1987). However, the Dobson and Lyles model (which derives from parasitic helminth population biology) only appears to produce this effect because it assumes that males are obligately monogamous when the polygyny skew $p = 1$. In reality, when the population sex ratio is skewed in favor of females, most primate males will facultatively behave polygamously, absorbing the unattached females in the population into their breeding groups (e.g., geladas: Dunbar 1977; for a more contentious example in gibbons, see Jiang at al. 1994); in contrast, the Dobson and Lyles model seems to make the erroneous assumption that these excess females will be unable to breed. Consequently it seems biologically unlikely that such a mechanism will be especially intrusive in the case of primates.

In fact there is currently little empirical evidence for Allee effects as a mechanism of population decline in primate populations. In populations that have recently either become extinct (e.g., the Xinglung rhesus monkeys: Zhang et al. 1989) or gone to the brink of extinction (e.g., the Hainan concolor gibbons, whose population fell to only seven or eight animals before recovering to twenty-one: Liu et al. 1989), the typical social structure of the species appears to have been maintained. The absence of such effects might reflect the characteristic flexibility of primates in both their social and their ecological patterns (see chapter 3).

Indeed, the most detailed study to date of the extinction of a primate population shows how socioecological flexibility might prevent potential Allee effects from taking hold, even though such flexibility may still not be enough to save the local population from extinction. When the long-term deterioration in the habitat of the Amboseli National Park (Kenya) led to unsustainable mortality from famine and predation in vervet monkeys, ultimately leading to local extinction (see above), an Allee effect that would have exacerbated the ongoing decline—namely a further increase in predation owing to excessively small group sizes—did not appear to emerge because groups that became too small fused with larger neighboring groups (Hauser, Cheney, and Seyfarth 1986; Isbell, Cheney, and Seyfarth 1991). Group fusion in these circumstances may be relatively common (macaques: Dittus 1986; howler monkeys: Pope 1998a), although the social and reproductive success of remnant individuals in their new group is often poor (Dittus 1986; Hauser, Cheney, and Seyfarth 1986).

SCALING OF EXTINCTION PROCESSES

Although the vulnerability of small populations is a universal feature across these four processes, the scaling of extinction risk to population size differs between them, which inevitably raises the question of their importance relative to each other. Lande (1993, 1995) has used theoretical models to evaluate this issue in some detail. He found that under the impact of demographic stochasticity persistence times increase exponentially with population size (fig. 7.5). In contrast, a population's response to both environmental stochasticity and periodic catastrophes (when these affect the survival and reproduction of all members of the population

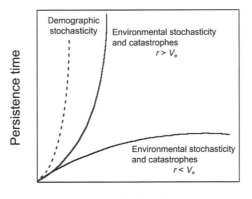

Population size

Figure 7.5 Covariation between population size and the risk of extinction from demographic and environmental stochasticity. The curves show how persistence time is influenced by population size for three kinds of perturbation. V_e is the variance in the rate of increase attributable to environmental fluctuations, and r is the mean rate of population growth. (Redrawn with the permission of Blackwell Science from Caughley 1994.)

equally) depends on the magnitude of the mean population growth rate, r, relative to the variance in the rate of population increase that is due to environmental fluctuations, V_e. When $r > V_e$, persistence times are affected by population size in much the same way as they are by demographic stochasticity, albeit less steeply; but when $r < V_e$ (i.e., environmental conditions are more variable), persistence times increase much less rapidly with increases in population size. In either case, very small populations are about equally vulnerable to both processes, while all other populations are more vulnerable to environmental stochasticity.

In addition, Lande (1995) points out that when small populations have survived the inbreeding depression caused by founder effects, environmental stochasticity will also be more likely to cause the population's extinction than any residual genetic stochasticity, simply because new deleterious mutations typically require one hundred generations or more to achieve fixation. Since the interval to the next environmental catastrophe will almost always be much shorter, especially in primates because of their slow life histories, populations will have crashed long before any deleterious mutations have taken sufficient hold to have a noticeable impact. Caughley (1994) makes a similar point about the current fashion for overemphasizing the importance of genetic factors in extinction.

Overall, then, environmental stochasticity seems to be a more serious threat to population survival than demographic stochasticity or genetic effects, especially in those primates with long generation times. These conclusions may also generalize to many other taxa, but they appear to be especially marked for primates.

7.3. Species Differences in Extinction Risk

Thus far, this chapter has considered how population size and extinction risk covary within species. For conservation planning, however, an issue of at least equal importance is how extinction risk varies between different taxa. This is a complex question because we might expect such variation to arise from a multitude of different sources impinging on the extinction process at different stages.

First, we might expect species to differ in vulnerability to extinction owing to differences in population size. Variation in population size is effectively a function of variation in population density and geographic distribution. The patterns and processes that determine these species traits were reviewed in chapter 5.

Second, for any given population size, some species may be more or less susceptible to population decline caused by habitat disturbance or hunting or other declining population processes. Part of this susceptibility is inevitably a function of geographic location: species that occur in areas of intensive habitat disturbance or hunting will be at greatest risk (e.g., African primates: Oates 1996b; see also sec. 10.2.2). But evidence suggests that within such areas interspecific variation in susceptibility is also marked. In addition, this variation can be highly specific: for example, some species are more vulnerable to agricultural practices than to forestry practices, and some are more vulnerable to hunting than to habitat disturbance. These patterns of interspecific variation, and the processes that underlie them, are reviewed in chapters 8 (esp. sec. 8.5) and 9 (secs. 9.5 and 9.6).

Third, once a population has declined, some species may be more vulnerable than others to the small-population processes that finally lead to extinction. This variation, which may also differ between small-population processes, is discussed below.

Fourth, how do all these facets of variability add up to determine interspecific variation in overall global extinction risk? Several studies have attempted to identify species traits that predict extinction risk at the global level. These also are discussed below.

Finally, it is important to remember that certain species traits might play a role at various stages in the process. For example, narrow dietary specialization might prevent a species from achieving high density (leading to a small initial population size), make a species more vulnerable to habitat disturbance (leading to a faster rate of population decline), and make a species more vulnerable to environmental stochasticity (leading to rapid extinction once the population is small). In addition, the same trait can make a species more or less vulnerable at different stages in the extinction sequence (e.g., body mass: see sec. 7.3.1). This creates further complexity in the extinction phenomenon.

7.3.1. Vulnerability to Small-Population Processes

The traits that determine a species' vulnerability to extinction once it has become small may be quite different from those that determine its vulnerability before then. This is well illustrated by comparing extinction rates in the Tana River primates: Syke's monkeys show lower extinction risk than crested mangabeys in forest patches of all sizes (see fig. 6.9), presumably because they occur in larger populations in those patches, but once population size is held constant they are at higher risk of extinction than crested mangabeys (fig. 7.2). In this section we seek to determine which traits might underlie interspecific differences in extinction risk once population size is held constant across taxa. There are four possible sources of such differences: they relate to population variability, life history, ecology, and social structure.

EFFECTS OF POPULATION VARIABILITY

Several authors have argued that if two populations of the same average size are compared, the population that shows the greater fluctuation in size is more likely to become extinct (e.g., Schoener and Spiller 1992). The empirical evidence is conflicting, however, with some studies supporting this assertion (e.g., British birds: Pimm, Jones, and Diamond 1988) and others finding no relationship (e.g., the same data set for birds: Tracy and George 1992; spiders: Schoener and Spiller 1992). The reasons for these discrepancies may reflect differences in the treatment of data and the analytical methods employed, although it is also possible that different taxonomic groups may show divergent patterns (Schoener and Spiller 1992).

With respect to primates, there is not enough information to determine whether there is a direct connection between population variability and extinction risk, although we can make some observations about patterns of population variability itself. First, population variability may be higher among species in "unstable" habitats such as strongly seasonal or successional environments (e.g., Neotropical primates: Robinson and Ramirez 1982). If this is true, then we might expect higher reproductive rates among primate species in such habitats as a result of selection for an ability to capitalize on favorable environmental conditions. Precisely this pattern is observed (see sec. 6.2.1). Second, Robinson and Ramirez (1982) suggested that frugivorous species may show greater population variability than sympatric nonfrugivores owing to the fluctuations that can occur in fruit production over successive years. However, this seems less likely given that frugivorous primates are well adapted to coping with the vagaries of fruit production from year to year (sec. 8.5.1).

Finally, population variability and thus extinction risk may be greater in smaller species because they are more vulnerable to environmental fluctuations in food availability (see sec. 3.1.1), as seen in some bird populations (e.g., Peters 1980; Cawthorne and Marchant 1980). Indirect evidence for this pattern in primates can be found across Neotropical communities, where only progressively larger species exist in areas of increasing rainfall variability (fig. 7.6).

EFFECTS OF LIFE HISTORY

The implications of large body size on primate life history strategies have already been discussed (sec. 3.1). Of particular importance is the fact that larger species have lower rates of reproduction but greater longevity. These two traits might act to increase or reduce, respectively, a population's vulnerability to extinction. Pimm, Jones, and Diamond (1988) argued that, compared with small-bodied species, large-bodied species of British birds were at lower risk of extinction in smaller populations but at higher risk in larger populations. However, a reanalysis of their data reveals no such crossover and suggests that larger species will always have an advantage over smaller species once population size is held con-

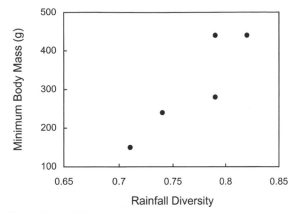

Figure 7.6 Effect of rainfall diversity on the lowest recorded primate body mass across Neotropical phytogeographic regions (rainfall diversity is calculated here as the mean of the differences of all possible pairwise comparisons of each monthly rainfall figure, where a high value indicates strong seasonality). Two high-altitude phytogeographic regions (southern Andes and northern Andes) are excluded here, since high-altitude regions are unlikely to be directly comparable to low-altitude regions. (After Pastor-Nieto and Williamson 1998.)

stant (Tracy and George 1992). This second pattern has also been confirmed for avifauna in four other island systems (Cook and Hanski 1995).

Studies of island shrew populations (Peltonen and Hanski 1991; Cook and Hanski 1995) also suggest that larger species are more resistant to extinction. In these studies, however, it is argued that smaller species are more vulnerable because of their higher metabolic rates and therefore their greater sensitivity to environmental stochasticity (specifically, temporal variation in food availability); this effect might be compounded by the positive scaling of fat stores to body size (see sec. 3.1.1). The finding that larger primates survive the gradual deterioration and contraction of foraging habitats better than smaller primates (during a period of forest submergence associated with a hydroelectric project: Peres and Johns 1991–1992) suggests that such processes might be important in primates.

Emmons (1984) also found that smaller species are more vulnerable to extinction than larger species because they are less able to cope with unfavorable periods of food scarcity (owing to metabolic constraints). However, Emmons proposed three alternative explanations: smaller species may be less able to travel long distances to seek out new food sources, more frequent losers in interspecific contest competition for food resources (sec. 4.3.1), or more likely to be dietary specialists (sec. 3.1.1). In Emmons's Neotropical mammal community, the last two suggestions appeared to have the greatest explanatory power, since smaller species in fact sometimes had larger home ranges than bigger species (although of course dietary specialization is also a determinant of population size, so the greater vulnerability of specialists is likely to be at least partially attributed to basic population size effects: see sec. 5.2.2).

Finally, although larger-bodied animals may have an advantage as a direct result of their body size, it is also (once again) possible that body size is a surrogate measure for another life history variable, such as reproductive rate (cf. sec. 5.2.2). In small populations, species with shorter generations are at greater risk of loss of heterozygosity because short generations accelerate the effects of genetic drift and inbreeding (Pope 1996). In addition, species with shorter generations are more vulnerable to demographic and environmental stochasticity because these can have a more profound effect on the age-sex class structure of their populations (Caughley 1994). For example, if a major famine results in no infants being born that year, the lack of recruitment to the cohort of breeding females several years later may have a disastrous impact on the population's ability to replace itself (Dunbar 1979).

EFFECTS OF ECOLOGY

In addition to the ecological effects of body size discussed above, dietary specialization can significantly increase a taxon's risk of extinction by reducing its ability to cope with ecologically stressful conditions. The theropithecines provide a spectacular example.

As grazers with a bulk-feeding strategy, the theropithecines as a group are highly specialized feeders that depend on the more digestible grasses found in low-temperature habitats. Dunbar (1993) modeled extinction risk in small-, medium-, and large-bodied *Theropithecus* baboons (the extant *T. gelada* and the extinct *T. darti* and *T. oswaldi*, respectively) and found that the larger species would have been far more vulnerable to extinction than smaller species because they would have found it more difficult to obtain enough grass in the foraging time available to them. As a result, this formerly widespread taxon of giant baboons became extinct throughout sub-Saharan Africa as global warming dramatically reduced its ecological base, leaving only the extant geladas isolated in their high-altitude habitat in Ethiopia (see sec. 7.4.1 for further details).

Differences in extinction risk between the Tana River primates (fig. 7.2) also appear to be directly related to dietary habits (with no consistent pattern in relation to body mass, dispersal pattern, or any other socioecological variable investigated: Cowlishaw, n.d.). For any given population size, red colobus monkeys show the highest extinction rates, probably because of their narrow dietary preferences. In contrast, crested mangabeys show the lowest extinction rates, presumably thanks to the ecological flexibility associated with their diet. Being frugivorous, this species is accustomed to variation in the availability of its food resources and has therefore developed a variety of coping strategies to deal with periods when food is scarce. These include variable territorial behavior, high mobility, and dietary flexibility.

EFFECTS OF SOCIAL STRUCTURE

Particular types of social structure may predispose a population to a high risk of extinction by making it more vulnerable to certain small-population processes. We have already discussed the role that Allee effects can play in small populations and how these may vary in severity depending on the social system (sec. 7.2.2). Here we focus on the ways social structure can influence population heterozygosity.

We noted earlier (secs. 6.4.1 and 7.2.2) that social structure can influence population heterozygosity through a number of mechanisms, including group fission, individual dispersal, and polygyny skew. We can provisionally conclude that taxa at greater risk of extinction from genetic effects in small populations are likely to be those with small group sizes, philopatric females, and a high polygyny skew (Chesser 1991; Jong, Ruiter, and Haring 1996), features that characterize many forest guenons (*Cercopithecus* spp.), the patas monkey (*Erythrocebus patas*), and perhaps the gorilla. In contrast, the peculiarities of the gibbon and callitrichid mating systems lead to levels of genetic diversity that may preadapt them for life in small populations. Pope (1996) points out that other factors such as variance in female lifetime reproductive output and intergroup differences in reproductive rates can also play an important role in determining heterozygosity (see table 7.2). However, convincing evidence that populations bear costs in any of these respects has yet to be provided for any primate population.

7.3.2. Vulnerability to Global Extinction

Although extinction is a complex process, several authors have attempted to find general correlates of the overall risk of extinction. Happel, Noss, and Marsh (1987) made the first attempt to identify the behavioral, ecological, and life history traits that might correlate with global extinction risk in primate taxa. Using an independent estimate of primate extinction risk, based on the original IUCN Red List codes (see sec. 10.2.2), they examined covariation between the degree of threat of extinction and female body mass, gestation length, group home range area, species geographic range area, maximum population density, and social system (monogamous vs. nonmonogamous). The authors found that only gestation length and geographic range area influenced extinction risk.

More recently, two analyses have examined changes in primate community structure after historical or anticipated patterns of selective extinction (subfossil lemur extinctions: Godfrey et al. 1997; future global primate extinctions: Jernvall and Wright 1998). These studies might allow us to obtain a broad overview of traits that might be associated with higher probabilities of extinction. In Madagascar, the lemurs that became extinct tended to have been larger and more terrestrial (or slow-climbing and suspensory), more folivorous, and more diurnal in their activity patterns than their congenors that survived to this day (Godfrey et al. 1997). At a global scale, the traits associated with potential future extinctions appear to be variable depending on the region in question: in particular, diurnal, large-bodied terrestrial species appear to be more vulnerable in Africa and Madagascar, and arboreal rain forest species tend to be more vulnerable in Asia and Madagascar (Jernvall and Wright 1998). The results of these two studies suggest that body mass, terrestriality, and diurnality are consistently associated with extinction risk in primates.

Finally, Harcourt and Schwartz (n.d.) have investigated the species traits that might correlate with the historical extinctions of primate species from the Sunda Shelf islands of Southeast Asia when these islands were isolated from the mainland. Their analyses, which incorporate methods that control for phylogeny, indicate that primate species were more likely to become extinct if they were larger, occurred at lower densities, occurred at lower latitudes (a possible indicator of low ecological flexibility: see sec. 5.1.2), and had smaller geographic ranges than closely related species that are still extant. Notably, there were no consistent effects that were due to diet, interbirth interval, or home range size.

The preceding studies suggest that a variety of traits might predispose a species to higher risk of extinction, although differences between their results makes it difficult to identify those traits that are likely to be most important. Moreover, the results of these studies must be interpreted with care owing to the problems associated with (1) the use of the original IUCN Red List codes (see sec. 10.2.2), and subsequent circularity associated with the use of geographic range area as both a category criterion and a predictor variable; (2) the lack of statistical controls for phylogeny (Happel, Noss, and Marsh 1987; Godfrey et al. 1997; Jernvall and Wright 1998), particularly when extinction risk appears to be phylogenetically clustered (see sec. 10.2.2); and (3) the lack of statistical controls for interrelatedness between predictor variables (all of the studies above), especially given that body mass, life history, diet, distribution, and abundance can all be correlated (see chapters 3 and 5).

One study has recently attempted to deal with all of these problems to identify key predictor variables for global extinction in primates. Using the new IUCN Red List codes (both for all threatened species and for those species threatened only by recent population decline), and employing statistical controls for phylogeny and interrelatedness between biological traits, Purvis et al. (n.d.) found that extinction risk in primates independently correlated with geographic range area, body mass, diet, and population abundance. High-risk species were those with small geographic range, large body mass, a diet at high trophic level, and low population density. Together these four variables accounted for 34–43% of the variance in extinction risk (for species in decline and for all threatened species, respectively); no further variance was explained by the other traits tested (island endemic status, gestation length, litter size, age at sexual maturity, interbirth interval, diurnality, home range area, and group size). Surprisingly, high extinction risk did not appear to be associated with slow reproductive rates, although it is possible that this effect was important but was represented in this analysis by body mass (the more direct measures of reproductive output may have been suppressed by body mass owing to the greater accuracy of the latter data). One final message of importance from this study is that less than 50% of the variance in extinction risk appears to be attributable to intrinsic biological species traits; most of this variance depends on the intensity of extrinsic anthropogenic activities, particularly habitat disturbance and hunting, which are the subject of the next two chapters.

7.4. Case Studies in Primate Extinctions

Based on the preceding material, we can start formulating some general rules regarding extinction processes in primate populations, although it remains to be seen how robust these rules might be in practice. In the meantime, we can learn a great deal about the extinction process from observing the event itself.

At present, empirical data describing primate extinctions are scarce, with only two documented contemporary cases. The first concerns a local extinction of vervet monkeys at Amboseli National Park (Kenya), owing to long-term natural deterioration in the habitat (a sequence of events we have already described in some detail: see secs. 6.2.2 and 7.2.2). The second concerns the extinction of a rhesus macaque population at Xinglung (China), at the northernmost limit of the species' range (Zhang et al. 1989). Historical documents suggest that macaques existed in northern China for many centuries, but by the 1960s the population at Xinglung (650 km away from the nearest population to the southwest) comprised only a single group of fifty to sixty individuals. Unfortunately, there is not enough evidence to distinguish the relative roles of extrinsic and intrinsic factors in the extinction of this population, which occurred in 1987 or shortly thereafter (although there appears to be little doubt that the decline in the population was at least partially driven by both habitat loss and hunting).

Given the paucity of information available on contemporary primate extinctions, we devote the rest of this chapter to accounts of primate extinctions reconstructed from the fossil and subfossil records.

7.4.1. Prehistoric Extinctions

Although the fossil and subfossil record offers a useful source of data on primate population extinctions, there are two caveats. First, although these accounts provide some insight into the processes that might lead to the decline of primate populations, they supply little information on what small-population processes cause extinction once the population is small. (To obtain further insight, it will be necessary to turn to theoretical population models: we return to this issue in sec. 10.2.3.) Second, much of what we know about prehistoric extinctions is based on inference and remains a matter of conjecture. Nonetheless, the two following accounts are extremely informative.

BABOONS

The baboons as a taxonomic group owe their origin to the radiation of large terrestrial cercopithecoid primates during the Pliocene. They derive from a widespread group of late Miocene large semiterrestrial species of the genus *Parapapio* that seem to represent the common ancestor of the mangabeys (genera *Cercocebus* and *Lophocebus*) and their relatives (drills and mandrills, genus *Mandrillus*) and the modern baboons (genera *Papio* and *Theropithecus*). By about 5 million years ago, there were two well-established genera present in African fossil sites, *Parapapio* and *Theropithecus,* apparently occupying contrasting niches: *Parapapio* a largely frugivorous woodland niche, the gelada a grazing open country niche (Jolly 1966; Lee-Thorp, van der Merwe, and Brain 1989; Benefit and McCrossin 1990; Fleagle 1999). About 4 million years ago these were joined by a new genus (the modern *Papio*) occupying much the same frugivorous woodland niche as *Parapapio*. With a mean adult body mass of about 22 kg (range 17–30 kg, $N = 4$ species), *Parapapio* were a little larger than living *Papio* (mean mass 21 kg, range 15–22 kg, $N = 4$ species), but fossil *Theropithecus* were considerably larger than either (50–96 kg, compared with 15 kg for the modern gelada) (data for fossil species from Fleagle 1999; those for extant species from Smith and Jungers 1997).

The subsequent history of these three groups can be seen as a process of speciation followed by decline and replacement. During the relatively wet phase of the later Pliocene, the dominant taxon was *Parapapio*. As the climate began to dry and cool, these species were gradually replaced by species of the genus *Theropithecus* whose range expanded with the emerging grasslands that typified the middle Pleistocene, achieving a widespread distribution (with relatively limited speciation) throughout the savanna grasslands of sub-Saharan Africa (Jolly 1966; Jablonski 1998). Fossil populations of *Theropithecus* are also known from the Mediterranean coast and even as far afield as northern India (Delson 1993; Delson and Hoffstetter 1993). Throughout the Plio-Pleistocene, the species of these two genera dominated the primate terrestrial communities of Africa. For reasons that remain obscure, *Parapapio* was replaced by *Papio* during the early Pleistocene after a period of coexistence. Nevertheless, the new *Papio* species remained confined to the more wooded ecotone between the grasslands and the adjacent forests (Jolly 1966).

In the end, the climatic warming of the past 200,000 years appears to have led to the extinction of almost all theopithecine populations because the temperate grasslands they depended on moved upward in altitude as global temperatures rose. The limited and dispersed distribution of high-altitude sites where these grasslands could persist meant that this habitat became increasingly scarce and fragmented. The ensuing series of extinctions ultimately resulted in the survival of only one relatively primitive small-bodied member of this remarkable genus, the gelada (*T. gelada*). This species was probably driven into the grasslands above 1,700 m on the Ethiopian plateau by the rapid rise in temperature (ca. 7°C in 50 years) associated with the Younger Dryas event about 10,000 years ago (Dunbar 1993, 1998b).

For two reasons, we can be fairly certain that climate has been the main driving force influencing the rise and fall of the genus *Theropithecus*. First, the occurrence of theropithecine fossils at sites on the North African coast is linked to the cycle of ice ages in Europe: Barbary macaques (*Macaca sylvanus*) occupied the Mediterranean coastal region during glacial maxima but were replaced by *Theropithecus* during the intervening periods when the macaques reinvaded Europe in the wake of the retreating ice fields (Pickford 1993).

Second, Dunbar (1993) used a model of theropithecine socioecology developed by Dunbar (1992b) to reconstruct the behavioral ecology and grouping patterns of the extinct theropithecine species (sec. 3.3.2). This showed that as the climate dried and the lowland grasses became less nutritious and digestible, the need to feed longer to compensate for reduced dietary quality would have reduced the time available for grooming and therefore led to group fragmentation. As a result, the theropithecines would have had trouble maintaining the minimal group sizes necessary to protect themselves from predators. This conclusion was reinforced by Lee and Foley (1993): they used data on the feeding efficiency of living geladas to model the energetics of the fossil species and concluded that the large-bodied taxa would have found it impossible to feed fast enough to meet their daily nutritional requirements as the cli-

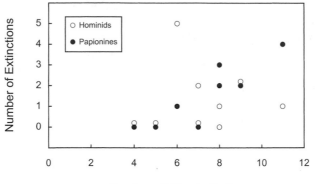

Figure 7.7 Number of species extinctions per 500,000-year period among the papionids and the hominids plotted against the number of climate cycles in that time period. (Data from R. Foley 1993, 1994.)

mate dried. Similar analyses were carried out for the papionids by Dunbar (1992d) using the systems model developed for *Papio* baboons by Dunbar (1992a) (see figs. 3.8 and 3.9). With climates for different fossil sites (and horizons within sites) estimated from paleoclimatic sources, the model indicated that maximum ecologically tolerable groups sizes would have been significantly lower at horizons and sites where papionids have not been found than at those sites were they are known to have existed. In other words, it would have been more difficult for the baboons to maintain cohesive groups large enough to ensure an acceptable degree of safety from predators.

R. Foley (1994) examined the extinction rates of African primates over the past 4 million years in relation to changes in climate documented in the deep sea cores. He used a 0.5 million year time unit to correlate the number of species last recorded in a particular time block with various measures of climatic stability for *Papio,* the Papionini as a whole, and the hominids. For all three groups, the frequency of extinctions correlated positively with low temperatures in both the current and the immediately preceding time frame, as well as with the number of climate cycles that occurred within that period (fig. 7.7). The cooler (hence drier) the climate, the more extinctions occurred. Foley interpreted these results as implying that species are unable to adapt quickly enough to sudden changes in vegetation precipitated by changes in climate.

Finally, Pickford (1993) has suggested that predation (notably by humans) may have contributed to the extinction of the East African theropithecines, especially after 300,000 years ago. They appear in very large numbers in the debris of *Homo erectus* butchery sites at Olorgesailie and elsewhere in Kenya from as early as 750,000 years ago (Leakey 1993). Shipman, Bosler, and Davis (1981) noted that young animals are represented with disproportionate frequency and interpreted this as evidence of deliberate hunting of young or weak animals. Although this pattern of mortality could just as easily reflect natural mortality in response to deteriorating environmental conditions, it remains plausible that hunting may have speeded up the attritional mortality imposed by climate change.

LEMURS

Madagascar and the Comoros Islands are known to have been home to at least twenty-two genera and forty-seven endemic species of Lemuriformes. Eight of these genera (fifteen species) went extinct during the past 1,000 years. These endemic species filled a wide variety of niches in a broad adaptive radiation (Godfrey et al. 1997). The following account of this extinction spasm is taken largely from Mittermeier et al. (1994) and Dewar (1984, 1997).

The eight lemur genera to become extinct (table 7.3) were, almost without exception, larger bodied than those genera existing today (fig. 7.8). *Archaeoindris*, the largest of the subfossil lemurs, reached an estimated mass of 200 kg (larger than an adult male gorilla) and dwarfs the largest living extant lemur, the indri, which weighs only about 6–7 kg (Smith and Jungers 1997). Whether the historical Malagasy primate fauna was genuinely this strongly skewed toward large body size remains unknown, but it is clear that all the extinct Malagasy species are related to living families, and all have been found in association with the subfossil remains of smaller living taxa.

Given their massive size, several of these extinct genera (notably *Megaladapis* and *Archaeoindris*) were likely to have had terrestrial lifestyles. In addition, anatomical proportions suggest that others may also have been terrestrial (e.g., *Archaeolemur*), with *Hadropithecus* filling a niche akin to that of the grazing gelada. In contrast, *Palaeopropithecus* probably had an arboreal lifestyle. The paleopropithecines have been dubbed the "sloth lemurs" owing to their apparent adaptation for hanging under branches. This form reached its zenith in the type genus, *Palaeopropithecus*. In contrast, the massive *Archaeoindris* probably occupied a niche similar to that filled by the giant ground sloth of the North American Pleistocene, while *Megaladapis* may have locomoted slowly and cautiously through the trees much like the modern koala.

Many of these species are known from sites distributed throughout Madagascar. Given such widespread distributions, and the wide diversity of habitats and niches these taxa are likely to have occupied, the sudden disappearance of all these species about 1,000 years ago is striking. It is

Table 7.3 The subfossil lemurs: Their ecology and life history

Superfamily	Family	Genus	Body mass (kg)	Locomotor style	Diet
Lemuroidea	Lemuridae	*Pachylemur*	10^1	Quadrupedal	Frugivore
Indrioidea	Archaeolemuridae	*Archaeolemur*	15–25	Quadrupedal	Frugivore
		Hadropithecus	15–25	Quadrupedal	Granivore
	Palaeopropithecidae	*Palaeopropithecus*	40–60	Suspensory	Folivore
		Archaeoindris	160–200	—	Folivore
		Babakotia	15–20	—	—
		Mesopropithecus	10	—	Folivore
Indrioidea (?)	Megaladapidae	*Megaladapis*	40–80	Vertical clinger	Folivore

Source: Data from Jolly 1986 and Mittermeier et al. 1994.

[1]Body weight estimate from Fleagle 1999.

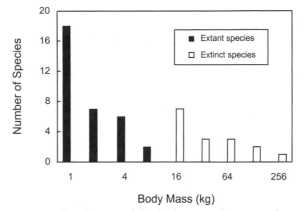

Figure 7.8 Frequency distribution of the body mass of extant and extinct Lemuri-formes of Madagascar. These data illustrate that only the smaller Lemuriformes survived the extinction processes of the past thousand years. (Data from Fleagle 1999.)

also notable that several other large animals (including elephant birds, the pygmy hippopotamus, and an endemic aardvark) also disappeared from Madagascar during this period.

What might have caused such a wave of extinctions? Although the climate of Madagascar is now believed to have been quite variable from the Pleistocene through to the present, there is no evidence to link local climatic changes to the extinctions of the lemurs (or indeed any other species that have disappeared from Madagascar in recent times, with the possible exception of a few aquatic bird species).

An alternative explanation implicates human intervention. The estimated dates of human colonization of Madagascar from Indonesia suggest that although the first arrivals landed not much earlier than A.D. 400, coastal Madagascar was widely settled by A.D. 1000. The archaeological record at several sites suggests that for a short period the people and the now extinct lemurs coexisted. Both habitat loss and hunting have been suggested as mechanisms that might have driven these lemur species to extinction. Habitat loss appears to have occurred through extensive burning (with a subsequent spread of grasslands) and, in later centuries, agricultural expansion. Hunting still occurs among some Malagasy cultures. and the extinct species, which were larger, slower moving, and more terrestrial than the extant lemurs, were likely to have been desirable and easily captured prey. Nonetheless, although human activities probably played the primary role in the lemur extinctions, it remains unclear precisely which ones were most important. In addition, these links are likely to be particularly complicated given that different species may have been affected by different processes, especially in different parts of the island.

In light of the present evidence, Dewar (1984, 1997) has suggested that livestock may have played a crucial role in these extinction processes, in two possible ways. First, as pastoralists moved with their livestock, their movements might well have been associated with burning and opportunistic hunting well into the interior of the island. Second, the livestock is likely to have had a major impact on the native plant communities, as well as competing with the indigenous herbivores. Moreover, both of these processes might be particularly important if feral populations of livestock spread across the country ahead of human movement. Livestock may therefore have been pivotal, because the transformation of native plant communities and the loss of native herbivores is likely to have precipitated a series of secondary extinctions associated with changes to the vegetation of the affected areas (sec. 7.2; cf. Owen-Smith 1988). If this is the case, hunting and other forms of habitat loss may have been less important than that previously postulated (although hunting may still have been the key anthropogenic pressure in the better preserved stretches of xerophytic bush and the rain forests of the east and northeast, where pastoralism was likely to have been minimal).

7.4.2. Lessons from Prehistoric Extinctions

Among the prehistoric extinctions, habitat disturbance (resulting from climate change for the baboons and human activity for the lemurs) appears to have been the primary factor precipitating extinction, although hunting may have exacerbated matters in both cases. This finding is consistent with the fact that habitat loss and fragmentation (together with hunting) would result in small isolated populations that would then be vulnerable to intrinsic small-population processes.

Nevertheless, some species clearly survived while others did not. Since most of the lemurs and baboons that went extinct were larger than their surviving contemporaries, it seems likely that body size was important: this would be consistent with the finding that body mass is a good predictor of extinction risk in contemporary primates (sec. 7.3.2). Similarly, Jablonski (1998) has recently noted that during Pleistocene-Holocene climate change in China the largest ape *Gigantopithecus* was the first to disappear; the somewhat smaller orangutan *Pongo* subsequently became extinct on the mainland but survived elsewhere, and the smallest apes (*Hylobates*) were relatively unaffected. Likewise in the New World, no extant atelines weigh as much as 10 kg (Smith and Jungers 1997), but two giant fossil ateline species have recently been discovered that weighed more than 20 kg: *Protopithecus brasiliensis* and *Caipora bambuiorum* (Hartwig 1995 and Cartelle and Hartwig 1996, respectively).

The fossil and subfossil case studies have two further lessons for us.

First, the extinctions among the lemurs of Madagascar illustrate the important fact that island populations are especially vulnerable to extinction (e.g., Smith et al. 1993b, Mace and Balmford 2000), for several reasons (sec. 4.1.1). Island populations are smaller than continental populations (owing to the physical limits of island size), less likely to benefit from rescue effects (owing to their isolation: see sec. 6.3), and in the case of oceanic islands, possess undersaturated communities that are especially vulnerable to introduced species and human colonization. The lemurs of Madagascar are not the only primates to demonstrate this pattern. The extinction of two of the three primate species that originally inhabited the Caribbean islands also occurred shortly after *Homo sapiens* arrived there some 4,500 years ago (Morgan and Woods 1986).

Second, although the small-population processes immediately responsible for fossil extinctions are difficult to identify, the systems models used to study the extinctions of *Theropithecus* and *Papio* fossil populations raise the possibility that Allee effects might come into play as population size declines to low values. Catastrophic collapse of the population may occur well before attritional mortality erodes group size completely, either because group sizes are too small to deter predators or because the social structure of groups is disrupted. This suggests that although the behavioral flexibility of primates might be sufficient to avoid Allee effects in many circumstances (sec. 7.2.2), we would do well not to write off this process as unimportant to primate population viability.

7.5. Summary

1. There are two key steps in population extinction: populations decline in size, and small populations then suffer from processes that cause extinction.

2. Some species naturally occur in small populations because of various constraints on abundance and distribution (chapter 5). Other populations may become small through natural fluctuations (sec. 6.2) or through the anthropogenic effects of habitat disturbance, hunting, introduced species, or secondary extinctions. Among primate taxa, only habitat loss and hunting currently appear to pose a significant threat to primate diversity.

3. Taxa that exist in small populations are more vulnerable to extinction through the processes of demographic stochasticity, environmental stochasticity, loss of heterozygosity (leading to both reduced ability to adapt to environmental change and inbreeding depression), and disruption of social structure (which may give rise to Allee effects).

4. Interspecific variation in extinction risk is primarily a function of variation in population size, vulnerability to declining population pro-

cesses, and vulnerability to small-population processes. The global extinction risk of a primate species is independently correlated with its geographic range area, its population density, its body mass, and its trophic level.

5. Species traits associated with high vulnerability to small-population processes include very small body size (high metabolic rate and relatively small fat stores) or very large body size (slow reproductive rates), dietary specialization (reduced ecological flexibility), short generation time (high risk of genetic drift) or long generation time (poor speed of demographic recovery), reduced longevity (the inability to ride out lean times), and a social structure characterized by high polygyny skew (high risk of genetic drift).

6. Population declines that led to the extinction of fossil baboons and subfossil lemurs appear to have been driven largely by habitat disturbance associated with climate change and anthropogenic disturbance, respectively.

8 *Habitat Disturbance*

In the preceding chapters we have described the processes that help determine population size and dynamics in primates and discussed their intrinsic role in primate population extinctions (chapters 5, 6, and 7). In this chapter and the next we examine the critical extrinsic forces that are driving primate populations to extinction: habitat disturbance and hunting.

Habitat disturbance poses a severe threat to the survival of many of the world's primate populations (Marsh, Johns, and Ayres 1987; Mittermeier and Cheney 1987) and has played an important role in fossil and subfossil primate extinctions (sec. 7.4.2). This chapter gives an overview of the key patterns and processes of habitat disturbance, the sorts of effects these processes have on primate populations, and the species traits that might underlie interspecific variation in the ability to cope with these disturbances.

8.1. Patterns of Habitat Disturbance

Since the vast majority of primate species live in tropical forests, our focus here is on those human activities that threaten these ecosystems and on how they affect forest structure in terms of habitat loss, fragmentation, and modification. Such processes can operate on a massive scale and often act in concert (e.g., fig. 8.1).

8.1.1. Causes of Disturbance

The first part of this section reviews the immediate causes of forest disturbance. We then consider the processes that drive these proximate mechanisms (their ultimate causation).

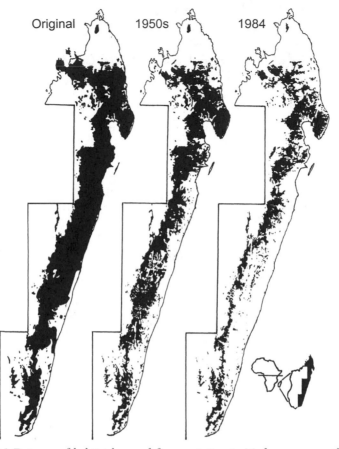

Figure 8.1 Patterns of habitat loss and fragmentation in Madagascar over the past thousand years. Distribution of eastern rain forest before the arrival of humans, during the 1950s, and in 1984. (Redrawn with the permission of Kluwer Academic/Plenum Publishers from Sussman, Green, and Sussman 1994.)

PROXIMATE MECHANISMS

Human activities that have a direct impact on tropical forest habitats can be classed under three broad headings. The first is agriculture, mainly in the form of shifting cultivation (slash-and-burn, or swidden, agriculture). Shifting cultivation is practiced by small-scale farmers and consists of clearing and planting crops in small patches of forest for two or three years until soil fertility is reduced or weeds encroach; the patch is then abandoned and left fallow to regenerate. Cultivation does not normally resume for at least ten years. At low population densities, this agricultural cycle contributes little to deforestation (Rijksen 1978). But increasing population pressure (leading to unsustainably short periods of fallow) and inappropriate agricultural methods (often practiced by immigrant farmers) have made shifting cultivation a major cause of tropical forest loss (Groombridge 1992). In contrast, permanent agriculture is believed to have been less important in global deforestation, although it has played a predominant role in some countries, such as Brazil. In the Atlantic Forest, sugarcane plantations were a major factor in deforestation (Ranta et al. 1998); more recently, government policies promoting settlement have resulted in the clearing of frontier Amazon forest by immigrating shifting cultivators, with the land later being purchased by cattle ranchers (Mahar and Schneider 1994). Another serious problem associated with the cultivation of forests is the use of fires to clear land: when they get out of control, they can spread into primary forest with devastating consequences (Neotropics: Cochrane and Schulze 1998; Southeast Asia: Kinnaird and O'Brien 1998).

Second, forestry can also make a significant contribution to deforestation (e.g., in Southeast Asia). Commercial forestry currently tends to be relatively less important in Africa and South America, although it is becoming increasingly significant as the tropical forests of Asia dwindle and alternative sources of timber are sought. Two basic silviculture techniques are practiced in tropical forests (Whitmore 1984): monocyclic systems, in which all commercial trees are removed in a single operation, and polycyclic systems, in which selected trees are removed in a series of felling cycles. Both may be classed as "selective logging." Since clear-cutting is not involved, these methods of forestry may have a relatively small impact on forest loss. But the extraction of only a few trees is normally associated with substantial peripheral damage through the process of logging: at Tekam (peninsular Malaysia), for example, removing less than 4% of the forest stand damaged over 50% of the trees (Johns 1988; Grieser Johns 1997). Consequently the impact of selective logging on the structure and future viability of existing tropical forests is likely to be much greater than its contribution to recorded deforestation(especially since these changes to forest structure also make these habitats more flammable: Cochrane and Schulze 1998).

Third, mining, hydroelectric projects, and other enterprises also contribute to forest disturbance. Although the direct impact of these projects may be relatively minor on the global scale, they can have two indirect effects of considerable importance. First, such projects can lead to large-scale ecological changes downstream: two new proposed dams along the Tana River (Kenya), for example, are expected to cause a drop in groundwater and to disrupt flooding and sedimentation cycles, with serious implications for the forest ecosystem and the primates downstream (Butynski 1995; see also Marsh 1986). Second, and of particular consequence, these ventures lead to extensive road building, opening up previously inaccessible forest areas and ultimately playing an instrumental role in promoting settlement and agriculture there (Groombridge 1992; Mahar and Schneider 1994; Chatelain, Gautier, and Spichiger 1996). This is a serious problem also associated with forestry projects. In the case of hydroelectric schemes, the problem is further exacerbated by the construction of towns to house the large workforces employed in building the structure (Kumar 1985). The role of accessibility in driving habitat disturbance is now well established: the distribution of secondary forest in western Madagascar, for example, is primarily influenced by distance from villages and secondarily by distance from roads (Smith, Horning, and Morre 1997). Its importance should not be underestimated. Indeed, accessibility also plays a key role in determining hunting pressure (as we discuss in chapter 9).

Finally, human activities also cause changes in climate, indirectly leading to large-scale forest disturbance. The viability of forest biomes is highly dependent on local climate, and the future distributions of these biomes are likely to be very different from those they now occupy if there is significant global climate change. Current estimates of the extent of climate change suggest that global temperatures will rise by between 1.5 and 4.5°C over the next century, albeit with considerable regional variation (IPCC 1992; Gates 1993; Houghton et al. 1996). Although its impact on the world's ecosystems remains difficult to predict with any certainty, in general we can expect most forest biomes to move poleward or to higher altitudes, with more arid grassland habitats replacing them in tropical latitudes. These effects can already be detected in temperate forest vegetation. Hamburg and Cogbill (1988) reported that the red spruce forests of the northeastern United States have retreated a considerable distance northward during the past 200 years. They calculate that the level of climatic warming observed in this area during the past 150 years (2.2°C in mean summer temperature) is equivalent to a very substantial 400 m altitudinal displacement in the spruce/hardwood boundary. This is equivalent in temperature terms to a northward displacement of about 200 km in the southern boundary for conifer species. The poleward shift of tropical forest belts may bring them into closer contact with regions where large-scale farming has already resulted in extensive land clearance. Forests attempting to colonize such zones will find it difficult to gain a foothold. Similar patterns among animal populations are now also becoming visible: the ranges of British birds have expanded northward by 19 km over a twenty-year period (Thomas and Lennon 1999).

ULTIMATE CAUSATION

The relative importance of these various mechanisms of deforestation differs from region to region (table 8.1). Agriculture (invariably shifting cultivation) usually plays a key role, but other processes are also very

Table 8.1 Proximate mechanisms of deforestation in three tropical countries (in the period 1981–88)

Process	Country Brazil	Indonesia	Cameroon
Total deforestation			
Forestry	2	9	0
Agriculture	89	80	100
Mining and related industries	<3	0	0
Hydroelectric production	4	0	0
Residual[1]	2	11	0
Agriculture			
Shifting	15	74	92
Permanent	85	26	8
Permanent agriculture			
Pastures	52	0	0
Permanent crops	5	10	63
Arable land	42	90	37

Note: Data given as percentages of forest loss (after Barbier et al. 1994).

[1]Includes other industries, infrastructure, housing, and fire loss.

important. In Southeast Asia, for example, commercial forestry is a highly significant threat. Most commentators agree that these mechanisms of forest disturbance have their ultimate origins in human population growth and resource consumption (see below). Although it is hard to determine which of these two forces is the more influential, that agriculture is often the primary mechanism in deforestation suggests that population growth may be the main problem. However, other factors such as poverty and inappropriate land tenure policies may exacerbate its effects.

The link between population growth and deforestation through shifting cultivation has been studied in a model of subsistence agriculture in the Ituri Forest (Democratic Republic of Congo) (Wilkie and Finn 1988). Zero population growth allowed indefinite utilization of the forest for shifting cultivation without reducing primary forest cover by more than 30%. In contrast, with 5% annual population growth, the number of primary forest patches is reduced, the number of forest clearings increases, the fallow period of the fields is shortened, and the area must be completely abandoned after about eighty years. Incorporating possible changes in the current land tenure system did not alter these results (although where large stretches of abandoned secondary forest already existed because of historical cash cropping, the expansion of subsistence agriculture can be absorbed up to a point: Wilkie et al. 1998a).

Several studies have attempted to identify, across countries, the key socioeconomic correlates of deforestation. Palo (1994) found that contemporary tropical forest cover progressively decreases with rising population density across a range of spatial scales in Latin America, Asia, and Africa (regression models: Palo 1994). Similarly, Green and Sussman (1990) determined that deforestation was more extensive in areas of high population density in Madagascar. Barnes (1990) found that in Africa the most reliable predictors of annual national forest loss were large initial forest area and large population size. (After these factors were accounted for, the additional effects of population density, per capita gross national product [GNP], total GNP, wood fuel consumption, charcoal consumption, and industrial wood production were negligible.) Using a more recent data set for African countries, Harcourt (1996) found that annual national percentage forest loss was strongly correlated with human density but that absolute area of annual national forest loss was correlated with national external debt. Harcourt suggested that the correlation with debt might reflect the fact that governments in developing countries adopt unsustainable forest management practices in order to service these debts. In conclusion, although these various studies are conducted at a range of spatial scales and across different geographic regions, they consistently find that human population pressure is the best predictor of tropical deforestation rates.

8.1.2. Types of Disturbance

Agriculture, forestry, and the other processes outlined above are directly responsible for global forest loss, fragmentation, and modification. In this section we describe the characteristic features of these three types of disturbance.

FOREST LOSS

It has recently been estimated that 8% of the world's tropical forests was lost in the decade 1981–1990 (World Resources Institute 1996). However, there is important regional variation in this figure, and care must be taken in interpreting these statistics (table 8.2). The annual percentage rate of deforestation in the tropical Americas, for example, is barely half that in Asia, but it destroys an area more than twice as large. At a finer scale, among the fifteen countries with the highest primate diversity, Brazil and Indonesia lose the largest absolute areas of forest annually (table 8.3) and together account for up to 45% of global tropical forest loss. Although large-scale conversion of tropical forests probably began about 1600 (after the introduction and intensification of livestock and cash crop farming), it is probably only in the past fifty years that tropical forest loss has really accelerated (Groombridge 1992). Deforestation

Table 8.2 Forest area and deforestation across major continental tropical regions

Region	Extent of tropical forest in 1990 ($\times$ 1,000 ha)	Annual percentage loss between 1981 and 1990	Estimated annual loss for 1991 ($\times$ 1,000 ha)
Africa[1]	540,669	0.73	3,947
Madagascar	15,782	0.79	124
Asia[2]	283,249	1.18	3,342
The Americas	959,704	0.74	7,101

Source: Data from World Resources Institute (1996).

Note: The figures for estimated annual loss in 1991 are based on the annual change in area over the period 1981–90.

[1]Includes Madagascar.
[2]Excludes temperate and middle East Asia.

rates in Ghana, Ivory Coast, and Uganda over this century support this argument (Barnes 1990).

The projections for future losses are subject to much debate but are invariably bleak (Myers 1994). The best estimates for future projections may be those developed by Barnes (1990), who fed national population growth projections into a multiple regression model that reliably predicted the annual area of forest loss in African countries based on population size. A selection of his results is shown in figure 8.2 (which includes six of the ten top African countries for primate diversity: table 8.3). The predicted patterns of deforestation are variable, reflecting the balance between the effects of dwindling forest area and increasing population size. It is worth noting, however, that a negative feedback mechanism may operate, with deforestation rates potentially slowing once forest area becomes very small (cf. Harcourt 1996). Precisely this pattern has been found in Madagascar: local forest loss appears to have slowed where the last remnants of forest are on slopes too steep to cultivate (Green and Sussman 1990).

In any analysis of deforestation rates, it is critical that historical deforestation also be considered. This provides an essential perspective on the preceding figures on contemporary deforestation rates: a 1% loss of forest becomes more serious when current forest cover is only a fraction of its original extent. Estimates of the remaining original forest in the key countries for primate diversity listed in table 8.3 suggest that, in most of the African countries, more than half the original forest may already have disappeared; in some, such as Nigeria and Uganda, less than 5% of the original forest cover is now thought to remain. Although some doubts remain about the precision of these figures (there is now some evidence that historical forest loss has not always been as severe as previously

Table 8.3 Natural forest area and deforestation rates for countries with high primate species richness

Region and country	Number of primate species	Area of natural forest in 1990 (× 1,000 ha)	Annual deforestation % rate between 1981 and 1990	Estimated area of forest loss in 1991 (× 1,000 ha)	Percentage of ancestral closed forest left	Total roundwood production in 1991–93 (× 1,000 m³)	Area of cropland in 1993 (× 1,000 ha)	Human population size in 1995 (× 1,000)	Per capita GNP in 1993 (U.S. $)
Africa									
Democratic Republic of Congo	31	113,275	0.61	691	59	43,252	7,900	43,901	264
Cameroon	29	20,350	0.57	116	48	14,483	7,040	13,233	820
Nigeria	23	15,634	0.71	111	14	114,704	32,385	111,721	300
People's Republic of Congo	22	19,865	0.16	32	62	3,438	170	2,590	950
Equatorial Guinea	22	1,826	0.37	7	50	613	230	400	420
Central African Republic	20	30,562	0.41	125	11	3,628	2,020	3,315	400
Angola	19	23,074	0.70	162	13	6,382	3,500	11,072	—
Uganda	19	6,346	0.92	58	7	15,099	6,770	21,297	180
Gabon	19	18,235	0.60	109	80	4,345	460	1,320	4,960
Madagascar	28°	15,782	0.79	125	37	8,600	3,105	14,763	220
Asia									
Indonesia	34°	109,549	1.00	1,095	67	185,426	30,987	197,588	740
The Americas									
Brazil	52°	561,107	0.61	3,422	—	268,879	48,995	161,790	2,930
Colombia	27	54,064	0.64	346	—	20,619	5,460	35,101	1,400
Peru	27	67,906	0.38	258	—	8,031	3,430	23,780	1,490
Bolivia	18	49,317	1.12	592	—	1,530	2,380	7,414	760

Source: Data on primate species richness from table 2.3; data on percentage of ancestral closed forest left from Sayer, Harcourt, and Collins (1992); all other data from World Resources Institute (1996).

° >30% of species occur only in that country.

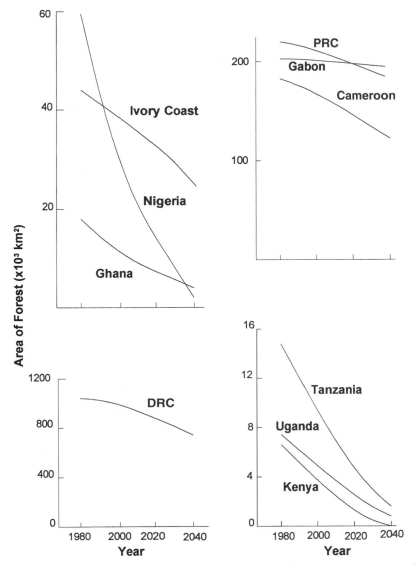

Figure 8.2 Projected patterns of forest loss over the next fifty years in a selection of African countries. Projections are based on the relation between population size and area of forest loss, using predicted patterns of future population growth. DRC is the Democratic Republic of Congo, PRC is the People's Republic of Congo. (From Barnes 1990 with the permission of Blackwell Science.)

thought: Fairhead and Leach 1998), they remind us that deforestation is not new. Comparable data do not exist for Latin America, although Asia may follow a pattern similar to that in Africa (the large extent of remaining forest cover in Indonesia is not typical for this region: Groombridge 1992).

Finally, what are the implications of these patterns for conserving primate diversity? In particular, is the rate of deforestation greater in countries with a high primate species richness? Analysis of the available data (table 8.3) suggests the following. First, as we would expect from island biogeography theory, countries with more forest have more primate species ($r_s = .52, p = .05$; see sec. 4.1.1). Although the annual percentage rate of deforestation is not related to total forest area ($r_s = .07, p = .80$), there is a strong trend for countries with more forest to lose a larger absolute area of forest annually ($r_s = .91, p < .001$). This trend may be mediated through both agriculture and forestry (area of forest loss is correlated with both area of cropland and roundwood production: $r_s = .58, p = .03$ and $r_s = .54, p = .04$, respectively). The ultimate cause of this pattern appears to be more strongly associated with the national population (human population size is correlated with both cropland and roundwood production: $r_s = .77$ and $r_s = .81$, respectively, $p < .001$ in both cases) rather than the national economy (per capita GNP is correlated with neither: $p > .60$ in both cases), although the impact of international markets on deforestation rates is difficult to ascertain from these data. Notably, there is no evidence that deforestation is slowing down in those countries where only a small proportion of the original forest cover remains: both the annual rate and the absolute area of forest loss are unrelated to the percentage of original forest remaining ($p > .50$ in both cases). In other words, we should not be lulled into a false sense of security because the percentage rate of annual forest loss is no greater in countries of high primate diversity: in absolute terms, these countries are losing much larger areas of forest annually, and there is no evidence (from these figures) that the area of forest loss will slow down as the fraction of forest remaining diminishes.

FOREST FRAGMENTATION

Habitat fragmentation poses a severe threat to primate populations around the world, with some of the most unusual primate communities now surviving only in highly fragmented habitats. A case in point is the Atlantic Forest of Brazil, which has suffered greatly from the combined effects of agriculture, forestry, and urbanization: it now exists only as a complex of remnant fragments constituting a mere 12% of the original 1 million hectares of forest (by contrast, 90% of the original 4 million hectares of the Brazilian Amazon is still standing: Brown and Brown 1992).

Moreover, as might be expected given the long history of global deforestation, the fragmentation of primate habitats by human activities is not new. The lion-tailed macaque has existed in a fragmented landscape since well before this century, thanks mainly to land conversion practices during colonial times (Kumar 1985). Similarly, the Mediterranean forests of the Barbary macaque in northern Africa have experienced successive waves of utilization and fragmentation since Roman times (Thirgood 1984; Fa et al. 1984).

Deforestation splinters any forest habitats into isolated fragments in a potentially complex pattern (Bascompte and Solé 1996; cf. Andrén 1994). A spatial model of fragmentation shows that the number of habitat patches may not begin to increase until forest loss is already well under way and peaks before deforestation is complete (fig. 8.3a). This is because habitat loss must be extensive before forest patches become completely isolated; but once high levels of deforestation are reached, all the forest is fragmented and any further loss destroys the fragments themselves. For similar reasons, the area of the largest fragment can show a dramatic discontinuity in the way it varies with the degree of deforestation (fig. 8.3b). These discontinuities suggest that there may be threshold levels of fragmentation beyond which primate population viability might decline dramatically. This possibility is explored further below (sec. 8.3).

The limited empirical evidence currently available suggests that the amount of fragmented habitat may be relatively small in relation to remaining continuous habitat, implying that it may be common for forests to be either lost completely or left intact. In the Brazilian Amazon in 1988, for example, forest fragments covered an area equivalent to only 7% of the area deforested (Skole and Tucker 1993). Similarly, in the Guiglo-Taï region of Ivory Coast in 1990, forest fragments constituted only 4.5% of the remaining area of forest (Chatelain, Gautier, and Spichiger 1996). This may be related to the fact that forests are largely eaten away at the edges by deforestation (e.g., along roadsides or around villages), so that only the fringes are fragmented.

In addition, there is evidence that as the number of fragments increases, so the average size of fragments declines (Costa Rica: Sanchez-Azofeifa et al. 1999), suggesting that once most forest has been lost the remaining isolates will be very small. Indeed, the frequency distribution of forest fragments is often strongly skewed toward smaller fragments (e.g., Los Tuxtlas, Mexico: Estrada and Coates-Estrada 1996; Pernambuco, Atlantic Forest: Ranta et al. 1998): in the Guiglo-Taï region of Ivory Coast 79% of unprotected forest areas have been lost, and of 984 fragments remaining in deforested areas most are less than 1 ha in size and fewer than 8% are larger than 10 ha (Chatelain, Gautier, and Spichiger 1996). There is also some evidence that fragments may not persist very

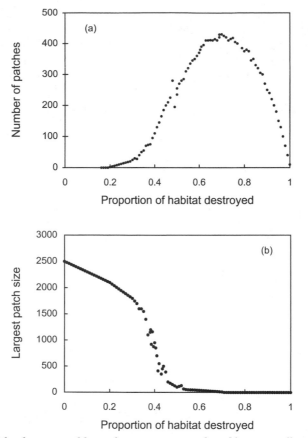

Figure 8.3 The dynamics of forest fragmentation predicted by a spatially explicit simulation model: (a) covariation between the number of fragments and the extent of habitat loss and (b) covariation between the size of the largest fragment and the extent of habitat loss. (Redrawn with the permission of Blackwell Science from Bascompte and Solé 1996.)

long, at least where there is ongoing deforestation: once again in Ivory Coast (Chatelain, Gautier, and Spichiger 1996) most fragments are young (84% were found in areas deforested between 1985 and 1990), while few are old (only 1% of fragments occurred in areas deforested before 1974). Although Ranta et al. (1998) found that fragments tended to be close to one another in the Atlantic Forest area (98% of the fragmented forest area is within 350 m of another fragment), this pattern may not be typical: at Los Tuxtlas, for example, isolation distances varied between 100 m and 8 km (Estrada and Coates-Estrada 1996). Unfortunately, small fragments may also be the most isolated (Ivory Coast: Chatelain, Gautier, and Spichiger 1996).

Finally, Bierregaard et al. (1992) report that patterns of tree mortality and recruitment in undisturbed forest differ between continuous and fragmented forest, particularly along forest edges where mortality may be very high owing to increased wind exposure (Lovejoy et al. 1984). Wind also dries out vegetation, making it more susceptible to fire. Murcia (1995) provides a recent review of such *edge effects* and the severity of the problem they pose. For example, Skole and Tucker (1993) estimated that the area of forest suffering from edge effects in Amazonia is far greater than the part that is either isolated or deforested. Although their penetration distance for such effects is probably an order of magnitude too large (cf. Murcia 1995), Ranta et al. (1998) have also found that the total area of edge zone exceeds that of interior habitat when edge effects operate at a relatively modest 60 m (in an area characterized by large numbers of small fragments in the Atlantic Forest).

Taken together, the preceding patterns have several implications. First, the small area of most fragments means that the size of primate populations in each fragment will also be small, so extinction will be a serious risk (see sec. 7.2), and that fragments will suffer severely from edge effects, since there will be more edge per unit area in smaller fragments. Second, that smaller fragments also tend to be the most isolated means those populations in greatest need of rescue effects (i.e., the smallest populations) are least likely to receive them. Third, in view of these area and isolation patterns, it seems likely that, even where a minority of primate populations do manage to persist, their long-term viability may be low owing to the gradual deterioration of forest fragments through edge effects.

FOREST MODIFICATION

Humans' use of tropical forests is not inevitably linked to deforestation. But even if enterprises such as shifting cultivation and selective logging do not always lead to large-scale forest clearance and fragmentation, they do tend to substantially modify the forest structure. The extent of this modification is difficult to assess, but if absolute deforestation is gauged through remote sensing as "the permanent depletion of the crown cover of trees to less than 10%" (World Resources Institute 1994), then the area of annual forest loss may be only a fraction of the area that is modified.

This modification primarily involves the replacement of primary forest with secondary forest. Secondary, or regenerating, forest consists of fast-growing, light-demanding species that rapidly encroach in areas where primary forest and its shady canopy cover have been lost (Archibold 1995). Although secondary forest appears in response to human-induced changes to the primary forest structure, we should also remember that it is a natural part of the ecosystem. It is particularly common at forest edges, especially along rivers and in forest gaps (which may be large or small depending on whether they are created through forces such as landslides and fires or branch falls and tree falls). Secondary forest patches are highly variable in their species composition but generally have a floral structure very different from that of primary forest. On Tiwai Island (Sierra Leone), for example, a comparison of the ten most common species in secondary and old forest reveals only one species in common (Fimbel 1994a). In addition, different types of disturbance may favor different types of secondary forest, depending on the amount of the original canopy lost and the composition of the remaining primary forest species. Agriculture and forestry can thus produce different kinds of secondary forest growth.

Although secondary forest is a natural feature of forest systems, it should be stressed that the long-term viability of regenerating forest largely depends on the persistence of pollinators and seed dispersers and the proximity of undisturbed primary forest (as a source of pollen and seeds: Rijksen 1978). If these roles are not filled, the long-term integrity of many modified forests may be destroyed. To the extent that primates act as pollinators and seed dispersers (sec. 4.4), forest patches may be pushed into a downward spiral as reduced forest cover either drives primates into extinction through stochastic population processes or forces them to move elsewhere.

Finally, climate warming may also lead to forest modification, although not through secondary plant growth. Climate changes are more likely to affect fruit production in tropical forests. Many of the tree species at Lopé (Gabon), for example, can set seed only if the minimum temperature at the appropriate time of year falls below 19°C (Tutin et al. 1997). With rapid climate warming in prospect, some marked changes can be expected in the fruiting patterns (and thus tree species composition) of tropical forests.

8.2. Effects of Habitat Loss

Primates appear to be particularly vulnerable to habitat loss. Primate taxa are more likely to be threatened with extinction by habitat loss than those of any other mammalian order, and among the primates it is the most prevalent source of species extinction risk (Mace and Balmford 2000). Island biogeography theory (IBT) predicts that a reduction in habitat area will inevitably lead to a reduction in species richness (according to the species-area curve: sec. 4.1.1). But the fact that many tropical countries have suffered extensive deforestation in the past, yet still retain high species richness, might be taken to mean that large areas of habitat may be lost without having a significant effect on primate biodiversity.

Unfortunately, it seems that this conclusion could not be more wrong. Studies of land-bridge islands and some national parks have shown that the extinctions associated with a loss of habitat often take place some time after habitat loss occurs (sec. 4.1.1; see also sec. 11.1.2). This is likely to reflect the fact that populations that become small and isolated may be able to survive for many years before they finally fall victim to the effects of small population size. Such processes are likely to be exacerbated by an absence of rescue effects and by the loss of ecological resources the population used in those areas of habitat that have disappeared.

Several authors have investigated the existence of time lags to extinction in contemporary tropical forest communities using IBT (Pimm and Askins 1995; Brooks and Balmford 1996; Brooks, Pimm, and Collar 1997; see also Pimm, Moulton, and Justice 1995). These studies are based on the simple relationship (proposed by Simberloff 1992) that the drop in the original number of species from S_o to S_n when the area of habitat declines from A_o to A_n can be estimated from

(8.1) $$S_n / S_o = (A_n / A_o)^z.$$

Provided that the original and new habitat areas and the original number of species are known, and a value for z can be estimated (see sec. 4.1.1), it is possible to predict the number of species remaining at equilibrium. Brooks and Balmford (1996) used this method to examine extinctions among the endemic birds of the Atlantic Forest, where 90% of the habitat has been lost but not a single bird species has become extinct. They found that the number of threatened endemic species in the Atlantic Forest was closely correlated with the number of species predicted to become extinct there (according to eq. 8.1). The same result has been obtained by Brooks, Pimm, and Collar (1997) for the endemic birds of insular Southeast Asia. Tilman et al. (1994) used metapopulation models to investigate time lags to extinction, or "extinction debts," using metapopulation models. They found that the proportion of species becoming extinct accelerates with increasing habitat loss. More worrying was the fact that the time lag between habitat loss and extinction in their tropical forest model was on the order of one hundred to four hundred years.

Cowlishaw (1999) recently used this approach to investigate extinction lags in African primates. The species richness of forest primates is strongly correlated with the extent of closed forest cover across African countries, but in support of the time lag hypothesis, the fit of this relationship is better for the historical forest cover than for present cover. Based on the observed link between species richness and forest area, the documented loss of forest in many African countries should have led to an average loss of 30% of the national primate fauna, typically constituting between one and eight species (table 8.4). But these losses have yet to be observed. Since the same model correctly predicts both the

Table 8.4 Primate extinction debts from historical forest loss in Africa

Country	Closed forest species currently present	Predicted losses of closed forest species	
		Optimistic scenario	Pessimistic scenario
Democratic Republic of Congo	26	0 (0.00)	1 (0.04)
Cameroon	27	3 (0.13)	8 (0.28)
Nigeria	22	5 (0.21)	8 (0.35)
People's Republic of Congo	20	0 (0.00)	1 (0.04)
Central African Republic	15	1 (0.08)	4 (0.24)
Angola	14	4 (0.26)	5 (0.39)
Uganda	15	5 (0.31)	7 (0.44)
Gabon	19	0 (0.00)	0 (0.00)

Note: Extinction debts estimated for African countries with high primate species richness, following table 2.3. Losses are given to the nearest whole number; figures in parentheses are the proportion of all closed forest species lost. The optimistic scenario assumes that historical forest loss has been overestimated or that some primate species may have already disappeared from the countries in question. Data from Cowlishaw 1999.

locations of endemic forest taxa threatened by habitat loss and the total number of forest primate species in Africa threatened by habitat loss (according to independent IUCN classifications; see sec. 10.2.2), it is unlikely that these predictions will turn out to be far wrong. Although local extinctions may not be considered as serious as global extinctions, many African species occur in only a handful of countries. Moreover, these estimated extinctions do not take into account the effects of hunting on primate populations. Given this omission, the magnitude of the extinction debts predicted by equation 8.1 is clearly a matter for grave concern.

Although IBT provides a useful guide to how the diversity of primate communities might become eroded after habitat loss, this body of theory is less useful in predicting which species are most likely to become extinct and what the ultimate mechanisms of that extinction are likely to be. Alternative models are required in these respects, and systems models (such as those presented in sec. 3.3.2) are one possibility. Dunbar (1998b) recently used such a model to predict how gelada baboons on the Ethiopian plateau would survive habitat loss as a result of climate change from global warming. The results suggest that the gelada's lower altitudinal limit will move upward by about 500 m for every 2°C rise in global temperature. Given the actual distribution of land surfaces at different altitudes on the Ethiopian plateau, this translates into an approximate halving of the land surface area available for geladas for each 2°C rise in temperature (fig. 8.4). The result will be a steady collapse in the species' total population size, with a 7°C rise in regional mean temperature being sufficient to reduce it from its current size of about

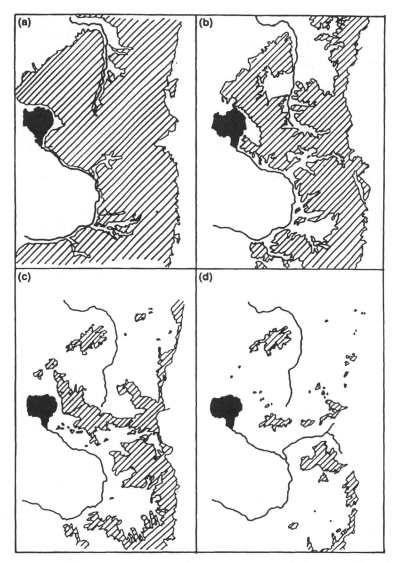

Figure 8.4 Decline in the geographical range (and thus population size) of the gelada baboon *Theropithecus gelada* with progressive increases in temperature through global climate change: (a) the species' current distribution on the Ethiopian plateau; projected distributions with (b) 2°C, (c) 5°C, and (d) 7°C rise in mean global temperature, based on the maximum ecologically tolerable group sizes predicted by systems models similar to those shown in figures 3.8 and 3.9. The solid black area is Lake Tana. (From Dunbar 1998b with the permission of Blackwell Science.)

250,000 to about 5,000. Although this would still be considered a reasonable population size from a conservation point of view, that these animals would be distributed around a large number of small isolated montane islands (with none having a population greater than 2,000) places them at great risk of small-population effects. Consequently the species' survival prospects would be considerably poorer than a simple numerical count would imply.

Finally, in the short term it possible that some primate populations might survive habitat loss by emigrating from the area under conversion. In some parts of the Ethiopian plateau, for example, large-scale conversion of the montane grasslands has forced gelada baboons to switch their foraging to steeper, uncultivated slopes; unfortunately, this is suboptimal habitat for the gelada, since the grass cover on these slopes tends to be sparser (Dunbar 1977). However, the success of that emigration depends on not only finding suitable habitat in the surrounding area but also finding suitable habitat that is not already occupied by a resident population of the same species (or another species that occupies the same niche). For example, since the expansion of agricultural activities on the slopes of Mount Kilimanjaro (Kenya) in 1982, there has been an emigration of males from the local population of olive baboons (*Papio anubis*) away from the mountain slopes and into the Amboseli basin, where they are competing with (and interbreeding with) the resident yellow baboons (*P. cynocephalus*) (Samuels and Altmann 1991).

8.3. Effects of Habitat Fragmentation

The effects of habitat loss and habitat fragmentation are intimately connected, since fragmentation is a natural consequence of loss. As deforestation takes place, animals whose ranges have been destroyed collect in the remaining fragments of pristine habitat. This influx can lead to isolates initially having high species diversity at unusually high abundances. But the competition and relative scarcity of resources in the fragment typically leads to a rapid decline, both through group emigration (e.g., tamarins and capuchins: Rylands and Keuroghlian 1988) and through direct mortality (where individuals are unable or unwilling to reach forest beyond the fragment). In this section we seek to identify the key features of habitat fragments that determine their capacity to support viable primate populations in the long term, and subsequently the ability of primate species to exploit new habitats that abut the remaining fragments of the original habitat.

8.3.1. Survival in Fragments

Once the fragment has reached its initial stable complement of species (those for which sufficient ecological resources are currently available), a variety of factors may be involved in determining long-term species richness and population persistence. Once again using island biogeography theory (IBT) as our guide, we initially focus on the effects of area, isolation, and age.

First, in accordance with IBT, larger fragments contain more species among both Neotropical frugivores and terrestrial mammals (across fragments of from 1 ha to more than 2,000 ha: Estrada et al. 1993; Estrada, Coates-Estrada, and Meritt 1994 respectively) and also among Neotropical primates (in comparisons between fragments of 10 ha and 100 ha: Lovejoy et al. 1986; Rylands and Keuroghlian 1988). The reasons larger isolates have more species overall have already been discussed (sec. 4.1.1). In the short term, habitat diversity may be most influential, although in the long term large population size will also be important.

Second, contrary to IBT, equal area Neotropical fragments that vary between 70 and 650 m in their degree of isolation did not show any association between isolation distance and primate species richness (Schwarzkopf and Rylands 1989). This finding suggests that this range of distances must therefore present either no barrier at all or an impenetrable barrier to movement, with the former being the more likely (given the observed movements of tamarins and capuchins between fragments). In contrast, with isolation distances of between 200 m and 8 km, Estrada, Coates-Estrada, and Meritt (1994) reported a strong negative correlation between isolation distance and the species richness of terrestrial mammals (independent of area). The most substantial drop in species richness was at the 4 km mark, although whether changes in primate species richness alone coincided with this mark was not reported. Another factor likely to play a role in isolation is the nature of the intervening habitat, including forest corridors. Rylands and Keuroghlian (1988) argued that primates experience greater isolation in fragments where the surrounding felled forest is burned (so that secondary forest grows more slowly).

Third, the role of fragment age is currently difficult to assess, since the fragments in the studies discussed here have been isolated for only two to five years (Schwarzkopf and Rylands 1988) and five to thirty-five years (Estrada, Coates-Estrada, and Meritt 1994). The short time frame of the former study is particularly problematic: owing to the effects of small population size, neither the 10 ha nor the 100 ha fragments may contain viable populations in the long term, even if sufficient ecological resources are available. The only factors that might mitigate this would be minimal fragment isolation and high terrestrial mobility. Both would enhance contact with surrounding populations, permitting immigration and therefore rescue effects and recolonization.

The combined effects of patch area, age, and isolation distance have been investigated for *Alouatta palliata* populations (Estrada and Coates-Estrada 1996). The population size of *A. palliata* in forest fragments at

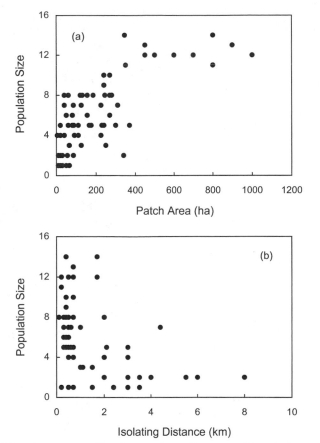

Figure 8.5 Population size of howler monkeys *Alouatta palliata* in forest fragments at Los Tuxtlas (Mexico) in relation to (a) area of forest fragment and (b) distance of forest fragment to nearest other fragment. Population size increases with area but declines with isolation. (After Estrada and Coates-Estrada 1996).

Los Tuxtlas was positively correlated with area and negatively correlated with age and isolation distance (fig. 8.5). Importantly, partial correlations were used to show that each of these parameters had an independent effect on population size, although the relationship with area was stronger than that with the other two variables.

In addition to the basic island biogeographic variables of area, isolation, and age, we might also expect habitat quality to be important, particularly for habitat specialists. The survival of tamarins in forest fragments, for example, appears to depend largely on the presence of secondary growth around the fragment edge (Rylands and Keuroghlian 1988; Schwarzkopf and Rylands 1989; see also Bernstein et al. 1976). In fact, tamarins did occur at higher density in a 100 ha forest isolate than in comparable 100 ha plots of continuous forest, although short-term refuge crowding may have contributed to this. In contrast, in forest patches along the Tana River, the number of red colobus and crested mangabeys is correlated with mean canopy height and, significantly, the shape of the patch (with patches of high area-to-perimeter ratio preferred—although the potentially confounding effects of patch area itself were not partialed out in this analysis: Medley 1993). Schwarzkopf and Rylands (1989) conducted a similar study for Amazonian forest fragments but controlled for area by comparing only similar-sized fragments (five 10 ha isolates). They found that primate species richness was greater in fragments that contained streams and more small trees and lianas but had a smaller percentage of large trees (although unfortunately the independent effects of these variables could not be teased out owing to the small sample size).

Finally, bear in mind that because of the spatial discontinuities in fragmentation processes (sec. 8.1.2), the effects of fragment area and isolation on primate populations may not correlate in any simple way with the degree of fragmentation (Boswell, Britton, and Franks 1998). Andrén (1994) has suggested that they should rather form a step function, with little influence when fragmentation is low but a very strong impact when remaining fragmented habitat falls below 30% of the original continuous habitat cover. Such a rule of thumb should be used with care, however, since critical threshold values, if they do exist, are likely to vary according to species and landscape types (Mönkkönen and Reunanen 1999; Andrén 1999).

8.3.2. Survival in the Matrix

Many primate species show great ecological and behavioral flexibility that, for some species in some circumstances, allows them to exploit the agricultural or urban environments that replace their original habitats. A critical requirement in this respect is that some small fragments of natural habitat must remain to provide a refuge for the population. In addition, species that utilize these new environments must be able to cope with their associated problems. For example, in Costa Rica squirrel monkeys suffer high mortality from both agricultural and urban processes, through both the direct and indirect effects of insecticide (i.e., poisoning and the loss of the arthropod prey base, respectively) and through electrocution on power lines (Boinski and Sirot 1997).

THE AGRICULTURAL MATRIX

Survival in the agricultural matrix may be largely dependent on the ability to obtain and subsist upon crops. Unfortunately, despite its economic importance, we know very little about crop raiding (Else 1991).

Table 8.5 Primate crop raiders

| Taxon | Number of species that raid crops | | |
	Yes	*No*	*Unknown*
Lorisiformes	0	0	11
Lemuriformes	1	0	20
Tarsiiformes	0	0	3
Callitrichidae	1	0	15
Cebidae	4	1	20
Cercopithecinae	35	2	5
Colobinae	6	6	11
Hylobatidae	0	2	4
Pongidae	3	0	1

Note: Species designated as experiencing pest control by Wolfheim 1983. Three species that do not experience pest control but that nevertheless raid crops are included as crop raiders here.

A preliminary comparative analysis of crop raiding across taxa indicates that although our basic knowledge is incomplete (very little information exists for prosimians, for example), primate species clearly vary in their propensity to raid crops (table 8.5). Most cercopithecine and pongid species eat crops, but colobine species are more evenly split and hylobatids avoid crop raiding altogether. Two patterns emerge from these data. First, folivores appear to be less common crop raiders than nonfolivores (the only howler monkey for which data exist is not a crop raider). Second, terrestrial species appear to be more likely to raid crops than arboreal species; the only cercopithecine species not reported as crop raiders are the diana monkey and the lion-tailed macaque (both strongly aboreal species). This may explain why, in Indonesia, plantations without mature trees are not used by gibbons, even though the leaf monkey *Presbytis rubicunda* will feed on the leaves of saplings there (Wilson and Johns 1982; see also Salafsky 1993).

Two recent studies provide detailed case studies of crop raiding that involve farmers cultivating on the edges of tropical forests in Uganda: Budongo Forest (Hill 1997) and Kibale Forest (Naughton-Treves 1998). At Budongo, 90% of farmers experienced damage from crop raiding, and 80% reported damage specifically by baboons. In fact, baboons were ranked highest of all crop raiding species in terms of both the damage they caused and how frequently they visited (wild pigs were ranked second, raiding 78% of farms, and vervet monkeys were third, raiding 33%). In comparison, at Kibale the three most important crop raiders, in terms of the number of farms raided, were redtail monkeys (88% of farms, 15% of field area damaged), baboons (72%, 24%), and bushpigs (72%, 15%). Other primates known to raid crops were, in order of importance: blue monkey, chimpanzee, guereza colobus, and redtail monkey at Budongo and chimpanzee, guereza colobus, and vervet monkey at Kibale. Curi-

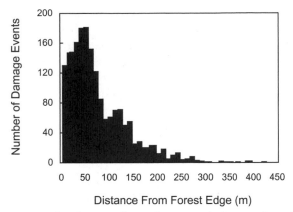

Figure 8.6 Frequency distribution of crop damage with distance from forest edge at Kibale, Uganda. Damage is recorded here for five species that together accounted for 82% of all crop raiding events: elephant *Loxodonta africana,* olive baboon *Papio anubis,* bushpig *Potamochoerus* spp., chimpanzee *Pan troglodytes,* and redtail monkey (*Cercopithecus ascanius*) (listed in order of magnitude of contribution, with elephants responsible for the greatest number of damage events). (Redrawn with the permission of Blackwell Science from Naughton-Treves 1998.)

ously, blue monkeys were present at Kibale but were not reported to raid crops.

At both sites, the frequency of crop raiding that farmers experienced was dependent on a variety of factors, including the crop currently cultivated and the proximity of the field to uncultivated habitat. Crop raiding tended to be more common in fields close to uncultivated habitat (fig. 8.6), probably because forest animals are reluctant either to stray far from the cover of trees or to visit distant fields once satiated from foraging on intervening fields. At Kibale over 90% of all raids took place within 160 m of the forest boundary; at Budongo farmers rarely suffered from crop raiding when there were two or more fields between their field and the forest edge. This emphasizes the important point that most crop raiding occurs only in a relatively narrow strip of farmland where cultivated areas are adjacent to wild habitats.

Variation in the intensity of crop raiding may reflect the relative costs and benefits of feeding on cultivated and noncultivated foods. At Kibale, chimpanzees, baboons, and redtails were more likely to forage on bananas in fields when alternative fruiting trees in the forest were not in fruit (although the abundance of forest fruit did not similarly affect primate appetites for maize: Naughton-Treves et al. 1998), and Arabian *Papio hamadryas* were more likely to raid crops in very arid areas where noncultivated foods are scarce (Biquand et al. 1992). Boulton, Horrocks, and Baulu (1996) found that the intensity of crop raiding by *Cercopithecus aethiops* in Barbados increased over a fourteen-year period owing to a reduction in food-rich areas (as the land area under cultivation decreased) while the monkey population remained constant. Conversely, if the costs increase (e.g., when the risk of predation by humans becomes too high), then crop raiding declines (Biquand et al. 1992). Maples et al. (1976) found that increased vigilance by farmers reduced raiding by *Papio cynocephalus*, halving both the number of baboons in each raid and the number of raids observed. Behavioral avoidance of farmers may also lead to changes in habitat use. Although gelada baboons rarely raid crops, they will spend more time lower down on the gorge slope when farmers are in their fields on the plateau top (Dunbar 1977).

THE URBAN MATRIX

Although it is still possible for some primates to survive within habitats fragmented by cultivation and forestry, the viability of primate populations in habitats fragmented by urban development is substantially more precarious.

In the initial stages of urbanization, traffic on the roads that separate forest fragments can be a major source of mortality as individuals or groups attempt to cross. Roadkills of red colobus monkeys at Jozani Forest Reserve (Zanzibar) may constitute an annual loss of 12–17% of the local population and therefore outweigh both habitat loss and hunting as a threat to population survival (Struhsaker and Siex 1996). Similar problems afflict populations of Angolan black-and-white colobus monkeys at Diani Beach (Kenya) (Julie Anderson, pers. comm.).

The effect of intensive urbanization on fragmented primate populations is even more severe. Indeed, Wolfheim (1983) lists only three primate species that can be commonly found in truly urban areas: the rhesus macaque, bonnet macaque, and Hanuman langur. All three species occur in India, where human population density is high but primates are often protected for religious reasons. The unusual combination of ubiquitous human settlement and protection from persecution may be pivotal in permitting urban survival for these species, but other factors might also be involved. Primary among these may be opportunities for crop raiding (each of these species is a common crop raider), since this can bring primates close to human dwellings. From this position, exploiting urban habitats may be only a relatively small step. It is notable that other primate species that have been reported to forage in urban areas, such as *Papio hamadryas* (Saudi Arabia: Biquand, Biquand-Guyot, and Boug 1989) and *Cercopithecus aethiops* (Malawi: King and Lee 1987), are also regular crop raiders.

Richard, Goldstein, and Dewar (1989) have gone so far as to propose that certain macaque species positively thrive in human habitats. These species might therefore be considered "weeds." These are to be distinguished from those species that simply survive well in secondary forest habitats, since secondary forest can arise from a variety of natural processes as well as from human activity (sec. 8.5.1). Although weed species must be able to make use of such vegetation, they must also be able to interact successfully with people. Richard, Goldstein, and Dewar (1989) suggest that this necessitates "curiosity, behavioral adaptability, an aggressive and gregarious temperament, and speed and agility on the ground." Their candidate weed species include the bonnet, long-tailed, and toque macaques as well as the rhesus macaque.

The species most strongly associated with urban areas is probably the rhesus macaque. It has been estimated that the towns and villages of India contain 60% of that country's total rhesus population (Southwick and Siddiqi 1968). However, there are differences between these two urban habitats that have important implications. The decline of village rhesus populations between 1959 and 1980 contrasts strikingly with the stability of the town populations and has been attributed to the increased defense of crops in villages (in towns, macaques scavenge on the streets and people have little interest in stopping this) and to the availability of refuges such as parks and temples in towns (in villages, such areas do not exist) (Southwick, Siddiqi, and Oppenheimer 1983). In semiurban habitats, rhesus macaque density increases in areas with a large number of roads (Shimla: Ross, Srivastava, and Pirta 1993), which probably reflects a combination of provisioning by local people and food stalls along the roads, although notably the same pattern is not seen in langurs, even though langurs are fed by humans in some areas and achieve higher birthrates in those areas where provisioning occurs (Srivastava and Dunbar 1996). Nevertheless, it should be stressed that urban areas are ultimately very poor habitats for the conservation of primate species.

8.4. Effects of Habitat Modification

Although habitat modification may have a less terminal impact on primate populations than its absolute loss, or even fragmentation, the effects of changes in floristic structure and composition can still be severe. This is particularly true where habitat modification goes hand in hand with loss and fragmentation, as it often does. In this section we examine what is known of small-scale disturbances, agriculture, and forestry on primate populations.

8.4.1. Small-Scale Disturbances

"Small-scale disturbances" include not only the traditional subsistence harvesting of naturally growing plants (especially for food and building materials), but also burning habitats and grazing livestock. Given the relatively low impact of this form of forest use, it might seem unlikely that it would be detrimental to primates. But if local people utilize a plant of ecological importance, a conflict of interest may occur. In forest patches along the Tana River, for example, some of the trees cut for canoes and honey, such as *Ficus sycamorus*, are important food resources for the crested mangabey and red colobus (Oates 1986a; Butynski and Mwangi 1994). In such small habitat patches, the loss of even one large tree can have a serious effect on the resident primate populations. Similarly, in the Amazon basin, *Chiropotes* spp. and woolly monkeys lose food resources through the destructive harvesting of trees such as *Couma macrocarpa* in the production of gums (Marsh, Johns, and Ayres 1987; Peres 1991).

Kinnaird (1992) provides a detailed account of conflict over the abundant palm *Phoenix reclinata* in the Tana River forests. Local people harvest all parts of the plant, for purposes such as food, building materials, tools, medicine, mat weaving, and basket making. Although some harvesting methods are relatively harmless to the palm (e.g., leaf collection), others are less so. The palm is also an important food resource for nonhuman primates, particularly the crested mangabey: in two forest patches, the palm constitutes 26% and 18% of the monkeys' annual diet. More important, the palm produces fruit during periods of food scarcity, contributing up to 62% of the diet in some months. Not surprisingly, there is a positive correlation between mangabey numbers and the abundance of *P. reclinata* (Medley 1993). Kinnaird considered whether human use of this palm threatens its availability to other primates by comparing forest patches with high and low levels of harvesting. Two patterns emerged: larger palm specimens are preferentially selected for harvesting, and high-intensity harvesting is associated with a change in the age structure of the palm population (larger specimens are less common than expected when harvesting pressure is most intense). Since only the larger palms are of reproductive age, this may affect the long-term viability of the palm population, with obvious consequences for the monkeys.

In general, human harvesting of natural resources seriously affects other primate populations only when it leads to major changes in habitat structure. This may be most common where burning and grazing are widespread. Fire is a natural feature of many ecosystems (Archibold 1995); consequently many plant species are fire adapted, and some plant communities require a particular burning regime for their maintenance. However, fire initiated and controlled by human agents may be inappropriate and extensive and may cause long-term damage to ecosystems (see especially Cochrane and Schulze 1998; Kinnaird and O'Brien 1998). Struhsaker and Leland (1980) noted that "bush" burning to make hunting pigs and duikers easier is a potential threat to the Zanzibar red colobus. Similarly, Starin (1989) suggested that red colobus monkeys are particularly vulnerable to fire in Senegal and Gambia. In the Moroccan fir forests of the Western Rif Mountains, Barbary macaques are threatened by deforestation associated with local people burning the understory to provide grazing for livestock (Mehlman 1984). The effects of the fire are then exacerbated by overgrazing, which prevents forest regeneration. The age structure of the forest is severely affected and its long-term viability called into doubt. In this case livestock may also directly compete for the same resources as the macaques: at Bou Jirrir (Morocco), domestic livestock feed on 78% of the tree and shrub species eaten by the macaques (Drucker 1984).

Changes in habitat structure are most likely to occur when local human population has increased or forest area has diminished. Both factors will lead to an increase in harvesting intensity per unit area of forest. For example, in the Uzungwa Mountains of Tanzania (home to the threatened red colobus subspecies *Procolobus badius gordonorum*), forest cover has progressively diminished and fragmented while local human populations have continued to grow. Consequently the effects of traditional harvesting on forest structure, particularly in pole cutting for house construction, have become severe (Rodgers and Homewood 1982). Indeed, in many areas where wood for burning and charcoal production was once collected from the dead litter of the forest floor, live branches are now removed or entire trees felled. Similar problems of tree felling and charcoal production jeopardize the last dwindling forest patches that support another threatened red colobus subspecies, *P. b. kirkii*, on the island of Zanzibar: by the early 1980s, the median size of forest patches was only 4 km^2 (Silkiluwasha 1981; Struhsaker and Siex 1996).

What do we know about the long-term impact of understory modification on primates? Harcourt and Fossey (1981) suggested that destruction caused by cattle in the Virunga Mountains (Rwanda) may have been responsible for demographic changes in the local gorilla population. These changes involved a decline in group size (presumably because gorillas in smaller groups compete less when foraging on the remaining food base: see Watts 1985) and consequently also a decline in the number of solitary males in the population (since most gorilla groups have one male). The Virunga gorillas also occurred at lower density where humans and cattle were more common, although whether this was due to forest modification was not clear. A more systematic account is supplied by Lawes (1992), who compared samango monkey populations in forest patches with and without recent signs of fire damage, woodcutting, debarking, and cattle grazing. Population density was consistently lower in disturbed forests, in two cases less than 50% of the density in undisturbed patches (fig. 8.7). Similarly, Medley (1993) found that habitat disturbance (caused by human activity, forest senescence, and damage from large mammals)

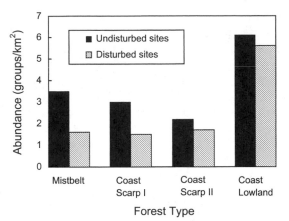

Figure 8.7 Effects of small-scale habitat modification (wood cutting, debarking, cattle grazing, and burning) on the abundance of samango monkeys *Cercopithecus mitis* in forests across Natal (South Africa), controlling for variation in forest vegetation type. (After Lawes 1992.)

reduced the number of crested mangabeys in forest patches along the Tana River, with the sympatric red colobus showing a similar but less consistent pattern.

Finally, identification of the precise mechanisms responsible for the reduced density of primates in areas of severe understory modification remains problematic. Changes in food availability may not always be the key factor, according to Ross and Srivastava (1994). In the Sariska Tiger Reserve (India), the density of Hanuman langurs was lower in areas of high disturbance (indexed by the presence of local people and the grazing of domestic animals), apparently as a result of reduced tree cover (the presumed effect of long-term understory disturbance). However, the abundance of shrubs and other plants in disturbed areas indicated that food availability was not the limiting factor; rather, the problem seemed to be a high risk of predation resulting from a scarcity of tree refuges. Support for this hypothesis came from the observation that the density of all-male groups (which have greater mobility than bisexual breeding groups that include many mothers and infants) were equally abundant in disturbed and undisturbed areas.

8.4.2. Agriculture

Agriculture can be broadly classified under two headings: permanent cultivation and shifting cultivation. Permanent cultivation generally leads to total habitat loss, although in some cases the forest understory is cleared but taller trees are left in place to provide shade for the crops planted below. In the short term, this canopy allows arboreal primates to continue using these areas. In the Western Ghats (India), for example, forests underplanted with cardamom are used by lion-tailed macaques, at least as a corridor to other forest areas, despite the removal of climbers and the midlevel trees of the canopy (Menon and Poirier 1996; Kumar 1985; Green and Minkowski 1977). Similarly, in Mexico forests underplanted with cacao and coffee are still used by howler monkeys, but not by spider monkeys (Estrada, Coates-Estrada, and Meritt 1994). Unfortunately, changes to the understory can prevent seedling recruitment, which leads to the senescence of the underplanted forest as the mature trees age and die and no younger trees replace them (Marsh, Johns, and Ayres 1987).

In contrast, shifting cultivation typically results in a mosaic of primary forest and patches of secondary forest at varying stages of regeneration. In these agricultural systems, primate populations have a greater chance of persistence since only a relatively small percentage of the land cover is actively cultivated at any one time.

Five studies of shifting cultivation systems (table 8.6) permit comparisons of species abundance in old forest and secondary forest (table 8.7). Their results indicate a great deal of variability in species responses across regions. In the Neotropics (two sites), primates appear to do badly in agricultural systems. Most primates occurred at lower density in the cultivated areas than in the nearby primary forest areas, with some taxa

Table 8.6 Five case studies of primate abundance in shifting cultivation systems

Region and site	Size of fields (ha)	Matrix still active?[1]	Size of patches (ha)	Age of regrowth (years)	Source
Africa					
Tiwai Island, Sierra Leone	1–4	No	25	5–12	Fimbel 1994a
Ituri Forest, Democratic Republic of Congo	0.5[2]	No	30[2]	5–40	Thomas 1991
Asia					
Batang Ali, Sarawak	?	No	?	30	Bennett and Dahaban 1995
The Americas					
Ponta da Castanha, Brazil	5	Yes	?	>10	Johns 1991
Selva Lacandona, Mexico	1–3	No	?	6	Medellin and Equihua 1998

Note: The area of each field (the size of field) and the total area of all fields combined at various stages of regeneration (the size of patch) are given, together with the approximate age of regrowth.

[1] If these areas are under cultivation, the agricultural matrix is described as "active."
[2] Field size from Wilkie and Finn 1988, patch size from Wilkie and Finn 1990.

Table 8.7 Patterns of primate abundance in shifting cultivation systems

Region and site	Greatest abundance observed in		
	Secondary forest	*Similar in both*	*Old forest*
Africa			
Tiwai Island	*Cercopithecus campbelli*	—	*Cercopithecus diana*
	C. petaurista		*Colobus polykomos*
	Cercocebus atys		*Procolobus badius*
	Pan troglodytes		
Ituri Forest	*Cercopithecus mitis*	*Cercocebus atys*	*Colobus angolensis*
	C. ascanius		*Procolobus badius*
	C. pogonias		
	Colobus guereza		
Asia			
Batang Ali	—	*Hylobates muelleri*	—
		Presbytis spp.	
		Macaca spp.	
The Americas			
Ponta da Castanha	*Callicebus moloch*	*Cebus albifrons*	*Saguinus fuscicollis*
	Saimiri sp.	*Pithecia albicans*	*S. mystax*
			Callicebus torquatus
			Cebus apella
			Alouatta seniculus
			Ateles paniscus
			Lagothrix lagotricha
Selva Lacandona	—	—	*Alouatta pigra*
			Ateles geoffroyi

Note: Variations in abundance are based, respectively, on: statistically significant differences in population densities (Batang Ali: Bennett and Dahaban 1995); statistically significant differences between the number of observed and expected sightings (Tiwai Island: Fimbel 1994b); survival ratios of abundance in secondary and old forest where the ratio shows a deviation of >0.50 from unity (Ponta da Castanha: Johns 1991; Ituri Forest: Thomas 1991); and species presence/absence from that habitat type (Selva Lacandona: Medellin and Equihua 1998).

completely absent from the former (e.g., *Ateles* was not seen in old fields at either site). In Asia (one site only) there were no substantial differences between forest types for either gibbons, langurs, or macaques (although the latter occurred at higher density in secondary forest, the difference was not statistically significant). In Africa (two sites), primates appear to do relatively well in agricultural systems. Here most species preferred secondary forest, although most colobines preferred old forest. *C. guereza* was an important exception in this case: this species also responds well to selective logging (sec. 8.4.3).

It is not clear whether these trends reflect genuine regional differences or simply variation between sites (differences between species at the same site will be discussed in sec. 8.5). The poor response of Neo-tropical species to agriculture might reflect historical or contemporary hunting at both the Brazilian and Mexican sites (see chapter 9); the im-pact of hunting in the Old World sites appears to have been minimal. Although the fields at Ponta da Castanha (Brazil) were also still under cultivation and surrounded by poor-quality forest, similar primate re-sponses were observed at Selva Lacandona (Mexico), where the condi-tions were more amenable (the fields were abandoned and embedded in a matrix of high-quality forest). This suggests that these other factors are unlikely to have been important. At the Asian site there may have been little difference in primate abundance in the secondary and primary for-est owing to the considerable age of the secondary forest there; after only six years at Selva Lacandona, there were still substantial differences between the vegetation in the old fields and in the surrounding forest (Medellin and Equihua 1998), but after thirty years it might be that any remaining differences in forest structure would be of negligible impor-tance. It is not clear why the African sites should have several species that showed a preference for old fields, but food resources may be an important factor. At Ituri, secondary forest is characterized by the pio-neer species *Musanga cecropioides* (Wilkie and Finn 1990; Thomas 1991): *Musanga* produces fruit throughout much of the year and is eaten by a variety of primates from prosimians to chimpanzees (Weisenseel, Chapman, and Chapman 1993; Hashimoto 1995).

The dynamic nature of species preferences for old and secondary for-est should also be stressed. Fimbel (1994a), working on Tiwai Island, has shown that the patterns of use of old and new forest can vary enormously across the year (fig. 8.8) and that these patterns tend to correlate with seasonal fluctuations in food availability. The temporary absence of *Cer-copithecus diana* from young forest coincides with an increase in fruit availability in old forest, and all guenons use young forest more inten-sively in the rainy season when old forest fruit production is low. The uniform absence of colobines from young forest may similarly be linked to the relative scarcity of leguminous trees in those areas (cf. sec. 5.2.1). However, it is notable that in those months when both habitats score equally for the abundance of fruit, the diana monkey still shows a clear preference for old forest while the other guenons do not. This suggests either that the diana monkey is more restricted in its dietary require-ments or that other factors are also involved.

Given the paucity of data describing the effects of shifting cultivation on primates, research in this area should be a high priority. The preced-ing studies suggest that hunting in and around areas of cultivation may play a crucial role in determining primate population persistence (see

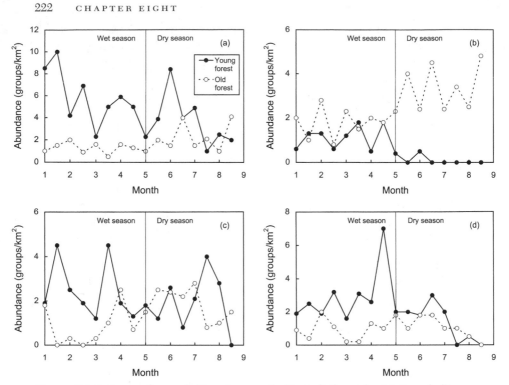

Figure 8.8 Patterns of use of old and young (cultivated) forest by (a) Campbell's monkeys *Cercopithecus campbelli*, (b) diana monkeys *Cercopithecus diana*, (c) spot-nosed monkeys *Cercopithecus petaurista*, and (d) sooty mangabeys *Cercocebus atys* on Tiwai Island, Sierra Leone. The plotted data are monthly densities (where month one is August) of sighted groups, averaged across a sixteen-month sample period. (After Fimbel 1994b.)

chapter 9). However, we can be sure that other factors related directly to habitat disturbance will also be important, especially the extent of cultivation, the distance to large tracts of primary forest, the rate of turnover of the fields (i.e., the length of the agricultural cycle), and the speed of forest regeneration in those fields. One last point to consider is that the old forest on Tiwai has itself regenerated following agriculture earlier this century. Since Tiwai supports high primate biomasses, it appears that the habitat heterogeneity that arises from small-scale cultivation need not be detrimental to primates in the long term (Oates et al. 1990).

8.4.3. Forestry

Since clear-cutting equates with forest loss, which can result only in population emigration or extinction, our concern in this section focuses on the effects of habitat modification from selective logging. We examine both the immediate and the long-term effects of logging on primate populations.

IMMEDIATE EFFECTS

Primates in areas where loggers are active are faced with a "stay or go" decision in response to the noise and associated disturbances. Which choice is preferred may depend on species-typical patterns of grouping, territoriality, and intergroup aggression (sec. 8.5).

For those species that remain, changes in behavior and ecology are inevitable. Johns (1985a, 1986) describes how pairs of white-handed gibbons at Tekam tried to stay as far as possible from the center of disturbance while remaining within their original territory (either within unlogged areas or behind the logging front as it swept through the forest). These pairs also reduced their calling frequency (a possible concealment strategy aimed at the loggers or at the neighboring groups whose territories they sometimes trespassed on), but a pair in an adjacent territory called more often (presumably because of increased territorial intrusion by groups avoiding the loggers). Similar responses were shown by banded leaf monkeys, which adopted a strategy of concealment and silence and remained largely hidden until late afternoon, after the logging teams had left. In contrast, a group of pig-tailed macaques spent more time in the area during and after logging than before it, possibly reflecting a highly opportunistic foraging strategy (sifting through the logging debris).

Once logging operations had finished, both species adopted new foraging strategies with a greater emphasis on poorer-quality foods, especially the abundance of young leaves in the regenerating forest (table 8.8). The langurs also commonly split into subgroups, apparently to maximize foraging efficiency on these new food sources. A change in activity budgets also ensued, with both taxa spending more time resting at the expense of other activities (fig. 8.9); this, it was argued, reflected the poor energy intake when animals are forced to feed disproportionately on leaves (cf. Mori 1979a). Finally, the size and the boundary positions of the home ranges were modified (table 8.8), and within the home range both species spent less time using the high canopy.

Although these data must be interpreted with care, since these observations were made on largely unhabituated animals (e.g., even before logging these gibbons spent only about 10% of their activity period feeding, compared with a norm of 22–55% in nine other gibbon populations: see Chivers 1984), some preliminary conclusions can be drawn. Johns (1985a, 1986) attributed these changes primarily to a loss of preferred food resources, which then initiated the described cascade of effects. However, other factors might also play a role. The opening of the canopy,

Table 8.8 Behavioral changes in *Hylobates lar* and *Presbytis melalophos* after logging

	Hylobates lar		*Presbytis melalophos*	
	Primary forest	Logged forest	Primary forest	Logged forest
Diet (% feeding time)				
Fig fruits	27	8	1	0
Liana fruits	17	0	1	0
Other fruits and seeds	33	52	58	42
Flowers	8	0	4	0
Young leaves	4	32	13	47
Shoots and petioles	8	4	21	11
Animal material	2	4	1	0
Ranging parameter				
Ranging area (ha)	18	13	15	19
Range overlap (%)	25	11	24	3
Old range used (%)	100	65	100	71

Source: Data are from the Tekam Forest Reserve, Malaysia (from Johns 1986).

the increased use of the lower canopy, and the fission of social groups into subgroups might all increase predation risk (Cheney and Wrangham 1987; Isbell 1994). This in turn could lead to both the observed increase in resting time (since primates may "rest" to be vigilant: Cowlishaw 1998) and the observed changes in diet and ranging patterns (to avoid dangerous areas: Cowlishaw 1997a). Such alternative explanations suggest that the causal link between habitat modification and behavioral and ecological responses might be more complex than initially envisaged. Given the uncertainties over the data, it is clear that more research is needed in this area.

The most ominous early response to logging is high infant mortality. Grieser Johns and Grieser Johns (1995) summarize data on infant:female ratios for species in the Tekam Forest Reserve at various stages during the logging process. The pattern revealed is one of almost 100% infant mortality during logging (table 8.9). This effect is common across a variety of species that differ widely in behavior and ecology. Just why logging operations should influence infant mortality in this way remains to be elucidated. Fortunately, within six months after logging, the ratio of infants to females began to improve (table 8.9).

LONG-TERM EFFECTS

In the long term, primates in logged forests must learn to cope with a habitat that has been abruptly modified in both its physical structure and its floristic composition. Coping strategies may involve marked changes

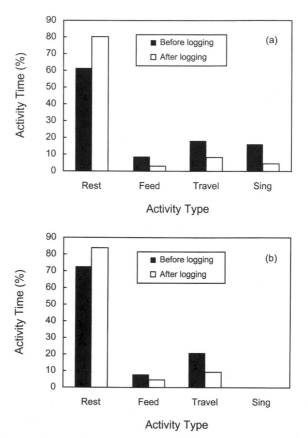

Figure 8.9 Time budgets of (a) white-handed gibbons *Hylobates lar* and (b) banded leaf monkeys *Presbytis melalophos* before and after selective logging at Sungai Tekam, Malaysia. (Data from Johns 1986.)

Table 8.9 Primate population demography after logging

	Infants per female				
		During	Years after logging		
Taxon	*Unlogged*	*logging*	0.5	6.0	12.0
Presbytis melalophos	0.41	0	0.13	0.14	0.16
Trachypithecus obscurus	0.31	0	0.09	0.20	0.09
Macaca fascicularis	0.25	0.18	0.18	0.25	No data
Hylobates lar	0.50	0	0.50	0.33	0.50

Source: Data are from the Tekam Forest Reserve, Malaysia (after Grieser Johns and Grieser Johns 1995).

in behavioral patterns, in addition to those already described, that persist over many years.

First, there is a common trend for groups to fission or become smaller in logged forests. At Kibale, *Colobus guereza* and *Cercopithecus mitis* tended to exist in smaller groups some years after logging (Struhsaker 1997). Red colobus (*Procolobus badius*) groups also appear to fission during foraging at this site. At Tekam, leaf monkeys (*Presbytis melalophos* and *Trachypithecus obscurus*) continued to forage in smaller groups eighteen years after logging (although the differences were not always statistically significant: Grieser Johns and Grieser Johns 1995). Second, interspecific interactions may also be affected by logging. At Kibale, the frequency of polyspecific associations of primate species in logged forest appears to be half that in primary forest, apparently because the antipredator benefits are outweighed by the foraging costs in an environment where normal patterns of food distribution have been severely disrupted (Struhsaker 1997). Third, in mouse lemurs, individuals are less likely to enter hibernation in logged forest, either because they find it difficult to achieve the body mass required to safely enter torpor in these habitats, or because the higher temperatures in logged forests inhibit torpid behavior, or because there are fewer large trees that provide suitable nest holes. Since mouse lemurs become torpid to save energy during the dry season, it is perhaps not surprising to find evidence that survival in these logged forests is substantially reduced as a result (Ganzhorn and Schmid 1998).

Although it is not easy to determine how these behavioral changes might influence population viability (the mouse lemur study is a notable exception), it is possible to say more about the link between various characteristics of the disturbance process and primate population abundance. In the rest of this section we examine this link with particular reference to six case studies (table 8.10), beginning with the two variables that may be of greatest importance in determining long-term primate abundance patterns in logged forest: the intensity of logging and recovery time since logging took place.

First, small-scale logging may have little long-term effect on some primate species. At Kibale, only one of seven diurnal species showed a significant decline in abundance in lightly logged forest (Skorupa 1986). Similarly, in lightly logged forest at Kirindy (western Madagascar), none of the seven lemur species present showed a decline in abundance; in fact in three cases there was a significant increase (Ganzhorn 1994, 1995; see also Smith, Horning, and Morre 1997). In contrast, at both these sites, the effects of high-intensity logging were quite different, with most species declining in abundance and some disappearing altogether (both diurnal and nocturnal species: for prosimians at Kibale, see Weisenseel, Chapman, and Chapman 1993). These results clearly suggest that the

Table 8.10 Six case studies of primate abundance in selective logging systems

Region and site	Logging intensity	Time since logging (years)	Source
Africa			
Kibale, Uganda	14–21 m³/ha	10–15	Skorupa 1986; Weisenseel, Chapman, and Chapman 1993;
Budongo, Uganda	20–80 m³/ha	2–50	Plumptre and Reynolds 1994
Madagascar			
Kirindy[1]	10% forest cover	2–6	Ganzhorn 1995
Asia			
Tekam, Malaysia	18 trees/ha, 3% of trees	18	Johns 1988; Grieser Johns and Grieser Johns 1995
Ulu Segama, Sabah	7% of trees	6–12	Johns 1992
The Americas			
Ponta da Castanha, Brazil	3–5 trees/ha	10	Johns 1991

[1]Data for lightly logged plots only.

severity of logging influences primate abundance. However, the connection between logging intensity and primate abundance is not universal: this association is absent across maroon leaf monkeys and white-handed gibbons at Ulu Segama (Johns 1992) and across blue monkeys, red-tailed monkeys, guereza colobus monkeys, and chimpanzees at Budongo (Plumptre and Reynolds 1994).

Second, the effects of recovery time on primate abundances in the Budongo Forest have been described by Plumptre and Reynolds (1994). At Budongo abundance increases with recovery time in most cases; for *Colobus guereza,* however, there is a consistent decline with length of time since logging. In contrast, no consistent patterns were found for the primates of Ulu Segama (Johns 1992). Once again there does not appear to be a universal effect.

Inconsistencies in primates' responses to logging seems to be characteristic across many of these sites. One reason for these inconsistencies is likely to be problems inherent in the research design: most of these data come from sites where different plots of primary and logged forest were surveyed at the same time. Consequently they fail to control for interplot differences in a range of important variables. Two of these are logging intensity and recovery time, whose effects we have already discussed. Another two are the area of forest subject to logging and the proximity of logged forest to primary forest. If the logged area is small relative to the home range area of a species, then it may be possible for many individuals to use both logged and unlogged forest during daily ranging patterns (e.g., chimpanzees in Kalinzu Forest, Uganda: Hashimoto 1995). If logged forest areas are distant from primary forest areas they will receive fewer immigrants (and therefore be less able to benefit from rescue effects or recolonization). For example, Johns (1991) describes how primate abundance in a 35 ha island of logged forest at Ponta da Castanha (surrounded by shifting cultivation and 500 m away from the nearest continuous forest) was lower than that witnessed in logged areas immediately adjacent to the unlogged forest.

Another reason for these inconsistencies is that species responses are at least partially influenced by changes in the availability and distribution of food resources in logged areas (Struhsaker 1997), and these can be highly variable even where logging intensity, recovery time, logging area, and isolation are all held constant. At some sites preferred food trees may also be valuable timber trees (e.g., *Chiropotes* spp. food trees in eastern Amazonia: Grieser Johns 1997); at other sites these trees may not be targeted but are still destroyed through associated damage (e.g., strangler figs are often attached to valuable timber trees: Leighton and Leighton 1983) or lost for other reasons (e.g., capuchin and howler monkey food trees are used as floats for rafting timber downstream in Amazonia: Marsh, Johns, and Ayres 1987); at still other sites, preferred food trees may be left intact.

What further complicates comparisons between sites is that some primary forests are dominated by trees that provide little food to primates. Where this situation exists, logging operations can actively improve the range of food resources available by breaking up the monodominant stands and promoting habitat heterogeneity. This appears to be why several primate species responded so well to logging at Budongo. Similarly, the growth of the colonizer *Musanga cecropioides* (which produces abundant large fruits) in logged areas of Kalinzu Forest has led to higher abundances of all primates in those areas (Howard, cited by Johns and Skorupa 1987).

Only one study provides a time series of the same area from before logging to several years after logging, thereby avoiding problems of sampling across sites with different disturbance characteristics. Grieser Johns and Grieser Johns (1995) describe trends in population density in the same forest plots at Tekam over an eighteen-year period following logging (fig. 8.10) in four different species across four different plots (three of which were free from hunting). Their results indicate that where hunting is absent the abundance of each species is increasing over time, but that there is a great deal of variation both within and between species.

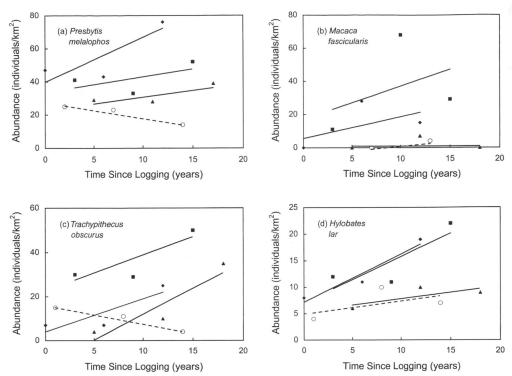

Figure 8.10 Changes in abundance with time since logging in the same plots at Sungai Tekam (Malaysia) for (a) banded leaf monkeys *Presbytis melalophos,* (b) long-tailed macaques *Macaca fascicularis,* (c) dusky leaf monkeys *Trachypithecus obscurus,* and (d) white-handed gibbons *Hylobates lar.* The plotted lines are least-squares regression lines; the dashed line indicates where hunting was also present. (After Grieser Johns and Grieser Johns 1995.)

8.5. Species Vulnerability Patterns

A common theme in the preceding material has been the bewildering variety of responses that different primates show to various forms of habitat disturbance. Although differences in extrinsic disturbance processes are likely to explain intraspecific variation in population responses (see above), intrinsic species traits are likely to play an important role in determining interspecific differences when disturbance processes are held constant. In this section we examine whether it is possible to identify the key species traits that underlie interspecific variation. First we ask to what extent human disturbance processes might mimic natural disturbance regimes, then we investigate which species traits might be associated with persistence in fragmented and modified habitats.

8.5.1. Disturbance Scales

Tropical forests are dynamic systems that naturally exhibit both temporal and spatial fluctuations in their ecology (Foster 1980; Jans et al. 1993; Archibold 1995). Temporal changes may occur cyclically (e.g., through seasonal flooding or fruiting) or at random (e.g., through hurricanes, landslides, or fires), whereas spatial heterogeneity may be associated with both geomorphology (e.g., through soil nutrient distribution) and the forest cycle itself (e.g., through gap dynamics). Hence, although some forests show less heterogeneity than others (indeed, monodominant stands do exist: see, e.g., Connell and Lowman 1989), most forest landscapes comprise an intricate patchwork of forest patches of different ages and species profiles. The Lopé Forest in Gabon, for example, illustrates this well with its highly seasonal periods of fruiting (Tutin et al. 1997) and its "complex mosaic of plant associations" comprising at least twenty forest types (White et al. 1995). We would expect species that have evolved in such dynamic ecosystems to be well adapted to their vicissitudes. Indeed, we would expect primates to perform better than most species in this respect, since their life spans are long enough to ensure exposure to a range of environmental extremes and their associated selective pressures.

In support of this argument, there is good evidence that primates can survive under a wide variety of ecological conditions in forest systems. For example, during periods of fruit scarcity, frugivorous primates are known to reduce the amount of fruit in their diet (e.g., gibbons: Leighton and Leighton 1983; long-tailed macaques: Berenstain, Mitani, and Tenaza 1986; chimpanzees and cercopithecines: Wrangham et al. 1991) and to migrate to areas where fruit is more plentiful (e.g., squirrel monkeys, white-faced capuchins, and woolly monkeys: Peres 1993b; orangutans: Leighton and Leighton 1983). Comparative studies of species' responses to fruit scarcity, in the context of keystone resource use, have been published for primate communities in Africa, Asia, and the Americas (see sec. 4.2.2). Similarly, many primate groups appear to be able to fission when food resources are naturally scarce (sec. 3.3). This capability appears to be put to common use in modified habitats where the distribution of food resources has changed and, potentially, where food resources have become more scarce in both logged forests (sec. 8.4.3) and cultivated forests (e.g., *Saguinus mystax, Pithecia albicans:* Johns 1991), as well as in naturally fragmented forests (e.g., *Cercopithecus cephus:* Tutin 1999). The latent ability of primates to survive even the most dramatic of natural disturbances is illustrated by the surprisingly limited effect that the Kalimantan Forest fires of 1982–1983 had on the local orangutan, gibbon, and long-tailed macaque populations (Berenstain, Mitani, and Tenaza 1986), although in the more severe Sumatran fires of 1997 primates fared less well (Kinnaird and O'Brien 1998).

Table 8.11 Primate species that show preferences for secondary forest

Common name	Species	Source
Colobus monkey	*Colobus guereza*	Thomas 1991
Blue monkey	*Cercopithecus mitis*	Thomas 1991
Redtail monkey	*Cercopithecus ascanius*	Thomas 1991
Crowned guenon	*Cercopithecus pogonias*	Thomas 1991
Vervet monkey	*Cercopithecus aethiops*	Kavanagh 1980; Chapman 1987b
Gorilla	*Gorilla gorilla*	Tutin and Fernandez 1984
Long-tailed macaque	*Macaca fascicularis*	Marsh and Wilson 1981
Pygmy marmoset	*Callithrix pygmaea*	Peres 1993b
Dusky titi	*Callicebus cupreus*	Peres 1993b
Collared titi	*Callicebus torquatus*	Peres 1993b
Red howler monkey	*Alouatta seniculus*	Peres 1993b
Black spider monkey	*Ateles paniscus*	Peres 1993b

Of key importance when considering a primate species' potential to respond to habitat disturbance (whether or not it is induced by humans) is the species' ability to utilize secondary forest. This is especially important where an anthropogenic disturbance process closely mimics a natural disturbance process (e.g., small fields in dispersed shifting cultivation systems can closely resemble the forest gaps that result from tree falls: see Medellin and Equihua 1998).

Many primates appear to do less well in secondary forest: Smith, Horning, and Morre (1997), for example, reported that all seven lemur species in a western Madagascar community occurred at higher densities in primary forest than in secondary forest. However, a substantial number of primate species do show a preference for secondary forest vegetation (see table 8.11; compare with table 8.7). In fact some species appear to be "secondary forest specialists" (e.g., the redtail monkey and guereza colobus: Thomas 1991) or "pioneer species" (e.g., howler monkeys: Peres 1997a). Boinski and Sirot (1997) have even suggested that a local extinction of squirrel monkeys in Costa Rica resulted from the gradual disappearance of their preferred secondary forest habitat as it matured into primary forest.

Preferences for secondary forest may at least partially reflect high rates of food production. Studies of plant populations in forests after logging suggest that plant growth can be strongly stimulated by selective removal of the vegetation. These studies have reported elevated leaf and fruit production after logging at Tekam (Johns 1988) and high rates of fruit and flower production at Ponta da Castanha (where most primates were more abundant in logged forest areas: Johns 1991). Ganzhorn (1995) also describes how logging in Kirindy Forest (Madagascar) led to an increase in fruit production and leaf quality (indexed by protein:fiber ratios) but a concomitant decrease in leaf biomass (since fewer trees were left to produce leaves). Nevertheless, it should be stressed that logging can also reduce the abundance and diversity of fruit resources, which may lead to a concomitant reduction in primate abundance (e.g., *Macaca nigra* in Sulawesi: Rosenbaum et al. 1998). This emphasizes the importance of target tree species identity and how it relates to food production and primate responses. At Kibale, for example, the basal area of primary forest food trees was reduced for *Cercopithecus mitis* following logging but was unchanged for *C. guereza*. Since the *C. mitis* population remained stable while the *C. guereza* population increased, it seems likely that productivity per unit of basal area must have increased or that alternative food sources were used (Skorupa 1986).

However, whether species that show preferences for secondary forest can sustain viable populations from using only such habitats remains to be answered. Thomas (1991) has suggested that although secondary forest may be a keystone resource for the survival of primates during periods of fruit scarcity, for many species the high energy burden of reproduction may still necessitate the use of primary forest at the peak of the fruiting season. Similarly, Weisenseel, Chapman, and Chapman (1993) have noted that although bushbabies may be attracted to some secondary growth foods (e.g., the fruits of *Musanga cecropioides*), the seasonal availability of food, together with limited availability of other essential ecological resources such as nest sites, might prevent bushbabies from existing permanently in such habitats (at least at the densities observed in primary forest). Most primate populations will probably benefit from access to both high-quality secondary forest and primary forest.

8.5.2. Habitat Fragmentation

Many ecosystems are also naturally fragmented. Blue monkeys occur in isolated forests in eastern South Africa (Lawes 1992), and the Tana River primates exist in groundwater forest patches created by the shifting meanders of the Tana River (Marsh 1986). This raises the possibility that some primate species may have evolved sufficient flexibility (or specificity) in their behavioral and ecological strategies to allow them to exploit patchy environments successfully. Tutin and White (1998) suggest that the expansive home ranges and large groups seen in chimpanzees and mandrills may be adaptations to forest habitats previously fragmented by intervening stretches of savanna. These behavioral traits allow these species to exploit such habitats today. This point serves to emphasize that although single fragments may be too small to support populations of a species, they can be embedded in a habitat mosaic in which species may persist.

Two studies have described primate species distributions in naturally fragmented habitat systems. In the first, Tutin and colleagues (Tutin et al. 1997; Tutin 1999) provide a detailed account of the use of thirteen natural forest patches (ranging from 0.5 to 11 ha in area and isolated by 0 to 450 m from continuous forest) by eight primate species at a savanna/forest border in the Lopé Reserve (Gabon). They report considerable variation in frequency of occurrence, such that in one 9 ha fragment only one species (*Cercopithecus cephus*) was resident over a seventeen-month period, but transient visits were also made by *Pan troglodytes*, *Lophocebus albigena*, *Colobus satanas*, *Cercopithecus nictitans*, and *Cercopithecus pogonias* during the same period (given in declining order of frequency; *Mandrillus sphinx* and *Gorilla gorilla* did not visit this fragment). However, when many of these taxa were present in fragments, their local abundance exceeded that in nearby continuous forest (specifically *M. sphinx*, *C. cephus*, *C. nictitans*, and *P. troglodytes*). These results suggest that fragments do not contain enough food to support most species throughout the year (across the thirteen natural forest patches, only *C. cephus* and *C. nictitans* were known as permanent residents), even though they are worth visiting at certain times. Relatively high abundances can be supported during these short periods, even though in fragments these species tend to eat rather fewer fruits and more leaves and insects than in continuous forest (fig. 8.11).

Why *C. cephus* and *C. nictitans* should be the only species to maintain a continuous high-density presence in fragments, while the remaining guenon *C. pogonias* copes particularly poorly with fragments (it is a rare visitor and occurs at extremely low densities when present), is unclear. It is possible that relatively small group sizes (and therefore small ranging areas) and high dietary flexibility might enable the first two species to be resident. Alternatively, or in addition, *C. pogonias* might suffer in competition with the other guenons in confined areas or depend on polyspecific associations with *C. satanas* and *L. albigena* for protection from predators.

In the second study, Cowlishaw (n.d.) used a modeling approach to estimate extinction rates in the Tana River forest patch system (see secs. 7.2 and 7.3). The results indicated that, in general, species with large populations in forest patches are less likely to suffer extinction in those patches (hence, in this system red colobus were always at greater risk of extinction than Syke's monkeys). In very small forest patches, however, population size effects appear to be less reliable predictors of extinction. In this case crested mangabeys are at greater risk of extinction than either red colobus or Syke's monkeys, despite similar or greater population size. One possible explanation is that mangabey groups require larger ranging areas (owing to their highly frugivorous diet) than groups of either of the other species. Finally, it appeared that a species' ability to withstand fluctuations in habitat quality are of great importance in determining population persistence in forest patches, even in large populations. This

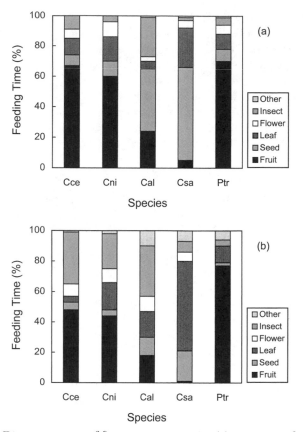

Figure 8.11 Dietary patterns of five primate species in (a) continuous forest and (b) a 9 ha forest fragment in the Lopé Reserve, Gabon. The five primate species are Cce, *Cercopithecus cephus;* Cni, *Cercopithecus nictitans;* Cal, *Cercocebus (Lophocebus) albigena;* Csa, *Colobus satanas;* and Ptr, *Pan troglodytes.* (After Tutin 1999.)

meant that the red colobus populations, characterized by low dietary flexibility, were most vulnerable to patch extinction.

Together the Tana River and Lopé studies suggest that small ranging areas and high ecological flexibility may be important determinants of a primate species' ability to persist in naturally fragmented systems in the long term. Unfortunately, less is known about those species traits associated with population persistence in habitats that have fragmented because of human activities. What we do know is largely qualitative, although we can draw some preliminary conclusions.

The most striking pattern that emerges is that a population is likely to survive in a fragment only when the area is at least as large as the natural

Table 8.12 Species residence in relation to group home range area, fragment area, and isolation date

Taxon	Home range (ha)	Number of groups in fragment				
		100 ha	10 ha			
		f1 1983	f2 1980	f3 1983	f4 1983	f5 1983
Ateles paniscus	>200	0	0	0	0	0
Chiropotes satanas	>200	0	0	0	0	0
Cebus apella	>100	1–>0	0	0	0	0
Pithecia pithecia	10	2–>1	0	0	0	0
Saguinus midas	30	4–>3	1–>0	1	0	1
Alouatta seniculus	?	5	1–>2	1	1	1

Note: Data shown for five different fragments: one of 100 ha (f1) and four of 10 ha (f2–f5). Fragment f2 was isolated in 1980, all other fragments were isolated in 1983. The group census is shown for both the number of groups at isolation and the number of groups following a second census in May 1985 (where both censuses report the same count, only one figure is given). After Lovejoy et al. 1986.

home range area of that species (table 8.12). Although species with larger ranges may be recorded in smaller patches (e.g., *Ateles geoffroyi:* Luna et al. 1987; *Pan troglodytes:* Chapman and Onderdonk 1998), this invariably reflects a temporary situation. Consequently, primates with large home ranges rarely occur in fragments (e.g., Rylands and Keuroghlian 1988; Bernstein et al. 1976), an observation that corresponds with the relatively high rates of extinction in small fragments observed in the species with the largest home range in the Tana River system (see above). We might also expect that species traits that correlate with large home ranges might also be associated with an inability to survive in fragments. This could explain why large-bodied frugivorous primates are often absent from forest fragments (e.g., mangabeys: Chapman and Onderdonk 1998), but folivorous primates that live in small groups are often present (e.g., howler monkeys: Bernstein et al. 1976; Chiarello and Galetti 1994). Similarly, Schwarzkopf and Rylands (1989) found that white-faced sakis existed only in the highest-quality fragments across five equal-area isolates, but that howler monkeys occurred in all fragments. Finally, species that have a preference for secondary forest might also do well in fragments, even where the fragment is smaller than the typical home range area, because fragments are often associated with extensive secondary forest growth (e.g., tamarins: table 8.12, sec. 8.3.1).

In comparison, a recent study of the correlates of survival of Australian rain forest mammals in fragments found that it was those species with high abundance in the surrounding habitat matrix that were more likely to persist in the fragments themselves (whereas body size, longevity, fecundity, trophic level, dietary specialization, and natural forest abundance were uncorrelated: Laurance 1991). It was concluded that this association might exist because taxa that are effective at dispersing between fragments are more likely to be able to cope with the ecological changes that occur in fragments. Similarly, Tilman et al. (1994; Tilman, Lehman, and Yin 1997) have argued that the weaker competitors in a community may show the best persistence in fragmented habitats simply because such species tend to survive in communities only by virtue of their greater powers of dispersal, and dispersal ability is likely to be crucial to long-term population persistence in such areas.

How far dispersal ability might determine primate survival in fragmented habitats, and the relative importance of other characteristics that have thus far received little attention but might also influence dispersal ability (e.g., the degree of terrestriality and territoriality), awaits a more systematic study of the primate species' traits that are associated with persistence in recently fragmented habitats.

Finally, because habitat fragmentation leads to the isolation of small populations, the genetic traits of species are likely to have important implications for long-term population viability. For example, Scheffrahn, Rabarivola, and Rumpler (1998) found that levels of genetic variation in *Eulemur macaco* were lower within fragmented habitats than in undisturbed habitats. Pope (1996) provides a case study for two highly endangered primate genera endemic to the Atlantic Forest, the lion tamarins (*Leontopithecus*) and the woolly spider monkeys or muriqui (*Brachyteles*). Both taxa now exist only in forest fragments (*Brachyteles* is now known to occur only in nineteen patches throughout its former range: sec. 10.2.3), although each would have once occurred in large stretches of continuous forest. Genetic studies of both taxa indicate that tamarins exhibit much lower levels of heterozygosity than muriquis (table 6.4). Since the Callitrichidae may have evolved under conditions of low heterozygosity (sec. 6.4), they may be better able to cope with the effects of fragmentation. In contrast, the extant populations of muriquis are not preadapted to conditions favoring homozygosity, and the future loss of genetic diversity through drift and inbreeding may well have severe effects on population viability in this species (see sec. 7.2.2).

In the medium term, however, two groups of muriquis in an 800 ha fragment isolated more than half a century ago (about seven generations) have continued to grow and appear to exhibit low rates of infant and juvenile mortality, even though the population receives no immigrants from elsewhere (Fazenda Montes Claros Forest [Brazil]: Strier 1991). Since the deleterious effects of inbreeding are typically expressed by high infant mortality, this suggests that muriquis in a population of this size may not begin to suffer from inbreeding depression for some time. Alternatively, Strier (1997) has argued that muriquis may be more resistant to fragmentation than expected because previous bottlenecks have already purged deleterious alleles from their populations. Clearly, the genetic effects of fragmentation on muriquis remain an unresolved issue.

8.5.3. Habitat Modification

Several authors have attempted to identify general principles in primate responses to habitat modification. Marsh, Johns, and Ayres (1987), for example, note that the primary rain forests of Southeast Asia are dominated by colobines, whereas in secondary forest mosaics macaque species predominate. They suggest that cercopithecines are better able to survive disturbance than colobines because they have more opportunistic diets and are more terrestrial; consequently they can more easily cope with both the changes in food availability and the loss of continuous canopy. However, this broad grouping conceals important interspecific variation between closely related taxa. For example, among the African colobines, *Colobus guereza* often increases in abundance following disturbance, but *Procolobus badius* becomes very scarce or even disappears completely.

In a study dealing with taxa at the species level, Johns and Skorupa (1987) sought to find life history and socioecological correlates that could be used to predict primate species' responses to habitat disturbance. First they calculated species *survival ratios* (abundance in disturbed forest divided by abundance in primary forest) from sites around the world where primate responses to disturbance had been studied. Then they correlated species' mean survival ratios with body size, dietary diversity (the percentage of a species' diet accounted for by the five and ten most used foods), and diet type. Their analyses detected no statistically significant association with either body size or dietary diversity but did find that survival ratios increased with the degree of folivory and decreased with the degree of frugivory (once the effects of body size and dietary diversity were removed).

Recently, Harcourt (1998) reexamined correlates of species survival ratios using methods that control both for intraspecific variation (by identifying species that tend to do consistently well or consistently badly in response to habitat modification) and for phylogeny (by using analytical methods that control for differences in the degree of relatedness between different species). Harcourt found no effect of diet type (or body mass, local abundance, global abundance and distribution, altitudinal range, or maximum latitude), but he did find that species were more likely to show higher survival ratios if they had smaller home ranges.

The inconsistencies between these studies might reflect biases in the first study that were absent in the second study (owing to the controls

Harcourt used in his analyses). However, it is difficult to draw any clear conclusions from either study, since both combine data on species' responses from different types of habitat modification (forestry and agriculture): in effect, they treat dissimilar modification types as if they were a single process. We have already seen how forestry and agriculture can have quite different impacts on tropical forest systems and how different species respond to these processes in different ways. We therefore focus here on the independent effects of agriculture and forestry on primate populations.

Only one study has systematically investigated interspecific variation in primate responses to shifting agriculture alone. In a study of the socioecological correlates of the relative use of regenerating forest by different primate species on Tiwai Island, Fimbel (1994b) found no correlation between survival ratios and adult body mass, group mass, group spread, group size, or degree of terrestriality. However, there was a strong association with diet: species with more fruit in their diets showed relatively higher abundances in secondary forest areas associated with shifting cultivation (fig. 8.12).

Much more work has been conducted on interspecific response to selective logging alone. First, in terms of behavior during logging operations, it has been argued that territoriality may be an important component of the species' response, with territorial species remaining but nonterritorial species leaving (Wilson and Johns 1982). However, territorial species have also been known to abandon territories (e.g., indris: Petter and Peyrièras 1974), and a great deal of intraspecific variability exists: for example, in some cases orangutans remain in areas where logging operations are active whereas in others they migrate up to 15–30 km to escape them (Wilson and Johns 1982 and Mackinnon 1974, respectively). Part of this variation might be explained by variation in the intensity of the logging operation, but it is also likely that it is simplistic to attribute the decision to stay solely to territoriality.

A more sophisticated socioecological explanation has been offered by White and Tutin (n.d.). At Lopé, the abundance of chimpanzee night nests rapidly declined at the commencement of logging (owing to avoidance of loggers), but gorilla nests remained unaffected. Because gorillas range in cohesive units and are not territorial, it is relatively easy for groups to cope with the intrusion of loggers by moving into areas of forest where other gorillas are present. In contrast, members of chimpanzee communities forage alone or in small parties but risk serious conflict if they invade the territories of neighboring communities. Since they are less able to move around the logging, the entire community may be displaced. If so, their fate is likely to be dire, since aggressive rivalry between neighboring communities commonly leads to the death of intrud-

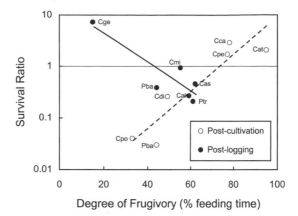

Figure 8.12 Persistence of primates following habitat modification by shifting cultivation or selective logging (at Tiwai and Kibale respectively: for details on study areas see tables 8.6 and 8.10). Persistence is estimated by survival ratios (abundance in disturbed forest divided by abundance in undisturbed forest): values above and below one indicate an increase or decline, respectively, in abundance following disturbance. These data indicate that interspecific variation in persistence is positively correlated with frugivory in cultivated areas but negatively correlated with frugivory in logged areas. Species are designated as follows: Cal, *Cercocebus (Lophocebus) albigena;* Cas, *Cercopithecus ascanius;* Cat, *Cercocebus atys;* Cca, *Cercopithecus campbelli;* Cdi, *Cercopithecus diana;* Cge, *Colobus guereza;* Cmi, *Cercopithecus mitis;* Cpe, *Cercopithecus petaurista;* Cpo, *Cercopithecus pogonias;* Pba, *Procolobus badius;* Ptr, *Pan troglodytes.* The plotted lines are least-squares regression lines. Data on persistence and diet are from Fimbel (1994a) for Tiwai cultivation and from Johns and Skorupa (1987), Rowe (1996), and Oates (1994a) for Kibale logging.

ing males (Goodall 1986). The unusually intense calling and drumming on tree buttresses at the commencement of logging strongly suggests precisely such conflict. Importantly, White and Tutin point out that this response may depend on the scale of logging. Where the logged area is much smaller than the territory of a chimp community, displacement may be less likely: in Budongo Forest, for example, logging areas are typically less than half the size of chimp territories, and chimp nesting patterns appear to be unaffected (Plumptre and Reynolds 1994).

Second, in terms of long-term interspecific abundance patterns in logged forest, Skorupa (1986) identified ecological groups that were vulnerable to selective logging. Using correlation matrix dendrograms, he identified what he termed "mature forest" core species at Kibale (Uganda). These species, which were especially adversely affected by logging, comprised *Lophocebus albigena, Cercopithecus lhoesti, Procolobus badius,* and *Pan troglodytes* (but not *Cercopithecus mitis, Cercopithecus ascanius,* and *Colobus guereza*). He found no such grouping for the Asian sites and species (*Hylobates lar, Macaca fascicularis, Presbytis melalophos,* and *Trachypithecus obscurus*) studied by Marsh and Wilson (1981).

However, these broad groupings also tend to mask finer patterns of differences in species response. To examine these finer differences, Skorupa (1986) carried out cross-species correlations of survival ratios against a range of socioecological and life history traits. He found that poor species survival was associated with large home range areas, broad group spread, and a high percentage of fruit, seeds, and flowers in the diet (fig. 8.12) (although the independent contributions of these three parameters were not ascertained). Skorupa suggested that these traits are typical of mature forest species and that these species are more vulnerable to extinction because of their wide resource monitoring patterns. These results are consistent with those obtained in the mixed-process studies described above: effects on home range area were reported by Harcourt (1998) and effects on levels of frugivory by Johns and Skorupa (1987). However, these consistencies might merely reflect the fact that these studies too were dominated by data from logged sites.

In conclusion, species persistence in modified habitats depends not only on the species in question but also on the type of habitat modification that has taken place. This finding is well illustrated by the observation that frugivorous species appear to survive particularly well in areas of shifting cultivation but particularly badly in areas of selective logging (fig. 8.12). Unfortunately, the mechanisms that underpin such different responses are currently poorly known.

8.6. Summary

1. Habitat disturbance in tropical forests arises through two routes: (a) directly through agriculture (permanent cultivation and the intensification of shifting cultivation) and forestry (selective logging), and (b) indirectly through those activities that enhance the accessibility of forests to colonization and exploitation (e.g., forestry, mining and hydroelectric projects) and, in the immediate future, global climate change. The primary driving forces behind these disturbance effects appear to be human population growth and consumption.

2. Tropical forest disturbance can lead to forest loss, fragmentation, and modification. Global forest loss was estimated at 8% between 1981 and 1990 (though with high regional variation), but this figure fails to account for substantial deforestation in previous years. Forest fragmentation has also been extensive, but the dynamic relationships between deforestation, fragment size, age, isolation, and edge effects are still poorly known. Forest modification may be the least damaging, since it is usually associated with secondary forest growth and thereby the retention of the forest's potential for long-term regeneration.

3. Habitat loss generally leads to the loss of populations and therefore species, usually through mortality or emigration. In countries that have experienced widespread deforestation, many primate species may currently exist only by virtue of a time lag between initial habitat loss and eventual population extinction.

4. The viability of habitat fragments for primate populations depends on the area, isolation, age, and quality of the fragments available. Primate populations are more likely to survive in fragments that are large in area, less isolated (although isolation effects may exist only at distances greater than 650 m), and young in age (particularly where edge effects are intense) and that have particular ecological characteristics (e.g., primary or secondary forest features, depending on the ecological requirements of the primate species in question).

5. The effects of habitat modification on primate populations are complex and appear to be largely dependent on the type and intensity of modification, the time since such modification took place, and the proximity and area of any remaining primary forest. Despite detailed accounts of the effects of habitat modification on primate abundance, little is known about the proximate mechanisms responsible for such effects, although food availability is likely to be a key factor. Several species appear to cope well in modified habitats, although they also appear to be secondary forest specialists. Habitat modification may not always lead to outright population loss, but it is usually associated at least with changes in primate community structure.

6. Broadly, interspecific variation in the ability to cope with habitat disturbance is likely to reflect at least partially how closely such processes mimic natural habitat dynamics. The species traits that might be associated with persistence in disturbed areas include (1) for survival in fragmented habitats, folivores with small home ranges, a high degree of mobility, and perhaps a history of low heterozygosity; (2) for survival in agricultural mosaics, a frugivorous diet; and (3) for survival in selectively logged forests, folivores with small home ranges. This clearly indicates that species that can cope well with one type of disturbance may be still vulnerable to another.

9 Hunting

Like habitat disturbance, hunting poses a serious risk of extinction for a considerable number of the world's primate populations (Mittermeier 1987a; Mittermeier and Cheney 1987). Moreover, it can be a substantially greater threat than many forms of habitat disturbance (Oates 1996b).

Human predation on primates has a long history. Shipman, Bosler, and Davis (1981) describe how giant gelada baboons (*Theropithecus oswaldi*) were butchered, and possibly hunted, by *Homo erectus* at one fossil site in East Africa between 400,000 and 700,000 years ago. More recently, prehistoric hunting has been implicated in the extinction of the orangutan in Java (Rijksen 1978) and, perhaps most notably, the extinction of fifteen species and eight genera of lemurs in Madagascar (sec. 7.4.1).

The most important reason contemporary humans hunt other primates is for food (e.g., Mittermeier 1987a). Subsistence hunting is widespread, but market demand for primate meat is generally a more serious threat. The threat of hunting is especially serious for primate populations in Africa and Latin America, since in Asia three of the predominant religions (Islam, Hinduism, and Buddhism) proscribe eating primate flesh (Southwick and Siddiqi 1994). Nevertheless, other Asian societies still hunt and eat primates (Mitchell and Tilson 1986; Kuchikura 1988), and the pattern of hunting across continental regions (table 9.1) suggests that hunting is as widespread in Asia as in Africa and Latin America. In Asia, however, more of the hunting is likely to be for medicinal ingredients (Jiang et al. 1991). It is also notable that, in each region, the smallest taxa (the Lorisiformes, Tarsiiformes and callitrichids) are rarely hunted while the largest (the pongids and cebids) are widely hunted. This appears to reflect a consistent preference among hunters for larger prey, a preference we discuss in some detail below.

After food, Mittermeier (1987a) suggests that the most important source of hunting mortality is the control of primates as agricultural pests (see sec. 8.3.2), particularly in Africa and Asia. Unfortunately this pressure is likely to become worse as cropland continues to replace tropical forests (Mittermeier and Cheney 1987). The intensity of such pest con-

Table 9.1 Distribution of hunting across primate taxa

| Taxon | Median extent of hunting across species | | | |
	Africa	Madagascar	Asia	The Americas
Lorisiformes	Rare (5)	—	Rare (1)	—
Lemuriformes	—	n.a. (12)[1]	—	—
Tarsiiformes	—	—	None (1)	—
Callitrichidae	—	—	—	Rare (11)
Cebidae	—	—	—	Widespread (24)
Cercopithecinae	Common (25)	—	Common (8)	—
Colobinae	Common (6)	—	Common (11)	—
Hylobatidae	—	—	Common (5)	—
Pongidae	Widespread (3)	—	Widespread (1)	—

Note: Hunting is categorized on a 4-point scale (from Wolfheim 1983), where hunting can be defined as none; rare: occurs in 1–24% of countries in the species range; common: occurs in 25–49% of countries; and widespread: occurs in ≥50% of countries. According to the criteria used, hunting includes killing primates for resources but not for pest control or live capture. The hunting pattern identified for each taxon in each region is the median across all species in that taxon-region group for which these data exist (the number of species is given in parentheses).

[1]Hunting is defined as n.a. (not applicable) in Madagascar because the majority of lemurs are restricted solely to this country; therefore even very localized hunting will lead to a "widespread" classification (in terms of the number of countries in that species' range where hunting occurs).

trol can be severe; Fitzgibbon, Mogaka, and Fanshawe (1995) estimate that the hunting and trapping of *Cercopithecus mitis* and *Papio cynocephalus* to reduce crop raiding around Arabuko-Sokoke Forest (Kenya) currently is several times the maximum sustainable harvest for these species.

The remaining incentives for hunting may generally make a relatively minor contribution to population decline, but under certain conditions they can play a more serious role for some target species. This is particularly true for the national and international trade in live primates and primate parts, as we will see below.

This chapter takes the following format. First, in order to understand the dynamics of subsistence and market hunting pressures, we introduce evolutionary foraging theory as a framework for interpretation. Second, we review some of the patterns and processes involved in the hunting of primates. Next, we investigate the domestic and international trade in primates and then examine the effects of hunting on primate behavior, ecology, and population biology. Finally, we briefly investigate the traits that might be associated with species' abilities to resist hunting pressure and the relative impact of hunting and habitat disturbance on primate populations.

9.1. Optimal Foraging Theory

Foraging theory provides a useful starting point for the study of human hunting behavior. Stephens and Krebs (1986) give a detailed introduction to the large literature on foraging theory in animals. In summary, optimality models can be used to explore the underlying determinants of hunting decisions. As with all models, simplifying assumptions are necessary in their formulation; such models remain instructive, however, since they provide an explicit and precise structure for hypothesis testing. Two basic models underlie much of this theory. Both derive from Holling's disk equation, which measures the rate of energy return in terms of two key time components (search time and handling time) that are considered mutually exclusive:

$$R = E_f / (T_s + T_h),$$

where R is the rate of energy gain and E_f is the energy gained during the total foraging time T_f (where T_f is the total time spent searching for [T_s] and handling [T_h] food items).

The two main optimality models are the prey choice model and the patch departure model. The prey choice model addresses the question whether, once a forager has encountered potential prey, it should stop to attack it or ignore it and search for more rewarding prey: in other words, the animal is viewed as being engaged in a series of "search or eat?" decisions. The patch departure model, on the other hand, is concerned with determining the point when it is most profitable to leave the current feeding patch to find the next: a "stay or go?" decision. We will not consider the second model here, since it has only rarely been applied to humans (although it might still play a useful role in explaining the use of different hunting areas around villages: e.g., Hames 1979; Kuchikura 1988; Alvard 1994).

Since we refer to the prey choice model repeatedly throughout this chapter, we will describe it in some detail. The model allows us to determine which animals a hunter will attack upon encounter, given that each potential prey species i can be characterized by its energy content (e_i), encounter rate (λ_i), and handling time (h_i). The hunter is assumed to be concerned to maximize his return rate per unit time spent hunting (his "profitability"). In the simplest case, the hunter encounters prey items belonging to only two species. If he is to maximize his return on hunting, then the prey choice model suggests that when he encounters the smaller species (low energy content, e_1) he should ignore it and continue searching for the larger species (high energy content, e_2) whenever

$$e_1 / h_1 < \lambda_2 e_2 / (1 + \lambda_2 h_2),$$

where the energy gain per unit handling time of prey species i (i.e., e_i/h_i) is defined as its profitability (for derivation, see Stephens and Krebs 1986). In effect, because the larger species provides a bigger return per unit time invested in hunting it, it is more profitable to take only individuals of the larger species and ignore individuals of the less profitable smaller species whenever they are encountered. Note that the model assumes the hunter has encountered an item of the smaller species: his decision to attack the smaller species is therefore independent of how often it is encountered (λ_1 does not enter into the inequality).

If there are more than two species, it is possible to rank all potential prey according to their profitability. Prey species should be added to the diet in order of their increasing profitability rank until

$$e_{n+1} / h_{n+1} < \sum \lambda_i e_i / (1 + \sum \lambda_i h_i),$$

where n is the number of prey species included in the diet. This inequality is termed the "prey algorithm." Once again the encounter rate of a species does not influence whether it is included in the diet, although the encounter rate of more profitable species is important. The model's prediction that hunters will either always accept or always reject prey is referred to as the "zero-one" rule. Deviations from this rule, in which some prey are inconsistently hunted, are termed "partial preferences" (Krebs and McCleery 1984). Finally, note that this model addresses only prey choice upon encounter; it does not make any predictions about the relative contributions of different species to the diet.

9.1.1. Optimality in Human Foragers

Applications of foraging theory to human behavior have met with promising results, although inevitably there are exceptions and complications with both the models' assumptions and predictions (Kaplan and Hill 1992). Based on the prey choice model, Hawkes, Hill, and O'Connell (1982) predicted that hunters should prey on only those animals whose profitability (e/h) exceeds the mean foraging return rate (E_f/T_f). For tropical forest hunters, this prediction has been explored among the Ache (Paraguay: Hawkes, Hill, and O'Connell 1982; Hill and Hawkes 1983), the Semaq Beri (Malaysia: Kuchikura 1988), and the Piro (Peru: Alvard 1993). In each case the observed hunting patterns are largely consistent with those predicted. For the Semaq Beri, all diurnal primates are profitable enough to be hunted, and all are observed in the diet (fig. 9.1). However, this case study also reports the existence of partial preferences, since primates are not always pursued on encounter, contrary to the zero-one rule. This is also commonly seen among other hunters (e.g., the Piro: Alvard 1994), where prey that should be ignored is sometimes taken. There are several reasons why partial preferences may be expressed.

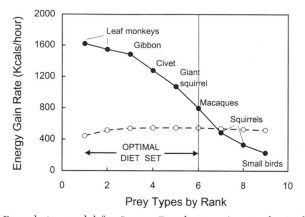

Figure 9.1 Prey-choice model for Semaq Beri hunters (peninsular Malaysia). The solid circles plot the profitability e/h (energy content per unit handling time) from hunting individual species; the open circles plot the cumulative mean foraging return rate E_f/T_f (total energy gained over total time foraging) as it changes with the addition of each prey species shown. The optimal diet set is composed of those species whose profitability exceeds that of the mean foraging return rate. Those species that do not exceed this figure (the smaller squirrels and birds) are rarely hunted by the Semaq Beri, as predicted. The leaf monkeys are *Trachypithecus obscurus* and *Presbytis melalophos*, the gibbon is *Hylobates lar*, and the macaques are *Macaca fascicularis* and *Macaca nemestrina*. (After Kuchikura 1988.)

First, partial preferences may reflect between-site variation in prey abundance. As hunters travel through different hunting areas, the density of more profitable (e/h) prey may decline such that, on encounter, a species of low profitability that was previously excluded from the diet may be included (following the prey algorithm). Hames and Vickers (1982) showed precisely this effect in Ye'kwana and Yanomamö hunters in Venezuela: where large prey had been depleted near villages, the hunters took smaller prey that were never hunted farther away.

Second, Alvard (1993) shows that this pattern is further complicated by "attack limited" hunters: where hunters are limited by the number of cartridges or arrows they have, they may prefer to save their weapons for the most profitable game and expend them on less profitable prey only when the likelihood of encountering profitable prey is low (e.g., when the hunters are almost back at their village). This pattern is shown by Piro hunters for prey such as capuchin monkeys (Alvard 1993). Moreover, this is consistent with the observation by previous authors that hunters are reluctant to shoot smaller primates: their ammunition is often limited because of the expense (e.g., Mittermeier 1987a; Peres 1991; Vickers 1991).

Table 9.2 Intensity of subsistence hunting across human societies

Region	Number of societies	Percentage of societies where subsistence hunting intensity is[1]			
		Low	Moderate	High	Very high
Africa	485	33	54	10	3
Madagascar	6	100	0	0	0
Asia	222	51	32	10	7
The Americas	125	13	32	31	24

Source: Data from the World Ethnographic Sample (P. Gray, pers. comm., after Murdock 1967). All societies listed occur within the tropics (23.5° latitude). These societies include those that are most populous in each region, although the data are not weighted for the number of people.

[1]Subsistence hunting intensity is defined as the percentage of all subsistence that is derived from hunting, where low intensity is 0–5%, moderate is 6–15%, high is 16–25%, and very high is 26%+.

A final reason for partial preferences is that the circumstances of the encounter (e.g., whether the prey detects the hunter first: Alvard 1993) can affect attack decisions. (Recall that the classic optimal foraging models were developed with "static" prey like fruits, leaves, and stationary invertebrates in mind. Avoidance behavior by the prey may violate some of the models' assumptions.)

Although model predictions may not always be met, the deviations in themselves are instructive, since they help us pinpoint other determinants of hunting behavior beyond time and energy. Undoubtedly conventional foraging models need to be modified if primates are hunted not for their energy content but for their other values, such as the monetary value of pelts or the medicinal properties of body parts. In the former case, the currency subject to gain rate maximization might be not energy but money. Nevertheless, since a predominant reason for hunting primates in traditional societies is to gain food, these models provide a useful starting point for studying hunting patterns.

9.2. Hunting Patterns

We have already seen that the extent of hunting varies between different primate taxa and geographical regions (table 9.1). This section examines in more detail both the complexity of variation in human hunting patterns and the processes that might underlie these patterns.

We begin with a broad geographical analysis of the relative contribution that hunting makes to subsistence economies in each of the major tropical regions where primates exist (table 9.2). This analysis indicates that hunting intensity tends to be low in Asia and Madagascar and high in the Americas. The factors that underlie these trends might include the availability of prey animals, the availability of alternative protein sources such as fish, the opportunities for livestock production, and the cultural traditions of the people. It is also important to remember that those cultures that do not hunt may still have an impact on local fauna if they buy wild meat from neighbors who do hunt.

An analysis of the relative contributions of different terrestrial vertebrates to the harvests collected by some of the societies in these regions suggests that mammals are the most important prey (table 9.3). Breaking down the mammalian harvest into constituent groups, it is obvious that three taxa predominate: primates, ungulates, and rodents (table 9.4). However, the relative importance of these three groups is variable. In addition, only a few species make up the bulk of the prey. The most important mammal, for example, tends to contribute about a quarter of the entire mammalian harvest (median: 24%). In only three cases is this a primate (Amerindians: *Cebus apella;* Semaq Beri: *Trachypithecus obscurus;* Siberut hunters: *Simias concolor*).

The prey choice model can provide some insight into what determines this variation in prey composition even where hunters are foraging in the same or similar assemblages. Since species profitability (*e/h*) and encounter rates (λ) determine prey choice in a given assemblage, it is probably variation in these variables that is critical. First, hunters in the same assemblage might harvest different prey because their use of different weapons leads to dissimilar handling times and encounter rates for prey. Second, even where hunters in the same assemblage use similar hunting techniques, the prey they hunt may differ because the hunters have different taboos or preferences for prey: these may in some cases be related to factors other than profitability and encounter rates. Third, the same hunter may show variable patterns of prey offtake contingent on spatial and temporal variation in prey profitability and encounter rates. All these issues are explored below. Finally, it is also likely that some variation in harvesting patterns between societies arises from differences in the prey species assemblages where the society hunts. For example, societies in the Americas probably hunt fewer mammals (and more birds and reptiles: table 9.3) because this region is relatively impoverished in terms of its large mammal fauna. Tilson (1977) has likewise argued that the low offtake of ungulates on Siberut Island (Indonesia) (table 9.4) is due to the local scarcity of these animals.

9.2.1. Hunting Methods

Hunting normally takes place during the day, but night hunting by torchlight or spotlight is also common in some areas, usually involving shotguns. In Bakossiland (Cameroon), for example, night hunting was

Table 9.3 Importance of different taxa in hunter harvests and market sales

Region and society	Mammals	Birds	Reptiles	Sources
Africa				
Efe and Lese, Democratic Republic of Congo	100	0	0	Wilkie 1989
Bubis, Equatorial Guinea	98	2	1	Colell, Maté, and Fa 1994
Fang, Equatorial Guinea[1]	100	0	0	Sabater-Pi and Groves 1972
Asia				
Semaq Beri, Malaysia	100	0	0	Kuchikura 1988
Siberut hunters, Indonesia	95	0	5	Tilson 1977
The Americas				
Amerindians, Neotropics	55	35	10	Redford and Robinson 1987
Colonists, Neotropics	68	16	16	Redford and Robinson 1987
Ye'kwana, Venezuela	30	56	14	Hames 1979
Yanomamö, Venezuela	65	27	8	Hames 1979
Tirio, Suriname	56	36	8	Mittermeier 1991
Carib, Suriname	95	0	5	Mittermeier 1991
African markets				
Rio Muni, Equatorial Guinea	95	1	4	Juste et al. 1995
Bioko, Equatorial Guinea	99	0	1	Juste et al. 1995
Bendel State, Nigeria	97	0	3	Anadu, Elamah, and Oates 1988

Note: Data are given as the percentage of carcasses.

[1] Based on percentage of informants for whom this was the most frequent meat diet.

Table 9.4 Importance of key mammal taxa in hunter harvests and market sales

Region and society	Primates	Ungulates	Rodents	Predominant[1]	Sources
Africa					
Efe and Lese, Democratic Republic of Congo	2 (2)	91 (1)	1 (3)	47	Wilkie 1989
Mijikenda and Sanya, Kenya[3]	23 (2)	6 (3)	59 (1)	36	Fitzgibbon, Mogada, and Fanshawe 1995
Bubis, Equatorial Guinea	15 (3)	37 (2)	45 (1)	27	Colell, Maté, and Fa 1994
Fang, Equatorial Guinea[4]	22 (3)	35 (2)	41 (1)	—	Sabater-Pi and Groves 1972
Asia					
Semaq Beri, Malaysia	97 (1)	0	2 (2)	74[p]	Kuchikura 1988
Siberut hunters, Indonesia	81 (1)	18 (2)	0	42[p]	Tilson 1977
Wong Garai, Indonesia	23 (2)	52 (1)	3 (3)	23	Wadley, Colfer, and Hood 1997
The Americas					
Amerindians, Neotropics	30 (1)	19 (3)	26 (2)	24[p]	Redford and Robinson 1987
Colonists, Neotropics	27 (3)	31 (2)	37 (1)	24	Redford and Robinson 1987
Ye'kwana, Venezuela	38 (1)	17 (3)	30 (2)	21	Hames 1979
Yanomamö, Venezuela	15 (3)	30 (1)	24 (2)	18	Hames 1979
Peruvians[2]	26 (2)	33 (1)	25 (3)	14	Bodmer et al. 1994
African markets					
Bakossiland, Cameroon	25 (3)	32 (2)	43 (1)	>11	King 1994
Rio Muni, Equatorial Guinea	23 (3)	43 (1)	31 (2)	30	Juste et al. 1995
Bioko, Equatorial Guinea	25 (3)	36 (2)	37 (1)	30	Juste et al. 1995
Bendel State, Nigeria	17 (3)	19 (2)	61 (1)	34	Anadu, Elamah, and Oates 1988

Note: Data are given as the percentage of carcasses unless otherwise stated; rank importance is given in parentheses.

[1] Percentage contribution made by the single most important mammal species in the harvest; a superscript *p* indicates that this species is a primate.

[2] Excludes small rodents.

[3] Most primates at the site were captured for reasons of pest control; the rodent data include two large insectivores.

[4] Based on percentage of informants for whom this was the most frequent meat diet.

practiced in about a third of all villages where hunting occurred (King 1994). Where possible, hunters will maximize their encounter rates by ambushing primate groups at locations they are known to frequent, such as fruiting trees (e.g., Iban hunters: Wadley, Colfer, and Hood 1997). Primates are usually trapped or shot. Shotguns, blowguns, and bow and arrow are the weapons most commonly used for shooting primates (the last two invariably firing poisoned projectiles), and traps may range from simple snares to elaborate cage traps (in the case of live-trapping) (Sabater-Pi 1981). Shooting appears to be the more common method, perhaps because it is hard to trap intelligent arboreal animals. This impression is supported by Muchaal and Ngandjui's (1999) finding that primates in southern Cameroon were captured more frequently in the dry season, when firearms replaced snares as the primary means of obtaining wild meat.

However, some primates involve more specialized hunting techniques. The drill is hunted with a combination of guns and dogs. The dogs are used to find drill groups, trail them as they flee, and eventually chase them up into trees; once the group is trapped, the hunters pick off individuals with guns (Gadsby 1990; Colell, Maté, and Fa 1994). Less sophisticated is the method the Hadza of Tanzania use to hunt *Papio* baboons: the hunters surround a sleeping site, dislodge the baboons by making a lot of noise and firing arrows at them, and finally club them to death as they attempt to escape the circle of hunters (Shipman, Bosler, and Davis 1981).

Many other primates are also hunted by groups rather than by individuals. Starin (1989), for example, describes communal hunts to control crop-raiding primates in West Africa. Similarly, when an Ache hunter encounters a group of capuchin or howler monkeys, he calls to other members of his hunting party to help him chase and capture them. The pursuit may be lengthy, and some hunters may climb into trees to flush the monkeys out; the capuchins are then shot only once they are stationary and the hunter has a clear view (Hill and Hawkes 1983). The Ache may also imitate calls to locate a troop and attract an animal to the hunter (as hunters are also known to do in West Africa: King 1994).

INTRODUCTION OF THE SHOTGUN

Hunting technique is dependent on the technology available. Perhaps the most important change for subsistence hunters has been the adoption of the shotgun, a particularly serious development given the potential killing power of these guns. According to both the Ye'kwana and the Yanomamö, the maximum killing distance for monkeys with traditional weapons is 17 m with blowguns and 25 m with arrows; in comparison, shotguns are effective at up to 45 m (Hames 1979). Moreover, shotguns are more likely to hit prey because of their wider projectile spread. Piro shotgun hunters need, on average, 1.3 shots per kill, whereas Machiguenga bow hunters in Amazonia require 30 shots per kill (Alvard 1995a). In terms of foraging decisions, this means that hunters armed with shotguns will be able to get within firing distance of a greater proportion of potential prey animals before they are alarmed and flee; as a result, they will have higher encounter rates with accessible prey (i.e., higher λ) than hunters without shotguns. This will increase the mean foraging return rate, which according to foraging theory will increase the minimum acceptable value of e/h for attacking a given prey species.

Hill and Hawkes (1983) explored this "acceptable profitability" prediction among the Ache hunters of Paraguay. Of the nine prey species most commonly hunted, the capuchin monkey had the lowest e/h value. Hill and Hawkes concluded that including capuchins in the diet was consistent with maximizing the rate of energy gain for Ache bow hunters but not for shotgun hunters. The model therefore predicted that only the former would prey on this species. Ache foraging patterns supported this prediction: bow hunters harvested 0.14 kg of capuchin per hunting hour (pursuing monkeys for 13% of foraging time) compared with 0.02 kg among shotgun hunters (2% of foraging time). Alvard (1993) provides more extensive data for Piro hunters (fig. 9.2); in this case it is clear that shotgun hunters consistently tend to ignore the smaller primate species that bow hunters will always pursue. Surprisingly, then, this suggests that the introduction of shotguns may actually reduce harvesting intensity for some species, at least among subsistence hunters.

For those species profitable enough to remain in the shotgun hunter's diet, however, the intensity of harvest may substantially increase. Hames (1979) found that the Ye'kwana, who hunt predominantly with shotguns, captured monkeys at a rate of three animals per 120 hours spent hunting; by comparison, the bow-hunting Yanomamö captured only one monkey per 120 hours of hunting in the same area (see table 9.4). In contrast, however, Yost and Kelly (1983) found that the Waorani of Amazonia prefer blowguns to shotguns when hunting monkeys because blowguns are silent (thus allowing several members of the same primate social group to be shot), poison darts are more effective (poisoned animals fall to the ground as their muscles relax), and shotgun shells are expensive. Consequently the median harvest of primates when using blowguns is over four times as great as when using shotguns (across the eight primate species they hunt). Yost and Kelly suggest that because the blowgun is better designed than the bow and arrow for catching arboreal prey, the adoption of the shotgun is likely to lead to an increase in hunting efficiency of arboreal prey only for hunters who previously used bows (see also Kuchikura 1988).

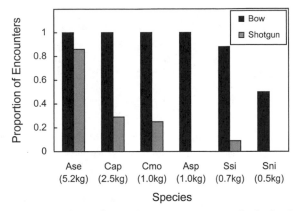

Figure 9.2 Frequency with which primate species were attacked after being encountered by Piro hunters (Peru) using bows and arrows or shotguns. Species are listed in order of body mass and designated as follows: Ase, *Alouatta seniculus;* Cap, *Cebus apella;* Cmo, *Callicebus moloch;* Asp, *Aotus* spp.; Ssi, *Saimiri sciureus;* Sni, *Saguinus nigricollis.* (Data from Alvard 1993.)

These studies indicate that when traditional hunters adopt shotguns the effects on the intensity of primate predation can be difficult to predict. Although shotguns can lead to a massive increase in hunting pressure, the presence and severity of the increase depend on the profitability of the prey and the weapon the shotgun replaces. Sustainability of harvests using shotguns is considered in section 11.2.2.

9.2.2. Food Taboos and Hunter Preferences

Food taboos can play an important role in determining hunting intensity in primate communities (Mittermeier 1987a; 1991). This is particularly true for those taboos that forbid the consumption of any primate, such as those practiced by Muslims, Hindus, and Buddhists. Yet most taboos apply only to particular species within a community, and their identity can vary between societies. Milton (1991), for example, found that each of four neighboring tribes of Amazonian Indians tabooed quite different species.

What determines taboos is difficult to ascertain. Around some villages in Ghana, monkeys are not hunted because they are believed to be the children of gods (Fargey 1992). In contrast, the Fang state that they do not hunt black colobus because its meat is dry and has a bitter taste and that they do not hunt chimps because of their close resemblance to humans (gorillas, however, are not exempt on these grounds: Sabater-Pi and Groves 1972.) Nevertheless, such proximate reasons may differ from the reason underlying the original adoption of the taboo: in many cases, the reasons offered for a taboo may be a culturally generated justification after the fact. Taboos may first arise from external constraints on hunting, such as the availability of alternative prey (i.e., the relative ranking of species according to the prey algorithm). Hence in Amazonia Peres (1990) notes that taboos against eating primates are practiced only where alternative food sources are abundant. Similarly, Waorani hunters appear to place taboos on animals that are difficult to catch (Yost and Kelly 1983).

An important point for conservation purposes is that the existence of personal food taboos does not guarantee reduced hunting pressure. In Africa, a hunter may have a taboo against eating a particular primate, but he may still kill those he meets and trade them with someone who does not share that taboo (e.g., Nigeria: Gadsby 1990; Democratic Republic of Congo: Aunger 1994). Alternatively, even when primate populations flourish owing to local taboos, hunters from outside the locality can enter the area to hunt and export the meat to other parts of the country where these species are eaten: precisely this process threatens the last stronghold of chimpanzees in Guinea (Ham 1998).

In addition, although certain taboos may be associated with particular cultural groups, within these groups individuals may show very different preference and avoidance patterns (e.g., Aunger 1994; Mittermeier 1987a). Indeed, the same person might practice different taboos at different times during his or her life: when women become pregnant in Bakossiland (Cameroon), for example, they are forbidden to eat either prosimians (for fear of giving birth to an ugly or deformed child) or *Cercopithecus erythrotis* (for fear of miscarriage) (King 1994).

The adoption and abandonment of taboos appears to be very flexible. Consequently, although some primates may be currently protected by taboos, we cannot assume that hunting will not take place in the future. A case in point is provided by Medway (1976), who reports that the Iban of Sarawak, who previously had a taboo against killing orangutans, have recently started hunting them for food (see also Rijksen 1978). The reasons for this change in attitude are not clear, although Medway suggests that scarcity of conventional game and greater access to shotguns may be implicated. Similarly, Akyem taboos are reportedly being abandoned as game animals become scarce in Ghana (Dei 1989), and Waorani taboos against eating peccary, deer, and tapir were abandoned once dogs could be used to help capture them (Yost and Kelley 1983). These latter examples confirm that the key foraging parameters identified by the prey choice model (profitability and encounter rates) may well contribute to the development of some food taboos.

It is also well known that verbal preferences for particular primates by hunters do not always match their harvesting patterns (Mittermeier 1991). For example, only 19% of Fang respondents say primates are their most common food, but 33% say primate meat is their preferred choice (with almost two-thirds of these having a preference for mandrills: Sabater-Pi and Groves 1972). Similarly, actual harvests rarely represent hunting preferences (where preferences are measured in terms of the number of attacks per encounter). Bodmer (1995a) measured hunters' preferences for eighteen mammal species (including nine primate taxa) in Peruvian Amazonia using Ivlev's index of selectivity I_s, which compares the utilization of a species in relation to its abundance. Bodmer found that hunters show greater selectivity for large prey but not for more abundant prey, although the volume of species harvested is correlated not with selectivity but with the reproductive rates of the prey species: in other words, although hunters prefer larger prey, they tend to harvest more prey with high r_m (presumably because their populations recover more quickly).

9.2.3. Temporal and Spatial Trends

Temporal variation in hunting is strongly linked to seasonality. A variety of factors might operate in this connection. First, seasonal patterns might result from variations in prey encounter rate, λ. The increase in hunting during the dry season in certain parts of West Africa (Anadu, Elamah, and Oates 1988; Juste et al. 1995) might be at least partially due to lower vegetation cover and improved access (no torrential storms), both of which are likely to increase prey encounter rates. Similarly, Waorani hunting of howler and capuchin monkeys falls off with lower encounter rates in both the wet season (when the monkeys are less active and are hidden by the noise of the rain) and immediately following it (when the monkeys are widely dispersed, since flowers and fruit are scarce) (Yost and Kelly 1983). In contrast, in Cameroon, hunting may increase during the wet season, possibly owing to the increased abundance of primates as they aggregate during the fruiting period (King 1994).

Seasonal changes in prey profitability (e/h) may also play a role. During the month when woolly monkeys are at their fattest, they are targeted by the Waorani to the extent that most other types of game are ignored (Yost and Kelly 1983; see also Peres 1991). Their larger body size increases their profitability both by maximizing e and by reducing h (fatter monkeys travel lower in the canopy and do not flee so quickly from hunters).

Other factors besides changes in hunting efficiency can contribute to seasonal hunting patterns. Predominant among these is the availability of alternative food sources. Kuchikura (1988) reports that hunting by the Semaq Beri was reduced during the hot season when more purchased food was available. In contrast, Dei (1989) reports how hunting among the Akyem (in Ghana) increased in the lean season before crop harvesting because of food scarcity. Notably, this increase was greatest among the poorest members of the society, although this may reflect not so much their own consumption as increased sale to markets (monetary income from "bush" harvests doubles during the lean season, and wealthier households make more purchases from markets during this period). Time restrictions may also be important. On Bioko Island (Equatorial Guinea) and in Cameroon, the agricultural tasks of the dry season are a major constraint since they reduce the time available for trapping (Colell, Maté, and Fa 1994 and Muchaal and Ngandjui 1999, respectively).

Spatial variation in hunting patterns may be more predictable than temporal variation. In particular, hunting intensity is often greatest close to agricultural settlements, where the incentive to hunt is at least partially driven by the need to control crop raiding. In these situations frequent or highly damaging crop raiders are the targets of hunting (including trapping), whether they are rodents (e.g., Dei 1989) or primates (e.g., Fitzgibbon, Mogaka, and Fanshawe 1995). In addition, being close to markets can encourage high levels of hunting (see sec. 9.3).

Nevertheless, the key determinant of most spatial variation in hunting is almost certainly accessibility. The shorter the travel time to a hunting area, the greater the gain in hunting per unit time expended. Consequently prey tend to become depleted along roadsides (e.g., the abundance of *Cercopithecus nictitans* declines with proximity to roads in northeast Gabon: Lahm et al. 1998) and around villages, particularly in the case of larger animals (e.g., the Neotropics: Alvard 1994; Hill et al. 1997; Alvard et al. 1997; Cameroon: Muchaal and Ngandjui 1999). Mittermeier (1991) also reported a scarcity of primates in the vicinity of two villages in Amazonia, where only *Saguinus midas* persisted within several kilometers of either settlement. Accessibility constraints explain why, for example, the number of traps set in the Arabuko-Sokoke Forest of Kenya (Fitzgibbon, Mogaka, and Fanshawe 1995) and the number of signs of hunters in the Mbaracayu Reserve of Paraguay (Hill et al. 1997) rapidly diminish with distance from the nearest forest access point (fig. 9.3). In these studies, the crucial distance beyond which hunting ceases varies between 6 km (Arabuko-Sokoke) and 12 km (Mbaracayu). This distance is in broad agreement with other studies: 2–8 km at Kibale Forest, Uganda (Naughton-Treves 1998), 8 km in western Madagascar (Smith, Horning, and Morre 1997), 10 km in both the Brazilian Amazon (Peres and Terborgh 1995) and the Serengeti (Arcese, Hando, and Campbell 1995), and 15 km in the Ituri Forest, Democratic Republic of Congo (Wilkie et al. 1998b).

Two factors are important in determining travel costs per unit distance. First, road networks open up previously inaccessible areas to hunters and make it easy to transport kills, thereby promoting market hunting

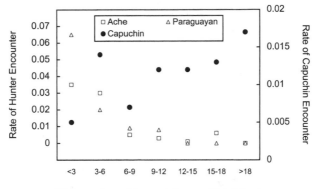

Figure 9.3 Probability of encounter with signs of hunters (of Ache or Paraguayan origin) and capuchin monkeys (*Cebus apella*) with increasing distance from the nearest access point for hunters into the protected area, in the Mbaracayu Reserve, Paraguay. As distance from the access point increases, the signs of hunters decline but the abundance of capuchin monkeys tends to increase. (After Hill et al. 1997.)

(sec. 9.3). The relatively healthy condition of primate populations in the southern forests of Bioko Island can be attributed to their isolation and to the absence of roads (Butynski and Koster 1994). Rivers serve a similar function, since they provide fast and easy access to forest areas for hunting and encourage the depletion of primate populations along river edges (see Freese et al. 1982). Second, more advanced transport can reduce travel costs and therefore increase accessibility. The adoption of the outboard motor in Amazonia has allowed hunters to forage in more distant areas of the forest where prey may be less depleted or less wary of hunters (Hames 1979).

9.3. Trade in Primates

So far this chapter has focused on the patterns and processes associated with the subsistence hunting of primates. However, among many hunters a substantial proportion of the primate harvest may be not for personal consumption but for sale at markets (e.g., Bioko: table 9.5). Such outlets, which provide a direct source of income, play an important role in determining hunting pressure. Market demand for primates becomes especially serious when it leads to professional hunters' harvesting large numbers of primates solely as a commercial enterprise. Whereas subsistence hunting for personal consumption may be sustainable at low human population densities, market hunting for commercial gain can be devastating

Table 9.5 Primate harvesting patterns on Bioko Island, Equatorial Guinea

Taxon	Body mass (kg)	Method of capture		Destination of prey		
		Trap	Gun	Sold	Eaten	Gift
Cercopithecus erythrotis	3.3	4	11	7	8	0
Cercopithecus pogonias	3.9	3	9	5	6	1
Cercopithecus nictitans	5.4	0	5	1	2	2
Cercopithecus preussi	5.5	4	0	3	1	0
Procolobus badius	10.0	0	1	0	1	0
Colobus satanas	12.5	0	3	2	1	0
Mandrillus leucophaeus	15.0	0	12	6	6	0

Source: Data from Colell, Maté, and Fa 1994.

(Mittermeier 1987a; King 1994). The trade that poses the most serious threat to primate populations is the meat trade, although trade in live primates or primate parts (such as colobus skins) can also present a very real risk of extinction for those species in high demand.

9.3.1. Trade in Primate Meat

The primate meat trade revolves around roadside stalls and local markets that often supply villages, towns, or cities with wild meat. Given the potentially large pool of consumers, the demand for this meat can be extremely high. The animals may be sold alive, as fresh carcasses, or more commonly smoked to preserve the meat for both transport to market and subsequent sale (e.g., King 1994). The route wild meat takes from the hunter to the consumer depends on the hunter (farmer or professional hunter) and the market (e.g., village or town), but it may involve a chain of intermediaries such as taxi drivers and stall owners (e.g., Bioko: Fa 1999). Although most wild meat trade is domestic, the international component can be substantial near national borders (e.g., between Nigeria and Cameroon: Gadsby 1990).

Market hunting may pose the biggest threat to the survival of many primate populations in West Africa, where "bushmeat" is extremely popular (Mittermeier 1987a). The annual value of the bushmeat trade in Nigeria alone has been estimated at between £15 million in the late 1960s and £150 million in more recent years (Anadu, Elamah, and Oates 1988). In other regions of the world, market hunting may be relatively less serious (primate meat is not considered worth buying or selling in some areas of West Kalimantan, for example: Wadley, Colfer, and Hood 1997). Nonetheless, commercial hunting of primates for meat currently threatens to become a significant threat to populations in many areas where they were previously hunted only on a subsistence basis.

The proportion of primate kills sold at market can be highly variable depending on the hunter and the region. Primate prey captured by Bubis hunters on Bioko is generally divided equally between market sale and personal consumption (Colell, Maté, and Fa 1994), although there is tendency for hunters to prefer selling larger prey (table 9.5), an apparently common pattern in western and central Africa (Fa 1999). In contrast, in southeastern Nigeria 90% of hunters do not eat any of their kill but hunt purely for local markets (Gadsby 1990). Gadsby also reports the ominous development of professional bushmeat traders who supply ammunition and high-quality weapons to local hunters in return for wholesale purchase of their kills.

As with subsistence harvesting patterns, a very small number of species make up most of the market, and these are rarely primates. In Equatorial Guinea, blue duikers and giant rats (Bioko), or blue duikers and African brush-tailed porcupines (Rio Muni), account for more than half of all market carcasses (Juste et al. 1995). Similarly, in Bendel State, Nigeria, three species (cane rat, giant rat, and Maxwell's duiker) account for 67% of the market and roadside carcasses for sale (Anadu, Elamah, and Oates 1988). This suggests that the contribution of primates to the bushmeat trade in these areas may be relatively small. In Nigeria this may partially reflect the food taboos of Muslim sections of the population. However, that a harvest is small does not necessarily mean it is sustainable (see sec. 11.2). With respect to primates, members of the genus *Cercopithecus* are consistently the most common taxon at these markets (Rio Muni: *C. nictitans;* Bioko: *C. erythrotis;* Bendel State: *C. mona*). Although many mammals show seasonal patterns of daily abundance in the markets of Equatorial Guinea, primates are relatively constant (Juste et al. 1995). The only primate that shows marked seasonality is *Cercopithecus neglectus*, which is seen in markets only during the first dry season of the year. The reasons for this temporal variation are unclear.

Market hunting can also affect the spatial patterns of hunting intensity. In the Ituri Forest, the resettlement of the Lese people along roadsides now means that the Efe communities, which trade their kills with the Lese, spend over nine months of the year hunting within 3 km of the road (Wilkie and Finn 1990). In southeastern Nigeria, Gadsby (1990) suggests that market hunting is more important to individuals in isolated villages than to those near roads. In isolated villages, the combination of a long journey to reach a road, the limited weight of agricultural produce that can be carried there, and the relatively lower value of agricultural produce per kilogram all combine to make it uneconomic for individuals to trade in anything other than meat. A similar pattern is seen around the Dja Biosphere Reserve (Cameroon), where the roads become impassable seasonally and bushmeat is the only rural product profitable enough to transport by foot (Muchaal and Ngandjui 1999). In contrast, as Gadsby notes, villages near roads can profitably trade in crops. The implications are that the primate harvest per hunter in isolated areas may be substantially greater than that near roadsides (although higher population densities close to roads means total hunting intensity can still be greatest in these areas). In Gadsby's study area there had been a reported shift over the preceding decades from subsistence hunting to market hunting as more roads were built through the region and local urban centers expanded.

Finally, the sustainability of market hunting in Equatorial Guinea was investigated by Fa and colleagues (Fa et al. 1995; Fa 1999). In both Bioko and Rio Muni, many primate species were sold at volumes beyond their predicted maximum sustainable harvests. This was particularly true in Bioko. In each region it appears that one primate taxon suffers particularly: in Bioko it is *Cercopithecus pogonias* (Fa et al. 1995) or *Mandrillus leucophaeus* (Fa 1999); in Rio Muni it is *Cercopithecus nictitans.*

9.3.2. Trade in Live Primates

The trade in live primates has both domestic and international markets. Within the country of origin, most live primates are traded as pets. In contrast, the international trade is largely driven by biomedical research, although trade in pets, zoo animals, and circus exhibits also contributes. However, live animals may also be traded for subsequent slaughter for both meat (sec. 9.3.1) and components for traditional medicines (sec. 9.3.3) on both domestic and international markets, especially in Asia (e.g., Vietnamese exports to China: Li Wenjun, Fuller, and Sung 1996; Li Yiming and Li Dianmo 1998).

Data on the severity of the domestic trade are difficult to obtain (Mittermeier 1991). Where this trade occurs as a by-product of subsistence hunting (e.g., when infants are captured as their mothers are killed), its impact might be limited to the selective targeting of mothers during normal subsistence hunting (Mittermeier 1987a; Teleki 1989), although demand for infants may intensify hunting pressure (e.g., gibbons: Wolfheim 1983). In contrast, where domestic trade in live animals is the result of demand for meat and medicinal ingredients, its impact is likely to be of the same magnitude as bushmeat hunting (sec. 9.3.1).

Data on the international trade are more widely available but are fraught with problems, since records for official exports (never mind illegal exports) are often incomplete. Moreover, the most recent attempt to synthesize what data are available is now fifteen years old (Mack and Mittermeier 1984). In the absence of more recent information, the following account of trade up to the 1980s is based on Kavanagh, Eudey, and Mack's (1987) review. For more recent regional summaries, see the following: Africa (Butynski 1996), Japan (Matsubayashi, Gogoh, and Suzuki 1986), United States (Held and Wolfle 1994).

These authors' primary finding is that the international primate trade progressively diminished between the 1960s and the 1980s. In terms of import patterns, the United States has consistently imported the greatest number of primates throughout this century. This reached its peak in the late 1950s when the development and production of the polio vaccine was at its maximum. Between 1968 and 1972, annual United States imports dropped from 127,000 to 20,000 individuals. The annual imports of the United Kingdom and Japan, the second and third largest importers, likewise declined from 30,000 to 8,000 between 1965 and 1975 (United Kingdom) and from 22,000 to 4,700 between 1972 and 1981 (Japan). These declines can mainly be attributed to two developments: technological advances combined with the increasing costs of primate subjects in vaccine development, and the widespread introduction of stringent laws to regulate traffic in live primates. The latter have been implemented both in major importing countries (e.g., the adoption of costly antirabies quarantine procedures in the United Kingdom and the prohibition of primate imports for the pet trade in the United States) and in source countries.

In terms of exports, the primary suppliers to the United States in the early 1970s were Peru, Colombia, India, and Thailand. All four countries imposed blanket bans on primate exports during the mid-1970s. The result was a reduction in the volume of primates traded, although this soon led to a switch in the identity of key primate species and the countries that exported them. By the early 1980s, the most important exporter of live primates to five of the world's largest importers was Indonesia, which produced almost twice as many primates as its nearest competitor, the Philippines. During 1978–1981 inclusive, only eleven countries contributed 1% or more of the imports recorded to the five major importing countries (Canada, Japan, the Netherlands, the United Kingdom, and the United States). Of these eleven exporting countries, 57% of the total trade was supplied by five Asian countries, with only 15% from two Latin American countries and 18% from three African countries (the remaining country, the United States, contributed 7% of the exports).

The Convention on International Trade in Endangered Species (CITES) has played an important role in regulating the live primate trade. CITES was established by an international agreement ratified in 1975, whereby signatories agreed to enact national legislation regulating the import and export of endangered animals and plants (including derivatives such as skins). If a species is threatened with extinction, it is placed in Appendix I of CITES, and international trade in that species is effectively banned. If a species might become threatened with extinction unless its trade is regulated, it is place in Appendix II of CITES, where its trade is carefully controlled (through the requirements of export and import permits) and monitored (annual trade reports are submitted to the

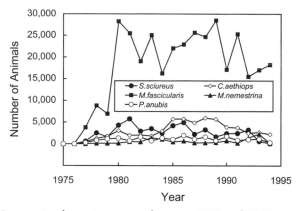

Figure 9.4 International species exports between 1975 and 1994 recorded by the CITES Secretariat for the five species that had the largest export volume between 1990 and 1994: vervet monkeys *Cercopithecus aethiops*, long-tailed macaques *Macaca fascicularis*, pig-tailed macaques *Macaca nemestrina*, olive baboons *Papio anubis*, and squirrel monkeys *Saimiri sciureus*. (Data supplied by the World Conservation Monitoring Centre with permission of the CITES Secretariat.)

CITES Secretariat). All primates are listed in the CITES Appendices, with most species in Appendix II.

Recent trends in the international primate trade recorded by CITES are summarized in figures 9.4 and 9.5. These figures describe the volume of trade for both species and countries that rank as the top five in volume during 1990–1994 inclusive. We make the following observations.

First, in recent years the volume of trade in long-tailed macaques has been almost an order of magnitude greater than that of the other four taxa that dominate the trade (vervet monkeys, squirrel monkeys, olive baboons, and pig-tailed macaques). This trade has emerged only since the late 1970s when the export of rhesus macaques from India was banned. Second, Indonesia and the Philippines share the bulk of the export trade (although the volume of exports from Indonesia has been in steady decline since the late 1980s). Since the mid-1980s the trade from Mauritius and China has gradually increased. Third, the United States remains the biggest importer of primates. In second place (with a trade volume approximately half that of the United States) comes Japan, followed by the United Kingdom, France, and the Netherlands.

An important point that these patterns stress is the dynamic nature of this trade, particularly with respect to the identity of species traded. The sobering lesson here is that a demand will always elicit a supply; national bans may simply shift the burden of demand from one species in one country to an alternative species elsewhere. Moreover, if one country continues to export primates legally, it provides a route through which

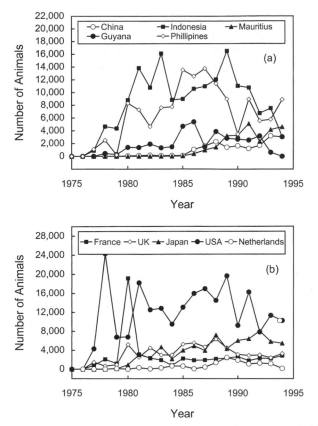

Figure 9.5 (a) Exports and (b) imports between 1975 and 1994 recorded by CITES Secretariat for those countries that had the largest trade volumes between 1990 and 1994. (Data supplied by the World Conservation Monitoring Centre with permission of the CITES Secretariat.)

primates illegally captured in other countries (and smuggled into the exporting country) can be sold. For example, after the 1974 national ban in Colombia, the trade in Colombian animals continued through Panama (Alderman 1989). Similarly, most of the chimpanzees exported from Sierra Leone come from Guinea and Liberia (Teleki 1989). An additional threat for chimpanzees is that as the international trade becomes increasingly restricted through legislation, pharmaceutical companies are establishing laboratories in source countries and obtaining chimps through the domestic market (Teleki 1989).

It is also important to appreciate that the acquisition of a single live primate does not simply reflect the loss of one individual from the wild population. There is substantial additional loss associated with mortality during capture, storage, and transport, and this may ultimately be the greatest source of population loss. In Sierra Leone, Teleki (1989) reports that five chimpanzees are killed for every infant captured. Moreover, only one infant in five survives the journey to the overseas buyer. Consequently the international trade in chimpanzee infants leads to the loss of twenty-four other chimpanzees for every infant successfully delivered. In terms of transport mortality alone, Kavanagh, Eudey, and Mack (1987) reported that about 20% of all primates imported to the United States in 1978–1979 were dead on arrival. Some species were clearly more hardy than others (macaques, for example, appear to suffer lower mortality than guenons), but it is not known why.

Finally, the degree of sustainability for this trade remains to be established. The astounding 90% crash in the size of rhesus macaque populations in India between 1959 and 1980 clearly indicates that even the most abundant and adaptable of primates can be seriously threatened by live-trapping for international markets (Southwick, Siddiqi, and Oppenheimer 1983). Ultimately, sustainability will be determined at least partially by the volume of offtake in relation to the size of the population. Hence exports from Indonesia may be relatively sustainable, since although Indonesia is the world's largest exporter it also has an immense area of tropical forest. In contrast, the Philippines exports half as many primates as Indonesia (1978–1981 inclusive) but has only 4% of that country's area of rain forest (World Resources Institute 1994).

9.3.3. Trade in Primate Parts

The market demand for primate parts has origins in both the medicinal trade and the ornamental trade. Although subsistence hunters may also use parts of their kills for medicinal and ornamental purposes (e.g., orangutans: Rijksen 1978), such small-scale personal utilization is unlikely to pose a serious threat to primate populations. Commercial demand for such products is much more serious. Indeed, in frontier trading posts between Vietnam and China, primates are sold in larger quantities than any other mammalian order, apparently primarily for medicinal purposes (e.g., Li Wenjun, Fuller, and Sung 1996; Li Yiming and Li Dianmo 1998). It is this commercial trade we focus on here.

Most hunting for medicinal purposes appears to occur in Asia (table 9.6), although primate parts are commercially sold for magical and medicinal uses elsewhere (e.g., chimpanzee skulls and prosimian skins in West Africa: King 1994). In Asia, parts traded may include the following (Wolfheim 1983): eyes (*Loris tardigradus:* love charms, cures for eye diseases); bones (*Macaca cyclopis:* aphrodisiacs); flesh, glands, and blood (*Trachypithecus johnii:* health tonics); brains (*Hylobates hoolock:* headache cure, Rabinowitz and Khaing 1998); gallstones (*Trachypithecus phayrei:* unspecified medicinal properties); and even the entire dried

Table 9.6 Primate species hunted for medicinal ingredients

	Percentage of species subject to hunting for medicines			
Taxon	*Africa*	*Madagascar*	*Asia*	*The Americas*
Lorisiformes	0 (7)	—	67 (3)	—
Lemuriformes	—	0 (13)	—	—
Tarsiiformes	—	—	0 (1)	—
Callitrichidae	—	—	—	0 (13)
Cebidae	—	—	—	4 (25)
Cercopithecinae	4 (25)	—	50 (10)	—
Colobinae	0 (6)	—	58 (12)	—
Hylobatidae	—	—	0 (6)[1]	—
Pongidae	0 (3)	—	100 (1)	—

Note: Calculated from data in Wolfheim 1983. The number of species for which information was available is given in parentheses. Cases where primates are used for magical ceremonies are not considered.

[1]Although recent reports do show that gibbons are also hunted for medicinal purposes (e.g., Eames and Robson 1993).

carcass (*Nycticebus coucang:* tonic, Martin and Phipps 1996). For any given species, little is known about the volume of this trade, its source, or its end point. However, Chinese markets may create a substantial demand for some species in certain regions, given the use of animal ingredients in traditional Chinese medicines and the vast and growing size of the Chinese population. Although some primate populations have reportedly been driven to extinction by hunting for medicinal purposes (e.g., several local populations of *Semnopithecus entellus* in India and *Trachypithecus phayrei* in Burma: Wolfheim 1983), the level of threat this trade currently poses to wild primates is unclear. However, Eudey (1987) identified several primate communities (such as those of the Guangxi region, China) as well as individual species (such as *Macaca cyclopis*) that over a decade ago were already seriously threatened by hunting for medicinal purposes.

The trade in ornamental primate products mainly involves skins, but a variety of other products are also traded, including skulls, necklaces made of teeth, stuffed individuals, and hands and feet (e.g., Amazonia: Mittermeier 1987a). The best-documented case of trade in primate skins is probably that of the black-and-white colobus monkey *Colobus guereza* (and to a lesser extent *C. angolensis* and *C. polykomos*). The following account is drawn from Oates (1977), Wolfheim (1983), and Mittermeier (1987a).

Guereza skins have traditionally been used by a variety of African cultures (such as the Kikuyu, Masai, Warusha, Watutsi, and Samburu) across a range of countries (including Kenya, Uganda, Rwanda, and Ethiopia). However, it is the intercontinental trade that has probably proved most destructive. Guereza skins have been traded out of Africa for many centuries. They were, for example, highly valued in Central Asia over five hundred years ago. In the mid-nineteenth century, the turnover in colobus skins on the European fur markets was colossal, with 1,750,000 skins being auctioned in London between 1871 and 1891. This trade continued into the twentieth century, although on a smaller scale and in quantities that varied over time with the whims of fashion. The most recent peak occurred in the early 1970s, when up to 10,000 skins a month were being sold in London, primarily to other European countries. The impact of this trade over the past two hundred years is thought to be responsible for the absence of colobus in many parts of East Africa, perhaps explaining this species' rather patchy distribution (see Oates, Davies, and Delson 1994).

A major recent outlet for colobus skins has been the tourist trade, particularly in Kenya and Ethiopia, where circular rugs (usually composed of five skins, although larger versions of up to seventy-five skins have been reported) were sold or exported to foreign markets. Surveys of tourist shops in these countries during the early 1970s revealed large numbers of rugs on display. During August and September 1972, twenty-four of sixty Kenyan tourist shops had skins in stock, and over a four-day period in 1974 a total of 982 skins were found on display in Nairobi alone. Dunbar and Dunbar (1975) estimated that in 1972 up to 200,000 guereza skins were on sale in Ethiopia. However, in all these studies the size of stocks not on display and the rate of turnover of stock proved difficult to ascertain. The impact of this trade on local populations was also difficult to determine, since the skins sold in one area might have been acquired in another; a substantial number of Kenyan skins, for example, are thought to have originated in Ethiopia. Since the 1970s, both Kenya and Ethiopia have outlawed the trade in colobus skins, and Oates and Davies (1994) do not cite this trade as a current threat to the viability of colobus monkey populations.

It is likely that the widespread distribution and the large populations characteristic of *C. guereza* played a crucial role in allowing its survival in the face of this substantial trade. Primate species that are less resilient would be unable to survive such intensive harvesting.

9.4. Effects of Hunting

In this section we examine how hunting can influence population size, population structure, and the behavior of individuals within populations; finally, we consider the consequences of hunting for the structure of ecological communities.

9.4.1. Population Size

Primate populations under continued unsustainable hunting pressure will rapidly and inevitably diminish in size. Once the population is very small, local extinction may be inevitable even once hunting has ceased (sec. 7.2). Local extinctions through hunting are known for even the most abundant species (e.g., rhesus macaques: Southwick and Siddiqi 1977), although they are likely to be most common among those species that already exist in small populations. Larger primates are therefore notoriously vulnerable to hunting, owing to both small population size (chapter 5) and hunter preferences (e.g., woolly monkeys: Peres 1990; lowland gorillas and chimpanzees: Tutin and Fernandez 1984).

Evaluations of the effects of hunting have been based largely on differences in population density between areas of high and low hunting pressure (e.g., Freese et al. 1982; Emmons 1984; Glanz 1991; Peres 1991; Puertas and Bodmer 1993) or in the same area during different periods (e.g., Sussman and Phillips-Conroy 1995). However, in such comparisons it is rarely possible to determine whether a change in abundance is the result of direct hunting mortality or due to the abandonment of an area (which might indirectly elevate mortality). Tutin and Fernandez (1984), for example, argue that both of these factors are likely to influence chimpanzee and gorilla densities in Gabon, where hunting is common. In addition, reports of reduced abundance may reflect not a genuine drop in population size but a failure to detect animals because they have become wary of humans (Hill et al. 1997).

Comparisons between sites are also easily confounded by between-site differences in natural abundance, particularly when the sites are several hundred kilometers apart. At least part of this variation is likely to arise through differences in habitat quality. Peres (1997b) has shown that intersite variation in the abundance of howler monkeys is a function of both habitat type (abundance increases with proximity to white-water rivers and structural heterogeneity of the forest canopy) and hunting pressure (abundance decreases as hunting pressure increases) (fig. 9.6). Finally, differences in abundance can occur even at the same site, owing to variation in local habitat type, season, time of day, and weather conditions. Hill et al. (1997) have shown that all these within-site factors can influence the abundance of prey populations in the Mbaracayu Reserve (Paraguay), in addition to the impact of hunting.

Nonetheless, in all the studies cited above, the abundance of most species did appear to be reduced in areas subject to hunting, and within these areas densities tended to be lowest where hunting intensity was greatest. Notably, hunting intensity frequently covaried with the accessibility of forest. In the Mbaracayu Reserve, *Cebus apella* densities are lower within 6 km of the nearest access point (Hill et al. 1997). Similarly, in the Arabuko-Sokoke Forest, *Cercopithecus mitis* densities are lower on the forest edge (where most trapping takes place) than in the forest center (Fitzgibbon, Mogaka, and Fanshawe 1995). At Mbaracayu, signs

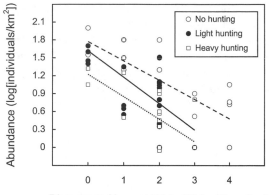

Figure 9.6 Covariation between local population density of howler monkeys *Alouatta* spp. and distance to white-water rivers, in areas of different hunting pressure across Amazonian forests. Abundance consistently declines with distance to river, but for any given river distance howler monkey density also tends to be higher where there is no hunting (open circles: dashed line), intermediate where there is light hunting (closed circles: continuous line), and lower where there is heavy hunting (open squares: dotted line). The plotted lines are least-squares regression lines. (After Peres 1997b.)

of hunters were also detected at low frequency for another 12 km (beyond the first 6 km distance), where the abundance of *C. apella* appeared to be relatively high, suggesting there may be no significant impact on prey abundance when hunting intensities are low (Hill et al. 1997).

It is possible that in some cases hunting is sustainable. If so, the population may be below carrying capacity but it will nevertheless continue to exhibit relatively constant numbers even while under hunting pressure. Indeed, that population density is lower in an area where hunting occurs does not necessarily imply that hunting is unsustainable (Robinson and Redford 1994). Whether primate harvests are sustainable is the subject of the following subsection.

SUSTAINABLE HARVESTS

The classic approach to this problem has been to compare offtake with recruitment (e.g., Caughley 1977; Caughley and Gunn 1996): if a population is to remain stable in numbers, then the harvest must not exceed the increment in growth, which is directly proportional to r_m. This makes r_m a trait of fundamental importance in determining whether a primate population will be able to recover from hunting. However, although this is a simple principle, it is complicated by the fact that population growth is usually density dependent.

The logistic equation for population growth (see sec. 6.2.2) has the useful property that it can be used to estimate the population size that will give the optimal yield (i.e., the maximum number of animals hunters can crop without leading to negative population growth). We do this by differentiating equation 6.3 and setting the derivative equal to 0. This yields $N = K/2$. In other words, a population size at half the carrying capacity shows maximum population growth and will allow us to maximize the offtake from the population. If this yield is constantly exceeded, however, the population will eventually become extinct.

Unfortunately, owing to the effects of environmental variation, populations rarely behave according to the logistic growth curve (sec. 6.2.2). In fact, stochastic models of population growth have generally found that population fluctuations reduce the maximum tolerable harvest below the deterministic maximum sustainable yield suggested by equation 6.4 (May 1994), emphasizing that caution should be employed in using deterministic models to set harvesting quotas. More recently, Lande, Engen, and Saether (1994, 1997) have shown that the time to extinction for a fluctuating population is maximized with a harvesting strategy characterized by zero offtake when the population is below K but maximal offtake when it is above K. Sustainable harvesting strategies may therefore exhibit an "on/off switch" nature, where the harvest rate must vary if yields are to be optimized and the risk of population extinction minimized.

So far, attempts to model hunting sustainability in wild primate populations (e.g., Crockett, Kyes, and Sajauthi 1996) have not used such sophisticated techniques. At present the most widely used method is the "population growth model" of Robinson and Redford (1991). This simple model bears some resemblance to the deterministic logistic model, although there are important differences. In this model the maximum potential production of a population per unit area is calculated from population density and r_m; however, maximum growth rate is assumed to occur at $0.6K$ rather than $0.5K$, and maximum harvest rates are modified by the amount of natural mortality that would be expected (larger harvests are allowed for short-lived species). Because primates are long-lived with unusually low reproductive rates (sec. 3.1), this model predicts that their maximum sustainable harvests are substantially below those of other large mammals. Furthermore, among primates larger species tend to have a lower potential harvest for the same reasons (fig. 9.7).

Recently, Slade, Gomulkiewicz, and Alexander (1998) developed an alternative model that allows for incorporating more detailed demographic data where this is available, including survival to age at first reproduction (a parameter they found to be of great importance). They noted that population growth rates estimated with their model were

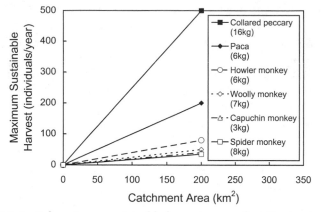

Figure 9.7 Potential maximum sustainable harvests across four Neotropical primate species (howler monkey *Alouatta*, woolly monkey *Lagothrix*, capuchin monkey *Cebus*, and spider monkey *Ateles*), in comparison with those observed for a Neotropical ungulate (peccary *Tayassu*) and rodent (paca *Agouti*), according to calculations using the Robinson and Redford model (1994). If the harvest exceeds the number shown for a given area, the harvest is predicted to become unsustainable. Note that the rodent and ungulate are about the same size as the primates listed, or larger, yet can still sustain a much larger harvest. Approximate body masses are estimated from Smith and Jungers (1997) for primates and Jorgenson (1995) for other taxa. (After Robinson and Redford 1994.)

more accurate than those obtained using Robinson and Redford's model; and that Robinson and Redford's model was liable to underestimate the probability that a species was being overharvested. However, Slade, Gomulkiewicz, and Alexander's model deals with "standard" rates of population growth rather than the "maximal" rates of population growth (r_m) that are the focus of Robinson and Redford's model. Consequently the choice of model for analyses of potential sustainability is not straightforward. In addition, the two models have common limitations. In particular, neither incorporates stochasticity or the influence of immigration and emigration; in addition, there is no satisfactory method of establishing the carrying capacity K.

Nevertheless, studies that have compared observed offtake with a theoretical maximum sustainable yield have consistently found that the hunting of primates for personal subsistence (Alvard et al. 1998), the bushmeat trade (Fa et al. 1995; Fa 1999), and the control of crop raiders (Fitzgibbon, Mogaka, and Fanshawe 1995) is not sustainable for many species. These findings are especially worrying because the figures for offtake in these studies are probably underestimates. This is likely for two reasons. First, in the case of market hunting, offtake estimates based on the number of carcasses at the market do not take into account the

Table 9.7 Effects of cropping on primate populations

Parameter and species	Date of cropping	Value at cropping			Date of recensus	Value at recensus
		Before	*After*	*Loss*		
Total population size						
Aotus nancymae (1)[1]	1981	71	29	−59%	1984	78 (+169%)
Aotus nancymae (2)	1984	78	12	−84%	1987	64 (+433%)
Saguinus mystax	1978	61	21	−66%	1982	45 (+214%)
Saguinus fuscicollis	1978	>63	>57	−10%	1982	59 (+104%)
Mean group size						
Aotus nancymae (2)	1984	3.7	—	—	1987	3.4
Saguinus mystax	1978	5.1	—	—	1981	5.2
Saguinus fuscicollis	1978	7.0	—	—	1981	5.5
Percentage of population comprising immatures[2]						
Aotus nancymae (2)	1984	33	—	—	1987	32
Saguinus mystax	1978	24	—	—	1981	46
Percentage of population comprising adult females[2]						
Aotus nancymae (2)	1984	35	—	—	1987	32
Saguinus mystax	1978	38	—	—	1981	25
Percentage of adult females breeding[2]						
Saguinus mystax	1978	50	—	—	1981	100

Source: Data for *Aotus nancymae* from the Tahuayo River, Peru (Aquino and Encarnación 1994). Data for *Saguinus mystax* and *S. fuscicollis* from the Yarapa River, Peru (Ramirez 1984).

[1]Data for successive cropping events in the same population of *Aotus nancymae*.
[2]Calculated from groups that were captured in their entirety.

kills that hunters keep for personal consumption (see table 9.5). Second, many animals are mortally wounded by hunters but manage to escape before they die (e.g., *Gorilla gorilla*: Harcourt, Stewart, and Inahoro 1989; *Lagothrix lagotricha*: Peres 1991); this may be particularly common with shotgun hunting, where the spread of shot is wide, but it may also occur when animals escape from snares (Noss 1998). Consequently, mortality in prey populations from hunting may greatly outstrip the actual number of animals captured, making sustainability a remote prospect.

Nonetheless, as theory would predict, there is evidence of sustainable harvesting for some primate populations. Studies of three Neotropical primate populations that have been commercially harvested to obtain live primates (table 9.7) indicate that effective recovery can occur within three to four years after the loss of several social groups from the population (*Saguinus mystax, S. fuscicollis*), even following successive harvests

(*Aotus nancymae*). However, the recovery pattern of S. *fuscicollis* was much slower than that of sympatric S. *mystax*, even though it was harvested at lower intensity. S. *fuscicollis* also exhibited poorer postcropping recovery than the more intensively harvested S. *mystax* at two other Peruvian sites (Glander, Tapia, and Fachin 1984). The reason for the greater vulnerability of this species is not clear, but since the two species often form polyspecific associations (sec. 4.3.1), it is possible that S. *fuscicollis* depends on a minimum abundance of S. *mystax* groups, perhaps because S. *mystax* provides S. *fuscicollis* with some protection against predators (Ramirez 1984).

Unfortunately, the mechanism of recovery in these studies is not always clear. This is important, because if recovery is the result of immigration by individuals from neighboring areas where there was no cropping (effectively a metapopulation source-sink relationship develops) rather than high rates of reproduction in the areas where cropping took place, it means that harvesting may be sustainable only in localized areas of a single continuous population rather than across the entire population as a whole. In the *Aotus* population, the authors were able to establish that both immigration and reproduction played a role in population recovery. Although it was difficult to determine their relative importance, immigration appeared to be more important after the second (1984) cropping (Aquino and Encarnación 1994). In the S. *mystax* population, the role of immigration was similarly difficult to ascertain, although the reproductive rates of females did increase following harvesting (table 9.7).

However, immigration and emigration cannot be a confounding factor in the population responses of *Cercopithecus aethiops* to cropping on the Caribbean island of Barbados. Here at least 10,000 monkeys were harvested over a fourteen-year period (1980–1994) from an introduced wild population (an average of 670 monkeys per year). Yet despite the intensity of the harvesting, the population grew by about 4% to almost 15,000 animals during the same period (Boulton, Horrocks, and Baulu 1996). In this case at least, there is clear evidence that extensive livetrapping can be conducted sustainably across a primate population.

Although the response to harvesting that these populations show could be interpreted with optimism, note that all these species are characterized by high ecological flexibility and small body size, both of which are associated with high r_m (see secs. 6.2.1 and 3.1). Consequently these are precisely the species we would expect to respond relatively well to cropping. Such patterns do not apply to most other primates (see sec. 11.2.2 for further discussion).

9.4.2. Species Population Structure

Hunting may precipitate substantial changes to population structure, which in turn may bring about genetic changes in the population. Modeling studies of ungulate populations demonstrate that a rapid loss of heterozygosity can result from inappropriate harvesting strategies (even where the total population size remains constant: Ryman et al. 1981) through changes in both the genetic effective population size (N_e) and, particularly, the generation interval (defined here as the average age of the mothers of the young born in any given year). The precise patterns that emerge are strongly dependent on the age-sex class selection of prey. In the following review, we examine how hunting may alter the age-sex composition and grouping patterns of a population and the consequences of such effects on social structure.

AGE-SEX CLASS COMPOSITION

Mortality from hunting is rarely distributed equally across all age-sex classes. This may be partly because some individuals are more conspicuous than others and therefore more easily detected and killed (e.g., males that give loud calls). Since most primates are group living, however, the hunter can usually choose his target from the entire group once the most conspicuous group member has been detected. Perhaps most commonly, the value of live infants encourages many hunters to target lactating females when hunting. The mothers are then eaten by the hunter or sold as bushmeat, while the infants, if they survive, are sold as pets (e.g., woolly monkeys: Peres 1991; chimpanzees: Teleki 1989). The result is a strongly male-biased population. For species in which males have the greater economic value, the reverse pattern can occur. Dunbar (1977) describes the effects of periodic selective hunting of prime age male gelada baboons by Galla tribesmen in Ethiopia to acquire male capes for use as ceremonial regalia. These harvests lead to the near disappearance of prime age adults from the population at the time of the harvest and so bring about dramatic changes in grouping patterns and social structure (see below).

The effects of live trapping on primate populations have been extensively investigated by Southwick and colleagues (e.g., Southwick and Siddiqi 1977, 1994). In the late 1950s, India exported more than 100,000 rhesus macaques annually for the international biomedical trade. During the 1970s the trade diminished, and it eventually ceased after India imposed an export ban in 1978 (sec. 9.3.2). Throughout this period, Southwick and colleagues made detailed observations on two rhesus macaque populations in Aligarh district, Uttar Pradesh (one unprotected population subject to trapping, the other semiprotected by local people). A comparison of the demography of the two populations (table 9.8) sug-

Table 9.8 Rhesus monkey demography in groups subject to heavy and moderate trapping

Demographic parameter	Heavy trapping	Moderate trapping
Natality (Percentage females giving birth per year)	76	90
Percentage immature animals in population	43	54
Infant losses per year	19	15
Juvenile losses per year	58	32
Adult losses per year	18	13

Source: After Southwick and Siddiqi 1977.

gests that live-trapping fell particularly heavily on juveniles, which trappers probably preferred because their smaller size made transport easier.

As a result, populations across India showed a marked deficit in the number of juveniles at this time. In 1959–1960, only 6% of the population was composed of juveniles (compared with 25% infants). After trapping stopped this deficit diminished so that by 1990–1991 juveniles accounted for 25% of the population (while the infant contribution remained the same). The loss rates in the semiprotected population shown in table 9.8 probably reflect those of natural mortality (although some limited trapping may also have occurred). The loss of adults in the semiprotected population was heavily biased toward males, probably indicating male dispersal rather than mortality; in the unprotected population, males and females were lost at equal rates. Note that the birthrate in the unprotected population was considerably lower than in the semiprotected population. This suggests that the disturbance associated with trapping alone may be sufficient to reduce the reproductive rates of females that are not trapped, perhaps through stress-induced reproductive suppression (sec. 3.3.1).

In contrast, Horrocks and Baulu (1988) found that males, particularly adults, were the most vulnerable to live trapping in a vervet population on Barbados, whereas females, particularly juveniles, were the least vulnerable. Consequently, although total population size remained constant over a six-year trapping period, the proportion of juveniles in the population, particularly female juveniles, increased. However, an increase in the actual numbers of juveniles, in addition to proportional changes in age-sex structure, suggests there may have been a concomitant increase in infant survivorship.

SOCIAL STRUCTURE

There are two basic reasons we might expect group size to change in response to hunting. On the one hand, if hunting pressure is high, it is plausible that groups will be smaller because mortality rates in each group will be higher. Precisely such an effect is observed in red colobus groups under heavy predation by chimpanzees (Stanford 1995). On the other hand, if group size is flexible in response to hunting, then we might expect groups to become either smaller (if this reduces the risk of detection by hunters) or larger (if this reduces the risk of attack by hunters).

The empirical evidence suggests that smaller groups are more common in areas of heavy hunting. Drill groups have become smaller and the large ephemeral congregations of these groups more uncommon in recent years in Nigeria (Gadsby 1990; Gadsby, Feistner, and Jenkins 1994). However, it remains unclear whether this change was due to hunting (and if so, how it was brought about) or to habitat disturbance, since both hunting and habitat disturbance are serious threats to drill populations throughout their range. Similarly, Watanabe (1981) found that where hunting pressure was high, snub-nosed langur (*Simias concolor*) groups were typically half the size of those where hunting was minimal, but once again it was difficult to control for confounding factors such as habitat disturbance and population density. In this case the variation in group size was also associated with differences in social system: the smaller groups were monogamous rather than polygynous. If hunting pressure was responsible for this difference, it seems that hunting also has the potential to introduce long-term genetic changes in populations though its effects on mating systems.

Smaller groups are not always associated with high hunting pressure, however. For example, Thomas (1991) reported that the group sizes of primates in the Ituri Forest (Democratic Republic of Congo) did not differ between areas where hunting was present and absent. In some cases groups may even become larger, although not necessarily as a defense against hunters or other predators. Dunbar (1977) describes how the selective hunting of adult male gelada baboons leads to an increase in the size of one-male units. Although large units are normally unstable (they are more frequently targeted for takeover: Dunbar 1984), the shortage of nonbreeding males that can take over and break up large groups results in group size drifting upward as females mature into their natal units and are unable to leave. This size increase may be further compounded by fusions with one-male units whose resident males have been shot. A further effect of the shortage of adult males is that subadult males now have the opportunity to mate in these units, often containing two or three times the number of females typical in areas where hunting does not occur. With time, the social structure of the population approaches normality as the males mature. But since the Galla shoot gelada males at eight-year intervals, the population has only a short period of tranquillity before the cycle starts again.

Finally, selective culling of groups or group members may also affect social structure. In the commercial cropping of live tamarins, entire groups are trapped, but then the adult breeding pair may be released back into the wild. Although removing the entire group may require trapping the fewest groups to reach the hunting quota (thereby reducing stress on other members of the population: Ramirez 1984), the removal of complete breeding groups is likely to lead to the erosion of population heterozygosity. In the short term, however, the choice of harvesting method used did not appear to affect the recovery of *Saguinas mystax* at one Peruvian site (Glander, Tapia, and Fachin 1984).

9.4.3. Behavioral Patterns

From the perspective of the primate, human hunters are as much predators as are raptors and carnivores. Consequently, primates might be expected to respond to the selective pressure of human hunting in the same way they respond to other predators, adopting behavioral and ecological strategies to minimize their risk of detection, attack, and capture.

Several authors have suggested that human predation affects patterns of vocalization. Watanabe (1981) reported that in areas of heavy hunting on Siberut Island, snub-nosed langur *Simias concolor* groups do not perform their conspicuous dawn loud calls, apparently to avoid detection by hunters. Similarly, Tenaza (1976) suggested that human hunting of gibbons *Hylobates klossii* on Siberut led offspring to separate from their mothers, and males to act as lookouts, during female song bouts. However, since interpopulation differences in patterns of singing behavior may be strongly influenced by other aspects of ecology and demography (see Cowlishaw 1992, 1996), it is difficult to confirm unequivocally that hunting pressure is responsible.

Nevertheless, Tenaza and Tilson (1977) described two *H. klossii* alarm calls (sirening and alarm trills) that were performed only after detecting human hunters, the animals' chief predators. Unlike their other vocalizations, these calls are as loud as territorial songs and can therefore be heard from neighboring territories. Tenaza and Tilson suggested that gibbons that have detected hunters are alerting not only other members of their group but also kin in neighboring groups. Vervet monkeys also have a special alarm call that is elicited only in response to unfamiliar humans (usually Masai herdsmen with their cattle). Although these herdsmen do not themselves appear to harass the vervets, their children may throw stones at the monkeys (Cheney and Seyfarth 1990).

Human predation can also affect other aspects of primate behavior. *Simias concolor* males have specialized displays that they perform only on encountering humans (Tilson and Tenaza 1976). However, this behavior is contingent on the presence of a mate with young: if offspring are present the male always performs this display while the female and young hide, but if the male is solitary he simply flees. At another site where heavy hunting took place, Watanabe (1981) noted that *S. concolor* groups were always quiet, would quickly flee or hide once they detected an observer, and rarely emitted alarm calls (see also Tilson 1977); in comparison, where hunting was rare, the groups were much noisier and did not flee from the observer. The efficacy of these tactics is perhaps reflected in the finding that hunters who visited previously unhunted areas captured far more monkeys than in traditional areas, although it is difficult to be certain about this since unexploited areas are also likely to have more prey available. Either way, the avoidance of alarm calling by primates in response to human hunters may be commonplace (e.g., diana monkeys: Zuberbühler, Nöe, and Seyfarth 1997).

Another antipredator behavior observed on Siberut Island concerns sleeping tree selection. *Hylobates klossii* appears to avoid sleeping in trees with lianas because indigenous hunters use these vines to climb the trees and shoot the primates at close range with poisoned arrows (Tenaza and Tilson 1985). This strategy also appears to be successful in reducing hunting pressure, since gibbons do not seem to be hunted as frequently as other primates despite their high local abundance.

9.4.4. Community Structure

The selective hunting of target species may dramatically alter primate community structure. The best available data to demonstrate this come from areas where subsistence hunting occurs. Since this hunting is generally associated with the targeting of larger taxa (those of higher profitability), the usual size distribution of species becomes skewed toward smaller species. If subsistence is not important, it becomes more difficult to generalize. If species are being killed to reduce crop raiding, then presumably it will tend to be the terrestrial nonfolivorous species that become depauperate in the community (although other species may also be inadvertently killed). In contrast, if species are being live-trapped, or killed for medicinal or ornamental purposes, then it is the species of economic value that will become scarce. Given the limitations of the data available, we focus here solely on the effects of subsistence hunting.

Studies in Latin America have found that larger primates are scarcer in hunted areas but that the abundance of smaller primates is less affected (e.g., Freese et al. 1982; Emmons 1984; Glanz 1991; Puertas and Bodmer 1993). The most detailed study to tackle this subject collated and compared abundance data from six Amazonian sites where hunting occurred and five where it did not (Peres 1990). Peres found that primate biomass was substantially reduced in hunted sites; a comparison for primates above and below 4 kg revealed that this difference was the result of reduced biomass of larger species—the biomass of smaller species was

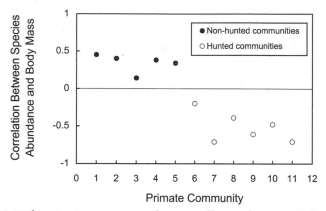

Figure 9.8 Within-site Spearman correlation coefficients between body mass and abundance for several Amazonian sites where hunting was present or absent. The communities listed from 1 to 11 are: Igarapé Açu, Urucu drilling site, upper Tefé drilling site, Cosha Cashu, and Açaituba (not hunted); Lago da Fortuna, Jaraquí, Riozinho, São Domingos, Tahuayo, and Ponta da Castanha (hunted). This analysis excludes the nocturnal owl monkey *Aotus* (although similar results are obtained when *Aotus* is included). (Data from Peres 1990.)

similar in both types of site. Further analysis of Peres's data permits an evaluation of the impact of hunting on body size–abundance associations. In nonhunted sites there is a consistent positive association between population density and body size. This appears to be because the three largest primates are folivores (*Alouatta*) or folivore-frugivores (*Ateles, Lagothrix*), and animals with folivorous diets are found to occur at higher abundances (sec. 5.2.2). In contrast, all correlation coefficients for hunted areas were negative (fig. 9.8); that is, larger species no longer occurred at high abundances. A comparison between coefficients at hunted and nonhunted sites is significant (one-way ANOVA: $F_{1,9} = 70.0, p < .0001$).

To determine the size threshold at which hunters most commonly cease attacking primates at these sites, we calculated the same correlation coefficients but for subsets of species of increasing body size: correlations were made for all taxa less than 8 kg, 7 kg, 6 kg, and so on. Once past the threshold in body size below which hunters usually ignore primates as prey, the correlation coefficients between hunted and nonhunted sites should be identical. The results of this analysis (fig. 9.9) suggest that 2 kg is the critical threshold. Below this body mass, hunting intensity appears to drop off, since the proportion of hunted sites with negative correlations is no more than that observed in nonhunted sites.

These results have to be interpreted with caution, owing to the problem of declining sample size at each successive step. In addition, they do not suggest a universal pattern, since even at the 2 kg size band there are

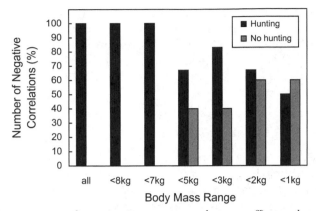

Figure 9.9 Percentage of negative Spearman correlation coefficients between body mass and abundance at sites in Amazonia with or without hunting, for different ranges of primate body mass (for list of sites see legend to fig. 9.8). Only when species of 5 kg or more are excluded from the comparison does the relation between body mass and abundance begin to show some resemblance between hunted and non-hunted sites, with convergence below 2 kg. This analysis excludes the nocturnal owl monkey *Aotus* (although similar results are obtained when *Aotus* is included). (Data from Peres 1990.)

more hunted sites with negative than positive correlations. Indeed, the body mass threshold is likely to vary between sites, since patterns of prey choice depend on the profitability and encounter rates of other available prey (at some of these sites 1–2 kg species are commonly hunted: e.g., Puertas and Bodmer 1993). Nevertheless, these analyses suggest that, in general, mortality from Amazonian hunters is substantial only among species that weigh more than 2 kg.

The possibility that intersite variation in ecological conditions may be a confound bedevils many comparisons. However, Sussman and Phillips-Conroy (1995) circumvented this problem by surveying two sites in Guyana and comparing abundance patterns with those obtained nineteen years earlier. They found that all primate taxa other than the smallest (*Saguinus*) exhibited a decline in population density (with the magnitude of the population decline ranging from 40% to 90%). They attributed this decline to sustained hunting pressure, although they were unable to rule out other possible within-site factors such as seasonality or long-term habitat disturbance.

Finally, it has been suggested (Emmons 1984; Bodmer, Fang, and Ibanez 1988; but see Puertas and Bodmer 1993) that where larger primates are depauperate, smaller species may become more abundant because the carrying capacity of smaller primates is partially determined by interspecific competition with larger primates (e.g., sec. 5.2.1). However, data from Freese et al. (1982) suggest that the abundance of small monkeys (*Saguinus* and *Callithrix*) shows no consistent variation with that of larger monkeys. A more parsimonious explanation may be found in ecological differences between sites. Nevertheless, the harvesting of one population might still potentially affect the persistence of another (e.g., sympatric *Saguinus* species: see sec. 9.4.1).

9.5. Species Vulnerability Patterns

The preceding section described the many effects hunting can have on primate populations. This section asks why populations of some species may be more resistant than others to the effects of hunting.

The logistic growth curve emphasizes the importance of the species trait r_m in determining how quickly a population recovers after harvesting (sec. 9.4.1). Correspondingly, mammals with low r_m show the greatest difference in abundance between heavily harvested and lightly harvested sites in Peru (Bodmer, Eisenberg, and Redford 1997). However, this is only part of the story.

If we are to understand why some species are more resistant than others, it is first necessary to identify those that are most likely to be targeted by hunters. We have already discussed the role of optimal foraging theory in determining patterns of prey choice (sec. 9.1) and the role of hunting technology, food taboos, and a variety of other factors in further refining that choice (sec. 9.2). Since hunters will generally attempt to maximize profitability (e/h), species that are above a certain body mass (high e) or that are easily captured (low h) will be most at risk of heavy hunting. Hence, to minimize h, for example, hunters in Africa may be more likely to attack colobines (which tend to be slower moving and less visually alert than sympatric cercopithecines) and avoid gorillas (since hunting gorillas is dangerous to the hunter) (Oates 1996b). Nevertheless, in some circumstances profitability will not be based on considerations of energy content per unit handling time; when primates are targeted for their organs or skins or as live specimens for the biomedical trade, financial gain per unit handling time may be the appropriate criterion (sec. 9.3).

However, those species that are profitable enough to always be attacked on encounter will not all experience the same level of offtake, simply because their encounter rates are likely to vary. The species that suffer high offtake will tend to be those that are encountered frequently (high λ). Within the profitable prey set, species that are abundant, noisy, have predictable ranging patterns (especially in small territories), and forage at a level in the canopy where they are easier to detect are likely to be at particular risk of hunting (Kuchikura 1988; Oates 1996b).

Because those species that suffer the highest offtake may also be the most abundant, there will not necessarily be a direct correlation between size of harvest and extinction risk (given that abundance correlates with population size and, potentially, reproductive rates; sec. 5.2). In fact it is quite possible that the species with the fewest individuals cropped may be at the greatest risk of extinction, since these species presumably already occur in small populations (evident from their low encounter rate). It is therefore important to discriminate between the actual number of individuals cropped and the proportion of the population that is cropped. The greater vulnerability of small populations, together with the tendency for large primates to occur in small populations, thus provides another explanation for why large primates appear to suffer so badly from hunting. Although all large primates may be at risk owing to their high profitability (Mittermeier 1987a), the relative risk within the profitable prey set is more likely to be determined by differences in population size than by any differences in body size.

This observation reminds us that the effects of hunting in a primate community result from an interaction between two complex sets of processes: the decision making of the hunter and the population dynamics of the prey. Fortunately we have models to describe both these processes: the prey choice model (sec. 9.1) and the logistic growth curve model (chapter 6), respectively. Although the interaction between these two processes is difficult to predict intuitively, especially in complex multispecies communities, Winterhalder and Lu (1997) have recently combined these two models to great effect. Their exercise, in which the hunting pressure a prey species experiences is the result of both standard prey choice decisions (assumed to follow the prey algorithm) and the number of hunters (the size of the hunter-gatherer population is assumed to obey a variant of the logistic growth curve), indicates the following conclusions.

First, in a single-prey system, the human and prey populations stabilize at sizes where the harvest of the prey species A is below the size that would provide its maximum sustainable yield. Second, in a two-prey system, prey species B is initially not sufficiently profitable for harvesting, but once the encounter rate of species A declines, species B is also incorporated into the diet. This expansion of the diet permits species A to recover to some degree. As a result, species B will disappear from the diet again. In their simulation species B was subsequently regained and lost on eight more occasions (contingent on the encounter rate between hunters and species A) before it finally entered the diet permanently. In other words, the exploitation of one species may buffer the exploitation of another, helping to prevent extinction in vulnerable populations (particularly those that are highly profitable but reproduce slowly). Third, in

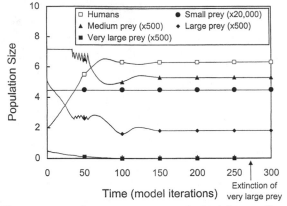

Figure 9.10 Changes in population size over time for both human foragers and their prey in a four-prey species system. Initially only two prey species (large and very large prey) are hunted. As the human population grows and the prey populations decline, medium-sized prey also enter the diet. Finally the human population reaches carrying capacity, but this occurs too late to prevent the extinction of the very large prey. At no point do the small prey enter the human forager diet. (After Winterhalder and Lu 1997.)

a four-prey system, prey switching becomes more complex but may no longer buffer prey populations sufficiently to prevent one of the more profitable prey species from becoming extinct (fig. 9.10). However, if the r_m of this species is doubled, then extinction no longer occurs, demonstrating the importance of this variable for prey species population dynamics.

These analyses emphasize the dynamic nature of hunter-prey interactions and suggest that attempts to predict a prey population's vulnerability to hunting without knowledge of the profitabilities and encounter rates of all other potential prey species in the community may be hopelessly inadequate. They also raise the possibility that different species characteristics might be influential in determining population viability at different times: the profitability of a given species (and the profitabilities and encounter rates of other species in the community: sec. 9.1) may be important initially in determining its viability, but once harvesting commences r_m may become crucial.

9.6. Hunting with Habitat Disturbance

It is almost inevitable that habitat disturbance will promote contact between humans and nonhuman primates and that hunting pressure will intensify as a result. In fact, the two processes often go hand in hand, with devastating consequences for primate populations. For example, Estrada, Coates-Estrada, and Meritt (1994) found that forest fragments with high levels of disturbance from logging, hunting, and removal of wood for fuel contained many fewer nonflying mammal species (including fewer individuals of *Alouatta palliata* and a complete absence of *Ateles geoffroyi*) than those fragments that were undisturbed. However, the critical issue for conservation management is which of these pressures has the greater impact on primate populations: without knowing this it becomes difficult to take effective conservation action. In this final section, we try to evaluate both the relative severity of threats posed by hunting and various forms of habitat disturbance and how far different species show different patterns of threat-specific vulnerability.

9.6.1. Relative Impacts by Threat Type

Oates (1996b) has argued that hunting is a greater threat than habitat modification, although not necessarily greater than habitat loss or fragmentation. He points out that many primates thrive in secondary forest areas but none are known to cope with heavy hunting pressure. Several other studies support this claim. Tutin and Fernandez (1984), for example, found that hunting pressure generally had a greater effect on gorilla and chimpanzee densities than habitat disturbance. Similarly, Chapman, Chapman, and Glander (1989) found that once spider monkeys were protected from hunting, they could thrive in disturbed regenerating forest.

Comparing hunting pressure with our three main categories of habitat modification (small-scale disturbances, agriculture, and forestry) adds further weight to this conclusion. First, the greater impact of hunting over small-scale disturbances is well illustrated by the effects of the rubber tapping industry on primate communities over the past century. There have been a number of local extinctions of woolly monkeys in *Hevea* habitats where rubber tapping occurred, even though the effect of tapping on the local habitat is minimal. These extinctions were the result of *Hevea* tappers' hunting the monkeys for food (Peres 1991). Peres suggests that the same may now happen in those habitats where *Couma* trees are ringed by tappers. *Couma* is an important source of fruit for woolly monkeys, and the ringing brings the tappers close to the monkeys. In this case habitat disturbance may exacerbate mortality, since many *Couma* trees do not survive ringing.

Second, Wilkie (1989) lists a variety of societies around the world that actively hunt in areas where shifting cultivation is practiced (a practice sometimes called "garden hunting": Jorgenson 1995). In the Ituri Forest, hunters acquire 59% of their total kills in secondary forest, although this forest makes up only 26% of the available habitat. Wilkie et al. (1998a,b) subsequently modeled the impact that human population growth (acting through the effects of shifting agriculture and hunting) had on forest structure and primate population dynamics in the Okapi Wildlife Reserve (Democratic Republic of Congo). Their results indicated that an expansion of shifting cultivation would have little influence on primary forest cover but that primate harvests might well become unsustainable in those areas where local hunters specialize on primate prey. Although these results may be unique to the Okapi Wildlife Reserve (the absence of any effects from agriculture here may have been solely due to the extensive areas of existing secondary forest, which absorbed agricultural activities), other studies confirm that hunting can have a dramatic impact on primate populations in cultivated mosaics. In Sulawesi, for example, *Macaca nigra* appears to do poorly in agricultural mosaics because of the associated hunting pressure (O'Brien and Kinnaird 1996; Rosenbaum et al. 1998). In Ivory Coast, chimpanzees become increasingly scarce as plantations and fallow land dominate the landscape, but this relationship may arise primarily through hunting, since the correlation is reduced once the effect of hunting intensity (indexed by the number of hunting trails) is removed by partial correlation (Marchesi et al. 1995).

Third, Grieser Johns and Grieser Johns (1995) have shown that the recovery trajectories of primates in logged forest compartments in Malaysia are severely compromised by hunting (fig. 8.10). Similarly, Harcourt (1980–1981) found that gorilla abundance in the Bwindi Forest, Uganda, did not differ between areas of logged and unlogged forest but did differ between areas that differed in evidence of human presence (the gorillas were more common where people were absent). Bennett and Dahaban (1995) noted that hunting increased significantly in an isolated area of forest in Sarawak (Borneo) once a road had been built by loggers. Hunting pressure continued to remain high after the loggers moved out, until the collapse of a bridge along the road once more made it hard to reach this remote forest (35 km from the nearest human habitation).

The connection between forestry operations and hunting has recently become a focus of attention as a potentially major force in primate extinctions (see Pearce and Amman 1995; Spinney 1998). Loggers not only make the forests more accessible (by constructing roads and trails), but they also provide local hunters with shotguns and cartridges. Moreover, they then use the logging trucks to transport meat from the local hunters to the nearby urban centers, thereby providing a direct route from the heart of the forest to a source of enormous urban demand. Consequently, even where logging operations are low impact, primate populations can be exterminated (e.g., Democratic Republic of Congo: Wilkie 1992; Gabon: Brugière and Gautier 1999).

9.6.2. Relative Vulnerability of Species

Relative differences between taxa in vulnerability to habitat disturbance and hunting has important implications for conservation. Perhaps the most reliable difference in vulnerability is related to body size. Since small species (typically <2 kg) are not profitable prey, they are rarely hunted. Consequently, although small species tend to be threatened by habitat disturbance only, medium- and large-bodied species are threatened by both habitat disturbance and hunting. This distinction is illustrated by Bodmer's (1995b) analysis for Loreto State (Peru). He estimated the relative annual losses from primate populations through habitat loss and hunting (including trapping for live export) and concluded that for large species (exact body mass not specified) the losses from hunting (45–89% of all individuals lost) were far greater than those from habitat disturbance (11–55% of all individuals lost) but that for small species the losses from habitat disturbance were much more serious (accounting for 85–96% of losses). Notably, trapping for live export (for biomedical research) contributed very little to these losses (no more than 2% of all hunting losses), irrespective of body size.

However, among the medium- and larger-bodied primates there may be further important patterns. In particular, species that do well in secondary forest (table 8.11) are almost always going to be more at risk from hunting than from habitat modification. Oates (1996b) provides several examples from Africa, including black-and-white colobus monkeys and the gorilla. In the Neotropics, howler monkeys can survive better than most primates in forest fragments, but their absence in fragments in some parts of the Atlantic coast suggests that they are not so resilient to hunting (Ferrari and Diego 1995).

Oates (1996b) evaluated the relative threats of hunting and habitat disturbance to folivorous African primates and found that geographical distribution played an important role in interspecific differences in vulnerability. In particular, East African species were more threatened by habitat disturbance than by hunting because hunting pressure is relatively low in this region. Within regions, marked differences also exist between taxa. In Gabon, Blom et al. (1992) estimated (1) that prosimians are less threatened by hunting than anthropoids and are generally more vulnerable to habitat loss than hunting, presumably because of their low body mass (and possibly also their nocturnal habits); (2) that most other primates in Gabon are at greater risk of extinction from hunting than habitat disturbance, probably because Gabon has suffered little historical forest loss (although *Colobus satanas* is most at risk from habitat loss, due to the taboos against its consumption); and (3) that cercopithecine species are most at risk from hunting, while the reduced risk among apes

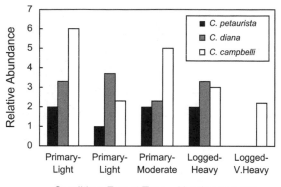

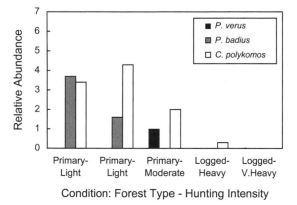

Figure 9.11 Relative abundance of West African guenons (*Cercopithecus petaurista,* *Cercopithecus diana,* and *Cercopithecus campbelli*) and colobines (*Procolobus verus,* *Procolobus badius,* and *Colobus polykomos*) at five sites in the Gola Forest, Sierra Leone, in relation to the local level of both selective logging and hunting pressure. Selective logging either has occurred (logged) or is absent (primary); hunting pressure is designated as light, moderate, heavy, or very heavy. (After Davies 1987.)

may also reflect local taboos (against eating species which so closely resemble *Homo sapiens:* Sabater-Pi and Groves 1972). However, both of these studies are based primarily on subjective impressions.

The most detailed and systematic evaluation of the relative effects of habitat disturbance and hunting has been carried out by Davies (1987). He surveyed five sites in the Gola Forest (Sierra Leone) that had experienced different regimes of selective logging and hunting (from primary forest with no hunting to logged forest with very heavy hunting). His results are shown in figure 9.11.

Both cercopithecine and colobine monkeys appeared to suffer more severely from hunting than from logging. The abundance of the three guenons in logged forest dropped dramatically where there was very heavy hunting, with only *Cercopithecus campbelli* remaining. The colobines appeared to be more sensitive to hunting than the guenons: this was especially true of *Procolobus badius,* which was not seen where hunting was of more than light intensity. Why *P. badius* should be so vulnerable and *C. campbelli* so resistant remains to be established. Since *C. campbelli* is the smallest monkey present (along with *C. petaurista*), but also the most abundant in primary forest, its resilience may arise from a combination of low profitability and large initial population size. It is also relatively cryptic in comparison with the other guenons (Kingdon 1997), and is therefore difficult for hunters to detect. In contrast, the reason for *P. badius's* vulnerability is less clear, although a combination of relatively large body size and group size might contribute (sec. 9.5).

9.7. Summary

1. Primates are hunted primarily for food (for personal subsistence and local markets), for products such as skins and organs (for ornamental and medicinal purposes respectively, often on international markets), to supply live subjects for biomedical research (for international markets), and to control crop raiding.

2. Subsistence hunting appears to be most widespread in the Neotropics and relatively uncommon in Asia and Madagascar. Among tropical hunters, mammals (mainly primates, rodents, and ungulates) tend to dominate the harvest. Hunting decisions by subsistence hunters may be best understood with reference to optimal foraging theory, which emphasizes the importance of prey encounter rates and prey profitability in determining prey choice.

3. Hunting intensity can vary in time (which can be due to seasonal variation in the encounter rates of prey, the profitability of prey, and the time available for hunting in the agricultural cycle), in space (which can be due to variation in local prey abundance, the accessibility of hunting areas, and the availability of transport), with hunting technology (shotguns, blowguns, and bow and arrow can each lead to different hunting patterns in different prey communities), and between human communities (e.g., through taboos).

4. The most important trade in primates (in terms of the threat it poses of widespread species extinctions) is likely to be the trade in primate meat, which is particularly severe in West Africa. By contrast, the international trade in live primates (which is regulated by CITES) poses a relatively less significant threat. This trade has declined since the 1960s but may still endanger those primate populations that are intensively targeted (previously the rhesus macaque, now the long-tailed macaque) or already endangered for other reasons (e.g., chimpanzees). The trade in

primate parts is less well known but may similarly threaten target populations; previously there has been serious demand from Western markets (for luxury items such as black-and-white colobus skins), but more recently most of this trade may be driven by the Asian markets (where body parts are used in traditional medicines).

5. Hunting in primate populations can reduce population abundance and size, but how far such hunting is sustainable remains unknown in most cases. Comparing theoretical models with observed harvests suggests that many species are hunted unsustainably. Where primates are harvested for live export there is some evidence of potential sustainability, but population recovery in these cases may often depend on immigration from outside the immediate area.

6. Hunting in primate populations can also modify the age-sex structure of the population (especially where a particular age-sex class is targeted: e.g., juvenile rhesus macaques and adult male gelada baboons); modify the behavior of individuals (e.g., by reducing the frequency of behaviors that attract hunters, such as loud calls); and modify the social structure of the population (e.g., in response to the demographic changes in age-sex composition). In addition, differential hunting intensity can cause marked changes in primate community structure, particularly favoring a preponderance of species that are less commonly hunted.

7. Some primate species do appear to be more vulnerable to hunting than others, but it is difficult to predict precisely the intensity or outcome of hunting on a primate population without knowledge of the other potential prey in the system. Nevertheless, as a general rule the species most vulnerable to hunting are those of large body size (high profitability) and low r_m (limited ability to recover from harvesting). Since both these traits are also associated with small population size, the risk such species face from hunting is often dire.

8. Hunting often occurs in conjunction with habitat disturbance, which brings people into direct contact with wild primate populations. Given that many forms of habitat disturbance can be relatively mild and that many primates persist reasonably well in modified habitats, it is not surprising that the effects of hunting are often more severe than those of habitat disturbance. This seems to be particularly true of logging operations, which establish a direct link between the forest and nearby urban centers for the market trade in bushmeat.

10 *Conservation Strategies*

I n this chapter and the following one we consider the ways wild primate populations can be protected from extinction. We first focus on conservation strategies and the principles that underlie them. The conservation tactics used to implement these strategies are the subject of the following chapter. We use "strategy'" to refer to a set of long-term objectives and "tactics" for finer-level methods used to achieve these goals.

10.1. Strategy Design Principles

Given their broad variability in scale and scope, it is difficult to generalize about conservation strategies. Politically, a strategy can be local, national, or international. Spatially, it may cover only a few hectares or vast stretches of habitat thousands of kilometers across. In temporal terms, most strategies operate over a five- to ten-year time frame, followed by reevaluation and revision in the light of successes and failures and other developments. A balanced approach to the conservation of both states and processes is usually adopted, and the strategy may operate at the level of individual taxa, communities, or ecosystems. The taxonomic unit of interest is normally the species, although subspecies and populations may occasionally be used. Finally, conservation strategies may focus on maintaining and enhancing wild populations (*in situ* conservation), captive populations (*ex situ* conservation), or both. Consequently there is no simple formula or structure for a strategy to follow; each is tailored to the problem it is designed to tackle.

Nevertheless, it is still possible to distinguish four principal components: (1) defining the strategic goals; (2) identifying the relevant units of conservation, which for primates typically include the species and community (cf. the "biospatial hierarchy": Soulé 1991); (3) evaluating these units and ranking them in order of their priority for action, taking into account the goals of the strategy and the limited time and resources

289

available; and (4) presenting realistic solutions (tactics) to the problems facing the targets, ones that have the potential for rapid implementation and whose outcome can be easily monitored. A variety of other components might also be present, including specific targets to evaluate the degree of success, a budget for costs, and research proposals to diagnose the precise nature of the threat, since, as Caughley and Gunn (1996) emphasize, an incorrect diagnosis can lead to wasteful or inappropriate action.

International conservation action is broadly guided by the "World Conservation Strategy" (IUCN, WWF, and UNEP 1980) and its successor, "Caring for the Earth" (IUCN, WWF, and UNEP 1991). The global strategy for primate conservation is headed by the IUCN/SSC Primate Specialist Group (PSG). This is one of several specialist groups operating under the Species Survival Commission (SSC) of the World Conservation Union (IUCN) that helps plan and coordinate international conservation in endangered taxa. Oates (1996a) defines the PSG's remit thus: "The PSG has set itself as a main goal the maintenance of the current diversity of the Order Primates, with a dual emphasis on: (1) ensuring the survival of endangered and vulnerable species wherever they occur; and (2) providing effective protection for large numbers of primates in areas of high primate diversity and/or abundance."

One route the PSG uses for achieving these ends is the publication of regional strategies, or "Action Plans" (Africa: Oates 1986a, 1996a; Asia: Eudey 1987; Madagascar: Mittermeier et al. 1992). (An Action Plan for the Americas is currently in preparation: see Rylands, Coimbra-Filho, and Mittermeier 1993; Rylands, Mittermeier, and Rodriguez-Luna 1997). These Action Plans set priorities among primate taxa for conservation action, identify regional primate communities, and make recommendations for conservation projects. In the case of Africa, the long-term continuity of the strategy has been emphasized with the publication of a second Action Plan ten years after the first (Oates 1986a, 1996a). This reviews the progress and achievements made since the original plan and makes recommendations for future development. Further information on the status and biology of threatened primates is provided in the IUCN Red Lists (see sec. 10.2.2) and IUCN Red Data books (Lee, Thornback, and Bennett 1988; Harcourt and Thornback 1990).

The PSG Action Plans have been supplemented with a global primate Conservation Assessment and Management Plan (CAMP) (Stevenson, Foose, and Baker 1992), developed by the IUCN/SSC Captive Breeding Specialist Group (CBSG) in collaboration with the PSG (see Seal, Foose, and Ellis 1994). Its purpose is to evaluate the global status of each primate taxon, both in the wild and in captivity, and determine its management needs. The CAMP report differs from the Action Plans in that it sets priorities among species for conservation attention according to extinction risk only (sec. 10.2.2); it operates at the subspecies rather than the species level (but cf. Mittermeier et al. 1992); it makes only broad categorical recommendations for action rather than specifying particular projects; and it is designed to develop a Global Captive Action Plan (see sec. 11.3).

A parameter of fundamental importance in any conservation strategy is the priority attributed to different projects. However, priorities depend on the goals of the strategy. Moreover, although there are a variety of reasons for valuing primate diversity, only those values that have a bearing on the strategy goals are considered in setting priorities. Were these goals to change, so too would the values and the priorities. The goals of the PSG have already been made explicit: the in situ preservation of species and communities. This focus is based on the view that successful in situ conservation is still thought practicable for primate taxa (e.g., Africa: Oates 1986b). However, should the role of ex situ conservation grow in importance and become a goal in these strategies, we might expect additional criteria, such as the viability of captive populations, to become part of its remit (e.g., Madagascar: Mittermeier et al. 1992). The important point here is that priorities are not absolute but relative and are designed with very specific goals in mind. Priorities should always be interpreted and applied in the context of the strategy within which they were identified.

In the following two sections of this chapter, we therefore investigate the philosophy and techniques used in exercises in evaluating conservation and setting priorities. First we examine this issue for taxonomic units of conservation, then we consider it for area-based conservation. In the final section of this chapter, we introduce some of the key components of the socioeconomic and political systems that provide the context in which conservation strategies must operate.

10.2. Setting Taxon Priorities

Since the time and finances available for conservation action are limited, it is essential that both these resources be used as efficiently as possible. The identification of taxa that are of high or low priority helps achieve this.

The first issue in setting taxon priorities is the choice of conservation unit. In principle, conservation strategies aim to focus on the "evolutionary significant unit" to maximize the preservation of biological diversity (e.g., Vogler and Desalle 1994). Therefore taxa that comprise distinct and well-differentiated sets of populations are normally the principal target for conservation action. Both species and subspecies fall into this category, although since species are normally better differentiated they are more commonly the focus. There are also pragmatic reasons behind the choice of conservation unit. The advantage of species is that they are conceptually understood by laypeople, legislators, and politicians (see also Caughley and Gunn 1996). The advantage of subspecies is that, as Strier (1997) notes, "Arguments about the importance of preserving relic populations representing two possible species may be more effective [at attracting public attention and funding] than those advocating the protection of a single species whose distribution has been severely fragmented into multiple populations."

At this stage of conservation planning, taxonomy obviously plays an important role in distinguishing species from subspecies (e.g., howler monkey: Froelich and Froelich 1987) and identifying appropriate conservation units (e.g., leaf monkeys: Rosenblum et al. 1997). Unfortunately, defining a taxonomic unit is rarely easy, for a variety of reasons (see sec. 2.1), making it extremely difficult to set priorities for conservation. PSG Action Plans therefore commence with a review of the taxonomic system they adopt. However, different plans use different conservation units. Mittermeier et al. (1992) consider the lowest-rank taxon available (subspecies or species). In contrast, Oates (1986a, 1996a) and Eudey (1987) concentrate on species, although both list subspecies where these are of special concern.

Once the conservation unit has been identified, priorities can be set. Surprisingly, despite the great importance of the criteria used to establish priorities, there is no agreed set of parameters for doing so. Nor is there a good understanding of the advantages and disadvantages of each possible parameter (Mace 1995a). The criteria used in existing primate conservation strategies are predominantly evolutionary uniqueness and extinction risk (secs. 10.2.1 and 10.2.2); less commonly, the association of the taxon concerned with other threatened primates is also considered (table 10.1). In all cases, the criteria are ranked numerically from low to high values (often on a simple scale of 1 = low to 4 = high), and the overall priority rating is obtained by adding scores on the different variables. Extinction threat is always defined using the widest range of rank scores, ensuring

Table 10.1 Taxon priority rating systems in Primate Specialist Group conservation strategies

Region	Priority rating criteria[1]				Taxonomic unit	Source
	1	2	3	4		
Africa	+	+	+	−	Species	Oates 1986a
	+	+	−	−	Species	Oates 1996a
Madagascar	+	+	−	+	Subspecies	Mittermeier et al. 1992
Asia	+	+	+	−	Species	Eudey 1987

[1]Priority rating criteria: (1) degree of extinction risk; (2) evolutionary (taxonomic) uniqueness; (3) association with other threatened primates; (4) existence in a protected area.

that it has the greatest weight in the final priority rating (so highly threatened species will always rank high overall).

Several other criteria could also be used (Mace 1995a). One would be "potential for recovery," which might incorporate factors such as the extent of remaining habitat, although this index might have limited value for primates, since at present all species arguably have a reasonable potential for recovery. Other possibilities are an "index of quality" of data used to establish the priority rating and a measure of the "viability of captive populations." This might also relate to recovery potential, if recovery requires reintroductions. For example, the drill, Africa's most endangered primate species, is poorly represented in captivity, and its future conservation is likely to require reintroductions (secs. 11.3 and 11.4). Yet despite the potential advantages of incorporating more information into priority setting, there are good reasons for maintaining an uncomplicated scoring system, particularly in terms of transparency and accountability (Williams 1999).

10.2.1. Evolutionary Uniqueness

By the criterion of evolutionary uniqueness, a taxon will be singled out for attention if its extinction leads to a disproportionate loss of primate diversity on a range of possible measures such as anatomical, ecological, or behavioral diversity. The information provided by taxonomy and phylogeny (sec. 2.1) is pivotal to this index. Most commonly, taxa with fewer living close relatives are attributed greater value because of their evolutionary uniqueness. Hence, for example, an endangered macaque (e.g., *Macaca nigra*) would be of lower priority than the endangered proboscis monkey (*Nasalis larvatus*), which represents a monotypic genus. Eudey (1987) codes both of these taxa similarly for extinction threat but assigns them to opposite ends of the range for taxonomic uniqueness.

One way of formalizing this quality might be to use the concept of *independent evolutionary history* (IEH), defined as the time since the taxon last shared a common ancestor with its closest relative (May, Lawton, and Stork 1995). For a phylogeny as well known as that of primates, it is relatively straightforward to estimate the number of years that constitute each species' IEH. Using Purvis's (1995) composite estimate of primate phylogeny, we have identified the highest and lowest IEH scoring primate species (table 10.2). Lemurs tend to score high (constituting three of the five highest scores), while members of the Cercopithecidae score particularly poorly (earning all five of the lowest scores). This pattern is not too surprising given macroevolutionary patterns in primate diversification (chapter 2).

However, several important points need to be taken into account. First, the value of IEH depends on the taxonomic status recognized for

Table 10.2 Rank evolutionary uniqueness of the highest and lowest rank
primate species

Taxon	Rank	Independent evolutionary history (IEH)[1] (million years)	Number of IEH estimates
Lepilemur mustelinus	1	18.6	1
Cebus apella	2	17.9	1
Varecia variegata	3	16.6	2
Pongo pygmaeus	4	14.5	2
Daubentonia madagascariensis	5	14.0	1
Cercopithecus mitis	199	0.25	1
Cercopithecus nictitans	200	0.25	1
Colobus angolensis	201	0.23	1
Colobus guereza	202	0.10	1
Colobus polykomos	203	0.10	1

Source: Data derived from Purvis's 1995 primate phylogeny.

[1]Number of years since sharing a common ancestor.

that taxon; if a species is subsequently split into several species, then new
calculations are necessary for each. Second, the values for IEH are often
based on single estimates of dates of diversification that will be prone to
error. Finally, the rate of evolution may differ greatly between taxa of
similar age depending on the intensity of the selective pressures they
have been exposed to; it is therefore possible that younger pairs of sister
taxa show less resemblance to each other than older pairs (sec. 2.1). This
last point stresses the importance of the different ways people value diversity (Williams, Gaston, and Humphries 1994). For example, the aye-
aye (*Daubentonia madagascariensis*) belongs to a monotypic family and
is one of the most unusual of primates, but it is only fifth in the IEH
priority list (table 10.2). Similarly, Goeldi's marmoset (*Callimico goeldi*)
belongs to a monotypic subfamily, yet it is entirely absent from the top
five. Consequently it is important to decide precisely what feature of
priznmate diversity is being valued (in this instance time since divergence rather than degree of change since divergence).

To complicate matters further, one school of thought argues that it is
not the older, more exceptional taxa that should be given priority but
rather the younger taxa, because the former are evolutionary dead ends
whereas the latter constitute the current diversification fronts providing
root stock for vigorous future evolutionary change (Erwin 1991; Brooks,
Mayden, and McLennan 1992). Hence species with the shortest IEH
should be given highest priority, such as the Galagidae and Cercopithecidae (Purvis, Nee, and Harvey 1995). Moritz (1995) suggests that dis-
agreement over the relative value of older and younger taxa reflects a

difference in objectives. Older taxa are valued more highly in the conservation of state and in preserving the current breadth of diversity, whereas younger species are valued more highly in the conservation of process and in promoting future diversity. However, the implicit assumption that diversification in the recent past is an indicator of future diversification is debatable. Hence we would not advise this rationale in the application of the evolutionary uniqueness criterion.

10.2.2. Extinction Risk

A taxon's degree of extinction risk, or *conservation status*, is the most important component of most exercises in setting priorities. Evaluation of this threat depends on our knowledge of the intrinsic and extrinsic processes that drive populations to extinction (chapters 7, 8, and 9). In this section we examine the classification systems that have been used to develop comparative indices of global extinction risk (together with the patterns of primate global extinction risk that emerge). In the following section we investigate the modeling approaches used to predict population extinction processes.

The simplest methods of determining conservation status equate extinction risk with a single characteristic that is likely to be associated with taxon viability. Rarity has commonly been employed as such a correlate, since by definition rare species tend to occur in small populations and small populations are most vulnerable to extinction. This approach has been further developed in the form of compound rarity (see sec. 5.3.2). However, rarity is a poor measure of risk since it is scale dependent and fails to incorporate the role of extrinsic forces in extinction threat.

Most systems of estimating threat rely on a suite of parameters, which usually adopt absolute rather than relative values and include both intrinsic and extrinsic correlates of extinction. In one of the earliest studies to adopt such an approach for primates, Wolfheim (1983) employed six variables to index extinction risk. Four were intrinsic species traits (distribution, abundance, body size, and habitat requirements), and two were extrinsic pressures (human predation and habitat loss). Species were considered highly endangered if they had a restricted geographic range, occurred at low population density, and were large-bodied habitat specialists that were hunted and also suffered from habitat loss. Although this index is admirably comprehensive, the differences in relative importance and the complex connections among these additive components make combining them into a single value of extinction threat problematic. Wolfheim (1983) ultimately resorted to subjective interpretation of the six parameters to arrive at a single value of extinction threat. Consequently the system also suffers from a lack of transparency and accountability.

The most widely employed index of conservation status is that devel-

Table 10.3 Criteria for extinction risk evaluation according to the Red List codes

Criteria	Critically Endangered (CR)	Endangered (EN)	Vulnerable (VU)
A. Declining population			
Population decline rate at least	80% in ten years or three generations[a]	50% in ten years or three generations[a]	20% in ten years or three generations[a]
using either			
1. population reduction observed, estimated, inferred, or suspected in the past or			
2. population decline projected or suspected in the future based on			
a. direct observation			
b. an index of abundance, appropriate for the taxon			
c. a decline in area of occupancy, extent of occurrence and/or quality of habitat			
d. actual or potential levels of exploitation			
e. the effects of introduced taxa, hybridization, pathogens, pollutants, competitors, or parasites			
B. Small distribution and decline or fluctuation			
Either extent of occurrence	<100 km^2	<5,000 km^2	<20,000 km^2
or area of occupancy	<10 km^2	<500 km^2	<2,000 km^2
and two of the following three:			
1. severely fragmented (isolated subpopulations with a reduced probability of recolonization, if once extinct)	= 1	≤5	≤10
2. continuing decline in any of the following:	at any rate	at any rate	at any rate
a. extent of occurrence			
b. area of occupancy			
c. area, extent, and/or quality of habitat			
d. number of subpopulations			
e. number of mature individuals			

3. fluctuating in any of the following:
  a. extent of occurrence
  b. area of occupancy
  c. number of subpopulations
  d. number of mature individuals
	>1 order of magnitude	>1 order of magnitude	>1 order of magnitude

C. Small population size and decline Number of mature individuals and one of the following two:	<250	<2,500	<10,000
1. rapid rate of decline at least	25% in three years or one generation°	20% in five years or two generations°	10% in ten years or three generations°

2. continuing decline and
  either a. fragmented or
      b. all individuals in a single subpopulation
	any rate
	all subpops ≦ 50
		any rate
		all subpops ≦ 250
			any rate
			all subpops ≦ 1,000

D. Very small or restricted Either 1. number of mature individuals or	<50	<250	<1,000
2. population is susceptible	(not applicable)	(not applicable)	area of occupancy <100 km² or number of locations < 5
E. Quantitative analysis Indicating the probability of extinction in the wild to be at least	50% in ten years or three generations°	20% in twenty years or five generations°	10% in one hundred years

Source: After Baillie and Groombridge 1996.

Note: Criteria are shown only for threatened categories. Taxa may also be classified as Low Risk (LR), with three further subdivisions: conservation dependent (cd), near threatened (nt), and least concern (lc) (in declining order of extinction risk). Taxa for which sufficient data are not available to determine their status according to this scheme are classified as Data Deficient (DD). Note that these criteria should not be applied without careful reference to the guidelines provided in the Red List.

°Whichever is the longer.

oped by IUCN (the World Conservation Union) in its Red Lists and Red Data Books (Mace 1995a,b). Initiated almost thirty years ago, the IUCN threat classification system has evolved through several stages. The system initially comprised three threatened categories (rare, vulnerable, and endangered), one "indeterminate" (threatened but not sufficiently known to determine which category it should be assigned to), one "insufficiently known" (inadequate data for classification), and one "not threatened." This classification scheme had a variety of problems, in particular its subjectivity (no numerical values were provided for any of the criteria), the absence of any time frame, and the exclusion of nonthreatened species from the Red Lists and Red Data Books.

In the early 1990s this system was subjected to a major overhaul, initiated by the work of Mace and Lande (1991). The primary purpose of this exercise was to reduce subjectivity and to make use of valuable information emerging from Population Viability Analyses (sec. 10.2.3). The aim was to produce a simple but flexible objective system with a strong scientific basis that also supplied a time scale over which extinction might occur and over which conservation action might be necessary. The new IUCN Red List system in its final form (detailed in Annex 2 of Baillie and Groombridge 1996) consists of categories for species that are extinct (Extinct in the Wild, Extinct), are threatened with extinction (Vulnerable, Endangered, Critically Endangered), are at Low Risk, are Data Deficient, or have not yet been evaluated. Assigning a taxon to any one of these classes is based on a detailed evaluation of the state of the global population (table 10.3). This system provides the basis for the most recent classification of global extinction risk in primates (the 1996 IUCN Red List: Baillie and Groombridge 1996), and its application has been tested in detail by Harcourt (1995, 1996) for the gorilla.

Whereas the CAMP report is based on an intermediate version of the revised Red List codes, the PSG Action Plans have not adopted either the old or the new IUCN system. Across these four plans, three novel evaluation systems have been used to ascertain both threat and priority for conservation action. This variation partially reflects the special needs and conditions specific to primate conservation in each geographic region, although they also reflect a reluctance to apply the Red List criteria owing to the detailed data required and the inevitability of resorting to guesstimates when providing such data (Mittermeier et al. 1992; Oates 1994a, 1996a). The criteria used in the various systems are compared in table 10.4 (following Mace 1995a). Population size is the only parameter common to all systems, although population trends and the threat of both habitat loss and hunting are also important variables. Rank values for each criterion are largely assigned subjectively, although numerical values are used for population size and geographic distribution (Mittermeier et al. 1992; Oates 1996a).

Table 10.4 Extinction risk rating systems

Region and publication	Extinction risk criteria[1]					Source
	1	2	3	4	5	
Africa						
Action Plan 1986	+	+	+	+	+	Oates 1986a
Red Data Book	+	−	+	+	+	Lee, Thornback, and Bennett 1988
CAMP Report	+	+	+	−	+	Stevenson, Foose, and Baker 1992
Action Plan 1996	+	+	+	+	+	Oates 1996a
Madagascar						
Red Data Book	+	−	+	+	+	Harcourt and Thornback 1990
Action Plan	+	−	−	−	−	Mittermeier et al. 1992
CAMP Report	+	+	+	−	+	Stevenson et al. 1992
Asia						
Action Plan	+	+	+	+	+	Eudey 1987
CAMP Report	+	+	+	−	+	Stevenson, Foose, and Baker 1992
The Americas						
CAMP Report	+	+	+	−	+	Stevenson, Foose, and Baker 1992

Note: This table includes only those IUCN publications that give comprehensive lists of species and status (excluding the Red Lists). It excludes evaluation in non-IUCN systems (e.g., Wolfheim 1983) and publications (e.g., Mittermeier et al. 1994).

[1]Threat rating categories: (1) Population size; (2) Population fragmentation; (3) Population trends; (4) Rarity (restricted distribution and/or low abundance); (5) Serious threat from habitat loss and/or hunting.

The conservation status of all primate species, according to the 1996 IUCN Red List, is listed in appendix 1. Below we discuss the patterns that emerge from these data.

SPATIAL PATTERNS OF RISK

Across world regions, there is a consistent trend for approximately half of all primate taxa to be threatened (table 10.5). An important pattern among the Old World primates is the relatively large number of high-risk taxa in Asia and Madagascar and the relatively small number in Africa. Similarly, comparing all recorded mammalian extinctions in Africa and Asia since 1600 reveals two extinctions in sub-Saharan Africa but double this number in Asia (Groombridge 1992; Smith et al. 1993a). Two explanations might be proposed. First, on a biogeographical basis, all the lemurs and a good proportion of the Asian taxa are island dwellers, which may put them at higher risk of extinction than the largely continental taxa of Africa (sec. 7.4.2). Second, in terms of differential human disturbance, Asia and Madagascar may be more severely affected by habitat loss (hunting is less of a problem in both areas: chapter 9). To discriminate between these two mechanisms, we compared like taxa in Africa and Asia (table 10.6).

Table 10.5 Geographic distribution of primate species global extinction risk

Region	Total number of species	Percentage of taxa in each Red List category[1]			
		LR	VU	EN	CR
Africa	65 (5)[2]	78	12	10	0
Madagascar	32 (0)	41	34	13	13
Asia	70 (14)	41	32	20	7
The Americas	90 (3)	64	21	9	6

Source: Data from Baillie and Groombridge 1996.

[1]Codes are abbreviated as follows: Low Risk (LR), Vulnerable (VU), Endangered (EN), and Critically Endangered (CR). VU, EN, and CR species are considered threatened.
[2]Numbers in parentheses are species classified as "Data Deficient." The values in the last four columns are given as the percentage of all species excluding those that are Data Deficient.

Table 10.6 Extinction risk of primate species by geographic region and distribution

Risk evaluation system	Threat status	Region	Number of taxa	Percentage of taxa with	
				Continental distribution	*Insular distribution*
CAMP (subspecies)	Critically Endangered	Africa	7	86	14
		Asia	30	57	43
Action Plans (species)	# 5–6 (high risk)	Africa	8	75	25
		Asia	25	48	52
Action Plans (species)	# 1–2 (low risk)	Africa	30	100	0
		Asia	18	62	38

Across lowest-rank taxa (species or subspecies), the number of critically endangered insular primate taxa in Asia is an order of magnitude higher than that in Africa (thirteen taxa vs. one). Asia also has more continental taxa in this category, but the difference between the two regions in this case is much less marked (seventeen taxa vs. six). We then made the same comparison between species using the PSG Action Plans threat classification system. The same trends were seen. In addition, we distinguished between high-risk and low-risk taxa. In the low-risk category, most taxa in both Africa and Asia were continental. In the high-risk category, the proportion of continental taxa declined and the proportion of insular taxa increased (compared with that seen in the low-risk category) in both Africa and Asia. These results support the hypothesis that insular primate taxa are more likely than continental taxa to be at high risk of extinction. They also support the hypothesis that Asia has more highly threatened taxa than Africa because it has more insular species.

Nevertheless, across both lowest-rank taxa and species, Asia contains more high-risk continental taxa. Turning then to the second hypothesis, comparing deforestation rates between Africa and Asia shows that Asia currently has half the area of rain forest yet exhibits a percentage annual loss of tropical forest almost twice that of Africa (table 8.2). This may well explain why Asia has more highly threatened continental species. Note that current habitat disturbance rather than previous disturbance may be most important, since Africa appears to have lost more of its original closed forest cover than Asia (Asia, $N = 12$ countries with 46% mean remaining forest cover; Africa, $N = 36$ countries, with 23% remaining forest: Groombridge 1992).

In summary, the severity of extinction risk among primates does vary across tropical regions. A proportion of this variation may reflect regional differences in both the insular faunal component and the severity of habitat disturbance.

TEMPORAL PATTERNS OF RISK

A temporal framework in patterns of global extinction is inherent in the latest 1996 Red List classification system, since each threat category is associated with a probability of extinction within a specified period (table 10.3). According to this system, if no conservation action is taken, we might expect seven Critically Endangered species to disappear within the next eighteen to sixty-three years, given that there are thirteen CR species, each is associated with a 50% probability of extinction within three generations (or a ten-year period, if longer), and the shortest and longest generation times are about six and twenty-one years (generation length is defined as the average age of breeding individuals, which can be estimated for the smallest and largest CR species using life history data from Rowe 1996). The distribution of these taxa may approximate two in Madagascar, two in Asia, and three in the Americas (table 10.5).

Smith et al. (1993b) suggest that changes in the Red List codes can be used to estimate variation in extinction risk over time. They present a comparative analysis of such changes across the major plant and animal taxa, including mammals, between 1986 and 1990 (during which time three Red Lists were issued). We have conducted a similar analysis for threatened primates from the three Red Lists issued from 1988 through 1994 (table 10.7). Although these data have to be interpreted with caution, given that the original Red List codes were quite subjective, some preliminary observations can nonetheless be drawn.

First, more taxa join the Red Lists over time than leave them (and of those few taxa that left the Red Lists in 1990, *Leontopithecus caissara* reappeared in 1994). Second, in all but one taxon (the hylobatids), most taxa have not changed in status over this seven-year period. In particular, none of the threatened Lorisiformes or pongids show a change

Table 10.7 Temporal dynamics of primate species extinction risk

	Threatened species[1]			Changes in conservation status		
Taxon	Joining lists (n)	Leaving lists (n)	Total number of taxa[2]	Increase in risk (%)	Decrease in risk (%)	No change (%)
Lorisiformes	1	0	5	0	0	100
Lemuriformes	2	0	24	33	0	67
Tarsiiformes	2	1	3	0	0	100
Callitrichidae	1	1	15	22	0	78
Cebidae	4	0	21	17	8	75
Cercopithecinae	4	1	21	6	6	89
Colobinae	1	2	15	15	5	59
Hylobatidae	0	0	5	60	0	40
Pongidae	0	0	4	0	0	100

Note: Temporal dynamics are estimated from analysis of the IUCN Red Lists, where data are combined across two periods (1988–1990 and 1990–1994). Data from IUCN 1988, 1990 and Groombridge 1993.

[1]We exclude subspecies from this part of the analysis, since changes in the Red List presence or absence of subspecies appears to be at least as strongly influenced by changes in taxonomy as by changes in conservation status.
[2]Total number of taxa refers to the number of taxa that are listed in the 1994 Red List.

in status over this time. Third, where changes have occurred, the vast majority show an increase in extinction risk (i.e., a transition from "rare" to "vulnerable" or from "vulnerable" to "endangered"). There are only three groups where a decline in extinction risk has been observed: the Cebidae, Cercopithecinae, and Colobinae. In only one of these cases does the number of declines equal the number of increases (the cercopithecines).

Although some of these changes are likely to be confounded by taxonomic revisions, these results provide a preliminary guide to temporal changes in primate extinction risk. Although the status of most taxa has not deteriorated, their situation has not improved either. This is borne out by the disproportionate increase in the number of species joining and by the number of increases in threat rating for those taxa already on the list. The substantial number of increases in risk among the hylobatids is consistent with our earlier identification of Asia as the most seriously threatened region (see above). However, the considerable variation seen between taxa in this analysis suggests that biological differences in vulnerability to extinction may also exist between groups. We investigate the potential existence of such differences below.

PHYLOGENETIC PATTERNS OF RISK

Analysis of the Red List codes across all mammals shows that the primate order contains significantly more threatened species than is typical for

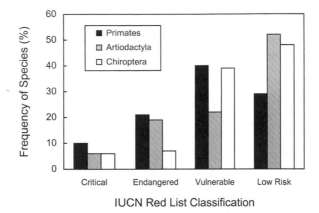

Figure 10.1 Red List classification of species of Primates, Artiodactyla (even-toed ungulates such as pigs, deer, and antelope), and Chiroptera (bats). Species classified as Extinct, Extinct in the Wild, Low Risk least concern, or Data Deficient are not shown. (Data from Baillie and Groombridge 1996.)

most other mammalian orders (Mace and Balmford 2000) (fig. 10.1). Similarly, McKinney (1997) has recently noted that primates are generally more extinction prone than other mammals and show a correspondingly short average species duration in the fossil record (see sec. 7.1): in other words, primates appear to be intrinsically more vulnerable to extinction than many other mammal groups. However, within the order the pattern of extinction risk among contemporary primates shows a pattern broadly similar to that seen in two other mammalian taxa (fig. 10.1): most threatened taxa are Vulnerable, and relatively few are Critically Endangered.

At a finer scale, important differences in extinction risk also emerge between primate groups. Compared with other mammalian families, the Galagonidae has significantly fewer threatened species than expected, while the Indriidae and Lemuridae have significantly more highly threatened species than expected (Mace and Balmford 2000). Similarly, a simple tabulation indicates that, compared with other primates, a relatively small proportion of Lorisiformes are threatened but that a relatively high proportion of lemurs, callitrichids, and hylobatids are Critically Endangered (table 10.8).

These differences may be influenced by spatial trends: for example, the large number of Critically Endangered lemurs and hylobatids is likely to be related to the insular distribution of the former taxon and the Asian distribution of the latter (see above). However, even within regions, marked differences exist between taxonomic groups (table 10.9): within both Africa and Asia, cercopithecine and colobine species experience higher extinction risk than lorisiform species.

Table 10.8 Taxonomic distribution of primate species extinction risk

Taxon	Total number of species	Percentage of taxa in each Red List category[1]			
		LR	VU	EN	CR
Lorisiformes	21 (4)[2]	88	12	0	0
Lemuriformes	32 (0)	41	34	13	13
Tarsiiformes	7 (5)	100	0	0	0
Callitrichidae	30 (1)	55	21	14	10
Cebidae	60 (2)	69	21	7	3
Cercopithecinae	56 (2)	67	19	13	2
Colobinae	36 (5)	42	32	19	6
Hylobatidae	11 (3)	50	25	13	13
Pongidae	4 (0)	0	25	75	0

Source: Data from Baillie and Groombridge 1996.

[1]Codes are abbreviated as follows: Low Risk (LR), Vulnerable (VU), Endangered (EN), and Critically Endangered (CR). VU, EN, and CR species are considered threatened.
[2]Numbers in parentheses are species classified as "Data Deficient." The values in the last four columns are given as the percentage of all species excluding those that are Data Deficient.

Table 10.9 Taxonomic distribution of primate species extinction risk by geographic region

Region and taxon	Total number of species	Percentage of taxa in each Red List category[1]			
		LR	VU	EN	CR
Global					
Lorisiformes	21 (4)[2]	88	12	0	0
Cercopithecinae	56 (2)	67	19	13	2
Colobinae	36 (5)	42	32	19	6
Africa					
Lorisiformes	18 (4)	100	0	0	0
Cercopithecinae	37 (1)	78	14	8	0
Colobinae	7 (0)	71	29	0	0
Asia					
Lorisiformes	3 (0)	33	67	0	0
Cercopithecinae	19 (1)	44	28	22	6
Colobinae	29 (5)	33	33	25	8

Source: Data from Baillie and Groombridge 1996.

[1]Codes are abbreviated as follows: Low Risk (LR), Vulnerable (VU), Endangered (EN), and Critically Endangered (CR). VU, EN, and CR species are considered threatened.
[2]Numbers in parentheses are species classified as "Data Deficient." The values in the last four columns are given as the percentage of all species excluding those that are Data Deficient.

The observation that extinction risk, and indeed extinction rate, is usually unevenly distributed across species within a taxon has already been made for birds and mammals (Bennett and Owens 1997; McKinney 1997; Russell et al. 1998; Mace and Balmford 2000). One suggestion is that species in smaller clades are more vulnerable to extinction. However, the evidence for this pattern in contemporary mammals is equivocal (Russell et al. 1998; Mace and Balmford 2000). Alternatively, previous authors have noted that certain traits may predispose species to extinction, so that certain taxonomic groups characterized by vulnerable traits will have higher rates of species extinctions: the results of these studies are described in section 7.3.2. Whatever the processes that underlie these patterns, it is clear that extinction events are unlikely to be indiscriminate across primates species.

10.2.3. Population Viability Analysis

Attempts to estimate the extinction risks of populations or species have focused on a form of modeling known as *population viability analysis* or PVA (Shaffer 1981: for reviews, see Boyce 1992; Burgman, Fearson, and Akçakaya 1993; Nunney and Campbell 1993; Lacy 1993–1994, Durant and Mace 1994). The aim of PVA is to determine the minimum size of population that will guarantee species survival according to some criterion (commonly 95% chances of survival as a viable breeding population for the next thousand years). Four basic types of viability analysis can be identified: subjective assessment, rules of thumb, analytical population models, and computer simulation models (Durant and Mace 1994).

Subjective assessment is the least satisfactory, since it is often based on personal predilections. To deal with this problem, rules of thumb have been developed that seek to identify some general principles to guide decisions. The best known is the 50/500 rule of Franklin (1980) and Soulé (1980), which suggests that populations with a genetic effective population size, N_e, of fewer than fifty in the short term or five hundred in the longer term are especially liable to extinction (see sec. 7.2.2). Another is the Dobson and Lyles (1989) suggestion that primate populations will collapse when adult female survival is less than 70% across the typical interbirth interval. However, these kinds of approaches have been criticized on the grounds that they ignore many crucial features of both the biology of the animals and the ecology of the local habitat. Analytical and simulation models attempt to get around these problems by estimating the extinction risks of individual populations directly.

Analytical models have used generalized demographic equations (such as the classic logistic equation or a Leslie matrix approach) to predict the future size of the population, using estimates of population-specific fecundity and mortality rates (or the intrinsic rate of increase,

r_m) and carrying capacity (K). Analytical models have been criticized on the grounds that they contain no measures of ecology and that it is often difficult to identify the appropriate parameter values to measure (Boyce 1992; Caughley and Gunn 1996). It is not always obvious, for example, how to estimate the parameters r_m and K in the logistic model. Accurate age-specific fecundity and survivorship schedules may also be difficult to determine for use in Leslie matrices. Such problems in parameter estimation can lead to grossly inaccurate estimates of extinction risk (Taylor 1995).

Simulation models circumvent some of these problems because they make use of a wider range of quantitative data on the species' ecology and life history to build a model that can be used to study the interactions between environmental parameters and the animals' functional responses. Such models can often involve feedback loops and may thus have complex nonlinear dynamics. One advantage of simulation models is that they allow us to explore the consequences of environmental stochasticity for a species' population dynamics, although they require more data. Simulation models are clearly the preferred method, if only because they are biologically more realistic and because they allow us to use sensitivity analysis to explore how far the outcome is dependent on the precise values of parameters.

A number of PVAs have now been carried out on primate populations (we review four below). Despite their often meticulous detail, it remains important to exercise caution in interpreting the results of such analyses. Their predictions are only as good as the model's ability to mimic reality and the quality of the data used. In particular, Caughley and Gunn (1996) point out that PVAs are invariably based on single-species models, whereas in real life species in biological systems are also affected by predators and ecological competitors. In addition, the PVA approach frequently assumes that the conservation problem is small population size, when we should perhaps be focusing on the extrinsic causes of small population size. Because PVA is designed to examine the consequences of low numbers rather than its causes, the status quo may often be accepted as the appropriate conservation management strategy when an answer to the question "Why did population size decline?" might produce very different recommendations.

Harcourt (1995) has forcefully pointed this out in the case of the mountain gorilla (*Gorilla g. berengei*). Three PVAs (Weber and Vedder 1983; Akçakaya and Ginzburg 1991; and Durant and Mace 1994) used a simple demographic model, a Markov model, and the POPGEN Leslie matrix model, respectively, to estimate demographic extinction probabilities. (The basic Leslie matrix approach is described in appendix 2.) The last was the most detailed in that it considered population subdivision into breeding groups and age structure as well as age- and sex-specific migration and reproductive rates. Even though it was the most pessimistic of the three, it still yielded a risk of demographic extinction after 750 years of just 10%.

Given these estimates, we would be led to infer that the gorilla is unlikely to go extinct in the immediate future. As Harcourt (1995) points out, however, these models ignore crucial features of the conservation biology of the gorilla, notably the effects of habitat loss. The Virunga gorilla population is surrounded by one of the highest-density human populations in Africa. Since country by country, the human population correlates significantly positively with the rate of deforestation, and since the human population was doubling every twenty-five years in Rwanda before the civil war, we can only expect the pressure on the Virunga habitat (and thus the gorillas themselves) to increase with time. Based on the current rate of deforestation, Harcourt (1995) calculated that it will take at most 100 years for all the forests in the region to be felled, at which point these forest-dwelling gorillas will inevitably become extinct. Similar calculations suggest that only three African countries (Central African Republic, Gabon, and People's Republic of Congo) have forest reserves that are likely to survive for the next 150 years at the current rates of deforestation. By implication, the gorilla will be extinct in at most 200 years. Concerns about the species' demographic viability are thus largely irrelevant. The real problem is deforestation.

These caveats aside, PVA can still be a valuable tool in appropriate areas of conservation management. One case in point would be the design of protected areas large enough to ensure that the extinction risk of the populations protected is at an acceptable level (Caughley and Gunn 1996; see sec. 11.1.2). A number of simulation software models now exist, including POPGEN (Durant and Mace 1994) and VORTEX (Lacy 1993), that can be used to run PVAs. In addition, PVAs can be extended to incorporate the broader features of the ecology of the study population, including both the habitat in which it is embedded and the social and economic aspects of the local human population. Such analyses are usually referred to as *population and habitat viability analyses* (PHVA) (Lacy 1993–1994). Though more diffuse in their approach, PHVAs have the advantage that they often require bringing together experts from different fields to discuss the situation pertaining to a given population. Rylands (1993–1994) provides an example where PHVA workshops have helped to identify essential conservation tactics in the case of the lion tamarins (*Leontopithecus* spp.), all of which are highly endangered. In the case of *L. rosalia*, nine specific proposals were identified that included the purchase of other forests outside existing reserves, measures to prevent fires and pollution within reserves, reforestation of degraded habitat within reserves, and the expansion of the existing reintroduction program.

TANA RIVER CRESTED MANGABEY

The Tana River crested mangabey (*Cercocebus galeritus galeritus*) occurs in riparian forest patches along the Tana River (Kenya). Although this system might function as a metapopulation (sec. 6.3), a previous study by Kinnaird and O'Brien (1991) ran a conventional PVA on this population using Lande and Barrowclough's (1987) analytical equations that incorporate the effects of variance in progeny number, unequal sex ratios, and fluctuating population size to determine the effective population size, N_e (and hence the risk of loss of genetic viability). They then combined the results of this analysis with Goodman's (1987) demographic model to calculate persistence time (i.e., expected time to extinction). Model parameters (numbers of breeding males and females, mean and variance in sex-specific lifetime reproductive output, mean generation length, fraction of individuals exchanged between groups per generation, probability of an infant's surviving to reproduce, and population growth rate) were estimated from population field data.

The model shows that persistence time is a linear function of population size, declining from 330 years in 1974 to 190 years in 1988 as a result of a putative decline in total population size. However, this does not guarantee that the population will survive for that length of time; stochastic events will result in the population's going extinct earlier than this in about half the cases! They therefore used simulation to determine probabilities of extinction within different time periods for persistence times up to 2,000 years (fig. 10.2). The results of this analysis suggest that, for a persistence time of 200 years, the chance that extinction will occur within the next century is 40%, clearly considerably higher than one would wish. To give a 95% probability of survival over the next 100 years, you would need a population of eight thousand individuals (an order of magnitude larger than the subspecies total) or you would need to extend the mean persistence time to 2,000 years.

This rather bleak prognosis is partly a consequence of the fact that the population is distributed along both banks of the Tana River. The river appears to act as a barrier to migration, thereby creating two effectively isolated subpopulations. Were the two subpopulations able to mix, the situation would be greatly improved, with survival being contingent on a proportionately smaller population size or lower persistence time. As the situation is at present, the effective population size on each side of the river may be as small as $N_e \approx 36$, a size that is unlikely to protect the species against loss of its genetic variation (but see Butynski and Mwangi 1994).

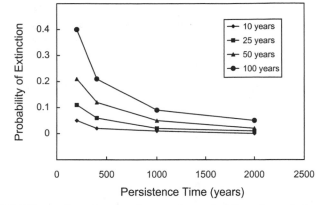

Figure 10.2 Effects of persistence time on the probability of population extinction within four specified time scales for the Tana River crested mangabey *Cercocebus galeritus galeritus*. A population with a persistence time of four hundred years would have only about a 3% chance of extinction in ten years (bottom line) but a greater than 20% chance of extinction in one hundred years (top line). (After Kinnaird and O'Brien 1991.)

BARBARY MACAQUE

The Barbary macaque (*Macaca sylvanus*) was widely distributed through-out the Mediterranean region during Pleistocene times, but now it exists only in relict populations in the Atlas and Riff Mountains of Morocco and Algeria. It also has a foothold on the European mainland in the form of a small managed population on Gibraltar. The Gibraltar population has been artificially provisioned since 1940; as a result, it has grown from a single group of approximately 10 animals in the late 1930s to about 150 animals in six groups by 1994 (Fa and Lind 1996).

Fa and Lind (1996) estimated that the minimum viable population size required to maintain 90% of the genetic diversity over the next century is 250–380 animals, about twice the size of the current population. With this as the best estimate for carrying capacity, they used the VOR-TEX model to simulate population growth rates over a hundred-year period under a range of thirteen demographic and life history scenarios based on variations in female fecundity rates, survival rates, and the probability that a disease catastrophe would decimate the population. Each population began with 100 animals, and a thousand replicates were run for each scenario.

The main findings were that no population ever reached the carrying capacity of 250 animals; that annual birthrates below 43% invariably resulted in extinction, even in the absence of catastrophes; and that even with the probability of a catastrophe as low as 0.1 per annum, extinction was inevitable unless birthrates exceeded 70% per annum. Since actual fertility exceeded this level in only about fourteen years between 1936 and 1994 for each of the two main groups (less than 25% of the time), the implications for the survival of an unmanaged population are at best bleak. The impact of disease was shown to be particularly important in this context: with a birthrate of 60% per annum, no population reached carrying capacity if there was an annual probability of a major disease outbreak higher than 0.04 (for outbreaks that reduced reproduction by 50% and survival rates by 35–50%).

WOOLLY SPIDER MONKEY

The woolly spider monkey, or muriqui (*Brachyteles arachnoides*), is a species of large South American monkey that is now confined to approximately nineteen scattered patches of Atlantic coastal forest in southeastern Brazil (Mittermeier et al. 1987; K. Strier, pers. comm.). Strier (1993–1994) carried out a PVA using VORTEX for one isolated population, with parameter values estimated from the population itself. The simulations looked at future changes in population size over a hundred-year period. Of particular importance in this case was the number and range of sensitivity analyses that Strier ran to assess the impact of parameters on population growth rates: these included variation in the impact of catastrophes on reproduction and survival rates, female age at first reproduction, natal sex ratios, carrying capacity, inbreeding effects, and environmental stability, as well as various combinations of these factors.

The main finding was that, under current demographic conditions, population growth rates were most strongly influenced positively by carrying capacity (indexed as the area of forest reserve) and female-biased natal sex ratios (the latter through its impact on recruitment to the cohort of breeding females: fig. 10.3) and negatively by environmental stochasticity. Female age at first reproduction had a significant effect (comparing values of seven versus eleven years for age at first reproduction), but its effect was simply to slow population growth rate (and thus reduce final population size by about 15% after a century). In addition, major catastrophes (assumed to occur once every fifteen years) influenced population growth rates only if they affected adult survival rates; their effects on birthrates had a negligible impact on population growth providing these extended only to the year in question. Finally, even though the total population size was only fifty-one animals, inbreeding risks were trivial compared with the impact of environmental factors.

The most obvious implication of the PVA is that the population's chances of surviving the next one hundred years will be maximized by increasing the habitat available and allowing the population to expand naturally. Long-term demographic lags play a role here too, because the

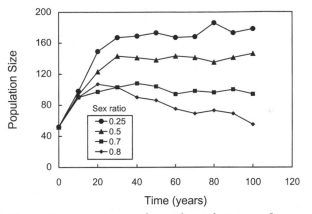

Figure 10.3 Fluctuations over time in the total population size for a woolly spider monkey population in Brazil's Atlantic Forest for different adult sex ratios (males per female), as determined from a PVA model (with all other variables held constant). The population will grow only providing there are at least two females for every male. (After Strier 1993–1994.)

impact of natal sex ratio starts to have its effect on population size only after twenty to thirty years.

SAMANGO MONKEY

The samango monkey, *Cercopithecus mitis,* is a small arboreal forest monkey common in wooded habitats along the eastern and southern coastline of Africa. Swart, Lawes, and Perrin (1993) developed a simple population demographic model based on the species' life history characteristics and the way these respond to population density. An innovative feature of this model is that it includes an Allee density-dependent effect on "fecundity" (the number of infants that survive to be recruited to the population as adults, a variable we will term *net fecundity* to avoid confusion with the birthrate). Swart, Lawes, and Perrin (1993) assumed that net fecundity falls at low population densities because infanticide by males invading new troops becomes more frequent under such conditions. The model's aim was to identify the threshold population density below which demographic collapse was inevitable. The parameters for the model were estimated directly from data on captive and natural populations reported in the literature. The model was age structured, with age- and sex-specific rates of mortality, maturation, migration, and reproduction that were also density dependent.

The simulations suggested that populations with medium to high densities (relative to the local habitat's carrying capacity) exhibited strong "equilibrium-seeking" behavior: populations within 50% of the equilibrium density (the local carrying capacity) take less than fifteen years to reach equilibrium again following perturbation (e.g., by an environmental catastrophe). This is primarily due to a feedback loop within the model that links birthrate and population density in a positive relationship. At densities below this threshold, however, the net fecundity function switches sign and, as a result of the negative feedback loop, recovery lags (the time taken to reach the original population size after a perturbation) rise dramatically; consequently, when population density is below 25% of the equilibrium value, the population is often driven to extinction.

This result mainly occurs because net fecundity is assumed to be a humped function of relative density (i.e., density relative to the carrying capacity): net fecundity declines as density exceeds carrying capacity because food resources come under increasing pressure, but it also declines when density falls significantly below carrying capacity because of the Allee effect. At relative densities between 0.5 and 1.0 times the carrying capacity, net fecundity is assumed to be stable and at maximum capacity. Sensitivity analyses in which mortality and net fecundity rates were varied indicate that net fecundity has a much higher impact than mortality. Nonetheless, varying the net fecundity rate by as much as 40% either side of the empirical value of 50% per annum does not alter the qualitative predictions of the model: the extinction risk is high whenever the population density is below about 25% of the carrying capacity.

Swart and Lawes (1996) extended this approach to examine metapopulation survival in the highly fragmented forests of the Kwazulu-Natal midlands of South Africa using a stochastic spatial population model. The metapopulation consists of about eight forest patches, none containing more than twelve troops of about twenty monkeys each. Migration between forest patches occurred either when males dispersed from their natal groups at maturity or, less often, when daughter groups moved to a new forest patch in search of a vacant territory after group fission (the likelihood of fission, however, was dependent on the size of the troop and the density of the population). By incorporating the effects of troop location (core vs. edge positions within forest patches) as well as the impact of stochastic environmental and demographic events into this model, they considered the way growth trajectories of individual troops within the metapopulation were affected by different levels of connectivity between the forest blocks. The model suggested that neither translocating troops nor providing connecting corridors would influence metapopulation survival over 200 years, but corridors could have a very strong effect on long-term survival over periods on the order of 800 to 1,500 years (see also sec. 11.4).

10.3. Setting Area Priorities

Area-based conservation priorities are less easy to set than those that are taxon based because the unit of conservation is harder to identify. Areas can be defined at the local, regional, continental, or global scale, although most strategies adopt a combination of these. The PSG Action Plans are generally continental in scope but operate at a combination of local and regional scales. Although they take into account political issues and constraints, they are independent of political boundaries. In contrast, several studies have considered as area conservation units either countries (e.g., for South America: Mittermeier and Coimbra-Filho 1977; Mittermeier, Kinzey, and Mast 1989; Rylands, Mittermeier, and Rodriguez-Luna 1997) or districts within countries (e.g., regions within Venezuela: Rodríguez and Rojas-Suárez 1996). The advantage of this approach is that the legislative and political decisions that can play such a pivotal role in conservation are often delimited by the boundaries of the units, but these units are often biologically meaningless in terms of identifying the limits of distinct ecosystems, communities, and populations. In most cases the decision on which approach to use is based on pragmatism and the aims and needs of each particular strategy.

Once the area scale has been decided, it is possible to define the set of areas available for consideration. The PSG Action Plans usually do this by adopting well-recognized biogeographical regions and then subjectively identifying key sites for local action within them. In all the plans, these local sites are most often identified by making use of preexisting protected areas (e.g., national parks). This means that the plans operate at what is likely to be the most tangible, practical, and useful area-based management unit. Nevertheless, more objective methods can also be used, which can help locate important sites outside preexisting protected areas. For example, once a region has been selected, it can be divided into an equal-area grid in which each cell is evaluated for potential conservation value. Such studies have been conducted both for South America (with a 500 by 500 km quadrat scale: Mittermeier and Coimbra-Filho 1977) and for Africa and Madagascar (with a 1° latitude-longitude scale: Hacker, Cowlishaw, and Williams 1998). This method forces quantitative evaluation of all possible area options, whereas the subjective method uses specialist knowledge to identify the key areas only. The former has the advantage that it forces all possible areas to be considered equally in an explicit fashion; its disadvantage is that it requires an extremely good database, and this is rarely available at finer scales. The paucity of suitable information for such analyses is illustrated by the prominent role given to surveys in the recommended conservation plans of the PSG Action Plans.

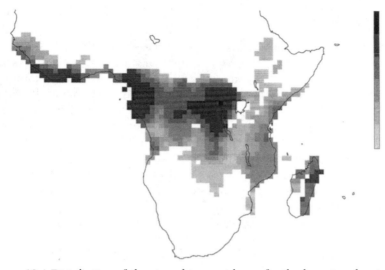

Figure 10.4 Distribution of threatened taxon richness for the lowest-rank primate taxa (species or subspecies) in Africa and Madagascar (at the 1° latitude-longitude grid cell scale). Threatened taxa are those classified as Vulnerable, Endangered, or Critically Endangered according to the IUCN Red List codes. The grid cell with the maximum value is shown in black, whereas the other nonzero scores are grouped into five classes (corresponding to the gray scale on the right), containing approximately equal numbers of grid cells. The map has been smoothed by taking each cell's score as the mean score of the surrounding cells. The five cells containing the Barbary macaque in North Africa are not shown. (Redrawn with the permission of Elsevier Science from Hacker, Cowlishaw, and Williams 1998.)

10.3.1. Currencies in Area Evaluation

Which area evaluation measure to use depends on the goals of the conservation strategy. Here we discuss five alternatives. The simplest measure, primate taxon richness, is usually indexed by the number of taxa, on either a continuous scale (e.g., Mittermeier and Coimbra-Filho 1977; Hacker, Cowlishaw, and Williams 1998) or a categorical scale (e.g., Oates 1986a; Eudey 1987). Taxon richness can also be defined by subgroups, for example, threatened species richness (fig. 10.4). Sites can then be ordered by their taxon richness score.

Second, taxon richness can be weighted by the evolutionary uniqueness of those taxa present (see sec. 10.2.1). According to these weightings, areas with a similar number of taxa may score differently depending on their taxon complement. The currency most commonly used to calculate these indices is termed *character richness*, since these indices are normally based on phylogenies, phylogenies are constructed on the basis of characters (genetic, morphological, behavioral), and character richness

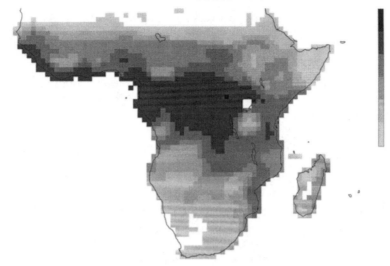

Figure 10.5 Distribution of character richness for the lowest-rank primate taxa in Africa and Madagascar (at the 1° latitude-longitude grid cell scale). Character richness is calculated as the number of intervening nodes between the two most phylogenetically distant taxa in each cell. A composite phylogeny was used so these patterns are likely to approximate "overall" character diversity. The grid cell with the maximum value is shown in black, whereas the other nonzero scores are grouped into five classes (corresponding to the gray scale on the right), containing approximately equal numbers of grid cells. The map has been smoothed by taking each cell's score as the mean score of the surrounding cells. The five cells containing the Barbary macaque in North Africa are not shown. (Redrawn with the permission of Elsevier Science from Hacker, Cowlishaw, and Williams 1998.)

indices are designed to attribute greatest value to that set of organisms that maximizes the variety of characters available (Humphries and Williams 1994; Williams and Gaston 1996). Inevitably, these indices can maximize the diversity of only those characters used to construct the phylogeny from which the index is calculated; which characters make the best currency depends on which aspects of diversity one wants to maximize (Williams, Gaston, and Humphries 1994). Geographical patterns of character richness have recently been described for the primates of Africa and Madagascar (fig. 10.5, Hacker, Cowlishaw, and Williams 1998). Note that taxon richness and character richness show a strong resemblance (compare figs. 2.3 and 10.5), but that Madagascar scores relatively poorly in character richness despite its unusual complement of primates (this is because the branch lengths of the phylogeny are not incorporated into this particular measure of character richness: cf. table 10.2).

The third common measure of area value is the number of rare (small distribution) taxa, or endemics, present. Importance is attached to such restricted-range taxa because rarity may be equated with extinction risk and these taxa are, by definition, rarely found in other areas. The collective restricted-range rarity of an assemblage may be scored on one of two scales. The first is categorical and uses the number of endemic taxa (e.g., Oates 1986a; Eudey 1987). The second is continuous: in this case the rarity of each taxon present is measured as the reciprocal of its range size and the overall score is calculated as the sum of reciprocal range sizes across all taxa present ("rarity-weighted richness": Williams et al. 1996). Geographical patterns of rarity-weighted richness have already been presented (sec. 5.1).

The fourth measure of area value is degree of threat to the ecological integrity of the area. Sources of threat may include agriculture, forestry, road building, or hunting pressure (Oates 1986a; Eudey 1987), although more indirect measures of threat may be used, such as human population density. This type of value is more difficult to quantify than the preceding three indices, not least because it is a predictive rather than a descriptive measure. For this reason it is usually estimated only for well-known areas at a fine-grain scale. Oates (1986a) and Eudey (1987) estimated a categorical degree of threat for areas of conservation action, but in this case key areas had already been selected so that threat had to be assessed only for a small subset of all possible areas.

Fifth, threat can also be equated with viability, such that areas with larger populations are given priority because they are likely to be more viable in the long term. Although this measure is also difficult to use in practice (not least because it requires good-quality data on population size: Turpie 1995), it can be extremely important in cases where conservation areas are being selected for a particular species. For example, it has been suggested that protected areas should be selected in the core of a species' range rather than on its periphery, since the core is more likely to contain viable populations (see sec. 5.3.1). Similarly, Dias (1996) has pointed out that a decision to conserve a sink habitat in a metapopulation will be a waste of time because the inevitable consequence will be its extinction as soon as its migrant-source population is cut off (see sec. 6.3).

10.3.2. Priority Ranking of Conservation Areas

Within any single area evaluation measure, priorities can be set simply by ranking areas in order of their scores (in the same way we ranked countries by primate species richness in sec. 2.2). The term "hot spots" has been used to refer to the top scoring 5% of areas; similarly, "cold spots" can be defined as the lowest scoring 5% (Prendergast et al. 1993; cf. Myers 1988, 1990). Hacker, Cowlishaw, and Williams (1998) have identified hot spots for taxon richness, rarity-weighted richness, and threatened taxon richness among African and Malagasy primates (fig. 10.6). Two important points emerge from a comparison of these alternative classifications. First, areas that score high in one currency may not score high in another. Although extinction risk is often equated with rarity, hot spots of rarity-weighted richness do not always coincide with hot spots of threatened taxon richness. Second, the proportion of taxa that are threatened increases in areas of high taxon richness. This pattern might arise because threatened taxa often require an area of high-quality habitat to maintain a viable population and these also tend to be areas of high taxon richness, or alternatively, because areas of high taxon richness are more severely threatened by human activity and therefore contain more threatened taxa.

Setting priorities for areas according to their currency scores will maximize value on an area by area basis. However, blanket protection of the highest-scoring hot spots may not be the most efficient way to maximize the value of an entire network of areas. For example, area by area, two adjoining areas might be selected because they both have high taxon richness; however, if the taxa in each area are identical and the goal is to maximize taxon richness across all areas, it would be more efficient to select another area with fewer species but fewer duplications. This is the principle of *complementarity* (e.g., Vane-Wright, Humphries, and Williams 1991). In these circumstances there is less emphasis on finding the top-scoring areas and more emphasis on finding that set of areas that achieves the conservation goal with maximum efficiency. A number of iterative algorithms have been developed to this end (e.g., Margules, Nicholls, and Pressey 1988). The most common goal is to represent all taxa with some specified frequency in a network composed of the fewest areas possible (e.g., Williams et al. 1996; see also Turpie 1995).

Hacker, Cowlishaw, and Williams (1998) investigated how many areas are required to represent all lower-rank primate taxa at least once in Africa and Madagascar (at the 1° grid cell scale) using a three-step algorithm: select those areas that have taxa limited solely to those areas, order all the remaining areas according to their rarity-weighted richness scores (where there is a tie, select the area with the greatest number of taxa), and pass backward through this sequence and eliminate redundant areas by removing those that lack unique taxa within the selected set. The resulting set of areas, the *near-minimum set,* is shown in figure 10.7. Comparing the efficiency achieved through the methods of complementarity and hot spot selection (table 10.10) clearly illustrates the strong performance of complementarity algorithms (see also Pressey and Nicholls 1989; Williams et al. 1996).

One problem with these methods is their dependence on univariate measures of value, although their advantage is that the decision process

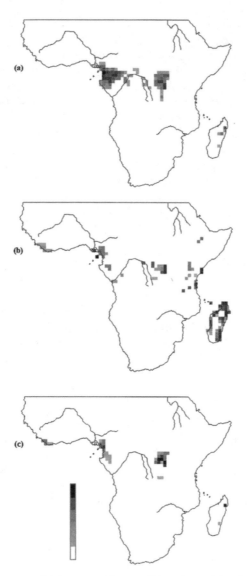

Figure 10.6 Hot spots of (a) taxon richness, (b) rarity-weighted richness, and (c) threatened taxon richness for the lowest-rank primate taxa in Africa and Madagascar (at the 1° latitude-longitude grid cell scale) (cf. figs. 2.3, 5.2, and 10.4, respectively). Grid cells with maximum values are shown in black, whereas the other nonzero values are grouped into four classes (corresponding to the gray scale at the bottom left), containing approximately equal numbers of grid cells. Note that the hot spot positions do not always precisely match the patterns shown in the preceding figures, since the hot spots are identified from the unsmoothed data. (From Hacker, Colishaw, and Williams 1998 with the permission of Elsevier Science.)

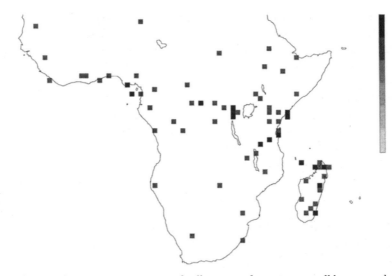

Figure 10.7 The near-minimum set of cells required to represent all lowest-rank primate taxa in Africa and Madagascar at least once (at the 1° latitude-longitude grid cell scale). (Data from Hacker and Cowlishaw, unpublished.)

Table 10.10 Comparison of hot spot and complementarity area selection methods

Evaluation	Primates and criteria	Number of cells	Number of taxa	Records per taxon[1]
Hot spots				
Taxon richness	All taxa	92 (5)	92 (45)	13.4
Threatened taxon richness	Threatened taxa	40 (5)	46 (51)	5.9
Complementarity				
Taxon richness	All taxa	69 (4)	205 (100)	2.4
Threatened taxon richness	Threatened taxa	41 (5)	90 (100)	1.7

Note: The figures in parentheses are the absolute values expressed as a percentage of the total number of cells/taxa present. After Hacker, Cowlishaw, and Williams 1998.

[1]Total number of records across represented taxa, divided by the number of represented taxa.

is explicit (e.g., Turpie 1995; Williams 1999). A simple method of taking additional factors into account is screening areas before selection; for example, if only highly threatened areas are of conservation interest, then all other areas can be filtered out before the identification of a near-minimum set (e.g., Williams 1999). Hence a combination of the four currency measures described above could potentially be used in a sequential screening process before setting eventual priorities based on a single currency.

In practice, exercises in setting area priorities are usually carried out on a specific subset of areas (e.g., protected areas for Afrotropical antelope: Kershaw, Williams, and Mace 1994; Kershaw, Mace, and Williams 1995). Oates (1986a) and Eudey (1987) identify a subset of sites for setting priorities for primate conservation in Africa and Asia. In Africa, this network of sites is subjectively chosen to represent all threatened taxa at least once and to be built largely around existing protected areas. Within this set, Oates used a summed four-criterion scheme: the number of high-priority primate species present, the imminence of threat to the ecological integrity of the area, the overall primate species richness, and the number of endemic primates present. Oates (1996a) has since simplified this system, adopting a single three-point priority rating scheme based on the number of species of conservation concern in the area and the degree of existing protection (although a very high rating is also assigned if the area is the only habitat of one or more endangered taxa).

The potential of local area-priority exercises for practical conservation management can be illustrated by recent attempts to identify conservation areas for lemurs in one region of western Madagascar (Smith, Horning, and Morre 1997) and for tropical forest in Pará State of Brazilian Amazonia (Veríssimo et al. 1998). In the first study, Smith and colleagues conducted an intensive survey of sites throughout the area to determine the ecological and land-use correlates of lemur species richness. Based on this information, they mapped predicted lemur species richness and then superimposed onto this map a second map showing the distribution of areas of low, moderate, and high risk of habitat disturbance. Combining these two maps allowed them to identify areas of high species richness at high risk of disturbance. On this basis the authors were able to show that the principal existing reserve in the area was poorly located (low lemur diversity, high risk of disturbance) and to make concise recommendations for the design of an alternative protected area system with a core zone (high diversity, low risk), a buffer zone, and a multiple resource use area.

In the second study, Veríssimo and colleagues estimated the areas of tropical forest at risk from logging by combining spatial data on tree species distributions, transport routes (rivers, roads), and sawmills with economic data on transportation costs and timber prices. They determined that 80% of 1 million km² of forest in Pará State is economically accessible for logging under current conditions, although 29% of forest is economically accessible only for mahogany and a further 30% is economically accessible only for twelve to twenty of the most profitable timber species. By combining their map of economic accessibility with maps of existing protected areas (many of which fall into areas of high economic accessibility) and priority conservation areas (areas of high species richness and rarity, many of which are not protected), the authors were able to develop a management proposal for the complementary zoning of conservation and timber extraction in this region.

10.4. Practical Considerations

Conservationists cannot be successful in their aims unless they take into account the interests of others who compete for the same natural resources. These other users often have goals that are incompatible with conservation. Consequently the successful development and implementation of conservation strategies requires an acute awareness of the sociopolitical and economic settings in which they are based. In this section we examine those aspects of human property systems and of traditional and developing societies that have important implications for conservation.

10.4.1. The Role of Property Ownership Systems

The conservation and management of primate populations and their habitats first requires identifying their owners. There are four basic ownership systems: open access, private property, common property, and state property (Turner, Pearce, and Bateman 1994). Each property system has its own advantages and disadvantages for conservation, and none alone can provide a universal best option (Rasker, Martin, and Johnson 1992). Indeed, it has been argued that none of these systems are adequate for wildlife conservation (Naughton-Treves and Sanderson 1995). This problem is further complicated by changes in ownership over time and the question of who has the right of ownership in the first place (Robinson 1993a). Nevertheless, these systems are practiced worldwide, and conservationists have little choice but to tailor their endeavors to them.

The first system, open access, is widely recognized to be the least desirable form of ownership for resource conservation. In open access systems, the resources are not owned by any one person but are shared by everybody. Because none of the users can control the offtake of others, there is a strong incentive for individuals to maximize their personal offtake before somebody else does. This system hence leads to spiraling overexploitation, termed the "tragedy of the commons" (Hardin 1968) (although the "tragedy of open access" would be more accurate, since traditional commons are a rather different kind of system: see below). In the remaining three systems, ownership is specified. The identity of the owner(s) has different implications for conservation in each case, although one problem that all owners share is the costs of defending resource use against exploitation by nonowners (Alvard 1998). Where these costs cannot be borne, the system may default to open access.

In the private property system, a single individual owns the resource. If resource conservation makes the greatest contribution to the owner's livelihood, then there is generally no conflict between his or her interests and that of conservation. If, however, there are alternative, more profitable, uses of the resource (e.g., the conversion of forest to cropland), then the practice of conservation largely depends on state intervention. Goodwill and the personal interest of the owner might also be important (Langholz 1996), but they should not be relied on. Consequently, even among those economists who strongly advocate private ownership for conservation, there are few who argue that it will succeed without state intervention (Rasker, Martin, and Johnson 1992). This intervention can include legislation or subsidies, although legislation requires effective enforcement and market incentives can be prone to failure, particularly where markets are changing rapidly.

In community and state ownership, several interest groups may be represented, although decisions over the use of resources may still be made by a single individual or a small subset of individuals depending on the political system in place. This may work both for and against conservation. Community ownership is often thought to be more efficient than state ownership because the local community is the direct user of the resource. As a result, community members have more information and are better able to carry out day-to-day management. Moreover, because they are direct users, they might be expected to have a greater vested interest in ensuring the long-term survival of that resource. However, if nonsustainable but highly profitable uses of that resource should emerge, state intervention may once again be required to make community ownership work for conservation purposes. For example, Caughley (1993) argued that elephants in the CAMPFIRE project might by now have been exterminated were it not that the communities involved are not allowed to overexploit the wildlife populations in their charge.

States are also under pressure to use the natural resources under their control in the most profitable way, although the state might also be expected to adopt a more farsighted and altruistic mode of resource utilization that might favor conservation (since the resource should benefit the country as a whole rather than a subset of the population). However, even then state ownership can fail because of lack of information, lack of financial resources, lack of political will, and corruption. These issues are particularly important since it is invariably the state that has responsibility for and ownership of the large protected areas that form the most important component of global conservation action. When government failure occurs, those areas can effectively come under any one of the alternative ownership systems (open access, private property, or communal property). Murphree (1996) argues that, owing to precisely this failure, most farmers in Africa ultimately end up as the owners of their land even though it is state land.

Yet whether or not state ownership exists, the state may still have a responsibility to intervene in private and common property systems (as previously described). This may cause an indirect form of management failure. For example, Anadu (1987; Areola 1987) blamed official apathy in colonial times and the current low priority rating of wildlife on the political agenda for the inadequate funding of conservation in Nigeria and the casual enforcement of the wildlife laws. One consequence of this is that today the last remaining gorillas in Nigeria are severely threatened by hunting even though hunting them has been illegal there since 1985 (Harcourt, Stewart, and Inahoro 1988). Although legislation can be an effective state tool (as in the case of the 1978 national ban on the export of live rhesus macaques in India: sec. 9.3.2), Struhsaker (1981, 1997) has lamented that legislation and policy rarely keep up with rapidly changing conservation problems. Similarly, Marsh (1988) has argued that an effective civil service is critical to successful conservation, although it rarely exists in developing countries. He suggests that one goal for conservationists in the developed world should be to raise the status of the profession in developing countries.

Finally, a component of fundamental importance in all management-ownership systems is owner security. Since the future is always uncertain, there will always be an incentive for resource owners to maximize present exploitation at the expense of future gains. Moreover, as uncertainty increases, so too does the incentive for maximizing immediate gains from exploitation (Alvard 1998; Milner-Gulland and Mace 1998; see sec. 11.2.2). Hence Marsh (1988) suggests that profitable unsustainable forestry is promoted by elected governments because they have only a short period of resource ownership before they must demonstrate to their electorate that they have improved the nation's economy. Similarly, farmers who become informal owners of state land have no long-term assurance of control over their fields (Murphree 1996). This problem of security is particularly profound in developing countries, where poverty and political instability can create very short time horizons (e.g., Ache hunter-gatherers in Paraguay: Milner-Gulland and Mace 1998). The implications of this problem are obvious: the argument that long-term gains can be made in conservation despite short-term losses is a weak incentive for conservation.

10.4.2. Conservation in Human Societies

Here we discuss whether conservation of primates and their habitats is more likely to be achieved in traditional societies or in developing societies. This question is important, since a great deal of conservation management currently revolves around the participation of traditional and developing communities (we shall return to this issue in section 11.2.1).

TRADITIONAL SOCIETIES

It is commonly argued that the subsistence practices of many traditional societies do not imperil biodiversity; indeed, some claim that these societies have an intrinsic conservation ethic (see Pimbert and Pretty 1995). This argument has sometimes been used by the people of these societies to justify their claims to land (Redford and Stearman 1993). These arguments are supported by the fact that in many regions of the world humans and other animals have coevolved in close proximity and the presence of people and their subsistence activities can enhance some aspects of local biodiversity.

An instructive example is the eviction of pastoralists and their livestock from their traditional grazing grounds in Nairobi National Park during the late 1960s (Western and Gichohi 1993). The park authorities had intended to promote an increase in the biomass of grazing wildlife by removing the competition it was believed they suffered from livestock. However, after the eviction, the grasslands that were formerly burned by the pastoralists and grazed by the livestock became rank and overgrown, and many of the herbivore populations migrated to the more heavily grazed pastoral rangelands outside the park. Similarly, low-intensity shifting agriculture can produce a mosaic of secondary and primary forest patches that can provide valuable resources to many primate populations (see sec. 8.5.1).

However, although humans may bring benefits in some circumstances for some species, there are several reasons it would be dangerous to assume that this will always be the case without reliable empirical evidence.

First, the animals that we see coexisting with people today are only those that have not already been driven to extinction (Balmford 1996).

Second, until recently many societies have carried out their modes of subsistence at relatively low population densities; but this is often no longer the case. For example, although small-scale shifting cultivation may not threaten primates, populations of agriculturalists that have recently undergone rapid expansion may not yet have had the opportunity to develop the traditions necessary to maintain agricultural sustainability. Just this problem exists with the Boki of Nigeria, whose subsistence practices pose a serious threat to the few remaining gorillas there (Harcourt, Stewart, and Inahoro 1989).

Third, along similar lines, many communities in developing countries are composed partially or entirely of migrants (most of whom left their homelands because of social, political, or economic problems: Charnley 1997). Unfortunately, migrants have rarely had time to develop sustainable practices in the use of unfamiliar natural resources; moreover, they often have little incentive to develop such practices, since they perceive themselves as transient users of those resources. For example, it is the migrant population at Okomu in Nigeria that is posing such a serious threat to the sustainable community utilization of this forest (Oates 1995).

Fourth, for a variety of reasons, many traditional societies no longer have the opportunity (or in some cases the inclination) to follow their traditional subsistence practices. For example, although some traditional dwellers in tropical forests appear to manage their forests carefully and sustainably (e.g., Fairhead and Leach 1996; Mackinnon 1998), others clearly will not do so if a substantial short-term profit can be made (the Ache: Milner-Gullard and Mace 1998). This problem may become particularly prevalent as new generations in these societies receive a Western education and show little interest in adhering to traditional values (Redford and Stearman 1993).

In conclusion, although we may be fortunate in that a conservation ethic may exist in some human societies, there is no reason to assume that this will always be the case. In fact, detailed studies designed to test the "ecologically noble savage" hypothesis have shown that apparent conservation by local people may simply be epiphenomenal (see sec. 11.2.2). Consequently we advise caution: unless we have clear evidence to the contrary, we should always assume that human activities may have some detrimental effect on local primate communities.

DEVELOPING SOCIETIES

Patterns of natural resource use in developing societies are particularly relevant to conservation for two reasons: at the macroscale, most tropical countries are composed of developing societies, and at the microscale, there is a strong trend toward integrating local conservation and development projects. A question of critical importance is whether developing and developed societies have a less severe impact on their natural environment than traditional societies (as Integrated Conservation and Development Projects, or ICDPs, appear to assume: sec. 11.2.1). In fact, although development certainly appears to alter traditional patterns of natural resource use, it does not appear to make it any less threatening, even though there is a wide understanding of the need to make development sustainable.

Moreover, the effect of development is highly unpredictable. For example, when Dardenolos, a previously isolated Amazonian community, was connected to the rest of the country by the building of a road in 1978, there was a marked change in hunting patterns. In 1978, before the road, thirty-five *Lagothrix lagotricha* were killed in a four-month period (along with eight *Chiropotes* and one each of *Cebus, Callicebus,* and

Callithrix); in 1980 no primates were reported killed in the same length of time. This decline in the harvest was also mirrored in total mammal offtake and occurred despite an increase in the size of the local human population. Although deforestation in this area had increased by 300% by 1980, this did not appear to be responsible for this change in hunting pattern. Rather, it seemed to be due to a combination of a switch toward more commercial hunting (and thus larger species), an increase in the consumption of domestic meat, the emigration of hunters as paid laborers, and an intensification of crop farming (Ayres et al. 1991).

The most detailed study of changes in resource use patterns with development comes from Godoy et al.'s (1995) study of the Sumu Indians of Nicaragua. Although it has been hypothesized that foraging specialization may increase with household income, these authors found no evidence that the number of types of animals a household hunted declined with increasing income. In fact, economic dependence on hunting and gathering may even show a quadratic relationship with income, peaking at middle-income households: poor households cannot afford to purchase and maintain guns and hunting dogs, while wealthy households can support themselves through more profitable agriculture and wage labor. Nevertheless, this should not be taken to imply that wealthy households have a lower impact on wild animal populations. In studies of the West African bushmeat trade, people are commonly found to spend more money on bushmeat as individual income increases (fig. 10.8), and this pattern appears to be repeated in both rural and urban areas (Martin 1983; Njiforti 1996).

Godoy, Wilkie, and Franks (1997) also investigated the effects of market integration on Neotropical deforestation rates in four Amerindian farming societies. Once again, they found a complex ∩-shaped relationship with income. This, Godoy and colleagues argued, arose because integration into the wage-labor market reduces deforestation but integration into the market for annual crops accelerates deforestation: since both strategies are adopted by members of these communities, a non-linear relation obtains (cf. Dardenelos above). However, the long-term stability of this pattern, and the effects of wage labor on forest habitats in other areas of the region, remains unclear.

Working on a much broader spatiotemporal scale than Godoy and colleagues, Marsh (1988) identified three stages in the development of the use of tropical forests. Initially, there is only traditional low-intensity forest use. This has a relatively low impact on primate populations (see sec. 8.4.1). Development begins with pioneering forest exploitation, which is characterized by intensive and unsustainable forms of land use that include logging, ranching, and permanent agriculture and lead eventually to the widespread opening up of forest areas. This phase is extremely

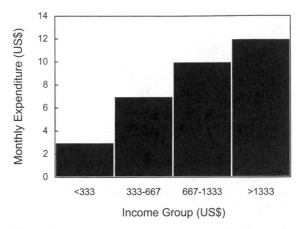

Figure 10.8 Effect of income per annum on purchase of bushmeat by people living in rural and urban communities in Nigeria. Note that urban and rural dwellers spent similar amounts on bushmeat. Income and expenditure are converted from Nigerian naira to United States dollars at the 1977 exchange rate (the period during which these data were collected). (After Martin 1983.)

damaging to primate populations (see secs. 8.4.2–3). Once the immediately available profits of the resource have been reaped, a more stable state emerges in which there is a diverse mosaic of land uses from intensive agriculture to protected conservation areas, with most timber now being drawn from managed plantations. The prognosis for primate populations in this sort of habitat may not be good, although this is the type of landscape where much conservation activity for primates already takes place and where, inevitably, even more is likely to take place in the future (see chapter 12).

Given these patterns, the question of how conservation can be successfully married to development requires careful thought. The World Conservation Strategy (IUCN, UNEP, and WWF 1980) advocates conservation and development as desirable and mutually dependent undertakings but recognizes that the two processes can conflict and therefore need to be carefully balanced. This point of view requires that, in some cases, development be subordinated to conservation, presenting a potentially very difficult dilemma. The successor document to the World Conservation Strategy (*Caring for the Earth:* IUCN, UNEP, and WWF 1991) puts greater emphasis on promoting development. This change in emphasis has elicited criticism from several conservationists, most notably Robinson (1993a,b), who rightly warns that "sustainable development requires conservation, but it is not the same process as conservation." We postpone further discussion of this troubling but fundamental issue to chapter 11.

10.5. Summary

1. Conservation strategies provide the guiding framework in which conservation action is undertaken. The principal components include stating goals, defining conservation units (usually species or geographical areas), setting priorities for conservation needs, and evaluating the tactical actions required for successful conservation. International primate conservation strategy is guided by the Action Plans produced by the IUCN/SSC Primate Specialist Group.

2. Taxon conservation units can be ranked using a variety of different criteria and weighting systems. There is no universally agreed system for setting priorities, but simplicity and transparency are advantageous. In primate strategy planning, two of the most important criteria have been evolutionary uniqueness and extinction risk. Primate taxa are ranked more highly if they are monotypic and threatened with extinction.

3. The IUCN Red List codes constitute the primary system by which the global extinction risks of all primate taxa have been evaluated. Approximately half of all primate taxa are threatened with extinction. Analyses of these codes indicate that primate taxa with distributions that are insular or in areas experiencing higher rates of forest loss tend to be more threatened. From 1988 to 1994 there was an increase in both the number of primate taxa threatened with extinction and the severity of threat for several of those taxa already threatened. Vulnerability to extinction may further vary with phylogeny (e.g., Lorisiformes tend to experience a low threat of extinction and pongids tend to experience a high threat).

4. Population extinction risk can be evaluated with population viability analysis (PVA). PVAs, in the form of computer simulation models (e.g., VORTEX), can provide detailed insights into the population processes that might ultimately lead to extinction. However, practical applications of PVAs may be limited by model and data quality and because they treat small population size as being the ultimate problem to deal with rather than merely the symptom (thus leading to the neglect of the extrinsic forces that caused the population to become small).

5. Area conservation units can also be ranked using a number of different criteria and weighting systems, including taxon richness, character richness, endemism, degree of threat, and viability. Although a set of areas can be identified solely based on high scores in such currencies (i.e., hot spots), it is often more efficient to identify a set of areas based on complementarity, so that areas in the set are chosen to complement one another and maximize the biological value of the network as a whole rather than any one area alone.

6. Before conservation strategies can be implemented, it is necessary to recognize the different sociopolitical property systems within which

primate taxa and their habitats exist. These are open access systems, private property systems, common property systems, and state property systems. Each is associated with a different set of costs and benefits for the development of successful conservation action. In addition, the time horizon of ownership security can be critical.

7. There is little evidence for an intrinsic conservation ethic in either traditional or developing societies, especially given the rapid rates of population growth and cultural change now being witnessed. Although human populations can have a range of neutral or even beneficial effects on the ecosystems in which they exist, such effects should not be assumed.

11 Conservation Tactics

onservation tactics are the short-term measures used to achieve the long-term goals of conservation strategies. These actions can be split into two broad categories: in situ, where the conservation action is deployed in the natural habitat, and ex situ, where this action takes place outside the habitat. Although conservation strategies normally focus on preserving species, the tactics are most commonly employed at the population level, since this is the most practical scale at which to make management decisions. A range of tactics are available for population management, depending on variables such as political instability and human population pressure (Soulé 1991).

The pattern of tactics advocated by the PSG Action Plans in their specific projects for conserving African and Asian primates is summarized in table 11.1. In situ conservation predominates, with surveys and protected area establishment and management being the favored recommendations. The surveys are designed to determine more clearly the distribution and status of populations, while the protected areas are designed to ensure their effective protection. A more significant contribution of ex situ conservation tactics is advocated in both the PSG Action Plan for Madagascar (Mittermeier et al. 1992) and the primate CAMP report (Stevenson, Baker, and Foose 1991). In this penultimate chapter, we examine in detail the primary tactics used to achieve the aims of primate conservation. First we examine the in situ tactics of protected areas and sustainable utilization. We then consider population management, namely ex situ captive breeding and its raison d'être of restocking and reintroductions.

There are two important tactics we do not cover here, largely because both tend to be intermediate steps toward more direct conservation action or play supplementary roles to such actions. The first is survey work, including monitoring: reviews of survey methods can be found in Defler and Pintor (1985), Brockelman and Ali (1987), Whitesides et al. (1988), and Buckland et al. (1993). The second is environmental education, in-

Table 11.1 Conservation tactics of the Action Plans for African and Asian primates

Recommended actions	African Action Plan 1986	Asian Action Plan 1987
Reintroduction	0	2
Protected area management	17	19
Protected area establishment	6	17
Species protection and management	0	4
Surveys	17	24
Research	1	4
Captive breeding	0	3
Trade controls	0	1
Legislation	0	0
Education and training	0	1
Social and human aspects of conservation	0	1
National conservation strategies	0	0
Political and intergovernmental action	0	0

Source: Data from Oates 1986a, and Eudey 1987. After Stuart 1991.

cluding awareness and training programs (e.g., Aveling 1987). Although there is little quantitative information available for determining either the impact of education and training programs on conservation or what determines their relative success or failure, there can be little doubt of their importance (for examples in primate projects, see Weber 1987; Wright 1992, 1997; Horwich 1998). Recent experiences in Democratic Republic of Congo and Rwanda have stressed the crucial role that locally trained staff can play: when war broke out in both countries, it was the local staff who kept the project running despite the lack of salaries, absence of senior management, and personal danger involved (Hart and Hart 1997 and Fimbel and Fimbel 1997, respectively).

11.1. Protected Area Systems

Protected areas are generally the preferred tactic of primate conservationists, since they have the potential to conserve both the patterns and processes of the natural ecosystem within which primates exist. The population densities of primate species are often higher in protected areas than outside them (e.g., howler monkeys: Crockett 1998; see also Caro et al. 1998; Norton-Griffiths 1998), and habitat loss can effectively cease in protected areas even when it continues unabated around them (e.g., in the Atlantic lowland tropical forests of Costa Rica: Sancho-Azofeifa et al. 1999). Although these results must be interpreted with care (because they might also arise owing to the designation of protected areas in localities of unusually high biodiversity or low economic potential), they do suggest that protected areas do have significant potential for conservation.

Nevertheless, protected areas have also been plagued by a diverse array of problems. In the past, the implementation and management of protected areas was usually established on the basis that traditional human habitation would be restricted (at least in theory). However, achieving this goal often meant forcibly evicting people from areas gazetted for conservation. These people, who had a tradition of using those resources, were now forbidden access to them and were rarely provided with any compensation for their eviction. Moreover, the expanding wildlife populations in the parks often imposed further costs on them through crop raiding, particularly by primates such as baboons (see sec. 8.3.2); crop raiding by buffalo and elephants may even cost human lives (Newmark et al. 1994).

The implications of this history are manifold. Because evicted people often have no alternative, they still use those resources within the protected area, albeit illegally. This can lead to problems of unsustainable hunting, forestry, and agriculture, if only because people who have no legal right to the resources have little incentive to use them sustainably. In addition, their treatment understandably generates resentment, with the consequence that local people may actively obstruct the running of the park (Hough 1988). Even where people were not originally evicted in creating the park, conflicts can occur when immigrants arrive in that region and want to live in protected areas. Substantial numbers may illegally occupy such areas (Pimbert and Pretty 1995), and many protected areas now exist only on paper (e.g., Madagascar: Green and Sussman 1990; the Neotropics: Peres and Terborgh 1995). This is exacerbated in that parks can never be guaranteed in perpetuity, even where they are effective. Where population pressure is excessive or political will is weak, formerly protected areas may simply be degazetted (e.g., sections of the Gunung Leuser National Park, which protects the last substantial population of Sumatran orangutans: Faust, Tilson, and Seal 1995). Alternatively, parks may be located in areas of low population density to avoid conflict with people but thereby fail to protect the priority conservation areas (China: Lan and Dunbar, n.d.).

The social problems involved in protected area design and management are discussed further below (sec. 11.1.3), and alternative forms of conservation action that can be used to alleviate some of these problems are addressed in section 11.2. Yet even where no serious social problems are associated with protected areas, there are many other difficulties that users of this tactic must overcome. One is the long-term viability of the primate populations in protected areas, given that ultimately many parks are likely to become islands in a sea of agriculture and forestry. Another

Table 11.2 Protected area systems across major continental tropical regions

Region	Total area protected ($\times$ 1,000 ha)	Percentage of land protected	Number of protected areas
Africa	90,899	3.1	285
Madagascar	740	1.3	16
Asia	42,525	1.4	548
The Americas[1]	67,506	3.9	391

Note: Data for totally protected areas only (IUCN categories I–III), from World Resources Institute 1996.

[1]South America only (Central America excluded).

problem is that predicted global climate change is likely to wreak severe changes on all ecosystems, so that many of the habitats currently existing in protected areas may completely change over the next fifty years. These biological considerations are also addressed below.

11.1.1. Distribution of Protected Areas

Africa currently has the greatest absolute coverage of protected areas, while the Americas have the largest proportional coverage (table 11.2). In contrast, Madagascar does poorly in both cases: the low absolute coverage is understandable given its size, but this makes the low proportional coverage an even greater cause for concern. In general, the extent of protected areas, in terms of both area coverage and the number of areas, increases linearly with country size (Ayres, Bodmer, and Mittermeier 1991); however, the slope of the regression is below one in both cases (0.89 and 0.35, respectively); in other words, larger countries have relatively less protected area in relation to their size than smaller countries.

Analysis of these patterns among the fifteen highest-scoring countries for primate species richness indicates that there is substantial variation in the proportion of land area under formal protection (table 11.3). This ranges from a minimum of zero (Equatorial Guinea) to a maximum of almost 9% in Colombia. No continental African nations do well, with four of the nine countries having less than 1% areal coverage. Further analyses of these data indicate that countries with large numbers of primate species also tend to have both large areas under protection and large numbers of protected areas (Spearman correlations: $r > .55$, $p < .05$ in both cases). In accordance with the findings of Ayres, Bodmer, and Mittermeier (1991), the slope of number of areas on country size is below one (0.70), but in contrast the slope of the extent of area is above one (1.5). Thus larger countries in this sample, which tend to have more primate species ($r = .49$, $p = .06$), also tend to have relatively larger areas of protected land. This is an encouraging finding.

Table 11.3 Protected area systems in countries of high primate species richness

Region and country	Number of primate species	Total area protected (× 1,000 ha)	Percentage of land protected	Number of protected areas
Africa				
Democratic Republic of Congo	31	9,917	4.4	8
Cameroon	29	1,032	2.2	7
Nigeria	23	2,226	2.4	13
People's Republic of Congo	22	127	0.4	1
Equatorial Guinea	22	0	0.0	0
Central African Republic	20	3,188	5.1	5
Angola	19	790	0.1	1
Uganda	19	876	4.4	7
Gabon	19	15	0.1	1
Madagascar	28[1]	740	1.3	16
Asia				
Indonesia	34[1]	14,397	7.9	97
The Americas				
Brazil	52[1]	20,423	2.4	149
Colombia	27	9,036	8.7	41
Peru	27	4,044	3.2	15
Bolivia	18	3,774	3.5	8

Note: Protected areas refers to totally protected areas (IUCN categories I–III). Data on primate species richness from table 2.3; all other data from World Resources Institute 1996.

[1] >30% of species occur only in that country.

Unfortunately, within countries the frequency distribution of reserve size can be strongly skewed toward smaller areas, such that the number of large protected areas is vanishingly small—for example, Madagascar (fig. 11.1) and Kenya (Armbruster and Lande 1993). However, this pattern is not always found across larger geographical regions (e.g., Amazon basin: Peres and Terborgh 1995; African regions: Siegfried, Benn, and Gelderblom 1998), raising the possibility that it might be scale dependent. Nevertheless, that such a frequency distribution might exist at the national scale in many countries raises serious concerns about the long-term viability of the primate populations in protected habitats. The problems associated with small protected areas are discussed below.

11.1.2. Maximizing Biological Value

A great deal has been written about the design of protected areas, mostly based on island biogeography theory (IBT; see sec. 4.1.1). Diamond (1975) proposed that protected areas that are surrounded by hostile habitats, such as agricultural fields, are analogous to land-bridge islands. If

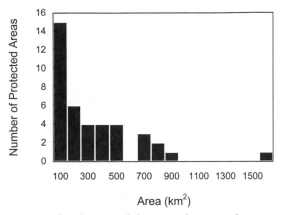

Figure 11.1 Frequency distribution of the size of protected areas in Madagascar. Most protected areas are relatively small. (Data from Mittermeier et al. 1992.)

true, then we have a serious problem, since the diversity of such areas at postisolation equilibrium is less than that in the same area before isolation (sec. 4.1.1). In an analysis of the mammal fauna of North American national parks, Newmark (1987) predicted that if the land-bridge model was correct, then the total number of extinctions should exceed the total number of colonizations, the number of extinctions should be inversely proportional to reserve size, and the number of extinctions should be directly related to reserve age. Each of these predictions was upheld.

Attempts to predict the magnitude of faunal collapse in protected areas after isolation have used island biogeography theory (cf. sec. 8.2). Soulé, Wilcox, and Holtby (1979) investigated the effects of future isolation on the large mammal fauna of the protected areas of East Africa. They predicted that "small parks will lose approximately 10 to 20% of their large mammal species in 50 years following isolation, and that a typical reserve will lose almost half of its large mammal species in 500 years." However, these findings have been criticized on methodological grounds by Western and Ssemakula (1981), whose alternative predictions suggest that a far greater proportion of the current fauna will survive isolation. Boecklen and Gotelli (1984) have also criticized such predictive exercises based on species-area curves, since the variance in the number of species explained by area is usually low (r^2 values rarely exceed .50: $N = 100$ studies), the z and $\log_{10}c$ parameter estimates are often very sensitive to individual cases (hence single points can have major effects on parameter values), and the predictions for the number of species in a given area from a species-area curve are often based on extrapolations beyond the existing range of values. Nevertheless, where studies can overcome these problems, their results may prove extremely useful.

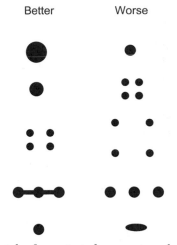

Figure 11.2 Design principles for protected area systems based on the principles of island biogeography theory. As a rule, protected areas are likely to conserve more diversity following isolation if they are large, continuous (or at least interconnected or closely spaced), and circular. (After Diamond 1975.)

DESIGN OF PROTECTED AREAS

The principles of island biogeography theory tell us that area and isolation are key determinants of species diversity but that after isolation the number of species in an area will decline with time (sec. 4.1.1). Unfortunately, nothing can be done to halt the passage of time, but we can take steps to maximize reserve size (solve the area problem) and increase colonization rates (solve the isolation problem).

Diamond (1975) proposed a series of design principles to achieve these ends (fig. 11.2). The application of these principles has been hotly debated, particularly in respect of whether to use a single large or several small (SLOSS) reserves in a protected area system. The SLOSS debate primarily revolves around the fact that several small reserves often contain more species between them than a single reserve of the same total size (Simberloff and Abele 1976), mainly because more habitats are sampled across multiple areas. In addition, when reserve area is plotted on an arithmetic scale, there comes a point at which any further increase in area does not lead to a significant increase in species richness (Western and Ssemakula 1981). Consequently, if the goal of protected areas is to maximize species richness, then several smaller reserves may be preferred. However, Diamond's principles were aimed at maximizing not current diversity but rather future diversity after isolation, and there is little doubt that extinction rates are reduced in larger areas because they house bigger populations. Consequently, the answer to the SLOSS debate depends largely on the conservation goal (Newmark 1986; Burkey 1989).

Where future extinctions are concerned, a further factor to consider in the SLOSS debate is species' vulnerability to environmental stochasticity. Where species vulnerability is high, the long-term survival of a population may be enhanced by subdividing that population into several small reserves. Although normally detrimental, this measure works on the assumption that the environmental fluctuations are at least partially disconnected in a set of different areas; consequently the extinction of one population from this process may not mean the loss of the entire species (McCarthy and Lindenmayer 1999). This issue reminds us that Diamond's (1975) design principles are based solely on island biogeography theory. An important limitation of these principles is that they do not consider a variety of biological factors that lie outside the immediate structure of IBT.

The question of optimal area size is not the only facet of Diamond's principles to have been challenged; debate has also focused on the costs and benefits of connectivity, specifically in relation to corridors (see Simberloff and Cox 1987; Beier and Noss 1998). This must be a serious consideration given the great distances between protected areas in many parts of the tropics (e.g., Africa: Siegfried, Benn, and Gelderblom 1998). Corridors may help to minimize isolation distances and therefore enhance population viability and species diversity. But despite the beneficial effect of increased migration rates, corridors can be associated with a variety of detrimental processes. Most notable are disease transmission, the spread of alien species, and the spread of fire. In addition, corridors can be especially vulnerable to degradation through edge effects and extractive use by local people (who can obtain easy access to these linear areas). Such access may significantly increase the threat to species that are targets of hunting (Noss 1987).

Nevertheless, a strong argument can still be made for corridors, particularly for mammals with low colonization rates (see sec. 4.1.1). Tutin et al. (1997) have stressed the importance of providing corridors for arboreal primates such as *Cercopithecus pogonias,* which appear able to use habitat fragments but are reluctant to cross open ground to reach them. Johns (1991) has argued that the existence of degraded forest corridors along watercourses at Ponta da Castanha played an important role in maintaining a healthy primate community in an area of selective logging and shifting cultivation (sec. 8.4). Similarly, lion-tailed macaques cannot maintain viable populations in forests underplanted with cardamom, but they can use the remaining forest canopy as a corridor to reach areas beyond the plantations (Green and Minowski 1977). Corridors also might be pivotal in maintaining the viability of samango monkey metapopulations in forest patches in South Africa (Swart and Lawes 1996). It is on the basis of evidence such as this that corridors have been recommended as a possible management tool for the Tana River primates (Oates 1996a).

The value of corridors partly depends on the dispersal patterns of the species they are designed for (Harrison 1992); species that disperse only short distances might make little use of corridors, while those that commonly disperse over long distances might use them frequently. To enhance the use of corridors, Harrison (1992) has proposed that their minimum width should equal the home range diameter of the species they are designed for, plus a further border to allow for edge effects. Effective corridors may be narrower, but only provided they are short enough to allow passage without the need for substantial foraging. The importance of corridor width can be illustrated by the howler monkeys at Los Tuxtlas (Mexico), which avoid the very narrow "live fences" (Estrada, Coates-Estrada, and Meritt 1994) but do use riverine vegetation (Estrada and Coates-Estrada 1996). However, these rules should not be assumed to apply universally (see Lindenmeyer and Nix 1993).

Finally, it is important to consider the long-term effects of global climate change on the design of protected areas. The poleward movement of forest belts does not of itself constitute a threat to primates, since we can assume they will simply move with the forests. The problem is that the reserves and national parks established to protect forests and their fauna may no longer coincide with the forests themselves (Peters 1988). Most forest reserves are only a few tens of kilometers in diameter, yet forest biome boundaries can be expected to move poleward by as much as 100 km for every 1°C of climate warming. It might seem appropriate, therefore, to place protected areas toward the pole side of forest biomes rather than the equator side to allow for possible poleward shifts in borders over the coming years.

Similarly, the lower altitudinal limits of forests can be expected to rise by about 200 m with every 1°C of climate warming. Although we cannot make any precise predictions about how specific habitats might change, it is sobering to note that, of the nineteen major national parks in sub-Saharan Africa that contain significant reserves of forest (as listed in IUCN 1971), five (26%) lie below 400 m in altitude and are thus at risk of losing their entire forest cover with a 2°C rise in global temperatures, while seven (37%) could lose their forest cover if temperatures rise by as much as 4°C (the upper limit on current predictions: Houghton et al. 1996). These habitat losses would, of course, be offset by the development of more extensive forest cover in high-altitude parks, but these would not generally be geographically close enough to absorb species from lowland habitats in decline.

11.1.3. Protecting Protected Areas

Independent of any erosion of primate diversity through isolation processes, protected areas can lose primate diversity through uncontrolled exploitation of the primate populations and habitats within them. To avert this, local authorities typically enact laws against using natural resources in protected areas. Antipoaching patrols then arrest offenders, who are punished by fines or prison sentences (see reviews by Milner-Gulland and Leader-Williams 1992; Leader-Williams and Milner-Gulland 1993). Fines are often more desirable, since these act as a "tax" on illegal activities, adding to the wealth of the state rather than reducing it. In addition, in developing countries, the threat of a prison sentence may be a less effective deterrent because the long-term future is less certain and so less valued than the immediate future. Yet these advantages must be balanced by the fact that guilty parties may rarely be able to pay fines. Since both fines and prison sentences have drawbacks, the best method to deter poachers is not by imposing high penalties on those detected, but rather by high rates of detection in the first place (Leader-Williams and Milner-Gulland 1993). In the rest of this subsection we examine how antipoaching patrols can be made more effective.

ACTIVE PROTECTION

The efficacy of antipoaching patrols has been well demonstrated in the Luangwa Valley (Zambia) by Leader-Williams, Albon, and Berry (1990). Survey teams working in those areas of the region where patrols were most active encountered both fewer poachers (per unit patrol effort) and more animals that were targeted by those poachers. It was subsequently estimated that effective park protection in this area required at least one man per 20 km². Similarly, the mountain gorillas of the Virungas suffered very little poaching when the density of park guards was at a level of one man per 2.6 km².

A detailed analysis of the efficacy of the antipoaching units in the Virungas has been conducted by Harcourt (1986; Harcourt et al. 1983). Harcourt compared the status of the mountain gorilla populations in the Democratic Republic of Congo section of the Virunga Conservation Area with that in the Rwandan section (where antipoaching units were much more frequent and efficient). In Rwanda, livestock were removed and gorilla groups monitored (for tourism purposes) so that animals caught in snares were quickly found and released. By 1981 the Rwandan section contained half as many snares as the Democratic Republic of Congo section, and the effects on the gorilla population were marked. Over an eight-year period, there were no dramatic overall changes in population size in either Rwanda or Democratic Republic of Congo, but population recruitment diverged radically. In Rwanda the number of immature animals increased by 17%; in Democratic Republic of Congo they *decreased* by 22%. Butynski, Werikhe, and Kalina (1990) similarly report a drop in poaching intensity after the introduction of well-equipped, trained, and disciplined antipoaching patrols in the Gorilla Game Reserve in Uganda.

Clearly, at sufficient intensity, antipoaching efforts can be highly effective at protecting primate populations. However, the precise value of the "safety threshold" figure is also partially dependent on the poaching pressure itself, which is likely to be a function of the density and socioeconomic status of the human population around the protected area. The number of patrols required may also be a function of their efficiency. Arcese, Hando, and Campbell (1995) have reported that using vehicles in patrols increases detection rates, especially in conjunction with foot patrols, and that financial rewards to antipoaching patrols for finding snares, making arrests, and impounding firearms can play a positive role both in boosting staff morale and in encouraging the confiscation of hunting equipment (the financial rewards may be particularly important when the social and personal costs to game guards of catching poachers are considered: Gibson and Marks 1995).

In the light of these findings, the solution to controlling poaching seems simple: increase the number of antipoaching patrols, provide vehicles for those patrols, and perhaps introduce a reward system for patrol staff. Although each of these measures requires a substantial financial input, the money may be well spent. Leader-Williams and Albon (1988) demonstrated a strong association between spending and the efficiency of conservation efforts: across nine African countries, government spending per square kilometer of conservation area in 1980 was positively correlated with the change in numbers of both elephant and black rhino populations between 1980 and 1984. The amount of money required to maintain stable populations during this period was estimated at $230 $km^{-2}yr^{-1}$ for black rhino and $215 $km^{-2}yr^{-1}$ for elephants. (All dollar amounts are United States dollars.)

These figures are undoubtedly at the top end of the range owing to the extraordinarily high value placed on rhino horn and elephant ivory at this time: effective protection of most large-mammal populations is likely to be considerably less expensive (Caro et al. 1998). Nonetheless, the financial resources required to underwrite even these levels of expenditure are rarely available in developing countries. In most countries, park protection and staff are severely underfunded. In the Okapi Wildlife Reserve (Democratic Republic of Congo), for example, only thirty-nine guards are employed (one per 352 km^2), and they lack suitable clothing and equipment and receive salaries of less than $0.20 per month (Stephenson and Newby 1997). In Brazilian Amazonia, only ten of thirty reserves have even a single guard, and across the entire region the coverage of guards is only one per 6,053 km^2 of protected area (Peres and Terborgh 1995). The result of such underfunding is that protection is often largely ineffectual.

In addition, once protected areas become dependent on antipoaching patrols for their security, the sudden withdrawal of funds can have a disastrous effect. At the Bawangalin Nature Reserve in China, budget cuts in 1991 and 1992 led to reduced forest protection, which resulted in the poaching of over half of the Hainan gibbon population there (eleven of twenty-one individuals: Southwick and Siddiqi 1994). This problem is likely to be particularly common in developing countries, where political instability can seriously affect long-term funding for protection (Oates 1996b).

PASSIVE PROTECTION

To protect areas without so great an investment, protected area systems might be spatially designed to enhance the efficiency of antipoaching patrols. Ayres, Bodmer, and Mittermeier (1991) have pointed out that, although the patrolling of protected areas is divided between perimeter and internal surveillance, the most important determinant of patrol effectiveness is the length of the perimeter. Consequently a single large reserve is better than several small reserves of the same aggregate area, since this shortens the total perimeter. In addition, round reserves should be preferred over square or rectangular reserves of the same area for the same reason. Minimizing the perimeter for a given area not only may improve the efficiency of antipoaching patrols, it may also reduce edge effects. Unfortunately, many existing protected areas have high perimeter-to-area ratios (e.g., African national parks: Siegfried, Benn, and Gelderblom 1998).

The relative efficiency of protected area systems can be compared across countries by regressing both the number and areal extent of reserves on country size (sec. 11.1.1) and then plotting the residual number of reserves against the residual total extent of reserves (following Ayres, Bodmer, and Mittermeier 1991). Countries with positive area but negative number residuals (type A) have a large amount of land protected in a small number of reserves, whereas countries with negative residuals in both parameters (type C) have a small amount of land in a small number of reserves. Both these country types have the potential to be cost effective in terms of area defensibility. In contrast, areas with both positive number and area residuals (type B: large number of large reserves) and countries with positive number but negative area residuals (type D: large number of small reserves) are likely to be less cost effective. However, the areal extent of reserves is also important. Overall, then, the ideal combination is A; the worst is D.

Conducting this analysis for countries of high primate diversity (fig.

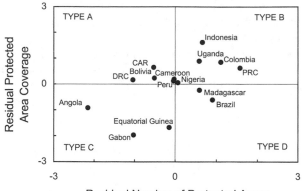

Figure 11.3 Covariation between the residual number of totally protected areas and the residual coverage of totally protected areas in countries with high primate diversity. CAR, DRC, and PRC are Central African Republic, Democratic Republic of Congo, and People's Republic of Congo, respectively. Types A, B, C, and D are described in the text. (Data from table 11.3.)

11.3) shows that Democratic Republic of Congo, Central African Republic, and Bolivia are doing well, but that Brazil and Madagascar are doing badly. There are two countries worth noting: Indonesia (with a relatively large area protected) and Gabon (with a relatively small area protected). In the first case, the large extent of area protected may be a natural consequence of the fact that much of it is relatively inaccessible; in the second case, Gabon has very little protected area because almost all of its vast area of intact forest is designated for future logging (Sayer, Harcourt, and Collins 1992).

More recently, Peres and Terborgh (1995) have made recommendations for designs based on perimeter area and accessibility. In an analysis of reserves in Brazilian Amazonia, they found that the proportion of each reserve accessible for hunting and other forms of extractive use (where a stretch of land was defined as accessible if it lay within 10 km of a road or navigable river) are at least 40% and reach up to 100% in the worst cases. Critically, larger reserves are more accessible, primarily because they encompass more rivers. Although larger reserves therefore require greater protection effort, there is no relation between reserve size and either the number of guards or the total number of personnel employed (and even the better-protected reserves are often poorly staffed and equipped). Peres and Terborgh argued that reserves should be designed to minimize accessibility and therefore the amount of active protection required. This could be achieved by using the edges of watersheds as their boundaries; in the absence of roads, this would mean that traffic in and out of the reserve could be controlled at a single point (and therefore with a minimum of staff), where the watercourse draining that watershed exits the reserve.

Unfortunately, although the design of protected areas might enhance protection, the ideal design may be constrained by existing political and habitat boundaries. In addition, even an ideal design cannot remove incentives for illegal use. To do this it is necessary to make the long-term existence of the protected area and its primates valuable to the local community. Increasing the opportunity costs of poaching will substantially reduce the incentive to poach. The following section addresses how this aim might be achieved. Note, however, that although incentives to stop poaching of primates may influence local people, they may have little direct effect on hunters from outside the area (Barrett and Arcese 1995).

11.2. Sustainable Utilization

Conservation has always been undertaken ultimately for human benefit. Nonetheless, in recent years there has been an increasing need for conservationists to justify the preservation of habitats and animal populations in purely economic terms. In this section we ask whether wildlife and wildlands can be conserved through sustainable use, either by consumptive harvesting or nonconsumptive tourism, given the economic pressures facing developing countries. First, however, we review an important approach to contemporary sustainable use programs.

11.2.1. Conservation and Development

Because the unsustainable exploitation of primate populations and their habitats by local communities is often believed to pose the greatest threat to their survival, a great deal of conservation action is now aimed at making these natural resources more valuable to local people in the long term. This goal is achieved by increasing the economic value of that resource, either by establishing new harvesting or tourism programs or by distributing more effectively the economic benefits of existing programs (e.g., the revenues from tourism that would previously have gone directly to the state are redirected to neighbors of the national parks). Such projects have the attraction that the improved economic benefits at the local level will bolster livelihoods and community development. This approach of emphasizing the link between biodiversity and the well-being of human populations has led to the emergence of Integrated Conservation and Development Projects (ICDPs).

The wisdom of this approach is still hotly debated (see Spinage 1998, 1999; Colchester 1998; Martin 1999), not least because the accumulated evidence suggests that humans show little natural inclination for sustainable use: what conservation does occur seems to be epiphenomenal rather than intentional (secs. 10.4 and 11.2.2). The approach has also been fraught with a variety of complex problems (which are discussed below). Nonetheless, harsh economic realities dictate that conservation will reap substantial benefits if ways can be found to ensure that this approach succeeds.

Projects that involve local people in conservation are generally termed community-based conservation projects (CBC). There are three basic types of community-based schemes (Gibson and Marks 1995). First, there are direct benefit-sharing schemes: local people gain directly from state-controlled conservation activities in the form of either cash income (e.g., from the sale of meat from safari-killed game) or employment as game scouts and tour guides. Second, there are indirect benefit-sharing schemes: local people gain indirectly if a share of the profits of conservation income is used on local development projects, such as building hospitals and schools. Finally, there are local empowerment schemes: local people are given responsibility for managing the resource, so that they have a vested interest in ensuring its long-term existence. CBC projects with a primary focus on primates include Ranomafana National Park in Madagascar (primarily an indirect benefit-sharing scheme: Wright 1992, 1997) and the Community Baboon Sanctuary in Belize (primarily a local empowerment scheme: Horwich 1998). These projects focus on lemurs and black howler monkeys, respectively (see table 11.4)

A variety of factors can influence the success of such projects. One key factor contributing to their failure is that the goals of the local community are often not the same as those of the conservationists who support the project (Brandon and Wells 1992; Barrett and Arcese 1995; Gibson and Marks 1995). Many community-based projects are designed to reduce damaging forms of resource use through sustainable alternatives, but the subsistence options are often not mutually exclusive. For example, agricultural production may be improved as an incentive to stop hunting and logging in neighboring forest, but logging and hunting may still be carried on during those seasons when there is little work in the fields. Precisely this problem has been identified in the attempt to combine conservation and development in the Okomu Forest Reserve in Nigeria, an important site for threatened primate species such as *Cercopithecus erythrogaster* (Oates 1995). Nevertheless, in some cases it is possible to identify appropriate links. In Rwanda, for example, local farmers may not be interested in protecting the montane forests for the mountain gorilla populations there, but they recognize the importance of these forests as watersheds for their cropland (Brandon and Wells 1992; see also Weber 1987).

Another important determinant of the success of these projects is

Table 11.4 Components of two primate community-based conservation schemes

	Ranomafana National Park, Madagascar	Community Baboon Sanctuary, Belize
Area	43,500 ha (core area, no people)	>4,662 ha (with people: see below)
Human population	25,000 people (in ninety-three villages) in 3-km-wide peripheral zone	>100 landowners (in seven villages) found throughout Sanctuary area
Target primate taxa	Twelve lemur species from nine genera (including two Critically Endangered and three Endangered species)	Black howler monkey *Alouatta pigra* (Low Risk, least concern: not globally threatened)
Research	Research stations built, biodiversity and human impact surveys and monitoring, studies of plant-animal interactions, animal ecology, population genetics, research training for local personnel	Monitoring of howler monkey population and other species (e.g., river turtles, migratory bird populations), howler monkey reintroductions, forest succession, human anthropology
Health	Demographic and health surveys and monitoring, hospital renovation and two new clinics built, health care services provided	Not applicable for this project
Conservation education	Four new schools built and seven renovated, village slide shows, rural libraries and museum established, monthly newsletter circulated	Conservation dinner lectures, T-shirts circulated, school talks, illustrated booklet circulated, guidebook written, radio and television programs broadcast, rural museum established
Economic development	Technical support provided in cultivation, forestry, husbandry, etc., native tree nursery established, native fish and crayfish aquaculture project started	Community garden developed
Ecotourism	Visitor center, campground, washrooms, and marked trail system established, brochures written, tourist guides trained	Tourist shelter built, camping areas and local family hostels established, guides trained, rural museum established, trail system marked out
Forest management	Research feeds into park management, monitoring of indicator species and water, Geographic Information Systems used to aid management	Individual management plans written for each landowner participating in the scheme to maximize habitat value to howler monkeys, hunting prohibited

Source: Data for Ranomafana from Wright 1997; data for Community Baboon Sanctuary from Horwich 1990, 1998. These lists are intended to provide examples of the activities undertaken in both projects and should not be considered comprehensive.

the distribution of the benefits that accrue from conservation. Hence Norton-Griffiths (1998) reports that wildlife has been conserved more effectively in Kenya where profits from tourism are distributed directly to landowners, whereas the wildlife has declined where revenues are captured by the tourist industry and higher-level administrative structures. Similarly, there has not been a long-term decline in illegal hunting in the LIRDP and ADMADE projects in the Luangwa Valley of Zambia (Lewis, Kaweche, and Mwenya 1990; Gibson and Marks 1995; Lewis and Phiri 1998), because the vast majority of community members receive minimal benefit from the program—certainly not enough to outweigh the income from hunting. In addition, even where conservation incentives might have balanced the monetary profit from hunting, they cannot easily compensate men for the respect and status that hunting brings them within their community (ADMADE: Gibson and Marks 1995).

These points emphasize that if such projects are to work, the short-term costs of management for long-term sustainable resource use must be affordable to all members of the community (Noss 1997). For these reasons, it has been proposed that those people who depend relatively more on the detrimental use of a resource should receive relatively greater benefits than the other community members for refraining from that use (Brandon and Wells 1992; Gibson and Marks 1995). This clearly makes approaches based on indirect-benefit sharing schemes (such as providing schools and hospitals) less viable. Indeed, the success of the CAMPFIRE scheme in the Masoka community, Zimbabwe, may reflect the fact that each household received its own substantial cash handout from the scheme in addition to enjoying communal benefits such as improved local schooling (Murphree 1996; Matzke and Nabane 1996).

Finally, we need to emphasize two additional problems. First, successful projects may attract immigrants from poorer areas, thereby placing potentially unsustainable pressure on a limited resource base (Brandon and Wells 1992). For example, at Okomu, immigrants have been drawn to the forest owing to the agricultural initiatives there, increasing the pressure from illegal hunting and logging (Oates 1995). In contrast, at Masoka, the original residents recognized that their personal cash benefits from the local conservation activities would be reduced if they had to be shared among more people; consequently the community turned away hopeful immigrants (Murphree 1996; Matzke and Nabane 1996). An alternative technique for dealing with immigration is to ensure that activities around the focal natural resources are established at least 25 km away (as in Ranomafana National Park, Madagascar: Wright 1997).

Second, these projects are generally labor and cost intensive. Barbier (1992) notes that the LIRDP project in Zambia is unlikely to become self-supporting before its fifteenth year. This is a serious problem because the whole philosophy of the approach—that wildlife "pays its way"—is not being met.

Undoubtedly, the next generation of community-based conservation projects should be an important tool for conservationists, although their role may be most valuable as just one facet of a mixed strategy alongside other conservation measures. Whether these projects are best seen as only short-term measures (e.g., Barrett and Arcese 1995) or as only long-term measures (Oates 1996b) remains to be seen.

11.2.2. Harvesting

Two issues are at the heart of sustainable utilization for primate conservation. One is the harvesting of resources in primate habitats (such as timber), the other is the harvesting of the primates themselves. The management of sustainable and profitable human extraction of resources from primate habitats, to ensure that the continued existence of these habitats can be justified on economic grounds, is a vast subject beyond the scope of this book. For recent reviews, we recommend Grieser Johns (1997) and Struhsaker (1997). Here we focus specifically on the exploitation of the primate populations themselves.

Although there is little doubt that hunting reduces the size of primate populations, it is much less clear how intense hunting has to be before it becomes unsustainable and drives a population to extinction. Although there is good evidence that primates are hunted and trapped at unsustainable levels, there is also no doubt that primates, like all other living organisms, have the potential to be harvested sustainably, albeit at low offtake levels. If unsustainable harvests can be converted to sustainable ones, or if sustainable harvests can be introduced to provide greater incentive for long-term habitat and population protection, then the cause of conservation may be significantly advanced. We have already discussed the principles of sustainable utilization and the potential for primate populations to sustain indefinite cropping (sec. 9.4.1). Here we focus on the evidence for potentially sustainable harvesting of primate populations in both subsistence and commercial hunting systems. Although the distinction between subsistence and commercial hunters is hazy, different processes may guide harvesting patterns in each case.

SUBSISTENCE HUNTING

The question whether traditional hunter-gatherers conserve their prey species has been hotly debated in recent times, largely because of the desire to employ traditional forms of resource use as the basis for community conservation projects and sustainable development. Evidence that harvest rates remain constant over long periods has been used to support the claim that subsistence hunters do hunt sustainably. The best example of such stability is provided by Vickers (1991), who shows that the number of kills of woolly monkeys and howler monkeys by Siona-Secoya hunters in Amazonia per man-hour spent hunting has not declined over a ten-year period. However, Vickers did report reduced harvest rates for woolly monkeys closer to the village.

Unfortunately, sustainable hunting is less likely to be due to active conservation measures undertaken by hunters than to be an epiphenomenal effect of low harvesting pressure owing to low hunter density (Alvard 1995a, Noss 1998), perhaps aided by the incorporation of alternative species into the diet as preferred prey become scarce (in accordance with the prey choice model: Winterhalder and Lu 1997). If hunters did practice an ethic of sustainable utilization, then we might expect that depleted populations would be allowed to recover to normal levels. Yet precisely the opposite is observed. Hames (1987) has shown that, across eleven subsistence hunting societies, hunters increase their hunting effort rather than reduce it as prey species become depleted. Similarly, when the Piro hunters of Amazonian Peru encounter profitable prey animals in depleted areas, they will attack them rather than ignore them (Alvard 1994).

Alvard (1993) suggested that "partial preferences" (sec. 9.1) might reflect conservation, since hunters might select prey only when they can be sustainably harvested (e.g., when prey occur at high abundance). However, the larger primates, which are the most vulnerable to hunting but also the most profitable to foragers (high e/h), were always attacked upon encounter by the Piro hunters he studied. It was only the smaller, less profitable, primates for which partial preferences were observed (fig. 9.2). The probable reasons for these preferences in the latter taxa have already been outlined (in sec. 9.1.1).

Alvard (1995a) also investigated whether Piro hunters attacking spider monkey and howler monkey groups selectively kill those members that are likely to make the smallest contribution to population growth (i.e., immature and old adults, the age-classes that have high mortality rates and that are postreproductive, respectively). Contrary to the principles of conservation management, adults were taken more often than immatures: 86% of the kills were adults for both species, in comparison with approximately 50% adult group membership (fig. 11.4). Similarly, the proportion of old adults killed was only 20% in both species. In addition, males and females were hunted in proportions similar to their abundance in the population: there did not appear to be selectivity for males. Alvard (1993, 1995a) concluded that the Piro are not trying to conserve their resource base but rather are maximizing the rate of energy gain during foraging, as predicted by optimal foraging theory (sec. 9.1).

In the light of these patterns, it might seem reasonable to expect that hunters who can increase their harvest rates through new technologies

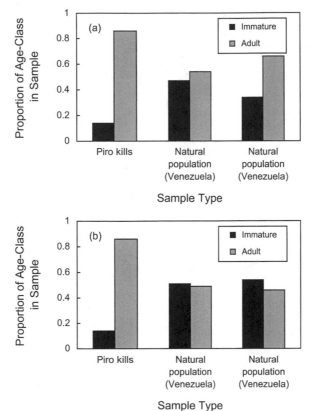

Figure 11.4 Patterns of selectivity by Piro hunters for different age-classes in (a) spider monkey *Ateles* and (b) howler monkey *Alouatta* populations. In both cases Piro hunters appear to target relatively more adults than immatures given the ratio of adults to immatures in two natural (unhunted) monkey populations. (Data from Alvard 1995a.)

will also increase their hunting pressure. However, although in Venezuela the Ye'kwana kill more primates as a result of using shotguns (over a 216 day period Ye'kwana hunters killed seventy-nine cebids using shotguns compared with the eighteen killed by the Yanomamö using bows in the same area), they also spend less time hunting (Hames 1979). Consequently, although the Ye'kwana are 343% more efficient when hunting than the Yanomamö, their per capita consumption of meat is only 16% greater. Consequently, the opportunity to increase exploitation does not mean it will be taken, although clearly this does not necessarily lead to the protection of primate populations; nor is it a pattern that might persist if market hunting became an incentive for utilization.

Alvard (1995b) similarly found that Piro shotgun hunters did not ob-

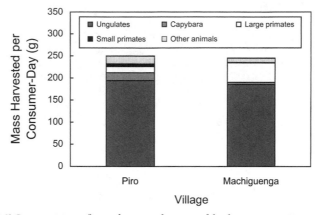

Figure 11.5 Composition of prey biomass harvested by hunters in a Piro community (Diamante) and a Machiguenga community (Yomiwato) in Peru. Both communities are found in the tropical forests in and around Manu National Park (Diamante is outside the park, Machiguenga is inside). (After Alvard 1995b.)

tain more meat per capita than the Machiguenga (who are bow hunters). The amount of meat harvested per consumer was identical in the villages of both communities, although the large primate biomass in the Machiguenga total was about double that in the Piro total (fig. 11.5). This difference in prey species composition probably arose because the Piro village is more than twice the size of the Machiguenga village, and therefore the populations of large primate prey in their hunting areas are already depleted. This interpretation is supported by the findings that large primates became progressively rare with increasing proximity to the Piro village and that the combined encounter rate of the four largest primates by hunters in the Piro area (0.09 per hour) was less than half that seen in the Machiguenga area (0.23 per hour) (Alvard et al. 1997). This finding suggests that a key determinant of potential sustainability is human population size and historical hunting pressure, while contemporary hunting technology may be relatively less important. This conclusion receives further support from Noss (1998), who describes currently unsustainable exploitation of traditional game species using traditional hunting methods by BaAka net hunters in the Central African Republic: the sole reason for unsustainability appears to be an increase in human population density.

Despite these findings, some studies do suggest that sustainable utilization by hunters can occur. The Ye'kwana and Yanomamö rotate hunting zones around their villages in such a way that areas that have been heavily depleted by hunting have a chance to recover. Hames (1979) describes how two once-rich hunting areas were abandoned owing to a decline in prey abundance while two more productive areas were opened up; these new areas had themselves been abandoned fifteen to twenty years before. Similarly, Alvard (1994) reports that Piro hunters spend less time hunting in depleted areas, where hunting success is low, and more time where hunting success is high. However, careful scrutiny suggests that these results are likely to be a simple by-product of optimal foraging rather than deliberate conservation practice (since if profitable prey are encountered in the depleted areas, the hunters will still pursue them).

A genuine conservation practice might exist on Bioko Island (Equatorial Guinea), where Bubis hunters reportedly avoid killing monkeys with infants, preferentially taking males or females without dependent young (Colell, Maté, and Fa 1994). Dei (1989) also reports that in Ghana Akyem hunters were previously bound by moral duty to hunt only the old and infirm animals and to leave the young and pregnant females; however, because of the current paucity of game, modern Akyem hunting is indiscriminate. These findings give some cause for optimism, but the weight of evidence clearly shows that conservation, if it occurs at all, is at best an epiphenomenon. Consequently, the development of conservation projects based on traditional hunting patterns in local communities is likely to be fraught with problems.

COMMERCIAL HUNTING

For many primate species, the commercial markets in primate meat, primate body parts, and live primates could in theory be supplied at low volume from populations that are harvested sustainably. There are two possible methods to achieve this.

First, these animals might come from primate "ranching" projects, where large populations can be supported in small areas through supplemental provisioning and where offtake can be carefully managed (Crockett, Kyes, and Sajuthi 1996). Although there are a variety of establishments around the world that breed live primates for medical research, large-scale in situ ranching has been successful only when substantial provisioning has been provided (Mittermeier et al. 1977). This makes the cost of ranching high. Although ranching might become a more viable option as wild primates become scarcer and market demand grows, it is also an approach that provides little incentive to conserve wild populations of primates and their natural habitats. Consequently the value of ranching in primate conservation strategies may be limited to reducing hunting pressure on wild populations.

Second, sustainably harvested primates might come from carefully managed wild populations. This approach has the advantage that it is potentially less expensive than ranching and that it might provide strong incentives for active conservation, since if the prey population disappears then so too does the income from that population. However, reports of the successful implementation of such projects are scarce. For example, Crockett, Kyes, and Sajuthi (1996) report one case in Peru where the live capture of tamarins and owl monkeys together with the harvesting of forest fruit provides a greater income to local people than conventional farming would; however, it is not clear whether this harvesting of wild primates is sustainable in the long term. Aquino and Encarnación (1994) present more detailed data on the live trapping of owl monkeys, suggesting that cropping can be sustainable provided there is a source of immigrants to enter the harvested population (sec. 9.4.1).

Unfortunately, all projects that advocate conservation based on financial profits through consumption must deal with the problem of economic discounting and the *discount rate* (the rate at which the value of an investment declines over time: Caughley 1993; Milner-Gulland and Mace 1998). A primate population represents capital, and the harvest from that population represents interest on that capital. If the financial gain through harvesting is less than the potential financial gain through interest on the equivalent sum of money invested elsewhere, then the economically rational course of action is to immediately and totally convert that primate population to cash and reinvest the capital so acquired. For harvested populations, the critical question therefore becomes the relative magnitudes of the population's intrinsic growth rate r_m and the economic discount rate d. Because primates have relatively low r_m (particularly large primates), the critical value of d (namely $d° = r_m$) does not have to be high before harvesting to extinction becomes the most rational economic action. Alvard (1998) illustrates this with a comparison between harvested populations of spider monkeys and peccaries: a decision to liquidate the peccary population would require access to an alternative cash investment with an interest rate of over 21%, but the decision to liquidate a spider monkey population would require the alternative investment's interest rate only to exceed 2%.

Lande et al. (1994) have investigated the effects of d on harvesting patterns and extinction risk in fluctuating populations. They concluded that even where $d < d°$, the profit from sustainable harvesting may be marginal. Once the costs of the harvesting operation have been taken into account, any economic incentive to harvest sustainably may vanish. In addition, if Allee effects operate in the harvested population, then the magnitude of $d°$ will be further reduced. Because of the impact of discounting, it seems unwise to base a conservation strategy for any wild population or habitat solely on economic profits from consumption (Caughley and Gunn 1996). This is especially true for primate taxa.

11.2.3. Tourism

An alternative to harvesting an animal population that might still provide a direct and sustainable cash income is tourism. Tourism has the potential to solve many conservation problems: it is a high-employment industry, showing rapid international growth, with good potential for sustainability provided it is carefully controlled. By 1996, travel and tourism constituted one of the largest industries in the world with an annual revenue on the order of $3 trillion; moreover, nature tourism is the fastest-growing sector of this industry (Giannecchini 1993). Tourism has the potential to contribute to conservation indirectly, through raising awareness and money for related conservation projects (e.g., orangutan rehabilitation centers: sec. 11.4.1), as well as directly, through conservation projects in which the threatened populations themselves are the tourist attraction (e.g., mountain gorillas: Harcourt 1986). Moreover, the financial benefits of tourism can be enormous. Since 1995, 12% of the revenue from two Ugandan mountain gorilla projects has been distributed to local communities, which have thus far received over $100,000 through the scheme (Butynski and Kalina 1998). Barnes, Burgess, and Pearce (1992) estimate that Khao Yai National Park (Thailand) has the potential to make $1 million a year, if its "viewing value" to tourists is captured in entry fees and other associated costs. Currently only a fraction of this revenue is collected.

The potential for successful sustainable tourism in primate conservation was epitomized by the Mountain Gorilla Project in the Parc National des Volcans (PNV) of Rwanda (Harcourt 1986). Tourism at this site was initially introduced in 1978 to make park protection, in the form of antipoaching units, self-sufficient (table 11.5). The foundations for the development of tourism were laid by the existing research base, which provided essential information on the behavior and ecology of the gorillas, and by a conservation awareness program conducted both within the local communities and nationwide. Between 1978 and 1984, the number of tourists that visited the park increased by 430%, and the increase in revenue was even greater. These profits were made even though the number of tourists that could visit the mountain gorillas was strictly limited (one visit of no more than six people per day per gorilla group). The resulting improvements in park protection yielded an increase in the recruitment rates of the mountain gorilla population during that period (sec. 11.1.3). Unfortunately, a variety of problems have since emerged in mountain gorilla tourism at PNV and other sites in the region, demonstrating that the achievement of long-term success in such projects is likely to be more difficult than first anticipated (fig. 11.6; review in Butynski and Kalina 1998; see below).

Nonetheless, for charismatic species such as the mountain gorilla, the potential for sustainable tourism remains promising. In contrast, those primate communities that do not boast such species may have relatively less potential for tourism. Widespread human disturbance can also reduce potential: a strong negative correlation exists, for example, between

Table 11.5 Main gorilla conservation activities at the Parc National des Volcans (PNV)

Year	Action	Summary
1976	Cattle and herders removed	All of the many cattle and herders that were in the park were removed. They had been causing considerable disturbance up to elevation of 3,500 m since the 1950s.
1978	Donor support enhanced	Donor support for conservation activities in and near the park greatly increased. Although published amounts are not available, the figure from 1978–1997 is likely to be well over U.S. $6 million.
1978	Lobbying increased	Lobbying government at various levels for improved support and management of the park expanded substantially as the numbers of conservationists, NGOs, embassies, and aid agencies involved in gorilla conservation increased.
1978	Security expanded	Park security activities were greatly expanded. The security force doubled, and the wardens and rangers were trained, well equipped, and supervised.
1978	Controlled tourism established	Controlled tourism on habituated gorillas began and generated approximately U.S. $4 million in gorilla viewing fees over the next thirteen years.
1979	Education program initiated	An environmental education program began among the people living in the vicinity of the park. This program reached hundreds of thousands of Rwandans.
1986	Veterinary program started	The Mountain Gorilla Veterinary Centre was established, primarily to enhance gorilla survival by monitoring the health of habituated gorillas and to administer emergency treatment. This facility has saved many gorillas from death.

Source: From Butynski and Kalina 1998 with the permission of Blackwell Science.

profits from game viewing and human population density across sites in the CAMPFIRE project (Murphree 1996). However, there is some evidence that even relatively unpromising primates do have the potential to be star attractions. In the Tangkoko DuaSudara Reserve in Sulawesi, 57% of visitors come to see a particular animal, and of these the first choice (74% of respondents) is either the spectral tarsier or the black macaque (Kinnaird and O'Brien 1996). Given the number of visitors to this reserve (over four thousand in 1993, with an increase of 250% in foreign visitors over the preceding three-year period), it is clear that even

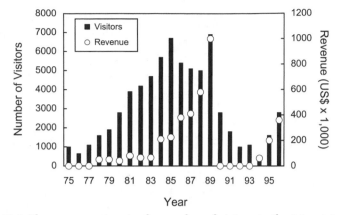

Figure 11.6 Changes over time in the number of visitors to the Mountain Gorilla Project in the Parc National des Volcans (Rwanda) and the revenue generated by those visitors. The decline in visitor numbers and revenue in the early 1990s reflects the political turmoil in that region. (After Butynski and Kalina 1998.)

primates such as macaques and nocturnal prosimians can capture the public imagination. Similarly, other primates such as red colobus monkeys (Zanzibar, Tanzania: Struhsaker and Siex 1996, 1998), howler monkeys (Horwich 1998), and a variety of species of lemurs (e.g., Ranomafana, Madagascar: Wright 1992, 1997) and macaques (e.g., Japan, Indonesia, and China: Koganezawa and Imaki 1999; Wheatley, Putra, and Gonder 1996 and Zhao 1996; respectively) can act as a focus for tourism.

Nevertheless, great care must be taken in introducing any tourism project, since sustainability is not always easy to ensure. One problem is that there is an obvious incentive to increase the number of tourists to maximize profits. This can lead to greater disturbance for the focal primate populations as tourist groups both get larger and visit more often (see below). It can also escalate environmental degradation. At Manuel Antonio National Park (Costa Rica), for example, a rapid growth in tourism and an associated development of hotels, restaurants, and other enterprises has led to substantial habitat loss and fragmentation, which may now pose a serious threat for the local squirrel monkey population (Wong and Carrillo 1996). Similarly, in Annapurna (Nepal), tourism led to deforestation as wood was harvested for heating and cooking for the visitors (Brandon and Wells 1992). Both of these processes can ultimately undermine the tourist industry by destroying the very resource it depends on (e.g., Western and Gicholi 1993).

One way to increase profits without increasing tourist numbers is to limit the number of visitors but charge more per visit. This approach naturally leads to a second problem. Tourists paying high fees expect spectacular wildlife experiences. In addition, such tourism requires good infrastructure and political stability. For many developing countries, these are absent: Wright (1997), for example, laments that at Ranoma-fana National Park (Madagascar), trail systems and park access points are inadequate, transport from town to the park is absent, and there are not enough hotels and dining facilities for visitors. Although backpackers and other low-cost tourists may be willing to visit countries that lack such features, the contribution they make to any sustainable tourism project will be relatively small unless they come in disproportionately large num-bers. This will, in turn, have knock-on effects that necessitate alternative conservation measures.

Finally, revenue from tourism can be highly sensitive to a variety of external factors that are largely unpredictable but not uncommon in many primate habitat countries. These include political and economic instability, travel restrictions, disease epidemics, international currency fluctuations, and military, terrorist, or criminal activities (Butynski and Kalina 1998). These problems suggest that even where successful tour-ism projects have been established, it might be advisable to have in place a variety of other conservation measures that would operate independent of tourist revenues.

IMPACT ON PRIMATE POPULATIONS

The most important issue of sustainable tourism projects is how much they disturb the animals themselves. Currently the empirical evidence for detrimental effects of tourism on primate populations is mixed. On the one hand, a comparison between areas where human traffic was rela-tively common and where it was absent in the Gunung Leuser National Park (Sumatra) found no difference in either the abundance or the be-havior patterns of pig-tailed macaques or orangutans (although other large mammals were affected: Griffiths and van Schaik 1993). Although the sample sizes were small, the results suggest that primates are more likely to habituate to the controlled presence of people than to leave the area. Grieser Johns (1996), for example, describes how chimpanzees in Kibale Forest (Uganda) slowly became habituated to tourist groups.

On the other hand, habituation responses cannot always be assumed. For example, although white-throated capuchins can habituate to tour-ists, squirrel monkeys in the same area appear not to (Boinski and Sirot 1997). Similarly, it has been reported that the roosting behavior of pro-boscis monkeys (*Nasalis larvatus*) along rivers declines in response to increasing human traffic (Griffiths and van Schaik 1993). Even where groups are habituated to researchers, contact with large numbers of tour-ists can be highly stressful. Kinnaird and O'Brien (1996) describe how habituated crested black macaques were much more likely to show nega-tive responses to tourists (e.g., fleeing or climbing into trees) as tourist

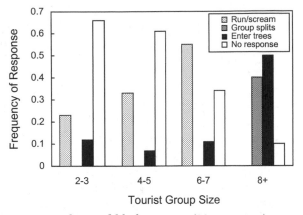

Figure 11.7 Responses of crested black macaque (*Macaca nigra*) groups to visits by tourist groups of different sizes in Tangkoko DuaSudara Nature Reserve (Sulawesi, Indonesia). Response patterns are averaged over three groups. (After Kinnaird and O'Brien 1996.)

group size increased (fig. 11.7). At the same site, spectral tarsiers delayed their departure from their day nests at dusk (potentially losing valuable foraging time before night fell) owing to the large crowds gathered outside, and on emerging they were apparently stunned by powerful camera flashes.

The greatest impact on primate populations is perhaps most likely to occur where tourists feed primates. Tourist provisioning has a major influence on normal foraging and ranging patterns in primates (e.g., *Macaca fuscata:* Koganezawa and Imaki 1996; *Macaca thibetana:* Zhao 1996). In addition, it also increases the risk of aggressive attacks on visitors, which must inevitably have serious consequences for the long-term viability of the tourist project (e.g., O'Leary and Fa 1993; Struhsaker and Siex 1996). The most extreme case of tourist provisioning is likely to be that of the Barbary macaques at Queens Gate on Gibraltar. These macaques can receive over a thousand visitors a day, and there are serious problems with overfeeding and obesity (O'Leary and Fa 1993). Moreover, Fa (1986) showed that birthrates correlated negatively with an index of tourist density for two of the Gibraltar troops ($r = -.85$ and $r = -.98$, respectively, $p < .01$). Similarly, a group of Japanese macaques at Nikko appeared to suffer an increase in mortality rates owing to its reluctance to shift its winter home range away from tourist feeding sites to lower altitudes where winter foods were more widely available (Koganezawa and Imaki 1999).

Interestingly, most cases of tourist provisioning tend to involve macaque species (red colobus are a notable exception: Struhsaker and Siex

1996), which might reflect their capacity to thrive close to human populations (sec. 8.3.2). If so, the question whether other primate species can cope successfully with such intensive human contact remains to be established.

The most serious threat that tourism presents to primate populations, particularly where close proximity is involved, is disease transmission. Primate are highly vulnerable to disease epidemics (sec. 6.2.2) and can easily contract human diseases. In 1966, for example, ten members of the Gombe (Tanzania) chimpanzee population fell seriously ill with polio, probably contracted from the human population on the southern boundary of the park after a similar outbreak there the previous month (Goodall 1986). Similarly, Barrett and Henzi (1997) describe a diarrheal disease that killed almost an entire troop of baboons, apparently contracted by foraging in an area contaminated by sewage effluent. Several authors have therefore expressed concern over disease transmission during contacts with tourists (e.g., O'Leary and Fa 1993; Struhsaker and Siex 1996; Butynski and Kalina 1998), especially given that transmission might also be bidirectional. Butynski and Kalina (1998) describe two disease incidents that occurred among the gorilla groups visited by tourists at PNV: several deaths occurred before successful treatment or vaccinations were given. If new diseases appear that are unrecognized or untreatable, the results could be catastrophic in such a small population.

Fortunately, careful management of tourists can mitigate tourist impact in most cases. Griffiths and van Schaik (1993) suggest that human traffic in protected areas should be permitted only along a small number of designated routes, retaining the surrounding area as a refuge from such disturbance. Where tourism is based around visits to specific animal groups, additional measures can be taken. These might include limiting both the size of the tourist groups and their proximity to the animals (Grieser Johns 1996; Kinnaird and O'Brien 1996). These points are well illustrated by the chimpanzees of Kibale Forest, where visitor distances of 10–20 m elicited negative responses five times more frequently than those over 40 m and where tourist groups of more than fifteen people elicited higher vocalization rates among the chimpanzees (Grieser Johns 1996). Butynski and Kalina (1998) have also suggested that, at least in the case of gorillas, sustainable tourism projects should be established only where there are reasonably large primate populations (so that the proportion of the population that experiences direct contact with tourists, and all the risks it entails, is as small as possible). Overall, provided the kinds of management steps listed above are adopted and monitored, sustainable tourism projects have the potential to make a substantial contribution to primate conservation.

11.3. Captive Breeding

Where wild populations are threatened with extinction but protecting the habitat and implementing sustainable use policies are not enough to ensure their survival, it may be necessary to intervene directly in population processes. This can involve two approaches. The first is *restocking* (supplementing an existing population to increase its size); the second is *reintroduction* (establishing a new population where the species previously existed but has since disappeared). A third approach, *introduction,* involves introducing a species into an area where it never existed before. However, owing to the disastrous effects introduced species can have on native ecosystems (sec. 7.2.1)—and primates are no exception in this respect (e.g., Konstant and Mittermeier 1982)—introductions are rarely used as a conservation tactic except where appropriate habitat can be found and other management options are restricted. Moving animals from one site to another is termed *translocation,* regardless of whether the animals transferred are drawn from wild populations or are captive bred.

Population management has the potential to make a substantial contribution to primate conservation. For example, Durant and Mace (1994) have demonstrated that active management of a model mountain gorilla population that has lost several of its social groups can significantly increase its persistence time (fig. 11.8). But population management is costly and time intensive, and may address only the symptom of a conservation problem (small population size) rather than its cause. In this section we consider the strategy of captive breeding for conservation and its limitations. In the following section we examine translocation programs designed to help with restocking and reintroductions.

11.3.1. Strategy

The basic philosophy of captive breeding for conservation is that a captive population (which may be scattered across a number of collections) is an integral component of the metapopulation currently managed by conservationists in the wild and that the wild and captive populations should be managed interactively to maximize population persistence time (Seal, Foose, and Ellis 1994). To achieve this, most conservation captive breeding programs aim to establish a self-sustaining population that will maintain at least 90% of the original heterozygosity for one hundred or two hundred years. Captive populations may also have another use: when a species becomes extinct in the wild, the captive population represents the only surviving population.

However, views on captive breeding as a tool in primate conservation have been mixed. On one hand, the IUCN Action Plans (Oates 1986a,

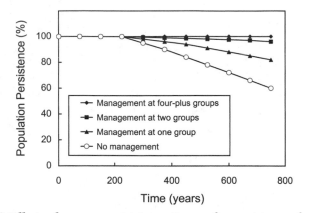

Figure 11.8 Effects of management intervention on the persistence of a mountain gorilla (*Gorilla gorilla beringei*) population, estimated with POPGEN (a population simulation model) based on one thousand simulations. In the model, the population is composed of ten social groups, each of which has an initial size of ten individuals and a ceiling of forty individuals and experiences demographic stochasticity. Management is initiated whenever the number of groups falls to the number indicated (where management is defined as intervention that ensures juvenile survival). Even when management is implemented with only one remaining group, the probability of population extinction is substantially reduced. If management is implemented earlier—at two, three, or four remaining groups—the probability of extinction is reduced further still. However, note that there is no further improvement in persistence time if management is implemented before the population falls to four groups. (After Durant and Mace 1994.)

1996a; Eudey 1987) decline to recommend captive breeding for any African taxa and endorse it only in four Asian taxa (*Simias c. concolor, Presbytis potenziani potenziani, Hylobates concolor hainanus,* and *Rhinopithecus bieti:* see table 11.1), and then strictly for the purposes of eventual restocking of wild populations (sec. 11.4). On the other hand, the recent Global Captive Action Plan for Primates (GCAP) outlines a far more extensive role (Stevenson, Baker, and Foose 1991). The purpose of the latter strategy, collaboratively developed by the IUCN Captive Breeding Specialist Group (CBSG), Species Survival Commission (SSC), and Primate Specialist Group (PSG), is to identify taxa whose conservation would be enhanced by captive breeding (Seal, Foose, and Ellis 1994).

As a rule of thumb, IUCN recommends that captive propagation of a taxon should occur when that taxon falls below a total of one thousand individuals (Seal, Foose, and Ellis 1994). However, as Caughley and Gunn (1996) note, once a wild population is so small that it requires supplementation with captive-bred animals, it may not be possible to remove enough animals for a captive breeding program without exacerbating the risk of extinction. This is a particular problem for the captive

Table 11.6 Summary of the Global Captive Action Plan (GCAP) for primates

| Taxon | Total subspecies recognized | Recommendations for percentage of all subspecies | | Current total captive population size ISIS[2] (1991) |
		To be held in captivity	To be treated as high priority[1]	
Lorisiformes	24	54	4	436
Lemuriformes	49	79	22	2,465
Tarsiiformes	4	25	0	24
Callitrichidae	53	53	13	2,760
Cebidae	119	42	9	1,798
African cercopithecines	106	32	7	2,225
Asian cercopithecines	23	35	13	1,100
African colobines	29	21	3	460
Asian colobines	66	45	35	407
Hylobatidae	27	48	30	930
Pongidae	9	78	56	3,212
Total	509	45	15	15,833

Source: Data from Stevenson, Baker, and Foose 1991.

[1]High priority captivity refers to species listed as "90%/100 years I" by the primate GCAP (see text for further information.

[2]International Species Information System (ISIS). Figures refer only to those taxa with captive recommendations.

breeding program recommended for the endemic Hainan gibbon (*Hylobates concolor hainanus*): at the time the program was proposed, the population totaled only thirty to forty individuals in four localities (Eudey 1987).

An alternative approach is not to allow wild populations to become so small before captive breeding programs are initiated. This is the philosophy on which the GCAP operates. In this case, all threatened species are recommended for captive breeding action, with the intensity of action directly comparable to the level of threat (indexed by the IUCN Red List codes: for details, see sec. 10.2.2). According to the GCAP strategy for primates, almost half of all primate taxa (45%) are recommended for captive breeding at a minimum level of twenty-five to one hundred individuals (table 11.6). A total of 15% are identified for urgent action in captivity (i.e., the establishment of a "population sufficient to preserve 90% of the average heterozygosity of the wild gene pool for 100 years as soon as possible": Stevenson, Baker, and Foose 1991). The distribution of these taxa indicates that pongids, Asian colobines, hylobatids, and Lemuriformes are most in need of such captive breeding programs (20% or more of all the taxa in each of these groups are recommended for urgent action); in contrast, African colobines, African cercopithecines, and cebids have the lowest proportion of taxa recommended for urgent action (table 11.6).

11.3.2. Constraints

Although this approach does potentially avoid the problem of exacerbating extinction risk in the wild, the blanket application of captive breeding to large numbers of threatened taxa has been strongly criticized owing to the limited success of such programs (Caughley and Gunn 1996; Snyder et al. 1996). Four main problems have been identified.

First, sophisticated maintenance and reproductive technologies are often required to breed wild animals successfully in captivity. For example, folivores are generally difficult to keep in captivity because of their dietary needs and specialized digestive systems (e.g., colobines: Oates and Davies 1994; *Brachyteles:* Pope 1998b; *Alouatta:* Horwich 1998). Similarly, breeding programs with the Philippine tarsier have been hampered by low offspring survival (Wright et al. 1987), while captive drills show very low birthrates (Gadsby, Feistner, and Jenkins 1994). In the latter case, low birthrates were attributed to the fact that most drills are kept as singletons or in pairs in substandard conditions, although these problems are now being addressed by improved coordination between collections. Yet even where wild-caught animals manage to breed successfully in captivity, there can be no assurance that breeding will then continue, as demonstrated by breeding failure in some second-generation captive-bred orangutans (Conway 1980).

Second, there is only so much space for captive breeding programs, and a trade-off must be made between maintaining a large enough captive population to ensure long-term viability in the face of small-population effects and providing enough room for captive breeding of all those species that require it (Ebenhard 1995). The small-population effects that most threaten captive populations are demographic stochasticity and loss of heterozygosity (sec. 7.2.2): a case study of the relative threats posed by these processes in a captive primate population is provided for gorillas by Mace (1988). In addition, large populations are required to avoid founder effects (although this may require only twenty to thirty individuals: Foose, Seal, and Flesness 1987). Maintaining large populations may be especially important for species with short generation times, since these species are at high risk from genetic drift. Inbreeding is also a problem, and many captive primates show high neonatal mortality (>19% mortality: Flesness 1987). The seriousness of inbreeding effects in captive primates has been demonstrated by Ralls and Ballou (1982), who showed that infant mortality rates were higher among inbred than noninbred animals in fifteen of sixteen primate colonies.

Currently, over fifteen thousand primates are held in captivity worldwide, according to the records of the International Species Information

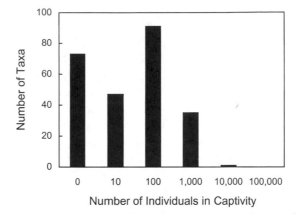

Figure 11.9 Frequency distribution of the size of captive populations of primate taxa (combines species and subspecies) on the International Species Information System (ISIS) database. This listing excludes zoological collections that are not participating in the ISIS scheme and all other captive animals (such as laboratory animals) not in zoological collections. This figure shows that a substantial number of taxa are not found in captivity; but among those that are in captivity the most common population size is below one hundred individuals. The one taxon with over one thousand individuals is the common chimpanzee *Pan troglodytes* (with two thousand captive specimens). (Data from Stevenson, Foose, and Baker 1992.)

System, ISIS (table 11.6). However, many captive populations are too small to maintain long-term viability. If a self-sustaining captive population is defined as a population "of over 100 individuals of which at least half were captive-bred," then in 1970 no captive population of "rare" primate species showed long-term viability (Flesness 1987). In 1979 only eight Red List primate taxa met this criterion. By 1984 this had increased to eight to eleven Red List taxa (22–30% of all Red List species held in captivity) from a total of thirty-seven to forty-six primate taxa that met this criterion (precise figures were not available owing to a combination of data sources). Analysis of the 1991 ISIS data on the number of individuals in each taxon recommended for captive action by the GCAP strategy ($N = 247$ species and subspecies) shows that seventy-three of these taxa are still not represented in captivity, while only half of those taxa that are represented occur in captive populations with more than thirty individuals. Overall, there are many small captive populations but very few that are large (fig. 11.9). Nevertheless, if it is crudely estimated that the global capacity for captive primates is about double the ISIS total, then it may be possible to maintain a captive nucleus of about 135 individuals for all those primate taxa identified as in need of urgent captive action (Stevenson, Baker, and Foose 1991), although less optimistic estimates for primates are given by Foose, Seal, and Flesness (1987; cf. Balmford, Mace, and Leader-Williams 1996).

Third, captive breeding is extremely costly in time and money. This is particularly true for the large primates that are slow reproducers. The captive maintenance costs per capita increase progressively with body size among mammals (Balmford et al. 1996). Moreover, the financial cost of maintaining captive populations in order to achieve the goal of retaining 90% heterozygosity over one hundred years increases with both body size and generation length in mammals (Balmford et al. 1996). This might discourage captive breeding for large, slow-breeding species. Yet even the small primates are extremely costly to breed in captivity. Kleiman et al. (1991) estimated that the cost of managing a captive population of 550 golden lion tamarins housed at one hundred institutions amounted to $911,875 in 1989 alone. (This estimate covers husbandry, food, veterinary care, recordkeeping, supplies and equipment, utilities, and housing.) Such high costs are a particularly important consideration where captive breeding is not an absolutely essential component of a conservation strategy, since the funding can divert support for more cost-effective forms of in situ conservation such as habitat protection.

Fourth, behavioral and genetic changes in captivity may preclude reintroduction. Foraging, antipredator, and social behavior may all be compromised if animals have lived all their lives in captivity, and the loss of heterozygosity may intensify vulnerability to disease (Snyder et al. 1996). Behavioral changes may be offset by rehabilitation programs (see below), but genetic changes are harder to deal with.

Given these problems, many believe that captive breeding for conservation should be used only where extinction would otherwise be inevitable (e.g., as a short-term measure to ensure the continued survival of a population, when this cannot be guaranteed in the wild during the period needed to bring the agent of extinction risk under control: Caughley and Gunn 1996). Nevertheless, some threatened primate species do breed well in captivity (e.g., the Barbary macaque: Fa 1987), and the wider application of captive breeding to supply the demand for primates for educational, entertainment, and biomedical purposes can still make a valuable contribution to in situ conservation by reducing the demand for wild live-caught primates (e.g., Oates 1994a).

11.3.3. Case Studies

The potential for effective captive breeding in primate conservation is exemplified by two projects in particular. The first is the captive breeding and rehabilitation facility for drills (*Mandrillus leucophaeus*) established within the geographic range of wild drills in southeastern Nigeria (Gadsby, Feistner, and Jenkins 1994). Since the drill is the most endangered primate in Africa (Oates 1996a), it is perhaps a justified candidate for captive breeding action. By undertaking captive breeding in the native range with a large semi-free-ranging group, it has been possible to reha- bilitate locally caught live-trapped animals; to enhance local conservation awareness of the drill and its habitat; to provide employment to local people (to maintain the center), who therefore receive a direct cash in- come from these animals; to produce captive-bred animals familiar with the natural habitat to which it is hoped they will eventually be reintro- duced; and to do all of these things relatively inexpensively. To date there have been no reintroductions, because the problems of habitat loss and hunting have not yet been brought under control. Nonetheless, the Nige- rian drill project demonstrates many of the positive contributions that captive breeding programs can make.

The second project is the Golden Lion Tamarin Conservation Pro- gram (Kleiman et al. 1986, 1991; Kleiman and Mallinson 1998). By the late 1970s the dwindling population of wild golden lion tamarins (*Leon- topithecus rosalia rosalia*) was estimated to number as few as 100–200 individuals (Coimbra-Filho and Mittermeier 1977). The Golden Lion Tamarin Conservation Program began with studies of the reproductive biology of the species in captive populations in the early 1970s. This re- search was then used to propagate the species in captivity with some success, leading to a rapidly growing captive population in the late 1970s. By the early 1980s field studies of wild golden lion tamarins were in- itiated to describe the behavioral ecology and population status of the remaining wild populations of this species. These field studies were ac- companied by habitat protection and restoration measures as well as con- servation education to gain the support of the local communities. The first reintroductions to protected habitat then took place in the mid- 1980s. By 1989 a total of 71 animals had been reintroduced. Of these, 27 have survived, and 26 more have been born (with an 81% survival rate). The estimated population in the wild is currently estimated to be 450 (of which about 64% live in protected areas), with an additional 550 in cap- tivity (Dietz, Dietz, and Nagagata 1994). The components of this project are detailed in table 11.7.

11.4. Restocking and Reintroduction

Restocking and reintroduction are the ultimate goal of captive breeding for conservation, although they may also be practiced using wild-bred animals. The use and the success of these methods in primate conserva- tion strategy have been limited to date, although their importance may grow in the future, particularly in the management of metapopulations in fragmented habitats.

Table 11.7 Components of the Golden Lion Tamarin Conservation Program

Heading	Activities
The captive population	Cooperative research, genetic and demographic management, and husbandry standards
The wild population	Field research on population status and behavioral ecology
The habitat	Protection, management, preservation, and restoration
Conservation education	Public relations, professional and community education
Reintroduction	Preparation, release, provision of critical resources, long-term monitoring and evaluation of success
Community ecology	Research on fauna and flora of the Atlantic coastal rain forests of Brazil

Source: From Kleiman et al. 1991 with the permission of the Zoological Society of London.

Thus far the IUCN Action Plans for African primates have not recommended either restocking or reintroduction of wild or captive-bred species (Oates 1986a, 1996a). However, at a localized scale translocations of wild African populations have been suggested, with recommendations for populations of both De Brazza's monkey and Tana River monkeys in Kenya to be moved from forest patches in unprotected areas to protected habitat (Mugambi et al. 1997 and Oates 1996a, respectively). The Asian Action Plan has recommended translocation only for the orangutan (Eudey 1987), and even this has proved highly controversial (Stuart 1991). Nevertheless, other restockings have taken place outside these recommendations (Wilson and Stanley Price 1994), such as the black-and-white ruffed lemur (Madagascar), black-handed spider monkey (Panama), golden lion tamarin (Brazil), Barbary macaque (Morocco), and lion-tailed macaque (India).

CRITERIA FOR TRANSLOCATIONS

Kleiman, Stanley Price, and Beck (1994) identify thirteen conditions under four headings that might constitute criteria for translocations (see also Stanley Price 1991; Caldecott and Kavanagh 1983). Table 11.8 applies these criteria to tamarin conservation. Four of them are worthy of special mention.

First, the availability of a sufficient stock for release is crucial, since the number of animals released into the new habitat is likely to be directly correlated with their success in establishing a new breeding population (e.g., Duncan 1997). Unfortunately, for many primate taxa it would

Table 11.8 Criteria for translocation scheme for lion tamarins

Criteria	Golden lion tamarin	Golden-headed lion tamarin	Black lion tamarin
Condition of species			
1. Need to augment the wild population	+	−	(+)
2. Available stock	+	+	−
3. No jeopardy to wild population			
Environmental conditions			
4. Causes of decline removed		−	−
5. Sufficient protected habitat	(+)	−	+
6. Unsaturated habitat	+	(+)	
Biopolitical conditions			
7. No negative impact for locals	−		
8. Community support exists	5	2	4
9. GOs/NGOs supportive or involved	+	+	+
10. Conformity with all laws/regulations	+		
Biological and other resources			
11. Reintroduction technology available	4	3	3
12. Knowledge of species' biology	5	1.5	3
13. Sufficient resources available	+	−	−
Translocation recommended	+	−	−

Note: The extent to which the different criteria are met for each species is designated with + for yes, (+) for possibly yes, − for no, and on a numerical scale of 1 (worst)–5 (best), with an empty space indicating no information available. After Kleiman, Stanley Price, and Beck 1994.

be unrealistic to expect reintroductions to occur on a very large scale. This may limit the potential use of this management tool. For example, Swart and Lawes (1996) modeled the effects of a restocking event of reasonable scale on the metapopulation of samango monkeys in Natal (South Africa). They found that restocking improves the probability of persistence only in the short term (fewer than five hundred years) and only in low-density populations (when restocking increases population size from 12.5% to 25% of carrying capacity; a further doubling has no additional effect) (fig. 11.10). Notably, in their analysis corridors were a much more effective management strategy than restocking for extending metapopulation persistence beyond five hundred years. The question of appropriate stock must also address genetic compatibility and the potential risks of both genetic pollution and outbreeding depression. Pope (1996, 1998b), for example, has expressed concern over translocations between the genetically distinct northern and southern populations of

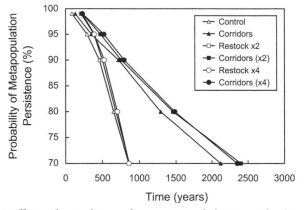

Figure 11.10 Effects of restocking and connectivity (habitat corridors) on the probability of samango monkey *Cercopithecus mitis* metapopulation persistence. In this stochastic spatial population model, the metapopulation is split between eight isolated forest patches. In the control condition, each patch is initially occupied at 12.5% of carrying capacity. Two possible restocking scenarios are shown, where restocking either doubles (Restock × 2) or quadruples (Restock × 4) the initial population size in the control condition. The three corridor scenarios demonstrate the effects of connectivity between patches in the control condition (Corridors) and the two Restock conditions (Corridors × 2 and Corridors × 4, respectively). For further information on this model, see section 10.2.3. (Data from Swart and Lawes 1996.)

muriquis in the Atlantic Forest. Similar concerns exist over the Sumatran and Bornean populations of orangutans.

One possible source of animals for translocation projects might be emigrants from isolated habitat fragments, particularly if these animals are unlikely to ever locate other groups in the same area. Such a suggestion has been made for isolated *Brachyteles* populations by Pope (1996).

Second, the risks involved in disturbing an existing wild population, particularly the introduction of diseases, must be carefully considered (Woodford and Rossiter 1994; Cunningham 1996). Rijksen (1978) describes how translocated orangutans introduced diseases from a local human population into a wild orangutan population. It is also possible for the translocated animals themselves to fall victim to disease: five deaths from an unknown disease occurred in one family group of golden lion tamarins that formed part of the release program for this species (Kleiman et al. 1986). However, although wild tamarins existed in the area, none existed specifically at the release site, so whether these animals caught the disease from conspecifics remains questionable.

Third, there is little to be gained from restocking or reintroduction programs if the original cause of decline has not been removed. A case in point is the Barbary macaque, a species threatened in the wild but with a large surplus captive stock (Fa 1987). Estimates of carrying capacities in habitats where the monkeys used to exist indicate that a reintroduction program might easily lead to the establishment of large populations in the wild (Fa 1994), but the problem of feeding competition from pastoralist livestock—implicated in their initial disappearance—remains to be remedied. In other cases where habitat suitability cannot be so easily assessed, probe groups and monitoring can be used to investigate reintroduction potential (Caughley and Gunn 1996).

Fourth, there is the question of financial resources. Translocation projects are notoriously expensive. Kleiman et al. (1991) estimated that each surviving golden lion tamarin reintroduced to the wild cost an average of $22,563 (costs from 1983 to 1989 inclusive)! Although Kleiman notes that these costs may decrease as program efficiency increases, she also concedes that expenses are always likely to be high, mainly because of the intensive pre- and postrelease training required for captive-bred animals (see below) and the isolation of the areas where the releases take place.

With operating costs like these, the question whether population manipulation is a cost-effective tool in primate conservation must be carefully addressed. There are two schools of thought. One argues that these vast sums of money would be better spent on in situ projects such as habitat protection, particularly since translocation projects can compete with these projects for funding (Mackinnon and Mackinnon 1991). The other school emphasizes the additional benefits that come from translocation projects that may not be directly financial but that still may make a significant contribution to conservation (Stuart 1991). For example, translocation projects can act as flagships to attract funding for related conservation projects that focus on habitat protection and education. The Golden Lion Tamarin Program has done precisely this (Kleiman et al. 1991; Dietz, Dietz, and Nagagata 1994).

Finally, the advantages and disadvantages of restocking and reintroduction techniques must ultimately be considered in terms of how successfully these programs achieve their aims. Beck et al. (1994) defined a successful reintroduction as one in which the wild population attains a size of five hundred individuals surviving independent of human intervention or showing long-term viability in a PVA. Only sixteen of the reintroduction projects reviewed by these authors (11%) were considered successful by this definition. No primate projects were on this list of successes. Perhaps the most important feature of the successful projects was that they released many more animals than the unsuccessful projects. They were also more likely to use local employment and undertake community education.

Wolf et al. (1996) undertook a broader analysis of the success of translocation projects, where success was defined simply as the establishment of a self-sustaining population (see also Griffiths et al. 1989; Wolf, Garland, and Griffith 1998). They found that among mammals the only reliably significant predictor for success was location of release: animals released into the core of the geographic distribution of the species fared better than those released on the periphery or outside the distribution. A variety of other variables, including life history parameters (e.g., body mass, age at first reproduction), ecological parameters (e.g., diet type, habitat quality), release parameters (e.g., wild or captive-bred animals), and conservation status, had no consistent effect. In the rest of this section we investigate the successes of primate translocation projects from captive and wild stock.

11.4.1. Translocation from Captivity

In the past, many primate translocation projects were developed for essentially ethical reasons rather than for conservation purposes. In these cases release programs have been seen as a solution to the wastage arising from the euthanasia of confiscated pets, former biomedical subjects, or wild animals that have been collected after being displaced by disturbances such as logging. Unfortunately, most of these early efforts at translocation were poorly planned and badly organized (Wilson and Stanley Price 1994). For example, one rehabilitation and release program for confiscated gibbons (*Hylobates muelleri*) in Sarawak had a 90% mortality rate in released animals (Bennett 1989, 1992). This failure may have been due to a variety of factors, including hunting, starvation, disease, and conspecific aggression. In such cases it must be seriously doubted whether it is ethically more acceptable to subject animals to a slow death in hostile habitat than a less painful death through euthanasia.

Better results were reported for a *Hylobates lar* release program by Tingpalapong et al. (1981), who released thirty-one former laboratory lar gibbons into closed forest in Thailand. All the adults had been wild caught, although five offspring were captive bred. This release program was more carefully planned in terms of the area (chosen because gibbons were already present but not in large numbers and the area was protected from hunting), postrelease provisioning of food and shelter, and systematic monitoring. The gibbons were released in combinations of family groups, pairs, and individuals. Of those released, two died at the release site, one was recaptured, twenty-four disappeared after varying periods (although in the interim some were seen feeding on natural fruits), and four joined wild groups. Notably, the most successful survivors were among the first to be released, suggesting that they may have taken up the few territorial vacancies available in the population.

A comparison between these two studies illustrates the importance of establishing translocation projects in protected areas and of providing some food and shelter for the animals during the release phase. Although such *soft releases* are financially more costly than *hard releases* (where no resources are provided for the animals), they are almost certain to be more cost effective in the long run simply because they will be more successful. For example, spider monkeys reintroduced to Barro Colorado Island survived at the site where food and shelter were provided but died at the other release sites (Konstant and Mittermeier 1982). The success of soft release projects may be enhanced by a training or "rehabilitation" phase to help the animals switch from a life of dependency on humans in a captive environment to independence in a natural environment. Rehabilitation involves actively coaching animals in the release program to develop lost or latent skills such as foraging, antipredator, and rearing behavior.

REHABILITATION PROJECTS

Primate rehabilitation projects fall into two categories: those that are part of ethical release programs and those that are part of conservation release programs. So far, most fall into the former category. Hannah and Mc-Grew (1991) provide a comprehensive review of rehabilitation projects for great apes (table 11.9) as well as giving a detailed account of one particular chimpanzee rehabilitation project in Liberia. Two lessons learned from the Liberian project were that female chimpanzees adapted more successfully to their new environment than males and that the highest mortality fell among those in the first subgroup to be released (individuals in subsequent subgroups were less likely to become lost, because they were in the company of animals familiar with the habitat). Hannah and McGrew (1991) also concluded that rehabilitation and release projects were likely to be more successful for orangutans than for chimpanzees because of the differences in their social systems. Communities of chimpanzees are highly territorial and will kill intruders (although individual females may be accepted peacefully: Treves and Naughton-Treves 1997), whereas orangutans are generally solitary and less aggressive (see also Borner 1985). Territorial aggression may also have been a problem for the gibbon projects described above (Bennett 1989).

Integrated rehabilitation and reintroduction projects for primate conservation are best illustrated by the Golden Lion Tamarin Conservation Program (sec. 11.3.3). The rehabilitation phase of this project is described in detail by Kleiman et al. (1986). Animals were systematically trained in how to feed, avoid danger, and locomote using new substrates. In the first case the tamarins were trained to find food embedded (hidden) in nonedible materials; adults tended to be less adept at finding

Table 11.9 Summary of chimpanzee rehabilitation projects

Release site	Date released	Number released	Age (years)	Background	Prerelease preparation	Provisioning	Follow-up	Adaptive behavior[1]	Outcome
Rubondo Island, Tanzania (2,400 ha)	1966–69	17	4–12	Wild born from zoos	No	No	No	1, 2 (?)	Reproducing population on island
Ipassa Island, Gabon (65 ha)	1968–72	8	4–8	Wild born from lab	No (?)	Yes	Yes	1, 2, 4, 6	Two escaped, others removed
Niokolo-Koba National Park, Senegal	1973–75	8	1–6	Wild born (6), captive born (2)	Yes	Yes	Yes	1, 2, 3, 5, 6	Moved to island in River Gambia
Baboon Island, Gambia (490 ha)	1979	9	1–13	Wild born (7), captive born (2)	Yes	Yes	Yes	1, 2, 4, 6	Added to group above
0.13 ha island, Florida	1975	8	4–11	Wild born (6), captive born (2) from lab	Yes	Yes	Yes	2 (crude)	?
Three islands, Liberia (6,27,28 ha)	1978 1983 1985	18 24 22	5–20+	Wild born from lab; some pets	Yes	Yes	Yes	1, 2, 3, 5	Reproducing populations on islands

Source: After Hannah and McGrew (1991).

[1]Adaptive responses: (1) eating foods, (2) nest building, (3) ant eating (without tool use), (4) ant dipping/termite fishing, (5) stone tool use (for nut cracking), (6) predatory behavior.

embedded foods than immatures, but those animals that were more efficient at finding embedded foods survived longer after release. In the second case, a poisonous toad was presented to the tamarins; although several tamarins attacked it and were subsequently poisoned by it (nearly fatally), this did not lead to an avoidance response on a second presentation. In the third case, the captive-bred animals disliked using natural flexible substrates (they were accustomed to the fixed perches in their enclosures) and seemed flummoxed by the problem of navigating through a complex three-dimensional habitat; the holding cages were subsequently modified to be more variable and naturalistic. After release, there were significant correlations between age and survival: younger animals were clearly more adaptable than older individuals. One way of countering this problem may be to release family groups, where older animals will have the opportunity to learn from younger ones.

Primates are difficult subjects for rehabilitation projects because many of their behaviors are learned, and standard learning periods may be many years. This is a particular problem for captive-bred animals that have never experienced a wild environment (Yeager 1997). Some subjects may initially settle into a normal wild existence but then abandon this existence to seek out and return to human company (chimpanzees: Treves and Naughton-Treves 1997). Where rehabilitation projects are the result of ethical rather than conservation interests, they may also do more harm than good to existing wild populations through disease transmission, feeding competition, and outbreeding depression (orangutans: Rijksen 1978; Yeager 1997). Nevertheless, Rijksen (1978) argues that although ethical rehabilitation projects should not generally be used to restock or reintroduce wild populations, they can still make an important contribution to conservation. By rehabilitating orangutans, for example, and allowing them to live a seminatural lifestyle around a rehabilitation center, it may be possible to promote public education, tourism, and research, benefiting both the orangutan and the rain forest environment as a whole.

11.4.2. Translocation from the Wild

Translocation projects involving wild primate stock have thus far been more often used to move a threatened population out of harm's way than to reintroduce or restock another threatened population. Partially as a result of such crisis management, many previous translocation projects have been badly planned and poorly monitored. Consequently several accounts exist of translocations where the outcomes either were failure or were unknown (Madagascar: O'Connor 1996; the Neotropics: Konstant and Mittermeier 1982; Horwich 1998). More recently, however, translocation projects have benefited from more careful planning and long-term monitoring (e.g., Ostro et al. 1999).

The results of a selection of translocation events, including both carefully planned and largely unplanned projects, are summarized in table 11.10. These data are encouraging since they demonstrate that translocations can succeed across a range of primate ecological niches with different diets and habits, including species that are arboreal or terrestrial, diurnal or nocturnal (*Daubentonia madagascariensis* is nocturnal), territorial or nonterritorial (*Alouatta pigra* is territorial: Ostro et al. 1999). In addition, the successful projects include taxa classified as Vulnerable (*Pongo pygmaeus*) or Endangered (*D. madagascariensis, Procolobus badius kirkii*). These data also indicate that restocking, reintroduction, and introduction schemes can all achieve success, although these outcomes appear to have been largely incidental to the rescue of the animals involved (only the translocation of *A. pigra* appears to have taken place as a specific conservation program to reestablish a viable population where an extinction had previously occurred: Ostro et al. 1999). Unfortunately, what determines success is less clear. In addition, many of these studies judge success on indirect measures over short periods when long-term success may differ. For example, since the "successful" baboon translocation, the new groups have suffered much higher mortality rates than the sympatric native troops (Kenyatta 1995). Nonetheless, it is possible to draw some preliminary conclusions from the evidence so far.

First, suitable release habitat is crucial. It must contain food and other key resources but not already be home to resident conspecifics (at least not with a population close to carrying capacity). These habitats should also be within the natural range of the species. All but two of the translocations listed used habitat that had recently suffered some disturbance or where hunting had previously taken place, since these sites had the potential to support healthy primate populations but the original population was below carrying capacity or completely absent (e.g., Strum and Southwick 1986; Andau, Hiong, L. K.; and Sale 1994; Vié and Richard-Hansen 1997). It is also important to ensure that the habitat disturbance has not been too extreme (the failure of the red colobus translocations may have partially reflected the small extent of remaining suitable habitat in the release areas: Struhsaker and Siex 1998) and that the area is protected from future human impacts.

Second, a soft release does not appear to be an essential ingredient for a successful translocation project, although in the two cases where provisioning was provided it clearly helped. The need for provisioning is likely to depend largely on the similarity between the source habitat and the new habitat. It may also depend on the season of release (if the release occurs in a lean season, provisioning may be vital, since the released animals will not know where to find keystone food resources), on the

Table 11.10 Primate translocation projects involving the movement of wild populations from one site to another

Species	Niche[1]	Translocation type	Release date	Soft release?	Number of animals[2]	Population response to translocation	Successful?	Source
Africa								
Papio anubis	T, Fr	Restock	1984	Yes	131 (3)	Settled with low mortality in 6 months	Yes	1
Procolobus badius	A, Fo	Reintroduce (Masingini)	1977–78	No	36 (?)	Increased to 56–64 individuals in 18 years	Yes	2,3
		Reintroduce (Kichwele)	1978	No	13 (?)	Declined to near-zero or zero levels in 18 years	No	2,3
		Introduce (Ngezi, Pemba)	1974	No	14 (?)	Declined to near-zero or zero levels in 18 years	No	2,3
Madagascar								
Daubentonia madagascariensis	A, In	Introduce (?)	1966	No	9 (n.a.)	Increased to 30 individuals in 30 years	Yes	4
Asia								
Macaca mulatta	T, Fr	Reintroduce	1983	Yes	20 (1)	Good survival and birth rates after 10 months	Yes	1
Pongo pygmaeus	A, Fr	Restock	1995	No	84 (n.a.)	Observed increase in number of nests afterward	Yes	5
The Americas								
Alouatta pigra	A, Fo	Reintroduce	1992–94	No	62 (14)	Good survival and birth rates after 6 months	Yes	6,7
Alouatta seniculus	A, Fo	Restock	1994–95	No	124 (29)	Majority survived and settled in general vicinity	Yes	8
Pithecia pithecia	A, Fo	Restock	1994–95	No	6 (1)	Majority survived and settled in general vicinity	Yes	8

Sources: (1) Strum and Southwick 1986; (2) Struhsaker and Siex 1998; (3) Silkiluwasha 1981; (4) O'Connor 1996; (5) Andau, Hiong, and Sale 1994; (6) Horwich 1998; (7) Ostro et al. 1999; (8) Vié and Richard-Hansen 1997.

[1] A, arboreal; T, terrestrial; Fr, frugivorous; Fo, folivorous; In, insectivorous.
[2] Number of animals is given as individuals, with number of social groups in parentheses. If species is solitary, listed as not applicable (n.a.).

restrictions on movement (if only a small area is protected, it might be prudent to encourage the released animals to stay where protection is most effective; otherwise translocated primates can move several kilometers from their release point: e.g., Vié and Richard-Hansen 1997), and on the dietary flexibility of the species (Strum and Southwick 1986).

Third, there is no strong pattern indicating that translocating large numbers of individuals will be any more successful than translocating only a few individuals. However, given the intrinsic vulnerability of small populations to extinction, we would always advocate that translocated populations should be as large as possible (given the availability of animals to translocate and the carrying capacity of the release habitat). The most successful of the Zanzibar red colobus translocations, for example, involved the greatest number of individuals (Struhsaker and Siex 1998). In addition, an appropriate age-sex composition is also important (an unfavorable age-sex composition might have been responsible for one of the red colobus translocation failures: Struhsaker and Siex 1998).

Fourth, the translocation of intact social groups may not always be essential. Strum and Southwick (1986) argued that translocated *Papio anubis* and *Macaca mulatta* groups retained their social integrity because they were moved as intact groups; however, this finding might equally be a function of strong female philopatry in cercopithecine primates. Non-cercopithecines that exhibit mixed-sex dispersal might show more fluid patterns of grouping (as might those species with a capacity for fission-fusion: sec. 3.3.1). For example, groups of *Alouatta seniculus* and *Pithecia pithecia* were translocated largely intact yet still split up after translocation (Vié and Richard-Hansen 1997). In contrast, one red colobus translocation was successful even though the released group represented only one-third of the original (Struhsaker and Siex 1998). Indeed, even among cercopithecines, partial groups can still translocate successfully (e.g., *Macaca mulatta*: Strum and Southwick 1986). The key factors in whether individuals should be translocated in socially intact groups is therefore likely to vary between species and probably also habitat (the latter might depend on the intensity of predation risk and therefore the minimum permissible group size).

Fifth, translocations can be costly in terms of both mortality and financial expense. Several of the projects listed experienced deaths from heat exhaustion (Andau, Hiong, and Sale 1994; Silkiluwasha 1981) or disease (Vié and Richard-Hansen 1997; Horwich 1998), but in all instances only a handful of individuals died and in other cases no deaths occurred at all (Strum and Southwick 1986). Mortality appears to be largely avoidable given sufficient planning and experienced project managers, especially since most deaths occur during capture and captivity. The financial expense may be more difficult to avoid, and it can be substantial: per animal, the orangutan translocation cost $175 (Andau, Hiong, and Sale

1994), the rhesus macaque translocation cost $150, and the baboon translocation cost $500 (Strum and Southwick 1986).

Finally, given the undesirability of introductions (sec. 11.3), it is perhaps surprising that these constitute two of the projects listed. However, these two cases might demonstrate those unusual conditions under which primate introductions can be justified. Both occurred on islands neighboring the source population; in the case of *D. madagascariensis*, the island was uninhabited (making subsequent protection of the introduced population much easier). Since the islands were neighbors to the source, they shared similar habitat types. Moreover, it is possible that these islands might once have contained populations of these primates that had become extinct (see sec. 7.4.2) but had not yet been recolonized (see sec. 4.1.1). The use of uninhabited offshore islands continues to be advocated for future lemur translocations (O'Connor 1996).

In conclusion, then, primate translocation schemes appear to hold a reasonable potential as a conservation management option in the future, provided suitable protected habitat exists. This conclusion is perhaps further substantiated by the observation that accidental primate introductions have so often proved successful—for example, long-tailed macaques on Mauritius and vervet monkeys in the West Indies (Konstant and Mittermeier 1982; Struhsaker and Siex 1998).

11.5. Summary

1. Key tactics in current primate conservation strategy can be grouped under the headings of protected area systems, sustainable utilization, and captive breeding, with the goal of restocking and reintroduction. Of these, the most important tactic is currently protected area systems.

2. The precise coverage of protected areas varies between region and country. In general, larger countries have relatively less area protected than smaller countries. Within countries, the size distribution of protected areas can be strongly skewed, with more areas being small and very few being large.

3. Evidence suggests that the long-term viability of populations in protected areas may be at risk from the effects of future isolation. A variety of measures have been proposed to circumvent these problems, including maximizing protected area size and connectivity. The latter may involve the use of corridors (although these are associated with potential costs as well as benefits). Alternative measures may be necessary to maximize present species richness in reserves. The problem of habitat loss as a result of global climate change remains to be addressed in protected area planning.

4. Protected areas can be defended from poaching in a variety of ways. Poaching can be reduced by intensive antipoaching patrols (which act

as a deterrent by increasing the risk of arrest). However, the financial investment this requires is prohibitive. Alternatively, the efficiency of existing antipoaching efforts can be maximized by designing protected area boundaries to minimize both the perimeter that has to be patrolled and the accessibility of the protected area to poachers.

5. Sustainable utilization is an alternative approach to primate conservation and may be used in conjunction with protected area systems to help deal with poaching. This tactic, which has the potential to contribute to both the local development of the human communities and the long-term conservation of the primate communities, has become very popular among conservationists in recent years. However, a variety of problems have emerged with this approach, suggesting that its potential may be more limited than previously assumed.

6. In principle it should be possible to harvest primate populations sustainably, albeit at low levels. Among traditional subsistence hunters, however, the available evidence currently suggests that any apparent conservation ethic is likely to be purely epiphenomenal. There is also a serious problem with the concept of sustainable commercial harvesting of primates: because of low reproductive rates, the sustainable yields will also be low, so commercial interests are likely to be best served by liquidating the stock (i.e., the primate population) and reinvesting the money elsewhere.

7. Tourism, a form of sustainable utilization that is nonconsumptive, appears to hold more potential than consumptive harvesting. But it may be possible only under certain conditions, and the interaction between tourists and wildlife must be carefully managed to avoid deleterious effects on the primate populations and their habitats. The most serious of these risks may be disease transmission from human visitors to nonhuman primates.

8. Captive breeding can play a useful role in restocking small populations that are otherwise threatened by extinction from small-population processes and reintroducing species to areas where previous populations have gone extinct. However, in both cases it is first necessary to ensure that the original source of threat or extinction has been effectively removed. The limitations associated with this captive breeding include the knowledge and expertise required for long-term successful breeding in captivity, the behavioral and genetic changes in captivity that might preclude subsequent release, the availability of sufficient captive breeding space, and the monetary and time costs.

9. Restocking and reintroduction programs (i.e., translocation programs), whether from captive or wild stock, are at present relatively uncommon in primate conservation. This may be because relatively few endangered primate populations can be saved from extinction through

such methods owing to ongoing problems in the wild and because even where the original problem might have been dealt with, there are potential difficulties with the availability of captive stock, disease transmission, and high financial costs. Nonetheless, translocation projects that have rescued local wild populations from almost certain extinction have proved successful, suggesting that this tactic might have a useful role in the future if suitable release habitat can be found.

12 *Conclusions*

The diversity and distribution of primate species have never been static. Primate populations have always been subject to continuing change as a result of a variety of ecological and climatic processes: this was as true before the emergence of *Homo sapiens* as it has been since (see chapter 2). Yet there can be little doubt that human activities have greatly accelerated the rate of detrimental change among primate populations. Human-induced habitat loss has had a profound effect on primates and continues to do so (chapter 8). The hunting of primates can pose an equally serious threat (chapter 9). Even though many primate species can persist in fragmented and modified habitats, the combination of habitat disturbance and hunting invariably has devastating effects on primate populations (sec. 9.6). As a result of such human activities, the imminent extinction of several primate species is rapidly becoming inevitable (sec. 10.2.2). Many species are unlikely to survive far into the first century of the new millennium (table 12.1).

Nevertheless, the successes achieved in conservation in the past few decades should not be underestimated. They have been tremendous, and current efforts continue to be so. The conservation of the golden lion tamarin is a case in point (sec. 11.3.3). The dramatic recovery of the rhesus macaque in India provides an another impressive illustration of the potential power of conservation action. Despite a 90% crash in the size of the species' population in India during the 1960s and 1970s, apparently driven by the international trade in live specimens, by 1990–1991 the population had recovered to 50% of its former size (Southwick, Siddiqi, and Oppenheimer 1983; Southwick and Siddiqi 1994). This recovery was largely brought about by a national export ban introduced in 1978. Similarly, Jiang et al. (1991) document the decline and recovery of an island population of rhesus macaques on the Nanwan Peninsula (Hainan, China). After human settlement some three hundred years ago, there was extensive deforestation and massive hunting (both to control

Table 12.1 Critically Endangered primate species

Africa	Madagascar	Asia	The Americas
None	Hairy-eared dwarf lemur,	Mentawai macaque,	Black-faced lion tamarin,
	Allocebus trichotis	*Macaca pagensis*	*Leontopithecus caissara*
	Golden bamboo lemur,	Tonkin snub-nosed monkey,	Golden-rumped lion tamarin,
	Hapalemur aureus	*Rhinopithecus avunculus*	*Leontopithecus chrysopygus*
	Broad-nosed gentle lemur,	White-rumped black leaf monkey,	Golden lion tamarin,
	Hapalemur simus	*Trachypithecus delacouri*	*Leontopithecus rosalia*
	Golden-crowned sifaka,	Silvery gibbon,	[no common name given],
	Propithecus tattersalli	*Hylobates moloch*	*Cebus xanthosternos*
			Yellow-tailed woolly monkey,
			Lagothrix flavicauda

Note: Half of these species are expected to be extinct within ten years or three generations (whichever is longer) from their date of evaluation (1996) unless conservation management can rescue their populations. Data from Baillie and Groombridge 1996.

crop raiding and to obtain medicinal ingredients): by 1958 all the forest had been cleared, and by 1964 there were only slightly more than one hundred monkeys left in five groups. Yet with the creation of the Nanwan Reserve in the same year and the provisioning of two groups, the population made a startling recovery. By 1987 there were approximately twelve hundred animals in twenty groups.

Although it is true that rhesus macaques are extremely versatile animals and often coexist well with people (sec. 8.3.2), there is evidence that the detrimental effects of habitat loss and hunting can be reversible for other species too. On Bioko Island (Equatorial Guinea), the indigenous primate populations appear to be larger now than they were earlier in this century, mainly because nearly 400 km² of agricultural and pasture land was abandoned and allowed to revert to secondary forest (Butynski and Koster 1994). That shotguns were confiscated from the civilian population in 1974 is likely to have been an additional contributory factor. These actions greatly enhanced the survival prospects of a variety of endangered primate species, including the drill and Preuss's guenon. Tragically, however, these taxa now appear to be at serious risk from the local bushmeat trade (sec. 9.3.1), a sobering reminder that conservation situations never remain static.

These case histories indicate that conservation action does have the potential to pull species back from the brink of extinction, and this gives us hope for the future. In this concluding chapter, our aim is to identify some of the most important issues and themes in the conservation of primates that have emerged from the preceding chapters.

12.1. The Past and Future of Primate Diversity

Since no primate extinctions have been recorded in the past four hundred years, some might assume that primates are resilient to human activities. Yet this conclusion would be a grave mistake, for three reasons.

First, evidence from the fossil and subfossil record over the past ten thousand years suggests that human communities can cause the extinction of primate species (see below). It also suggests that the lack of extinctions during recent centuries may partially reflect the fact that many vulnerable species had already disappeared (Balmford 1996). Second, the attritional decline of primate populations over the past hundred years or so has culminated in the small and fragmented distributions now seen in many contemporary species; as a consequence many species are now at serious risk of extinction (sec. 10.2.2). Indeed, even if the current levels of habitat disturbance and hunting were to cease today, the existence of extinction lags means we should still expect a substantial number of extant species to face inevitable extinction unless effective conservation action is taken (sec. 8.2). Third, new research reveals that if we take a slightly longer time frame (five hundred years), at least one primate species has gone extinct in recent history: it now seems that *Xenothrix mcgregori,* a cebid from Jamaica, went extinct shortly after European contact (Monkey fossils 1997). Other cases may also have been overlooked. Indeed, it has been estimated that probably fewer than half of the extinctions among birds that resulted from human activities in the central Pacific are known to science (Pimm, Moulton, and Justice 1995).

Nevertheless, it appears that primates have experienced fewer extinctions than many other taxa over the past four hundred years. The reasons for this, however, appear to be quite straightforward and offer little hope of protecting primate diversity in the future. First, most mammal extinctions during this period have occurred on islands (Groombridge 1992; Smith et al. 1993a). Island species have borne the brunt of these extinctions because their populations are smaller, more isolated, and—in the case of oceanic islands—especially vulnerable to introduced species and human activities (sec. 7.4). It seems likely that primates have escaped only because their distributions are predominantly continental. This conclusion is borne out by the observation that the single primate extinction now recorded during this period involved an island species (see above) and by the fact that earlier subfossil primate extinctions have been predominantly among island taxa (sec. 12.1.1). Unfortunately, this pattern of distribution is unlikely to help primates in the future: the number of continental animal extinctions has increased over the past two hundred years (Smith et al. 1993a), and this trend is set to continue.

Second, most mammal extinctions in the past four hundred years have taken place either through the effects of introduced Eurasian species or through overexploitation (Caughley and Gunn 1996). Owing to their arboreal tropical forest niches and their continental distributions, primates appear to be largely unaffected by exotic introductions (sec. 7.2). The reasons no primate species have been driven to extinction from overexploitation are less clear but include the following possibilities. First, most primates are relatively small mammals that are both less profitable to hunt (sec. 9.5) and more likely to occur in large populations (sec. 5.2.2) than many of those species that have been hunted to extinction (e.g., Stellar's sea cow). Second, most primates occur in the continental tropics, where recent contact with large populations of settling Eurasian communities has been relatively limited, at least compared with extinctions by overhunting elsewhere (e.g., the bluebuck in South Africa and the Mexican silver grizzly bear in North America). In these zones of contact with large populations of armed settlers, species were hunted to extinction either because of economic conflict (e.g., the Falkland Islands wolf) or for the commercial trade in meat and pelts (e.g., the quagga). Given that most indigenous human populations in the tropics are now expanding, possess firearms, are in conflict with primates over crops (sec. 8.3.2), and hunt primates intensively for their meat (or other products; sec. 9.3), it seems that primates now face a similarly serious risk of extinction from hunting. This risk is borne out both by evidence of primate extinctions from overexploitation in the subfossil record (sec. 7.4) and by the identification of hunting as one of the primary forces driving population decline in many contemporary primate species (secs. 9.4.1 and 9.6).

The forces that have favored the survival of primate species over the past four hundred years may no longer be a source of protection. Moreover, there is the very serious threat of habitat disturbance that has now taken place on a continental scale and threatens mammalian diversity in an unprecedented fashion. In addition, as more species and habitats are lost in the face of human activities, the frequency of secondary extinctions can also be expected to rise. In summary, the historical survival of primate species notwithstanding, primates are now under siege from powerful forces that currently seem likely to drive a massive wave of extinctions in the new millennium unless conservation is successful.

In the remainder of this section we draw attention to some of the forces that are likely to play a pivotal role in primate population declines but that are poorly understood: accessibility, migration, and secondary extinctions. Better comprehension of these forces will allow more effective conservation planning and action in the future.

12.1.1. The Role of Accessibility and Migration

Where human populations have contact with primate communities, there is a serious danger that the local primate populations will suffer through habitat disturbance and hunting. Any process that leads to more contact between humans and primates therefore poses a greater threat to long-term primate population persistence. Higher rates of contact primarily come about in two ways: through greater accessibility to primate populations and habitats (the supply side of the equation) or through increases in local human population density and consumption (the demand side of the equation). We have already touched on these sorts of processes. On one hand, habitat disturbance and hunting tend to be less widespread and occur at lower intensity in areas of limited accessibility (secs. 8.1.1 and 9.2.3). On the other hand, increases in population growth can lead to higher rates of habitat loss (sec. 8.1.1) and hunting (sec. 11.2.2). The impact of logging projects on the exploitation of primate populations illustrates both these issues: accessibility to primate populations is improved through the construction of logging tracks, and the size of the human population utilizing that resource dramatically increases through the presence of loggers and their connections with the local urban markets (sec. 9.6.1).

Given the importance of accessibility in habitat disturbance and hunting, it is surprising that the processes that influence accessibility, particularly road-building projects, do not receive more attention in conservation planning and management. Although road construction can be associated with a multitude of other economic changes whose impact on primate populations can be difficult to predict (sec. 10.4.2), there can be little doubt that roads will always have the potential to introduce long-term deleterious and largely irreversible changes to neighboring primate populations and habitats.

The influence of local human population growth on habitat disturbance and hunting is perhaps better appreciated than the role of accessibility. But population growth due to immigration from other areas is rarely distinguished from growth due to higher birthrates and lower mortality in the same locality. The difference between these two processes is enormously important. Immigration commonly leads to human communities' exploiting natural resources with which they have had no previous experience and therefore no opportunity to develop cultural traditions that might help promote sustainability. The unsustainable use of resources is exacerbated because most migrant populations have at best only short-term tenure of the land they occupy and so have little incentive to manage their resources to maximize long-term sustainability. Consequently migrant populations can have a particularly severe impact on primate populations and habitats (sec. 10.4.2). In contrast, we tend to find that human populations with a long history of existence in an area have often developed cultural mechanisms to promote sustainability (if they had not, either they or the natural resources they depended on would have disappeared by now).

Take, for example, taboos in Madagascar. The golden crowned sifaka *Propithecus tattersalli* has long been protected from local hunting by taboos, but it is not protected from hunting by immigrant gold miners (Meyers 1996). Similarly, in Sierra de Santa Marta (Mexico), the indigenous Popoluca Indians do not hunt monkeys for cultural and practical reasons (cultural because they consider them funny and curious animals akin to people, practical because they prefer to hunt at night with dogs). In contrast, immigrants to the area hunt and kill monkeys to use them as fishing bait or to sell infants as pets (López, Orduña, and Luna 1988).

History indicates that where human populations have had a long period of coexistence with primate populations the rate of primate extinctions can be relatively low (e.g., *Homo, Pan,* and *Gorilla* have coexisted for 60,000 years at Lopé, Gabon: Tutin and Oslisly 1995). In contrast, extremely high rates of extinction have been witnessed in primate communities that have only recently experienced contact with human populations: a wave of primate extinctions seems to have coincided with the arrival of humans in Madagascar and, perhaps, the Caribbean (sec.

7.4). Similarly, the disappearance of Neotropical primate megafauna took place at the end of the Pleistocene (sec. 7.4.3) at about the time the Americas were first colonized (and a wave of megafaunal extinctions swept the continent: Martin and Klein 1984).

In contemporary times, migrant populations are now recognized as playing a vital role in driving deforestation in many countries around the world (e.g., Ivory Coast: Sayer, Harcourt, and Collins 1992; Indonesia: Rijksen 1978; Wadley, Colfer, and Hood 1997). Moreover, migrant populations are seen as a serious threat to the success of community-based conservation projects: examples include the Okomu Forest Reserve (Nigeria: Oates 1995), the Okapi Wildlife Reserve (Democratic Republic of Congo: Stephenson and Newby 1997), and the Dzanga-Ndoki National Park and Special Reserve (Central African Republic: Noss 1997). This is due not solely to the problems outlined above but also to the fact that immigration leads to greater ethnic diversity, which can complicate conservation planning if different ethnic groups have different traditions of resource management and planning (Noss 1997). Ironically, the problem of migrancy is likely only to worsen, because environmental degradation forces populations that are not using their natural resources sustainably to move on.

Although migration from one rural area to another is our most serious concern here, migration from rural areas to urban centers should not be overlooked. Although the movement of populations into towns and cities might seem to alleviate the demand placed on primates and their habitats in rural areas, such an effect should not be assumed: urban dwellers with high incomes can place an intolerably high burden on natural resources in the surrounding countryside (e.g., through an increased demand for primate bushmeat: Martin 1983). Another important point is that it is not only human populations that can sweep into new areas but also cultural and technological traditions. When new technologies arrive in a society through cultural diffusion, the deleterious impact on local natural resources may be similar to that of the arrival of migrant populations (e.g., the adoption of shotguns by hunter-gatherers: secs. 9.2.1 and 11.2.2). The impact of new technologies will depend on whether the previously sustainable patterns of hunting were limited by the existing technologies or by cultural practices and demand.

In light of the threat posed by migration, it seems prudent that we should attempt to obtain a better understanding of migrant population movement (cf. Charnley 1997) and its impact on primates and their habitats, and that we incorporate this information into the development of primate conservation strategies where appropriate.

12.1.2. Secondary Extinctions and Habitat Decay

The dynamics of historical and current trends of habitat loss are relatively well understood (sec. 8.1). Much less is known about the patterns and processes of future habitat loss. This is of particular concern because processes that have already been set in motion may lead indirectly to extensive habitat loss in the future even though they do not cause habitat loss now. We have already discussed the serious problem of climate change on the viability of the world's forests (see chapter 8). Here we look at the possibility that habitat disturbance or hunting may lead to gradual habitat decay and eventual loss. There are several mechanisms through which this process might occur.

First, changes in forest structure following habitat disturbance can make forests more vulnerable to a variety of natural environmental catastrophes that will prevent recovery in the long term. Logged forests, for example, are more susceptible to fires, which can kill 10–80% of the living biomass (Cochrane and Schulze 1998; Nepstad et al. 1999). Moreover, the fires that burn in logged or cultivated areas often penetrate into neighboring undisturbed forest, especially in the dry season. Most ominously, once an area has burned it becomes increasingly flammable and each successive fire becomes more damaging. In addition to the problem of fire, disturbed forests can also suffer from arrested regeneration. For example, regeneration in selectively logged forest at Kibale (Uganda) is failing, possibly owing to disturbance by elephants (Struhsaker 1997; Grieser Johns 1997) or by a local absence of aggressive colonizing tree species (Chapman et al. 1999).

Second, habitat disturbance such as selective logging can cause extensive mortality among plant species (even those not targeted for extraction: sec. 8.1.1). Forest fragments also exhibit much higher rates of tree mortality and damage than comparable areas of continuous forest (Laurance et al. 1998), primarily through edge effects (sec. 8.1.2). Consequently small population effects in diminished plant populations might lead to their extinction over several generations (which, for many tropical forest tree species, may occur only over centuries).

Third, the extinction of keystone pollinators and seed dispersers, owing to either habitat disturbance or overexploitation, may prevent forest recovery or continued forest viability. Most seriously, a drop in the abundance of a keystone species might have the same effect as its extinction, if the decline prevents that species from performing the services it provides at the required levels. This means that even when a species is harvested sustainably, the long-term viability of its habitat may be in danger (Robinson 1993a).

The major problem in evaluating the potential impact of long-term habitat decay on primate populations is that the time windows of most studies of primate persistence in disturbed habitats or in exploited animal communities are relatively short compared with the life span of many of the tree species in those habitats (chapters 8 and 9). Primate populations thus might survive under these conditions in the short term but fail to survive in the long term because of the gradual disappearance of key ecological resources. Effectively, in all these cases, the primates will be victims in a chain of extinctions (sec. 7.2). The collapse of the Amboseli baboon and vervet populations provides a case in point. Here events that began during the 1950s and early 1960s did not show up in terms of their effect on the monkeys until as much as two decades later.

Given the time scales over which these secondary extinctions are likely to take place, it is difficult to gauge what their magnitude might be, or even whether they are likely to occur at all. Nevertheless it is clear that we must tackle this problem if we wish to ensure that the conservation management steps taken now are appropriately designed to maximize long-term habitat and population viability.

12.2. Diagnosing Populations in Trouble

If the conservation management of primates in declining populations is to be successful in spite of limited time and resources, it is imperative that the cause of each decline be correctly identified (Caughley and Gunn 1996). Otherwise, inappropriate steps might be taken that cannot be rectified before the extinction of the target population becomes inevitable. The complexity of species patterns in abundance and distribution in systems perturbed by humans means that mistakes are easy to make.

The system of forest fragments in eastern Minas Gerais (Brazil's Atlantic Forest) provides a good illustration of this complexity. In this system, marmosets are more common than either capuchins or howler monkeys, and these in turn are more common than muriquis, while titi monkeys and tamarins are absent from the locality altogether. This interspecific variation appears to reflect differential effects of several intrinsic species traits and extrinsic threat processes. The following explanations have been suggested (Ferrari and Diego 1995). First, the marmosets are most abundant because they thrive in secondary forest and are too small to be profitable to hunt. Second, muriquis and howler monkeys are probably both hunted equally, but howler monkeys are more resistant to hunting because they occur in larger populations in the forest fragments (by virtue of their folivorous diets). Third, capuchins are small enough to be of marginal value when hunting, but they are no more common than howlers in this area because their numbers are controlled to minimize crop raiding. Fourth, titi monkeys and tamarins are absent altogether owing to the local dryness and seasonality of the site. If correct, these explanations indicate that a different form of conservation management is likely to be required for each species. Moreover, the type of management necessary will be difficult to identify without knowledge of the key threat for that species.

The purpose of this section is to highlight the problems that must be overcome in diagnosing the conservation needs of threatened primate populations.

IS THE POPULATION DECLINING?

First it is necessary to establish whether the population is in fact declining. Evidence for population declines can come from either a drop in abundance or a contraction of geographic range. So far, most research in primate conservation biology has been aimed at evaluating changes in abundance associated with threats that are assumed to cause population decline (see chapters 8 and 9). We will therefore focus on population abundance here. In many cases a drop in abundance might be recorded and mistakenly interpreted as a population in decline. In fact, abundance might drop because of changes in habitat use, such as avoiding areas owing to habitat disturbance (sec. 8.4); because of changes in detectability, resulting from either incidental changes in behavior following disturbance (Johns 1985c) or the active avoidance of observers owing to hunting pressure (sec. 9.4); or because of changes in survey technique (a great deal of the apparent variation in the size of the Tana River crested mangabey and red colobus populations over the past decade, for example, has been attributed to variation in survey technique: Butynski and Mwangi 1994). The influence of these potentially confounding factors should always be considered.

WHAT IS THE MECHANISM OF DECLINE?

Once it has been established that a drop in abundance is the result of a decline in population size, it becomes imperative to identify the proximate mechanism that caused this decline. Given that primate populations may always be in flux owing to natural variation in the availability of food, the intensity of predation, and the prevalence of disease (sec. 6.2), it is always possible that a decline in abundance in a previously disturbed or hunted habitat may be entirely independent of the disturbance or the hunting (which might have had a negligible impact) and instead may result from an unrelated natural process (such as a particularly long dry season). If the source of population decline was wrongly diagnosed in this instance, valuable time and financial resources for conservation might be wasted.

If the decline is caused by human activities, it is still necessary to determine the mechanism that is responsible if conservation action is to be taken. In the case of declining populations in areas of habitat disturbance, the focus of research has been the role of changes in food availability on primate populations (sec. 8.4). This has been a sensible starting point, given that food is a critical limiting factor on the size of primate populations (chapters 5 and 6). Nonetheless, there are many other potential influences that may be of at least equal importance (see also Johns 1985b).

First, there is the problem that food may still be available but physically out of reach owing to the loss of arboreal pathways (e.g., Ganzhorn 1987). Second, interspecific competition may intensify if preferred foods diminish in availability or the animals are forced to switch to alternative foods. Such competition can have a substantial influence on primate populations (sec. 4.3). Third, these same changes may also promote disease transmission: increased ground feeding, for example, may enhance the transmission of gut parasites through fecal contamination of food. Given the prominence of disease in primate population crashes (sec. 6.2.2), its role may be worthy of special attention. Fourth, changes in the thermal environment are also likely to ensue from habitat modification, potentially leading to excessive heat loading and concomitant stress on time budgets (Stelzner 1988; Dunbar 1992b): if foraging or social time is compromised, the viability of social groups—and thus entire populations— may be at risk (sec. 7.4).

However, the most important of alternative impacts may be predation risk, given that predation can override food resources as a determinant of habitat use in primate populations (Cowlishaw 1997a). Changes in vegetation structure associated with habitat disturbance are almost certain to lead to changes in predation risk. For example, the disappearance of aerial pathways forces primates to travel lower in the canopy or on the ground, where predators can more easily reach them. Predators will usually modify their behavior accordingly: both Philippine eagles (Southeast Asia) and harpy eagles (Amazonia) appear to be more common in logged forest and to capture more prey when hunting there (Grieser Johns 1997). Primates in turn appear to respond to these changes in risk (see sec. 8.4.3). In plantations in Madagascar, small lemurs such as *Microcebus rufus* tend to be absent where the shrub layer is discontinuous, since this prevents individuals from foraging unless they travel across open ground (Ganzhorn 1987). Similarly, *Macaca nigra* prefers those disturbed habitats where fruits are higher in the canopy (Rosenbaum et al. 1998).

In other cases, where primates have no option except to use disturbed habitats, population decline may result from high predation rates. In logged forest in Kibale, monkeys experience higher rates of predation by both crowned eagles (because of the broken canopy) and chimpanzees (because of the loss of arboreal escape routes) (Grieser Johns 1997). Similarly, vervet monkeys forced to modify their home ranges as a result of deteriorating habitat quality suffered higher mortality from predation because the new terrain was unfamiliar to them (Isbell, Cheney, and Seyfarth 1990; Young and Isbell 1994). These effects may be exacerbated by other changes in behavior forced upon primates foraging in these habitats. In particular, primates are likely to experience higher predation when grouping patterns change (through probable changes in foraging costs). Many primate species form smaller groups when foraging in areas disturbed by forest fires (Berenstain, Mitani, and Tenaza 1986), shifting agriculture (Johns 1991), and selective logging (Grieser Johns and Grieser Johns 1995) and may also form polyspecific associations less frequently (logged forest: Struhsaker 1997). These changes are likely to elevate predation risk. In situations like these, it is possible that even native predators could cause the extinction of primate populations (Sweitzer, Jenkins, and Berger 1997).

MIGHT THE DECLINE LEAD TO EXTINCTION?

Clarifying the mechanism of decline is likely to require understanding the structure and dynamics of the primate population in question (chapter 6). It would be difficult to draw a connection between a decline in population and associated ecological changes (such as reduced food availability and increased predation risk) without knowing how those ecological changes influenced BIDE processes (sec. 6.2.1). Although the slow rates of reproduction characteristic of primates can hamper the detection and study of changes in population dynamics (Struhsaker 1997), the value of such insight is enormous. Knowledge of population structure and dynamics will also allow conservation biologists to determine whether the observed population decline had led to a significant increase in extinction risk and how rapidly recovery might otherwise take place. Where changes in population size are relatively small or transient, it may be appropriate to redirect conservation attention to more pressing problems.

The identity of sources and sinks in habitats is particularly important in this context (sec. 6.3). If population declines from habitat disturbance or hunting have occurred in sink populations, then their importance is marginal (given that the population was not viable in the first place). In contrast, declines that occur in a source population are of grave concern, since the local extinction of that population could lead to multiple extinctions in dependent sink populations in surrounding areas that have not suffered either habitat disturbance or hunting. In some cases habitat disturbance and hunting may convert one part of a source population to a sink population (e.g., hunting around villages: Alvard et al. 1997): provided the rest of the source population is untouched and large enough that it does not experience a significant increase in extinction risk, this switch from source to sink may not warrant conservation action.

WHAT THREATS UNDERPIN THIS MECHANISM?

Once the mechanism and severity of population decline has been identified, it can merely be a matter of working back through the chain of events that culminates in that mechanism to discover which human activities initiated the process. In many cases the source of threat will be obvious (e.g., a reduction in the number of food trees lost during selective logging). In other circumstances the ultimate source of threat may be less clear.

In the past, for example, there has been extensive contraction and fragmentation of the geographic ranges of species such as the woolly spider monkey (Mittermeier et al. 1987), chimpanzee (Teleki 1989), and orangutan (Rijksen 1978), but the relative importance of hunting and habitat disturbance in these historical changes has remained unclear. Similarly, in contemporary times primate populations can be threatened by a number of different human activities: examples include lion-tailed macaques threatened by forest fragmentation with underplanting (India: Menon and Poirier 1996), orangutans threatened by shifting agriculture and selective logging (Sumatra: Rijksen 1978), and lemurs threatened by a combination of shifting agriculture, livestock husbandry, and burning (Madagascar: Smith, Horning, and Morre 1997).

This problem has recently been emphasized by Oates (1996b), who believes that the threat of habitat modification has been overestimated and that hunting within modified habitats is the key threat to many primate populations. That many primates can thrive in secondary vegetation (sec. 8.5.1) certainly lends credence to this argument, as do those studies that have attempted to compare directly the effects of habitat modification and hunting on primate populations (sec. 9.6).

By identifying the threat that precipitates the mechanism driving the primate population decline, it becomes possible to manage the threat both at the interface between the threat mechanism and the primate population (e.g., by replanting food trees in logged forest) and at the source of the threat itself (e.g., by addressing world demand for tropical hardwoods). We explore these issues further in section 12.3.

LOGISTICAL CONSTRAINTS

Clearly, diagnoses in this detail require time and money—both likely to be in short supply. It is perhaps for this reason that many studies of declining primate populations typically try only to establish whether a drop in abundance has occurred in association with a particular threat. These studies often make three key assumptions: that the drop in abundance reflects a declining population, that the population is declining as a direct result of the associated threat, and that the decline represents a significant increase in extinction risk. Our concern is that these assumptions are dangerous for precisely the same reason that many of these studies do not attempt more analyses: the resources available for conservation are limited. If these resources are not directed appropriately, then conservation will suffer. Caughley (1994) and Caughley and Gunn (1996) discuss several cases where problems were misdiagnosed and conservation action failed as a result. Consequently, although we recognize that detailed diagnoses are expensive and time consuming, especially in the case of primates where population dynamics occur slowly and on a large scale, we strongly advocate that wherever possible research on threatened primate populations should attempt to address directly the issues we have raised here.

In addition, an important challenge for conservation biologists is to develop rapid and inexpensive methods to help achieve this goal. Some of the first steps have already been taken with the development of techniques such as Rapid Biodiversity Assessment surveys, which have recently been developed and refined for specific application to primate communities (Madagascar: Sterling and Rakotoarison 1998). Future steps might include the rapid identification of high-risk species according to characteristic species traits, which would allow conservation research to determine more quickly those taxa likely to be most in need of help (sec. 7.3).

12.3. Effective Conservation Action

Once the ultimate threat and proximate mechanisms of decline have been diagnosed, it is possible to devise and implement conservation action. The threat of extinction from habitat disturbance or hunting can be tackled at various points. Most commonly, conservation management takes place at the interface between the threat process and the threatened primate population, where the mechanism that drives the extinction operates. However, both habitat disturbance and hunting are largely driven by human population growth and consumption (see chapters 8, 9, and 11). Although action at the sociopolitical level is beyond the scope of most conservation projects, these are problems that remain of fundamental importance to the successful long-term achievement of conservation goals. Another immense sociopolitical problem that conservation biologists must continue to grapple with is funding for conservation. Greater financial support from wealthier countries is likely to be needed if poorer countries are to conserve their primate communities success-fully. One obvious solution would be to write off these countries' loans, since it is the interest on these debts that has left most developing coun-tries bereft of any funds to invest in health, education, infrastructure, and (only once all these are satisfied) conservation. In addition, the money already devoted to foreign aid needs to be spent with much greater efficiency and efficacy (Mittermeier and Bowles 1994; Struh-saker 1997).

The primary tactics employed in strategies for primate conservation have already been discussed in detail (see chapter 11). In this section we focus on two particular areas of management that will be critical to the future of primate conservation, namely the management of primates in fragmented habitats and the sustainable utilization of primate popula-tions.

12.3.1. Management in Fragmented Habitats

Current trends in habitat disturbance indicate that primate populations will increasingly be required to survive in fragmented habitats. Fortu-nately, many primates do have the potential for long-term survival in such environments over hundreds and thousands of years (e.g., the Tana River red colobus in East Africa, samango monkeys in southern Africa, Barbary macaques in North Africa, and lion-tailed macaques in India: sec. 8.5.2). However, we know surprisingly little about the processes that underpin survival in these cases. We have little understanding of how far a habitat can be fragmented, and in what pattern, before causing a significant in-crease in the risk of primate population extinction (secs. 8.3.1 and 8.5.2). Similarly, our knowledge of the range and efficacy of conservation steps for managing such problems is still in its infancy. Our ignorance in these matters is a serious concern, since the survival of many fragmented pri-mate populations is likely to require significant management inter-vention.

An issue of fundamental importance is the control of local extinctions in the subpopulations that compose the fragmented population: given their size, they will be at serious risk of extinction from small-population effects (sec. 7.2). Management plans will need to ensure that the rate of population extinctions does not exceed the rate of colonizations: in other words, the fragmented population—now effectively a metapopulation— must be maintained at a healthy equilibrium (sec. 6.3). To achieve this, migrants from one subpopulation must be able to rescue declining sub-populations or else recolonize empty patches elsewhere. This can be achieved through habitat corridors (sec. 11.1.2) or through translocation projects (sec. 11.4). Given the considerable expense of the latter, corri-dors are always likely to be the preferred option (see sec. 12.4 below).

To ensure that the subpopulations of a fragmented population are as well connected as possible, an extensive corridor system should be developed. To do this cheaply and effectively, the use of nonpristine habitats as corridors should be encouraged. That many primates are able to persist in secondary forest as well as in habitats disturbed by activities such as underplanting, shifting agriculture, and selective logging emphasizes the potential value of such habitats as corridors between more suitable habitat patches (secs. 8.4 and 8.5.1). Ganzhorn (1987; Ganzhorn and Abraham 1991) has shown how old *Eucalyptus* plantations, which can be viable habitats for some lemur species, can play a valuable role as corridors between (or even act as borders for) protected areas in Madagascar. Precisely this approach is now being used to develop corridors between fragmented populations of black lion tamarins in the Atlantic Forest (Mamede-Costa and Gobbi 1998).

Disturbed habitats can also be actively managed to promote their value to primate populations. Grieser Johns (1997), for example, documents the sorts of procedures that might be adopted during selective logging operations in order to maximize the retention of key ecological features for primate populations. These include preserving 5–10% of the forest concession intact, conserving keystone food resources in logged areas, and planting food trees for postharvest enrichment. The choice of tree species used for stabilizing soils in areas heavily damaged by logging can also be guided by the feeding requirements of particular primate species (e.g., *Albizia falcateria* for slow lorises at Tekam, peninsular Malaysia: Grieser Johns 1997). Horwich (1998) similarly outlines the habitat management steps that shifting cultivators in Belize have introduced in order to conserve black howler monkeys *Alouatta pigra*. More active management still can be considered in a variety of forms, including creating appropriate sleeping sites (e.g., squirrel monkeys: Boinski and Sirot 1997; and building nest boxes for hole-nesting primates such as black lion tamarins: Mamede-Costa and Gobbi 1998), and constructing bridges to connect forest fragments separated by roads (e.g., black lion tamarins: Mamede-Costa and Gobbi 1998; howler monkeys: Horwich 1998; black-and-white colobus monkeys: Julie Anderson, pers. comm.).

The potential value of such habitats in primate conservation raises the importance of the habitat matrix as a whole in fragmented landscapes. Not only does the matrix have the potential to help maintain the viability of fragmented populations at a local scale, it may also connect isolated national parks and other protected areas on a regional scale (e.g., Africa: Siegfried, Benn, and Gelderblom 1998). In many cases matrix populations might constitute only sink populations dependent on the source populations that persist in the remnants of intact or protected habitat. Nevertheless, these sink populations might still play an invaluable role for long-term primate persistence in two ways. First, the sink populations will buffer the small source populations against small-population effects: if a source population crashes owing to a series of male births or a localized disease outbreak, it will recover more quickly if individuals in the sink population can replenish it through back-migration. Second, individuals in the sink populations are likely to be crucial in ensuring the regeneration of the forests in the matrix through pollination and seed dispersal (sec. 4.4).

The role of primates in matrix forest regeneration (sec. 4.4) will also be important to local human communities, since both agriculture and forestry depend on forest regeneration. Lambert (1998) recently demonstrated the immediacy of these links: she found that more than a third of all forest trees at Kibale had seeds dispersed by primates, and 42% of those that were primate-dispersed had some direct utility to local people, including wood (e.g., for fuel, poles, or furniture), food, medicine, and fodder. In addition, local communities might be able to sustainably harvest these primate sink populations as another source of food. However, this will probably be feasible only in very specific conditions (see below): in most cases source populations may not be able to sustain sink populations depending on the intensity of harvesting (Wilkie et al. 1998a; cf. Muchaal and Ngandjui 1999). In fact the unsustainable exploitation of primates for meat is likely to be the biggest obstacle in efforts to conserve viable primate populations in the matrix (sec. 9.6).

Another problem to be faced in such systems will be crop raiding (sec. 8.3.2). Since a matrix approach to primate conservation will inevitably increase contact between farmers and primates, agricultural communities may be forced to control primate numbers through heavy hunting, which might threaten even the source populations. However, solutions are possible. For example, important forest fruiting trees could be actively conserved in edge habitats (to encourage potential crop raiders to stay in the forest), while the cultivated foods most desirable to crop raiders could be planted at least 500 m from the forest edge (Naughton-Treves et al. 1998). Perhaps most important, given that farmers will put up with heavy crop raiding by other people's livestock because there are institutional mechanisms (i.e., customary laws) that ensure compensation, it seems likely that crop raiding by primates would be tolerated if effective compensation mechanisms could be found (Naughton-Treves 1998).

12.3.2. The Limits of Sustainable Harvesting

A fundamental characteristic of primates is that they reproduce slowly: primates reproduce at only one-quarter to one-half the speed of other mammals of identical body size (sec. 3.1.1). Since reproductive rates underpin the maximum sustainable yield of a population (sec. 9.4.1), this makes them especially poor candidates for sustainable harvesting. Even when primates are harvested at lower rates than rodents or ungulates, harvests are less sustainable (Equatorial Guinea: Fa et al. 1995; Kenya: Fitzgibbon, Mogaka, and Fanshawe 1995; Peruvian Amazonia: Bodmer 1995a).

Although prey switching by hunters when prey reach low densities may help buffer exploited primate populations against extinction (sec. 9.5), the small size of the remaining populations and the slow reproductive rates of primates (which will keep the populations small for some time thereafter) will make extinction solely from small-population effects a serious risk. This problem is illustrated by recent evidence from Bioko Island (Fa 1999, Fa, Yuste, and Castelo, n.d.), where hunting appears to have intensified since 1989 (when a new government lifted restrictions on the civilian ownership of shotguns). Here the larger diurnal primate species (already heavily harvested for sale at market in 1990 at what were often estimated to be unsustainable levels: Fa et al. 1995; Slade, Gomulkiewicz, and Alexander 1998) became increasingly scarce in the wild over the course of the decade 1986–1996. As these species became increasingly scarce, they became more expensive in the market. In contrast, smaller prey species (including four birds and one squirrel that were absent in 1990) were more common in the market in 1996. Estimates of the population size of each primate species for the entire island indicate that the populations now remaining are likely to be dangerously small.

The unusually slow rate of reproduction characteristic of primates suggests that conservation through sustainable harvesting is likely to be a remote prospect in most instances. A handful of case studies suggest there might be some potential for sustainable yields, but this appears to apply only to a very specific combination of species and scales, namely species that reproduce relatively quickly and that are harvested in localized areas of the population so that immigration can promote recovery (sec. 9.4.1). In addition, given the problems associated with the tragedy of the commons and economic discounting (secs. 10.4.1 and 11.2.2), it is clear that the sustainable harvesting of primate populations will always be risky under most conditions.

Nevertheless, although sustainable utilization will rarely be a useful tactic to promote primate conservation, the unsustainable harvesting of many primate populations remains a problem to be solved. In many primate communities solutions will be required that do not depend on the effective establishment and implementation of protected areas (sec. 11.1). There are three possibilities.

First, hunters can be encouraged to change their hunting patterns. At Tahuayo (Peru), the sustainability of primate harvests could be promoted by asking hunters to target males preferentially and to reduce their exploitation of species that are currently being harvested unsustainably (including the primates) while hunting sustainably harvested species at current levels (most of the ungulates and rodents) (Bodmer et al. 1994). However, this entails a substantial economic cost to the hunter, who would have to be compensated in some way. A further problem with this approach is that even where primates are not targeted by hunters they can still suffer significant mortality if snares are used to catch ungulates and rodents (e.g., gorilla populations in Rwanda: Plumptre et al. 1997; Hall et al. 1998). Hence snaring must be controlled if this option is to be pursued, and this may be difficult since snares are cheap to use and easy to obtain (Lewis and Phiri 1998).

Second, given that the most serious source of overexploitation in primate populations is often commercial hunting for the bushmeat trade, controls on this domestic trade might be introduced. Wilkie et al. (1998b) suggest that it should be possible to control market hunting locally by focusing efforts on the handful of traders in the business (rather than the numerous hunters and consumers). The primary control step they suggest is taxing the transport of bushmeat, since this minimizes opportunities for corruption (such as laundering confiscated carcasses) and can be implemented at minimum cost (provided only a small number of transport routes exist). The traders will attempt to absorb these costs either by increasing the sale price to customers (which will reduce demand for bushmeat) or by reducing the price they are willing to pay to hunters (which will reduce their incentive to hunt and thus increase their incentive to find alternative sources of income). In both cases, the intensity of harvesting of all bushmeat taxa should decline.

Third, alternative sources of meat could be introduced to alleviate the demand for bushmeat (e.g., Martin 1983; Wilkie et al. 1998b; Fa 1999). Such alternative meat sources must be cheap in relation to bushmeat and produced in sufficient quantity to meet demand. Existing preferences for wild meat over domestic meat suggest that the domestication of small game, such as cane rats and giant rats, might be better than a total switch into true domestic livestock such as cattle (e.g., Njiforti 1996). Although cattle, pigs, and rabbits have the advantage that they are more efficient meat producers than many potential bushmeat species, the technical problems of breeding such species in tropical climates, together with the devastation that exotic species can wreak on native ecosystems (sec. 7.2), means that the domestication or semidomestication of native bushmeat species would be preferable. The successful domestication of the cane rat (*Thryonomys* spp.) in western and central Africa indicates that this approach might be successful (Jori, Mensah, and Adjanohoun 1995).

Ultimately, solutions to the problem of subsistence and market hunting for primate meat may require a combination of these approaches.

For example, Wilkie et al. (1998b) suggest that taxation and alternative meat sources should go hand-in-hand, with taxes set high enough to help overcome cultural preferences for bushmeat. However, given that a switch from bushmeat to domestic meat may have an adverse effect on the economy, taxation should initially be introduced at low levels and gradually raised to provide time to find alternative sources of income, including livestock production. The money raised from taxes might, for example, be used to compensate isolated communities that would be unable to market livestock competitively because of the travel costs involved.

12.4. Finding Unique Solutions

There has been a tendency for conservation management to follow fashions. In the early days of conservation, protected areas tended to be the preferred approach. More recently, sustainable utilization and community conservation have come to the forefront. Unfortunately, conservation problems are too complex and too variable to yield to one simple universal solution. In fact, most conservation problems are likely to benefit from a mixture of tactics that are carefully chosen and integrated based on the precise nature of the problem being dealt with (hence also the importance of correct diagnoses). Ultimately, each conservation problem will have its own unique solution.

In this book we have tended to stress the primary tactics of protected areas, sustainable utilization, captive breeding, and translocation (see chapter 11). The future will require a careful mixture of these approaches, in addition to others such as environmental education and training, and biodiversity survey work and monitoring. Moreover, there is a greater need than ever to develop innovative solutions to conservation problems, at both the strategic and tactical levels. For example, trust funds are now seen as one fresh approach to maximizing the long-term benefits of financial support to conservation projects. This is particularly the case where large sums of money are invested by the donor community over short periods (e.g., Mittermeier and Bowles 1994; Oates 1996b): Struhsaker (1997) details how this approach would have benefited the conservation of Kibale Forest, and Butynski and Kalina (1998) describe how a trust fund based on gorilla tourism in Uganda is now generating sufficient funding for park management. Similarly, while conservationists have rapidly forged collaborative projects with local communities, they have been slower to develop such projects with private enterprises and corporations, even though these can play a pivotal role in the unsustainable exploitation of habitats and wildlife (Milner-Gulland and Mace 1998).

The right balance between approaches in a mixed conservation strategy may not be easy to achieve. One case in point would be the balance between consumptive and nonconsumptive utilization (e.g., the bush-meat trade and tourism): given the problems of sustainable harvesting in primate populations (sec. 12.3.2), it is likely that nonconsumptive utilization will always provide the better option. However, tourism may not be an option in very remote areas or in countries with little infrastructure for tourism (sec. 11.2.3). In these cases sustainable utilization of primate habitats (rather than primates) might be advocated. In particular, Grieser Johns (1997) presents a persuasive argument for the potential of community forestry in conservation practices.

However, community-based conservation projects must also be implemented with more care in the future, given the complexities inherent in such an approach. Becker and Ostrom (1995) suggest that communities need to satisfy the following criteria if conservation management is to have any chance of working: the participants should be relatively homogeneous in their asset structure, information, and preferences; they should share functional systems of reciprocity and trust; and they should not discount the future at a high rate. The number of communities that fulfill these criteria may be relatively small, necessitating alternative approaches to community-based conservation in many cases. Fortunately, other property systems do exist that can also be successfully exploited for conservation management (sec. 10.4.1).

These examples illustrate the different mixtures of in situ conservation tactics that might be developed. Another case in point is the balance we should adopt between in situ and ex situ methods of conservation. Balmford, Leader-Williams, and Green (1995) have demonstrated that although the two approaches produce similar rates of population growth in target mammalian taxa (including *Leontopithecus rosalia* and *Gorilla gorilla*), the per capita costs of ex situ conservation are much greater than those of in situ conservation. Given that in situ conservation can also protect other species that coexist in the same habitat as the target species, and that ex situ conservation has limited practical use for future reintroduction unless the habitat of the target species is protected (or can be restored), primate conservation should probably continue with its current emphasis on in situ methods wherever possible. However, ex situ conservation may still prove invaluable where a primate taxon is declining rapidly, the forces driving that decline cannot be controlled at present, but opportunities might exist for reintroductions in the future. Ex situ conservation might also constitute an inexpensive and useful basis for community-based sustainable tourism projects where the captive propagation program occurs within the natural range of the taxon. The drill project in Nigeria (sec. 11.3.3), with its focus on drill conservation within its geographic range, exemplifies these strengths.

In addition, the future will require new techniques in conservation management, together with some difficult trade-offs. Some of the specific methods that might be envisaged for the conservation of both disturbed habitats and exploited populations have already been discussed (secs. 12.3.1 and 12.3.2). The trade-offs are perhaps more difficult to predict. One might arise from the possible use of livestock production to alleviate the demand on primates in the bushmeat trade (see below). If this method is implemented, it is likely to mean sacrificing areas of forest for conversion into pastures. Although undesirable, the alternative of a more extensive but empty forest would be worse (Robinson 1998).

Finally, developing successful solutions to conservation problems in the future will also require a deeper understanding of the socioeconomic dynamics of the human populations that are threatening the primate populations or that might otherwise constitute the solution to a conservation problem. The need for such insight has become especially clear in recent years as a result of the rapidly increasing number of attempts to conserve wildlife through utilization. Brandon and Wells (1992), for example, describe many integrated conservation and development projects (sec. 11.2.1) that have failed because of lack of preparation and failure to identify the root cause of the conservation problem.

Although research has begun to address the links between household income, markets, and development (sec. 10.4.2), we have barely started to understand the way they influence how intensively primates and their habitats are utilized. We also need research into the noneconomic values of activities that threaten primate populations. For example, in Bakossiland (West Cameroon), hunting is considered recreational as well as economic, and hunters are respected in their communities for their skills and achievements (King 1994). In addition, we need to understand the role of corruption in socioeconomic systems more clearly and learn how to act on it more effectively. (For the impact of corruption on the conservation of the orangutan, see Rijksen 1978; Environmental Investigation Agency 1998.) Elucidation of these practices will allow conservation interventions to be more effective.

A key facet of local socioeconomic systems is the range of subsistence options available to individuals in communities (Noss 1997). In the same way that hunters will switch between different prey when a particular species becomes unprofitable (following the prey-choice model: chapter 9), hunters will abandon hunting altogether and switch to an alternative income-generating enterprise when hunting becomes a relatively poor source of income. For example, in Bayango (Central African Republic), people commonly switch back and forth between hunting, fishing, and farming on one hand and, on the other hand, employment with local mining, logging, and conservation projects. There are two implications to this. First, such opportunism means that people are less willing to accept short-term costs for long-term sustainability because sustainability is not highly valued when the community does not depend solely on the target resource for its survival (e.g., the future loss of primates from forests may not be perceived as a problem when duikers and rodents will still be available). Second, but more encouraging, if the resource is not indispensable, then conservation management has all the more potential to be compatible with the normal activities of the local communities.

We have emphasized here the importance of developing unique and appropriate multiapproach solutions to each conservation problem, where each solution is sensitive to the local socioeconomic conditions. We would also like to highlight the importance of ensuring that any such solutions are inexpensive, easy to implement, and require little maintenance. This is especially important for primate conservation, where most habitat countries cannot afford substantial or long-term investments in conservation projects and where conservation projects are otherwise handicapped by political instability (e.g., Oates 1996b).

We have already discussed the value of designing protected areas to minimize the need for expensive active protection measures (sec. 11.1.3). A similar philosophy can be applied to all areas of conservation practice. For example, in disturbed habitats, it might be argued that, if fragmented primate populations are to be maintained, then corridors will be a cheaper management tactic than restocking programs (secs. 12.1.2 and 12.4); furthermore, if corridors are required, it will be cheaper to construct them from existing gallery forest, plantations, and secondary forest mosaics (which arise from selective logging and shifting agriculture) than from protected pristine habitat (sec. 12.3.1). Similarly, if the bushmeat trade is to be controlled through taxes, it is best to collect them in a way that requires minimal personnel and equipment (sec. 12.3.2). Developing and implementing such inexpensive, low-maintenance approaches will make an enormous contribution to the goal of primate conservation.

APPENDIXES

APPENDIX 1

Primate Species and Conservation Status

Species list 1	Common name	Species list 2	IUCN Red List codes	
LORISIDAE				
Arctocebus calabarensis	Angwantibo	*Arctocebus calabarensis*	LR nt	
		Arctocebus aureus	LR nt	
Euoticus elegantulus	Western needle-clawed bushbaby	*Euoticus elegantulus*	LR nt	
Euoticus inustus	Eastern needle-clawed bushbaby	*Euoticus pallidus*	LR nt	
		Galago matschei	LR nt	
Galago alleni	Allen's bushbaby	*Galago alleni*	LR nt	
Galago granti	Grant's bushbaby	*Galagoides granti*°	DD	
Galago moholi	Southern lesser bushbaby	*Galago moholi*	LR lc	
Galago senegalensis	Northern lesser bushbaby	*Galago senegalensis*	LR lc	
		Galago gallarum	LR nt	
Galagoides demidoff	Demidoff's bushbaby	*Galagoides demidoff*	LR lc	
		Galagoides orinus°°	DD	
		Galagoides rondoensis°°	DD	
		Galagoides udzunguensis°°	DD	
Galagoides zanzibaricus	Zanzibar bushbaby	*Galagoides zanzibaricus*	LR nt	
Loris tardigradus	**Slender loris**	**Loris tardigradus**	**VU**	**A1cd**
Nycticebus coucang	Slow loris	*Nycticebus coucang*	LR lc	
Nycticebus pygmaeus	**Pygmy slow loris**	**Nycticebus pygmaeus**	**VU**	**A1cd**
Otolemur crassicaudatus	Thick-tailed greater bushbaby	*Otolemur crassicaudatus*	LR lc	
Otolemur garnettii	Garnett's greater bushbaby	*Otolemur garnetti*	LR nt	
Perodicticus potto	Potto	*Perodicticus potto*	LR lc	

CHEIROGALEIDAE			
Allocebus trichotis	**Hairy-eared dwarf lemur**	LR 1c	**CR** **A1c, B1 + 2abc**
Cheirogaleus major	Greater dwarf lemur	LR 1c	
Cheirogaleus medius	Fat-tailed dwarf lemur	LR 1c	
Microcebus murinus	Gray mouse lemur	LR 1c	
Microcebus myoxinus**			**VU** **B1 + 2abc**
Microcebus rufus	Brown mouse lemur	LR 1c	
Mirza coquereli[1]	**Coquerel's dwarf lemur**		**VU** **A2cd, B1 + 2abc**
Phaner furcifer	Fork-marked lemur	LR nt	
LEMURIDAE			
Hapalemur aureus	**Golden lemur**		**CR** **A2cd**
Hapalemur griseus	Gray gentle lemur	LR 1c	
Hapalemur simus	**Broad-nosed gentle lemur**		**CR** **A2cd**
Lemur catta	**Ring-tailed lemur**		**VU** **A1c**
Lepilemur mustelinus	Sportive lemur	LR 1c	
Lepilemur dorsalis			**VU** **A2cd, B1 + 2c**
Lepilemur edwardsi		LR 1c	
Lepilemur leucopus		LR 1c	
Lepilemur microdon		LR 1c	
Lepilemur ruficaudatus		LR 1c	
Lepilemur septentrionalis			**VU** **A2cd**
Eulemur coronatus	**Crowned lemur**	LR 1c	**VU** **A1cd, B1 + 2bc**
Eulemur fulvus	Brown lemur		
Eulemur macaco	**Black lemur**		**VU** **A1cd**
Eulemur mongoz	**Mongoose lemur**		**VU** **A1c, C2a**
Eulemur rubriventer	**Red-bellied lemur**		**VU** **A2c**
Varecia variegata	**Ruffed lemur**		**EN** **A1cd**
INDRIDAE			
Avahi laniger	Woolly lemur	LR 1c	
Avahi occidentalis**			**VU** **A2cd**
Indri indri	**Indri**		**EN** **A1c + 2c**

(continued)

Appendix 1 (continued)

Species list 1	Common name	Species list 2	IUCN Red List codes	
Propithecus diadema	**Diademed sifaka**	**Propithecus diadema**	**EN**	**A1cd**
Propithecus verreauxi	**Verreaux's sifaka**	**Propithecus verreauxi**	**VU**	**A2cd**
Propithecus tattersalli	**Golden-crowned sifaka**	**Propithecus tattersalli**	**CR**	**A2c, B1 + 2bcd**
DAUBENTONIIDAE				
Daubentonia madagascariensis	**Aye-aye**	**Daubentonia madagascariensis**	**EN**	**A2cd, C2a**
TARSIIDAE				
Tarsius bancanus	Western tarsier	Tarsius bancanus	LR 1c	
Tarsius pumilus	Pygmy tarsier	Tarsius pumilus	DD	
		Tarsius dianae	DD	
Tarsius spectrum	Spectral tarsier	Tarsius spectrum	DD	
		Tarsius pelengensis**	DD	
		Tarsius sangirensis**	DD	
Tarsius syrichta	Philippine tarsier	Tarsius syrichta	LR cd	
CALLITRICHIDAE				
Callimico goeldi	**Goeldi's monkey**	**Callimico goeldi**	**VU**	**A1c**
Callithrix argentata	Silvery marmoset	Callithrix argentata	LR 1c	
		Callithrix leucippe**	**VU**	**B1 + 2c**
		Callithrix saterei**	DD	
		Callithrix nigriceps**	**VU**	**B1 + 2c**
Callithrix humeralifer	Tassel-eared marmoset	Callithrix humeralifer	LR 1c	
		Callithrix chrysoleuca**	**VU**	**B1 + 2c**
Callithrix jacchus	Common marmoset	Callithrix jacchus	LR 1c	
		Callithrix aurita	**EN**	**B1 + 2abcde, C2a**
		Callithrix flaviceps	**EN**	**B1 + 2abcde, C2a**
		Callithrix geoffroyi	**VU**	**B1 + 2b, C2a**
		Callithrix kuhlii	LR 1c	

		Callithrix penicillata	LR 1c	
Cebuella pygmaea	Pygmy marmoset	*Callithrix pygmaea*	LR 1c	
Leontopithecus chrysomelas	**Golden-headed lion tamarin**	***Leontopithecus chrysomelas***		**EN B1 + 2abcde, C2a**
Leontopithecus chrysopygus	**Black lion tamarin**	***Leontopithecus chrysopygus***		**CR B1 + 2abcde, C2a**
		Leontopithecus caissara		**CR B1 + 2abcde, C2a, D1**
Leontopithecus rosalia	**Golden lion tamarin**	***Leontopithecus rosalia***		**CR B1 + 2abcde, C2a**
Saguinus bicolor	Bare-faced tamarin	*Saguinus bicolor*	LR 1c	
Saguinus fuscicollis	Saddleback tamarin	*Saguinus fuscicollis*	LR 1c	
Saguinus imperator	Emperor tamarin	*Saguinus imperator*	LR 1c	
Saguinus inustus	Mottle-faced tamarin	*Saguinus inustus*	LR 1c	
Saguinus labiatus	White-lipped tamarin	*Saguinus labiatus*	LR 1c	
Saguinus leucopus	**White-footed tamarin**	***Saguinus leucopus***	VU	**A1c, B1 + 2c, C2a**
Saguinus midas	Red-handed tamarin	*Saguinus midas*	LR 1c	
Saguinus mystax	Moustached tamarin	*Saguinus mystax*	LR 1c	
Saguinus nigricollis	Black-and-red tamarin	*Saguinus nigricollis*	LR 1c	
Saguinus oedipus	**Cotton-top tamarin**	***Saguinus oedipus***		**EN B1 + 2abcde, C2a**
		Saguinus geoffroyi	LR 1c	
Saguinus tripartitus	Golden-mantled saddleback tamarin	*Saguinus tripartitus*	LR 1c	
CEBIDAE				
Alouatta belzebul	Red-handed howler monkey	*Alouatta belzebul*	LR 1c	
Alouatta caraya	Black-and-gold howler monkey	*Alouatta caraya*	LR 1c	
Alouatta fusca	**Brown howler monkey**	***Alouatta fusca***	VU	**A1c**
Alouatta palliata	Mantled howler monkey	*Alouatta palliata*	LR 1c	
		Alouatta coibensis[2]	LR 1c	
Alouatta seniculus	Red howler monkey	*Alouatta seniculus*	LR 1c	
		Alouatta sara	LR 1c	
Alouatta villosa	Guatemalan howler monkey	*Alouatta pigra*	LR 1c	
Aotus azarae	Southern owl monkey	*Aotus azarai*	LR 1c	
		Aotus infulatus	LR 1c	
		Aotus miconax	VU	**A1c, B1 + 2c**

(continued)

Appendix 1 (*continued*)

Species list 1	Common name	Species list 2	IUCN Red List codes
Aotus trivirgatus	Northern owl monkey	*Aotus nancymae*	LR 1c
		Aotus nigriceps	LR 1c
		Aotus trivirgatus	LR 1c
		Aotus brumbacki	**VU B1 + 2c**
		Aotus hershkovitzi	DD
		Aotus lemurinus	**VU B1 + 2c**
		Aotus vociferans	LR 1c
Ateles belzebuth	**Long-haired spider monkey**	**Ateles belzebuth**	**VU A1c**
Ateles fusciceps	**Brown-headed spider monkey**	**Ateles marginatus**	**EN B1 + 2abcde**
		Ateles fusciceps	**VU A1c, B1 + 2abcde**
Ateles geoffroyi	Black-handed spider monkey	*Ateles geoffroyi*	LR 1c
Ateles paniscus	Black spider monkey	*Ateles paniscus*	LR 1c
		Ateles chamek	LR 1c
Brachyteles arachnoides	**Woolly spider monkey**	**Brachyteles arachnoides**	**EN B1 + 2abcde, C2a**
		Brachyteles hypoxanthus°°	**EN B1 + 2abcde, C2a**
Cacajao calvus	**White uakari**	**Cacajao calvus**	**VU A1c**
Cacajao melanocephalus	Black-headed uakari	*Cacajao melanocephalus*	LR 1c
Cacajao rubicundus	Red uakari	(synonymous with *C. calvus*)	
Callicebus brunneus	Brown titi monkey	*Callicebus brunneus*	LR 1c
Callicebus caligatus	Chestnut-bellied titi monkey	*Callicebus caligatus*	LR 1c
Callicebus cinerascens	Ashy titi monkey	*Callicebus cinerascens*	LR 1c
Callicebus cupreus	Red titi monkey	*Callicebus cupreus*	LR 1c
Callicebus donacophilus	Bolivian gray titi monkey	*Callicebus donacophilus*	LR 1c
Callicebus dubius	**Titi monkey**	**Callicebus dubius**	**VU B1 + 2c**
Callicebus hoffmannsi	Hoffmann's titi monkey	*Callicebus hoffmannsi*	LR 1c
Callicebus modestus	Titi monkey	*Callicebus modestus*	LR 1c
Callicebus moloch	Dusky titi monkey	*Callicebus moloch*	LR 1c
Callicebus oenanthe	**Andean titi monkey**	**Callicebus oenanthe**	**VU B1 + 2c**

Scientific name	Common name	Scientific name	Status	Threat
Callicebus olallae	Beni titi monkey	*Callicebus olallae*	DD	
Callicebus personatus*	**Masked titi monkey**	**Callicebus personatus**		VU A1c
*Callicebus torquatus***	Yellow-handed titi monkey	*Callicebus torquatus*	LR 1c	
Cebus albifrons	White-fronted capuchin	*Cebus albifrons*	LR 1c	
Cebus apella	Brown capuchin	*Cebus apella*	LR 1c	
		Cebus xanthosternos°°		CR B1 + 2abcde, C2a
Cebus capucinus	White-throated capuchin	*Cebus capucinus*	LR 1c	
Cebus olivaceus	Wedge-capped capuchin	*Cebus olivaceus*	LR 1c	
		Cebus kaapori°°		VU A1c, B1 + 2c
Chiropotes albinasus	White-nosed saki	*Chiropotes albinasus*	LR 1c	
Chiropotes satanas	Black saki	*Chiropotes satanas*	LR 1c	
Lagothrix flavicauda	**Yellow-tailed woolly monkey**	**Lagothrix flavicauda**		CR B1 + 2abcde, C2a
Lagothrix lagotricha	Common woolly monkey	*Lagothrix lagotricha*	LR 1c	
Pithecia aequatorialis	Equatorial saki	*Pithecia aequatorialis*	LR 1c	
Pithecia albicans	Buffy saki	*Pithecia albicans*	LR 1c	
Pithecia irrorata	Bald-faced saki	*Pithecia irrorata*	LR 1c	
Pithecia monachus	Monk saki	*Pithecia monachus*	LR 1c	
Pithecia pithecia	White-faced saki	*Pithecia pithecia*	LR 1c	
Saimiri boliviensis	Bolivian squirrel monkey	*Saimiri boliviensis*	LR 1c	
Saimiri oerstedii	**Red-backed squirrel monkey**	**Saimiri oerstedii**		EN B1 + 2abcde, C2a
Saimiri sciureus	Common squirrel monkey	*Saimiri sciureus*	LR 1c	
Saimiri ustus	Golden-backed squirrel monkey	*Saimiri ustus*	LR 1c	
Saimiri vanzolinii	**Black squirrel monkey**	**Saimiri vanzolinii**		VU B1 + 2c, C2a

CERCOPITHECIDAE
(subfamily CERCOPITHECINAE)

Scientific name	Common name	Scientific name	Status	Threat
Allenopithecus nigroviridis	Allen's swamp monkey	*Allenopithecus nigroviridis*	LR nt	
Cercocebus albigena	Gray-cheeked mangabey	*Lophocebus albigena*	LR 1c	
Cercocebus aterrimus	Black mangabey	*Lophocebus aterrimus°*	LR nt	
Cercocebus galeritus	Agile mangabey	*Cercocebus galeritus*[3]	LR nt	
		Cercocebus agilis[3]		

(continued)

Appendix 1 (continued)

Species list 1	Common name	Species list 2	IUCN Red List codes
Cercocebus torquatus	White-collared mangabey	Cercocebus torquatus	LR nt
		Cercocebus atys°°	LR nt
Cercopithecus aethiops	Vervet monkey	Cercopithecus aethiops[5]	LR 1c
Cercopithecus ascanius	Red-tailed monkey	Cercopithecus ascanius	LR 1c
Cercopithecus campbelli	Campbell's monkey	Cercopithecus campbelli	LR 1c
Cercopithecus cephus	Moustached monkey	Cercopithecus cephus	LR 1c
Cercopithecus denti	Dent's monkey	(synonymous with C. wolfi)	
Cercopithecus diana	**Diana monkey**	**Cercopithecus diana**	VU A1c + 2c
Cercopithecus dryas	Dryas monkey	Cercopithecus dryas	DD
Ceropithecus erythrogaster	**Red-bellied monkey**	**Cercopithecus erythrogaster**	VU A1c
Cercopithecus erythrotis	**Red-eared monkey**	**Cercopithecus erythrotis**	VU A1c
		Cercopithecus sclateri	EN B1 + 2c
Cercopithecus hamlyni	Owl-faced monkey	Cercopithecus hamlyni	LR nt
Cercopithecus lhoesti	L'hoest's monkey	Cercopithecus lhoesti	LR nt
Cercopithecus mitis	Blue monkey	Cercopithecus mitis	LR 1c
Cercopithecus mona	Mona monkey	Cercopithecus mona	LR 1c
Cercopithecus neglectus	De Brazza's monkey	Cercopithecus neglectus	LR 1c
Cercopithecus nictitans	Spot-nosed monkey	Cercopithecus nictitans	LR 1c
Cercopithecus petaurista	Lesser spot-nosed monkey	Cercopithecus petaurista	LR 1c
Cercopithecus pogonias	Crowned monkey	Cercopithecus pogonias	LR 1c
Cercopithecus preussi	**Preuss's monkey**	**Cercopithecus preussi**	EN A1cd + 2c
Cercopithecus salongo	Salongo monkey	(synonymous with C. dryas)	
Cercopithecus solatus	**Sun-tailed monkey**	**Cercopithecus solatus**	VU B1 + 2a, C1
Cercopithecus wolfi	Wolf's monkey	Cercopithecus wolfi	LR 1c
Erythrocebus patas	Patas monkey	Erythrocebus patas	LR 1c
Macaca arctoides	**Stump-tailed macaque**	**Macaca arctoides**	VU A1cd
Macaca assamensis	**Assamese macaque**	**Macaca assamensis**	VU A1cd
Macaca cyclopis	**Formosan macaque**	**Macaca cyclopis**	VU A1cd

Macaca fascicularis	Long-tailed macaque	*Macaca fascicularis*	LR nt
Macaca fuscata	**Japanese macaque**	***Macaca fuscata***	**EN A2cd**
Macaca maurus	**Moor macaque**	***Macaca maura***	**EN A1cd**
Macaca mulatta	Rhesus macaque	*Macaca mulatta*	LR nt
Macaca nemestrina	**Pig-tailed macaque**	***Macaca nemestrina***	**VU A1cd**
		***Macaca pagensis*°°**	**CR A1cd + 2c**
Macaca nigra	**Celebes macaque**	***Macaca nigra***	**EN A1acd**
		Macaca nigrescens°°	LR cd
Macaca ochreata	Booted macaque	*Macaca ochreata*	DD
		***Macaca brunnescens*°°**	**VU C1**
Macaca radiata	Bonnet macaque	*Macaca radiata*	LR 1c
Macaca silenus	**Lion-tailed macaque**	***Macaca silenus***	**EN C2a**
Macaca sinica	Toque macaque	*Macaca sinica*	LR nt
Macaca sylvanus	**Barbary macaque**	***Macaca sylvanus***	**VU A1c, C1**
Macaca thibetana	Tibetan macaque	*Macaca thibetana*	LR cd
Macaca tonkeana	Tonkean macaque	*Macaca tonkeana*	LR nt
		Macaca hecki°°	LR nt
Mandrillus leucophaeus	**Drill**	***Mandrillus leucophaeus***	**EN A1acd + 2cd + C1 + 2a**
Mandrillus sphinx	Mandrill	*Mandrillus sphinx*	LR nt
Miopithecus talapoin	Talapoin	*Miopithecus talapoin*	LR 1c
Papio anubis	Olive baboon	*Papio anubis*°	LR 1c
Papio cynocephalus	Yellow baboon	*Papio cynocephalus*°	LR nt
Papio hamadryas	Hamadryas baboon	*Papio hamadryas*	LR nt
Papio papio	Guinea baboon	*Papio papio*°	LR 1c
Papio ursinus	Chacma baboon	*Papio ursinus*°	LR nt
Theropithecus gelada	Gelada baboon	*Theropithecus gelada*	LR nt

CERCOPITHECIDAE
(subfamily COLOBINAE)

Colobus angolensis	Angolan black-and-white colobus	*Colobus angolensis*	LR 1c

(continued)

Appendix 1 (*continued*)

Species list 1	Common name	Species list 2	IUCN Red List codes
Colobus badius	Red colobus	*Procolobus badius*[4]	LR nt
		Procolobus pennantii[4]	
		Procolobus preussi[4]	
		Procolobus rufomitratus[4]	
Colobus guereza	Guereza	*Colobus guereza*	LR 1c
Colobus kirkii	Zanzibar red colobus	(synonymous with *P. pennantii*)	
Colobus polykomos	Western black-and-white colobus	*Colobus polykomos*	LR nt
Colobus satanas	**Black colobus**	**Colobus vellerosus****	VU A1c + 2c
		Colobus satanas	VU A1c
Nasalis larvatus	**Proboscis monkey**	**Nasalis larvatus**	VU A2c
Presbytis aurata	**Ebony leaf monkey**	**Trachypithecus auratus**	VU A1c
Presbytis comata	**Sunda leaf monkey**	**Presbytis comata**	EN A1c, C2a
Presbytis cristata	Silvered leaf monkey	*Presbytis fredericae*°°	DD
		Presbytis hosei	LR 1c
		Trachypithecus cristatus	LR nt
Presbytis entellus	Hanuman langur	*Semnopithecus entellus*	LR nt
Presbytis francoisi	**François's leaf monkey**	**Trachypithecus francoisi**	VU A1cd + 2cd, C2a
		Trachypithecus laotum°°	DD
		Trachypithecus poliocephalus°°	EN A2c, C2a
		Trachypithecus delacouri**	CR A1d, C2a
Presbytis frontata	White-fronted leaf monkey	*Presbytis frontata*	DD
Presbytis geei	Golden leaf monkey	*Trachypithecus geei*	DD
Presbytis johnii	**Nilgiri leaf monkey**	**Trachypithecus johnii**	VU A1d
Presbytis melalophos	Banded leaf monkey	*Presbytis melalophos*	LR 1c
		Presbytis femoralis	LR nt
		Presbytis thomasi	LR nt
Presbytis obscura	Dusky leaf monkey	*Trachypithecus obscurus*	LR 1c

(continued)

	Common name			
Presbytis phayrei	Phayre's leaf monkey	*Trachypithecus phayrei*	DD	
Presbytis pileata	**Capped leaf monkey**	***Trachypithecus pileatus***		**VU A2c**
Presbytis pontenziani	Mentawai leaf monkey	*Presbytis pontenziani*		**VU A1c + 2c**
Presbytis rubicunda	Maroon leaf monkey	*Presbytis rubicunda*	LR 1c	
Presbytis vetulus	**Purple-faced leaf monkey**	***Trachypithecus vetulus***		**VU A1c**
Procolobus verus	Olive colobus	*Procolobus verus*	LR nt	
Pygathrix avunculus	**Tonkin snub-nosed monkey**	***Rhinopithecus avunculus*[6]**		**CR C1, E**
Pygathrix brelichi	**Brelich's snub-nosed monkey**	***Rhinopithecus brelichi*[6]**		**EN C2b**
Pygathrix nemaeus	Douc langur	*Pygathrix nemaeus*		**EN A1cd**
Pygathrix roxellana	**Chinese snub-nosed monkey**	***Rhinopithecus roxellana*[6]**		**VU C2a**
		***Rhinopithecus bieti*[6]**		**EN C2a**
Simias concolor	**Pig-tailed langur**	***Simias concolor*[7]**		**EN A1cd + 2c**
HYLOBATIDAE				
Hylobates agilis	Agile gibbon	*Hylobates agilis*	LR nt	
Hylobates concolor	**White-cheeked gibbon**	***Hylobates concolor***		**EN A1cd, C2a**
		Hylobates gabriellae	DD	
		Hylobates leucogenys	DD	
Hylobates hoolock	Hoolock gibbon	*Hylobates hoolock*	DD	
Hylobates klossii	**Kloss's gibbon**	***Hylobates klossii***		**VU A1c + 2c**
Hylobates lar	Lar gibbon	*Hylobates lar*	LR nt	
Hylobates moloch	**Javan gibbon**	***Hylobates moloch***		**CR A1c, C2a**
Hylobates muelleri	Müller's gibbon	*Hylobates muelleri*	LR nt	
Hylobates pileatus	**Pileated gibbon**	***Hylobates pileatus***		**VU A1cd + 2cd**
Hylobates syndactylus	Siamang	*Hylobates syndactylus*	LR nt	
PONGIDAE				
Gorilla gorilla	**Gorilla**	***Gorilla gorilla***		**EN A2cd**
Homo sapiens	Human	*Homo sapiens*		

Appendix 1 (continued)

Species list 1	Common name	Species list 2	IUCN Red List codes
Pan paniscus	**Pygmy chimpanzee**	**Pan paniscus**	**EN A2cd**
Pan troglodytes	**Common chimpanzee**	**Pan troglodytes**	**EN A2cd**
Pongo pygmaeus	**Orangutan**	**Pongo pygmaeus**	**VU A1cd, C1**

Note: This appendix provides two alternative species lists for contemporary primate diversity, together with the conservation status of these taxa. Species list 1 is based on Corbet and Hill (1991); the family names and the common English names are also given according to this taxonomy. Species list 2 is based on Groves (1993), but with additional modifications (see below). Groves's taxonomy differs from that of Corbet and Hill in several respects: certain species names are revised (e.g., *Presbytis aurata* becomes *Trachypithecus aurata*); certain species are split into two or more taxa (e.g., *Arctocebus calabarensis* becomes *Arctocebus calabarensis* and *Arctocebus aureus*); and certain pairs of species are combined into one (e.g., *Cacajao rubicundus* is considered the same species as *Cacajao calvus*). Overall, Groves's taxonomy is more speciose than that of Corbet and Hill.

Conservation status is based on the most recent evaluation of global extinction risk in primates (Baillie and Groombridge 1996); the IUCN Red List codes that form the basis of this evaluation are explained in chapter 10 (see table 10.3). This evaluation follows Groves's taxonomy, but with some modifications: some species not recognized by Groves were recognized by Baillie and Groombridge and have therefore been added to Groves's list here (these are marked with asterisks; see below), while others recognized by Groves either have been renamed (seven cases), have been omitted (one case), or do not appear to have been recognized (two sets of cases) by Baillie and Groombridge (these are marked with numbers; see below).

Bold type highlights species classified as threatened by Baillie and Groombridge (1996): see final column.

°Species not recognized by Groves (1993) but recognized by Baillie & Groombridge (1996) and Corbet and Hill (1991).

°°Species not recognized by either Groves (1993) or Corbet and Hill (1991) but recognized by Baillie and Groombridge (1996).

°°°Species that are recognized by Groves (1993) and Baillie and Groombridge (1996) but appear to have been inadvertently omitted from Corbet and Hill (1991).

[1]Listed as *Microcebus coquereli* by Groves (1993).

[2]*Alouatta coibensis* appears to have been inadvertently omitted from the IUCN Red List; its two subspecies (on the same list) are classified as "Endangered" and "Critically Endangered."

[3]The Red List appears to consider *Cercocebus agilis* a subspecies of *Cercocebus galeritus.*

[4]The Red List appears to consider *Procolobus pennatti, P. preussi,* and *P. rufomitratus* as subspecies of *Procolobus badius* (note, however, that all three subspecies are classified as "Endangered" in the same listing).

[5]Listed as *Chlorocebus aethiops* by Groves 1993.

[6]Listed as genus *Pygathrix* by Groves 1993.

[7]Listed as *Nasalis concolor* by Groves 1993.

APPENDIX 2
Leslie Matrices

Leslie (1945) introduced a convenient way of estimating the future size and composition of a population using data from life tables. His approach was based on the mathematical technique known as matrix algebra. Since Leslie matrices form the basis for many simulation models of population dynamics (including several of those discussed in this book), we summarize the basic principles as an aid to understanding how these models work.

Leslie showed that the future age structure of a population can be determined from its current age structure by iterating the following matrix equation:

$$(\textbf{A2.1}) \quad \begin{pmatrix} m_0 & m_1 & m_2 & \cdots & m_{x-1} & m_x \\ p_0 & 0 & 0 & \cdots & 0 & 0 \\ 0 & p_1 & 0 & \cdots & 0 & 0 \\ 0 & 0 & p_2 & \cdots & 0 & 0 \\ \cdot & \cdot & \cdot & \cdots & \cdot & \cdot \\ \cdot & \cdot & \cdot & \cdots & \cdot & \cdot \\ 0 & 0 & 0 & \cdots & p_{x-1} & 0 \\ 0 & 0 & 0 & \cdots & 0 & p_x \end{pmatrix} \times \begin{pmatrix} n_{00} \\ n_{10} \\ n_{20} \\ n_{30} \\ \cdot\cdot \\ \cdot\cdot \\ n_{(x-1)0} \\ n_{x0} \end{pmatrix} = \begin{pmatrix} n_{01} \\ n_{11} \\ n_{21} \\ n_{31} \\ \cdot\cdot \\ \cdot\cdot \\ n_{(x-1)1} \\ n_{x1} \end{pmatrix}$$

Equation A2.1 consists of a transition matrix $\textbf{M}$ with m rows and x columns (on the left) that specifies the birthrates in each age-class (m_x) and the probabilities of an individual's surviving from each age-class x to join the next age-class $x + 1$ (p_x), plus two column vectors that specify (on the left of the equality sign) the age structure of the population at time $t = 0$ and (on the right of the equality sign) its age structure in the following time interval (time $t = 1$), where n_{xt} is the number of individuals in the xth age class at time t. The second column vector is obtained by summing, across each row of the transition matrix, the product of each cell in that row and the corresponding cell in the first column vector. In other words, n_{x1}, the number of individuals in the xth age class at time $t = 1$, is

$$n_{x1} = \sum_x M_{ix} n_{x0},$$

where M_{ix} is the cell corresponding to the intersection of the ith row and the xth column in the transition matrix. The size of the population at time t is given by the sum of the column vector

$$N_t = \sum_x n_{xt},$$

where N_t is the total size of the cohort at time t. Separate analyses can be carried out for each sex.

Since the diagonal row in the transition matrix gives the proportion of each age-class that survives to join the next age-class in the following time interval, the size and composition of a population at any future time can be determined by iterating equation A2.1 across as many time intervals as is required. At each iteration the n_{x1} column vector becomes the n_{x0} column vector for the next cycle.

APPENDIX 3

Primate and Conservation Organizations

The following is a selection of organizations for people interested in primates and their conservation. These organizations are active across a broad sweep of issues (those focusing on specific species, geographic regions, or conservation problems have not been included here owing to space limitations). The Web sites and postal addresses are given for each of the organizations (which are in alphabetical order under each heading). Readers interested in becoming involved in primate conservation should visit these Web sites for more information. We also recommend visiting the Primate Info Net (http://www.primate.wisc.edu/pin), which includes links to a variety of other primate conservation programs (http://www.primate.wisc.edu/pin/conserv.html).

Organizations Supporting Primates and Primate Conservation

American Society of Primatologists
Web site: http:/www.asp.org
Contact: Steven J. Schapiro, Treasurer ASP
            Department of Veterinary Sciences
            University of Texas
            M. D. Anderson Cancer Center
            Route 2, Box 151-B1
            Bastrop, TX 78602-9733
            USA

International Primate Protection League
Web site: http://www.ippl.org/index.html
Contact: International Primate Protection League
            P.O. Box 766
            Summerville, SC 29484
            USA

            International Primate Protection League
            116 Judd Street
            London WC1H 9NS
            UK

417

International Primatological Society
Web site: http://indri.primate.wisc.edu/pin/ips.html
Contact: Dr. Richard W. Byrne
            School of Psychology
            University of St. Andrews
            St. Andrews, Fife, Scotland KY16 9JU
            UK

Jane Goodall Institute
Web site: http://www.janegoodall.org/index.html
Contact: Stewart Hudson, Executive Director
            The Jane Goodall Institute
            P.O. Box 14890
            Silver Spring, MD 20911-4890
            USA
            [There are many national offices besides the USA,
            including several African countries.]

Primate Conservation, Inc.
Web site: http://www.primate.org
Contact: Primate Conservation, Inc.
            1411 Shannock Road
            Charlestown, RI 02813-3726
            USA

Primate Society of Great Britain
Web site: http://www.ana.ed.ac.uk/PSGB/home.html
Contact: Dr. H. C. McKiggan, Information Officer
            School of Psychology
            University of St. Andrews
            St. Andrews, Fife, Scotland KY16 9JU
            UK

Organizations Supporting On-the-Ground Conservation Action (Involving Primates or Primate Habitats)

Conservation International
Web site: http://www.conservation.org
Contact: Conservation International
            2501 M Street NW, Suite 200
            Washington, DC 20037
            USA

IUCN (The World Conservation Union)
Web site: http://www.iucn.org/themes/ssc/sgs/sgs.htm#mammals
Contact: IUCN-US, Washington Office
            1630 Connecticut Avenue NW, Third Floor
            Washington, DC 10009
            USA

            IUCN World Headquarters
            Rue Mauverney 28
            CH 1196 Gland
            Switzerland
            [There are many national offices. If the national address is un-
            known, write to the World Headquarters]

IUCN/SSC Primate Specialist Group
Web site: None available at present
Contact: Dr. Russell A. Mittermeier, Chair
            Conservation International
            2501 M Street NW, Suite 200
            Washington, DC 20037
            USA

Fauna and Flora International
Web site: http://www.ffi.org.uk
Contact: Fauna and Flora International
            Great Eastern House
            Tension Road
            Cambridge CB1 2DT
            UK

Wildlife Conservation Society
Web site: http://www.wcs.org
Contact: Wildlife Conservation Society
            2300 Southern Boulevard
            Bronx, NY 10460-1099
            USA

WWF (World Wide Fund for Nature)
Web site: http://www.wwf.org
Contact: WWF United States
            1250 24th Street NW
            Washington, DC 20037-1175
            USA

WWF International
Avenue du Mont-Blanc
CH 1196 Gland
Switzerland

Zoological Society of London
Web site: http://www.zsl.org
Contact: Zoological Society of London
Regent's Park
London NW1 4RY
UK

Organizations Supporting the Field of Study of Conservation Biology

Society for Conservation Biology
Web site: http://conbio.rice.edu/scb
Contact: Alice Blandin, Membership Information
University of Washington
P.O. Box 351800
Seattle, WA 98195-1800
USA

REFERENCES

Abbott, D. H. 1984. Behavioural and physiological suppression of fertility in subordinate marmoset monkeys. *Am. J. Primatol.* 6:169–186.

Abbott, D. H.; Keverne, E. B.; Moore, G. F.; and Yodyinguad, U. 1986. Social suppression of reproduction in subordinate talapoin monkeys, *Miopithecus talapoin.* In *Primate ontogeny,* ed. J. Else and P. C. Lee, 329–341. Cambridge: Cambridge University Press.

Aiello, L. C., and Wheeler, P. 1995. The expensive tissue hypothesis. *Current Anthropol.* 36:199–211.

Akçakaya, H. R., and Ginzburg, L. R. 1991. Ecological risk analysis for single and multiple populations. In *Species conservation: A population-biological approach,* ed. A. Seitz and V. Loeschke, 73–87. Basel: Birkhauser.

Alberts, S. C., and Altmann, J. 1995. Balancing costs and opportunities: Dispersal in male baboons. *Am. Nat.* 145:279–306.

Alderman, C. L. 1989. A general introduction to primate conservation in Colombia. *Primate Cons.* 10:44–51.

Alexander, R. D. 1974. The evolution of social behaviour. *Ann. Rev. Ecol. Syst.* 5:325–383.

Allee, W. C. 1931. *Animal aggregations: A study in general socioecology.* Chicago: University of Chicago Press.

Altmann, J. 1979. Age cohorts as paternal sibships. *Behav. Ecol. Sociobiol.* 6:161–169.

Altmann, J.; Alberts, S. C.; Haines, S. A.; Bubach, J.; Muruthi, P.; Coote, T.; Geffen, E.; Cheesman, D. J.; Mututa, R. S.; Saiyalel, S.; Wayne, R. K.; Lacy, R. C.; and Bruford, M. W. 1996. Behavior predicts genetic structure in a wild primate group. *Proc. Natl. Acad. Sci. USA* 93:5797–5801.

Altmann, J.; Altmann, S. A.; Hausfater, G.; and McCuskey, S. A. 1977. Life history of yellow baboons: Physical development, reproductive parameters and infant mortality. *Primates* 18:315–330.

Altmann, J.; Hausfater, G.; and Altmann, S. A. 1985. Demography of Amboseli baboons. *Am. J. Primatol.* 8:113–125.

Altmann, J.; Schoeller, D.; Altmann, S. A.; Muruthi, P.; and Sapolski, R. M. 1993. Body size and fatness of free-living baboons reflect food availability and activity levels. *Am. J. Primatol.* 30:149–161.

Altmann, S. A. 1998. *Foraging for survival.* Chicago: University of Chicago Press.

Alvard, M. S. 1993. Testing the "ecologically noble savage" hypothesis: Interspecific prey choice by Piro hunters of Amazonian Peru. *Human Ecol.* 21:355–387.

———. 1994. Conservation by native peoples: Prey choice in a depleted habitat. *Human Nature* 5:127–154.

―――. 1995a. Intraspecific prey choice by Amazonian hunters. *Current Anthropol.* 36:789–818.

―――. 1995b. Shotguns and sustainable hunting in the Neotropics. *Oryx* 29:58–66.

―――. 1998. Evolutionary ecology and resource conservation. *Evol. Anthropol.* 7:62–74.

Alvard, M. S.; Robinson, J. G.; Redford, K.; and Kaplan, H. 1997. The sustainability of subsistence hunting in the Neotropics. *Cons. Biol.* 11:977–982.

Anadu, P. A. 1987. Progress in the conservation of Nigeria's wildlife. *Biol. Cons.* 41:237–251.

Anadu, P. A.; Elamah, P. O.; and Oates, J. F. 1988. The bushmeat trade in southwestern Nigeria: A case study. *Human Ecol.* 16:199–208.

Andau, P. M.; Hiong, L. K.; and Sale, J. B. 1994. Translocation of pocketed orangutans in Sabah. *Oryx* 28:263–268.

Andelman, S. 1986. Ecological and social determinants of cercopithecine mating patterns. In *Ecological aspects of social evolution,* ed. D. I. Rubenstein and R. W. Wrangham, 201–216. Princeton: Princeton University Press.

Andrén, H. 1994. Effects of habitat fragmentation on birds and mammals in landscapes with different proportions of suitable habitat: A review. *Oikos* 71:355–366.

―――. 1999. Habitat fragmentation, the random sample hypothesis and critical thresholds. *Oikos* 84:306–308.

Andresen, J. E. 1999. Seed dispersal by monkeys and the fate of dispersed seeds in a Peruvian rain forest. *Biotropica* 31:145–158.

Andrewartha, H. G., and Birch, L. C. 1954. *The distribution and abundance of animals.* Chicago: University of Chicago Press.

Andrews, P. J., and Aiello, L. C. 1984. An evolutionary model for feeding and positional behaviour. In *Food acquisition and processing in primates,* ed. D. J. Chivers, B. A. Wood, and A. Bilsborough, 429–466. New York: Plenum Press.

Aquino, R., and Encarnación, F. 1994. Owl monkey populations in Latin America: Field work and conservation. In Aotus: *The owl monkey,* ed. J. F. Baer, R. E. Weller, and I. Kakoma, 59–95. San Diego: Academic Press.

Arcese, P.; Hando, J.; and Campbell, K. 1995. Historical and present-day antipoaching efforts in Serengeti. In *Serengeti II: Dynamics, management and conservation of an ecosystem,* ed. A. R. E. Sinclair and P. Arcese, 506–533. Chicago: University of Chicago Press.

Archibold, O. W. 1995. *Ecology of world vegetation.* London: Chapman and Hall.

Areola, O. 1987. The political reality of conservation in Nigeria. In *Conservation in Africa: People, policies and practice,* ed. D. Anderson and R. Grove, 277–292. Cambridge: Cambridge University Press.

Arita, H. T.; Robinson, J. G.; and Redford, K. H. 1990. Rarity in Neotropical forest mammals and its ecological correlates. *Cons. Biol.* 4:181–192.

Armbruster, P., and Lande, R. 1993. A population viability analysis for African elephant (*Loxodonta africana*): How big should reserves be? *Cons. Biol.* 7:602–610.

Atkinson, I. 1989. Introduced animals and extinctions. In *Conservation biology for the twenty-first century,* ed. D. Western and M. Pearl, 54–75. New York: Oxford University Press.

Aunger, R. 1994. Are food avoidances maladaptive in the Ituri Forest of Zaire? *J. Anthropol. Res.* 50:277–310.

Aveling, R. J. 1987. Environmental education in developing countries. In *Primate conservation in the tropical rainforest,* ed. C. W. Marsh and R. A. Mittermeier, 231–262. New York: Alan R. Liss.

Ayres, J. M.; Bodmer, R. E.; and Mittermeier, R. A. 1991. Financial considerations of reserve designs in countries with high primate diversity. *Cons. Biol.* 5:109–114.

Ayres, J. M., and Clutton-Brock, T. H. 1992. River boundaries and species range size in Amazonian primates. *Am. Nat.* 140:531–537.

Ayres, J. M.; de Magalhaes Lima, D.; de Souza Martins, E.; and Barrieros, J. L. K. 1991. On the track of the road: Changes in subsistence hunting in a Brazilian Amazonian village. In *Neotropical wildlife use and conservation,* ed. J. G. Robinson and K. H. Redford, 82–92. Chicago: University of Chicago Press.

Baillie, J., and Groombridge, B. 1996. *1996 IUCN Red List of threatened animals.* Gland, Switz.: IUCN.

Balmford, A. 1996. Extinction filters and current resilience: The significance of past selection pressures for conservation biology. *Trends Ecol. Evol.* 11:193–196.

Balmford, A.; Leader-Williams, N.; and Green, M. J. B. 1995. Parks or arks: Where to conserve threatened mammals. *Biodiversity Cons.* 4:595–607.

Balmford, A.; Mace, G. M.; and Leader-Williams, N. 1996. Designing the ark: Setting priorities for captive breeding. *Cons. Biol.* 10:719–727.

Barbier, E. B. 1992. Community-based development in Africa. In *Economics for the wilds,* ed. T. M. Swanson and E. B. Barbier, 103–135. London: Earthscan.

Barbier, E. [B.]; Bockstael, N.; Burgess, J.; and Strand, I. 1994. The timber trade and tropical deforestation in Indonesia. In *The causes of tropical deforestation,* ed. K. Brown and D. W. Pearce, 242–270. London: UCL Press.

Barnes, J.; Burgess, J.; and Pearce, D. 1992. Wildlife tourism: Community-based development in Africa. In *Economics for the wilds,* ed. T. M. Swanson and E. B. Barbier, 136–151. London: Earthscan.

Barnes, R. F. W. 1990. Deforestation in tropical Africa. *J. Afr. Ecol.* 28:161–173.

Barrett, C. B., and Arcese, P. 1995. Are integrated conservation-development projects (ICDPs) sustainable? On the conservation of large mammals in sub-Saharan Africa. *World Development* 23:1073–1084.

Barrett, L., and Henzi, S. P. 1997. Environmental determinants of body weight in chacma baboons. *S. Afr. J. Sci.* 93:436–438.

———. 1998. Epidemic deaths in a chacma baboon population. *S. Afr. J. Sci.* 94:441.

Barton, R. A. 1996. Neocortex size and behavioural ecology in primates. *Proc. Roy. Soc. Lond.,* ser. B, 263:173–177.

Barton, R. A.; Byrne, R. W.; and Whiten, A. 1996. Ecology, feeding competition and social structure in baboons. *Behav. Ecol. Sociobiol.* 38:321–329.

Barton, R. A., and Dunbar, R. I. M. 1997. Evolution of the social brain. In *Machiavellian intelligence II,* ed. A. Whiten and R. Byrne, 240–263. Cambridge: Cambridge University Press.

Basabose, K., and Yamagiwa, J. 1997. Predation on mammals by the chimpanzees in the montane forest of Kahuzi, Zaire. *Primates* 38:45–55.

Bascompte, J., and Solé, R. V. 1996. Habitat fragmentation and extinction thresholds in spatially explicit models. *J. Anim. Ecol.* 65:465–473.

Bauchop, T., and Martucci, R. W. 1968. Ruminant-like digestion of the langur monkey. *Science* 161:698–700.

Bearder, S. K. 1987. Lorises, bushbabies and tarsiers: Diverse societies in solitary foragers. In *Primate societies,* ed. B. B. Smuts, D. L. Cheney, R. M. Seyfarth, R. W. Wrangham, and T. T. Struhsaker, 11–24. Chicago: University of Chicago Press.

———. 1999. Physical and social diversity among nocturnal primates: A new view based on long term research. *Primates* 40:267–282.

Bearder, S. K., and Doyle, G. A. 1974. Ecology of bushbabies *Galago senegalensis* and *Galago crassicaudatus,* with some notes on their behaviour in the field. In *Prosimian biology,* ed. R. D. Martin, G. A. Doyle, and A. C. Walker, 109–130. London: Duckworth.

Bearder, S. K.; Honess, P. E.; and Ambrose, L. 1995. Species diversity among galagos with special reference to mate recognition. In *Creatures of the dark: The nocturnal prosimians,* ed. L. Alterman, M. K. Izard, and G. A. Doyle, 331–352. New York: Plenum Press.

Beck, B. B.; Rapaport, L. G.; Stanley-Price, M. R.; and Wilson, A. C. 1994. Reintroduction of captive-born animals. In *Creative conservation,* ed. P. J. S. Olney, G. M. Mace, and A. T. C. Feistner, 265–286. London: Chapman and Hall.

Becker, C. D., and Ostrom, E. 1995. Human ecology and resource sustainability: The importance of institutional diversity. *Ann. Rev. Ecol. Syst.* 26:113–133.

Begon, M.; Harper, J. L.; and Townsend, C. R. 1986. *Ecology: Individuals, populations and communities.* Oxford: Blackwell.

Beier, P., and Noss, R. F. 1998. Do habitat corridors provide connectivity? *Cons. Biol.* 12:1241–1252.

Benefit, B., and McCrossin, M. L. 1990. Diet, species diversity and the distribution of African fossil baboons. *Kroeber Anthropol. Soc. Papers* 71–72:77–92.

Bennett, E. L., and Dabahan, Z. 1995. Wildlife responses to disturbances in Sarawak and their implications for forest management. In *Ecology, conservation and management of Southeast Asian rainforests,* ed. R. B. Primack and T. E Lovejoy, 66–86. New Haven: Yale University Press.

Bennett, J. 1989. Gibbon rehabilitation in Sarawak. *Sarawak Gazette* 116:14–22.

———. 1992. A glut of gibbons in Sarawak—Is rehabilitation the answer? *Oryx* 26:157–164.

Bennett, P. M., and Owens, I. P. F. 1997. Variation in extinction risk among birds: Chance or evolutionary predisposition? *Proc. Roy. Soc. Lond.,* ser. B, 264:401–408.

Bentley-Condit, V. K., and Smith, E. O. 1997. Female reproductive parameters of Tana River yellow baboons. *Int. J. Primatol.* 18:581–596.

Berenstain, L.; Mitani, J. C.; and Tenaza, R. R. 1986. Effects of El Nino on habitat and primates in east Kalimantan. *Primate Cons.* 7:54–55.

Bernstein, I. S.; Balcaen, P.; Dresdale, L.; Gouzoules, H.; Kavanagh, M.; Patterson, T.; and Neyman-Warner, P. 1976. Differential effects of forest degradation on primate populations. *Primates* 117:401–411.

Bierregaard, R. O.; Lovejoy, T. E.; Kapos, V.; Santos, A.; and Hutchings, R. W. 1992. The biological dynamics of tropical rainforest fragments. *BioScience* 42:859–866.

Biquand, S.; Biquand-Guyot, V.; and Boug, A. 1989. The status of the hamadryas baboon in Saudi Arabia. *Primate Cons.* 10:28–30.

Biquand, S.; Biquand-Guyot, V.; Boug, A.; and Gautier, J.-P. 1992. The distribution of *Papio hamadryas* in Saudi Arabia: Ecological correlates and human influence. *Int. J. Primatol.* 13:223–243.

Birkinshaw, C. R., and Colquhoun, I. C. 1998. Pollination of *Ravenala madagas-cariensis* and *Parkia madagascariensis* by *Eulemur macaco* in Madagascar. *Folia Primatol.* 69:252–259.

Bittles, A. H.; Radha Rama Devi, A.; and Rao, N. A. 1990. Inter-relationships between consanguinity, religion and fertility in Kearnataka, south India. In *Fertility and resources,* ed. J. Landers and V. Reynolds, 62–75. Cambridge: Cambridge University Press.

Blackburn, T. M.; Brown, V. K.; Doube, B. M.; Greenwood, J. J. D.; Lawton, J. H.; and Stork, N. E. 1993. The relationship between abundance and body size in natural animal assemblages. *J. Anim. Ecol.* 62:519–528.

Blackburn, T. M., and Gaston, K. J. 1997. A critical assessment of the form of the interspecific relationship between abundance and body size in animals. *J. Anim. Ecol.* 66:233–249.

Blackburn, T. M.; Gates, S.; Lawton, J. H.; and Greenwood, J. J. D. 1994. Relations between body size, abundance and taxonomy of birds wintering in Britain and Ireland. *Phil. Trans. Roy. Soc. Lond.,* ser. B, 343:135–144.

Blackburn, T. M.; Lawton, J. H.; and Gregory, R. D. 1996. Relationship between abundances and life histories of British birds. *J. Anim. Ecol.* 65:52–62.

Blackburn, T. M.; Lawton, J. H.; and Pimm, S. L. 1993. Non-metabolic explanations for the relationship between body size and animal abundance. *J. Anim. Ecol.* 62:694–702.

Blom, A.; Alers, M. P. T.; Feistner, A. T. C.; Barnes, R. F. W.; and Barnes, K. L. 1992. Primates in Gabon—Current status and distribution. *Oryx* 26:223–234.

Bodmer, R. E. 1995a. Managing Amazonian wildlife: Biological correlates of game choice by detribalized hunters. *Ecol. App.* 5:872–877.

———. 1995b. Priorities for the conservation of mammals in the Peruvian Amazon. *Oryx* 29:23–28.

Bodmer, R. E.; Fang, T. G.; and Ibanez, L. M. 1988. Primates and ungulates: A comparison of susceptibility to hunting. *Primate Cons.* 9:79–83.

Bodmer, R. E.; Eisenberg, J. F.; and Redford, K. H. 1997. Hunting and the likelihood of extinction in Amazonian mammals. *Cons. Biol.* 11:460–466.

Bodmer, R. E.; Fang, T. G.; Moya I. L.; and Gill, R. 1994. Managing wildlife to conserve Amazonian forests: Population biology and economic considerations of game hunting. *Biol. Cons.* 67:29–35.

Boecklen, W. J., and Gotelli, N. J. 1984. Island biogeographic theory and conservation practice: Species-area or specious-area relationships? *Biol. Cons.* 29:63–80.

Boesch, C. 1997. The emergence of cultures among wild chimpanzees. In *Evolution of social behaviour patterns in primates and man,* ed. W. G. Runciman, J. Maynard Smith, and R. I. M. Dunbar, 251–268. New York: Oxford University Press.

Boesch, C., and Boesch, H. 1981. Sex differences in the use of natural hammers by wild chimpanzees: A preliminary report. *J. Human Evol.* 10:585–593.

———. 1989. Hunting behaviour of wild chimpanzees in the Tai National Park. *Am. J. Phys. Anthropol.* 78:547–573.

Boinski, S. 1987a. Mating patterns in squirrel monkeys: Implications for seasonal sexual dimorphism. *Behav. Ecol. Sociobiol.* 21:13–21.

————. 1987b. Birth synchrony in squirrel monkeys (*Saimiri oerstedi*): A strategy to reduce neonatal predation. *Behav. Ecol. Sociobiol.* 21:393–400.

Boinski, S., and Sirot, L. 1997. Uncertain conservation status of squirrel monkeys in Costa Rica, *Saimiri oerstedi oerstedi* and *Saimiri oerstedi citrinellus*. *Folia Primatol.* 68:181–193.

Borner, M. 1985. The rehabilitated chimpanzees of Rubondo Island. *Oryx* 19:151–154.

Borries, C.; Launhardt, K.; Epplen, C.; Epplen, J. T.; and Winkler, P. 1999. DNA analyses support the hypothesis that infanticide is adaptive in langur monkeys. *Proc. Roy. Soc. Lond.*, ser. B, 266:901–904.

Boswell, G. P.; Britton, N. F.; and Franks, N. R. 1998. Habitat fragmentation, percolation theory and the conservation of a keystone species. *Proc. Roy. Soc. Lond.*, ser. B, 265:1921–1925.

Boulton, A. M.; Horrocks, J. A.; and Baulu, J. 1996. The Barbados vervet monkey (*Cercopithecus aethiops sabaeus*): Changes in population size and crop damage, 1980–1994. *Int. J. Primatol.* 17:831–844.

Bourlière, F. 1985. Primate communities: Their structure and role in tropical ecosystems. *Int. J. Primatol.* 6:1–26.

Boyce, M. S. 1992. Population viability analysis. *Ann. Rev. Ecol. Syst.* 23:481–506.

Brandon, K. E., and Wells, M. 1992. Planning for people and parks: Design dilemmas. *World Development* 20:557–570.

Brain, C. 1992. Deaths in desert baboon troop. *Int. J. Primatol.* 13:593–599.

Brain, C., and Bohrmann, R. 1992. Tick infestation of baboons (*Papio ursinus*) in the Namib Desert. *J. Wild. Dis.* 28:188–191.

Brandon-Jones, D. 1996. The Asian Colobinae (Mammalia: Cercopithecidae) as indicators of Quarternary climatic change. *Biol. J. Linn. Soc.* 59:327–350.

Brockelman, W. Y., and Ali, R. 1987. Methods of surveying and sampling forest primate populations. In *Primate conservation in the tropical rainforest*, ed. C. W. Marsh and R. A. Mittermeier, 23–62. New York: Alan R. Liss.

Bronikowski, A. M., and Altmann, J. 1996. Foraging in a variable environment: Weather patterns and the behavioural ecology of baboons. *Behav. Ecol. Sociobiol.* 39:11–25.

Brooks, D. R.; Mayden, R. L.; and McLennan, D. A. 1992. Phylogeny and biodiversity: Conserving our evolutionary legacy. *Trends Ecol. Evol.* 7:55–59.

Brooks, T., and Balmford, A. 1996. Atlantic Forest extinctions. *Nature* (Lond.) 380:115.

Brooks, T. M.; Pimm, S. L.; and Collar, N.J. 1997. Deforestation predicts the number of threatened birds in insular Southeast Asia. *Biol. Cons.* 11:382–394.

Brown, J. H. 1984. On the relationship between abundance and distribution of species. *Am. Nat.* 124:255–279.

————. 1995. *Macroecology*. Chicago: University of Chicago Press.

Brown, J. H., and Kodric-Brown, A. 1977. Turnover rates in insular biogeography: Effect of immigration on extinction. *Ecology* 58:445–449.

Brown, J. H.; Mehlman, D. W.; and Stevens, G. C. 1995. Spatial variation in abundance. *Ecology* 76:2028–2043.

Brown, K. S., and Brown, G. G. 1992. Habitat alteration and species loss in Brazilian forests. In *Tropical deforestation and species extinction,* ed. T. C. Whitmore and J. A. Sayer, 119–142. London: Chapman and Hall.

Brugiére, D., and Gautier, J. P. 1999. Status and conservation of the sun-tailed guenon *Cercopithecus solatus,* Gabon's endemic monkey. *Oryx* 33:67–74.

Bshary, R., and Noë, R. 1997. Anti-predation behaviour of red colobus monkeys in the presence of chimpanzees. *Behav. Ecol. Sociobiol.* 41:321–333.

Buckland, S. T.; Anderson, D. R.; Burnham, K. P.; and Laake, J. L. 1993. *Distance sampling.* London: Chapman and Hall.

Burgman, M.; Fearson, S.; and Akçakaya, H. R. 1993. *Risk assessment in conservation biology.* London: Chapman and Hall.

Burkey, T. V. 1989. Extinction in nature reserves: The effect of fragmentation and the importance of migration between reserve fragments. *Oikos* 55:75–81.

Busse, C. 1977. Chimpanzee predation as a possible factor in the evolution of red colobus monkey social organisation. *Evolution* 31:907–911.

Butynski, T. M. 1988. Guenon birth seasons and correlates with annual rainfall and food. In *A primate radiation: Evolutionary biology of the African guenons,* ed. A. Gautier-Hion, F. Bourlière, J.-P. Gautier, and J. Kingdon, 284–322. Cambridge: Cambridge University Press.

———. 1989. Comparative ecology of blue monkeys (*Cercopithecus mitis*) in high- and low-density subpopulations. *Ecol. Monog.* 60:1–26.

———. 1995. Primates and hydropower. *Swara,* September–October, 28–30.

———. 1996. International trade in CITES Appendix II African primates. *Afr. Primates* 2:5–9.

Butynski, T. M., and Kalina, J. 1998. Gorilla tourism: A critical look. In *Conservation of biological resources,* ed. E. J. Millner-Gulland and R. Mace, 294–313. Oxford: Blackwells.

Butynski, T. M., and Koster, S. H. 1994. Distribution and conservation status of primates in Bioko Island, Equatorial Guinea. *Biodiversity Cons.* 3:893–909.

Butynski, T. M., and Mwangi, G. 1994. Conservation status and distribution of the Tana River red colobus and crested mangabey. Report to the Kenya Wildlife Service.

Butynski, T. M.; Werikhe, S. E.; and Kalina, J. 1990. Status, distribution and conservation of the mountain gorilla in the Gorilla Game Reserve, Uganda. *Primate Cons.* 11:31–41.

Byrne, R. W.; Whiten, A.; Henzi, S. P.; and McCullogh, F. M. 1993. Nutritional constraints on mountain baboons (*Papio ursinus*): Implications for baboon socioecology. *Behav. Ecol. Sociobiol.* 33:233–246.

Caldecott, J. O. 1980. Habitat quality and populations of two sympatric gibbons (Hylobatidae) on a mountain in Malaya. *Folia Primatol.* 33:291–309.

Caldecott, J. O., and Kavanagh, M. 1983. Can translocation help wild primates? *Oryx* 17:135–139.

Cant, J. G. 1980. What limits primates? *Primates* 21:538–544.

Caro, T. M., and Laurenson, M. K. 1994. Ecological and genetic factors in conservation: A cautionary tale. *Science* 263:485–486.

Caro, T. M.; Pelkey, N.; Borner, M.; Campbell, K. L. I.; Woodworth, B. L.; Farm, B. P.; Kuwai, J. O.; Huish, S. A.; and Severre, E. L. M. 1998. Consequences of different forms of conservation for large mammals in Tanzania: Preliminary analyses. *Afr. J. Ecol.* 36:303–320.

Cartelle, C., and Hartwig, W. C. 1996. A new extinct primate among the Pleistocene megafauna of Bahia, Brazil. *Proc. Natl. Acad. Sci. USA* 93:6405–6409.

Carthew, S. M., and Goldingay, R. L. 1997. Non-flying mammals as pollinators. *Trends Ecol. Evol.* 12:104–108.

Caughley, G. 1977. *Analysis of vertebrate populations.* Chichester, Eng.: Wiley.

———. 1993. Elephants and economics. *Cons. Biol.* 7:943–945.

———. 1994. Directions in conservation biology. *J. Anim. Ecol.* 63:215–244.

Caughley, G., and Gunn, A. 1996. *Conservation biology in theory and practice.* Oxford: Blackwell.

Caughley, G., and Sinclair, A. R. E. 1994. *Wildlife management and ecology.* Oxford: Blackwell Science.

Cawthorne, R. A., and Marchant, J. H. 1980. The effects of the 1978/79 winter on British bird populations. *Bird Study* 27:163–172.

Ceballos, G., and Brown, J. H. 1995. Global patterns of mammalian diversity, endemism, and endangerment. *Cons. Biol.* 9:559–568.

Chalmers, N. R. 1968. Group composition, ecology and daily activities of free living mangabeys in Uganda. *Folia Primatol.* 8:247–262.

Chapman, C. A. 1987a. Flexability in diets of three species of Costa Rican primates. *Folia Primatol.* 49:90–105.

———. 1987b. Selection of secondary growth areas by vervet monkeys (*Cercopithecus aethiops*). *Am. J. Primatol.* 12:217–221.

———. 1989. Primate seed dispersal: The fate of dispersed seeds. *Biotropica* 21:148–154.

———. 1995. Primate seed dispersal: Coevolution and conservation implications. *Evol. Anthropol.* 4:74–82.

Chapman, C. A.; Chapman, L.; and Glander, K. E. 1989. Primate populations in northwestern Costa Rica: Potential for recovery. *Primate Cons.* 10:37–44.

Chapman, C. A.; Chapman, L. J.; Kaufman, L.; and Zanne, A. E. 1999. Potential causes of arrested succession in Kibale National Park, Uganda: Growth and mortality of seedlings. *Afr. J. Ecol.* 37:81–92.

Chapman, C. A., and Onderdonk, D. 1998. Forests without primates: Primate/plant codependency. *Am. J. Primatol.* 45:127–141.

Chapman, C. A.; Wrangham, R. W.; and Chapman, L. J. 1995. Ecological constraints on group size: An analysis of spider monkey and chimpanzee subgroups. *Behav. Ecol. Sociobiol.* 36:59–70.

Charles-Dominique, P.; Cooper, H. M.; Hladik, A.; Hladik, C. M.; Pages, E.; Pariente, G. F.; Petter-Rousseaux, A.; Petter, J. J.; and Schilling, A. 1980. *Nocturnal Malagasy primates: Ecology, physiology and behaviour.* New York: Academic Press.

Charnley, S. 1997. Environmentally-displaced peoples and the cascade effect: Lessons from Tanzania. *Human Ecol.* 25:593–618.

Charnov, E. L. 1991. Evolution of life history variation among female mammals. *Proc. Natl. Acad. Sci. USA* 88:1134–1137.

———. 1993. *Life history invariants.* Oxford: Oxford University Press.

Charnov, E. L., and Berrigan, D. 1993. Why do female primates have such long life-spans and so few babies? Or life in the slow lane. *Evol. Anthropol.* 2:191–194.

Chatelain, C.; Gautier, L.; and Spichiger, R. 1996. A recent history of forest fragmentation in southwestern Ivory Coast. *Biodiversity Cons.* 5:37–53.

Cheney, D. L.; Lee, P. C.; and Seyfarth, R. M. 1981. Behavioural correlates of non-random mortality among free-ranging female vervet monkeys. *Behav. Ecol. Sociobiol.* 9:153–161.

Cheney, D. L., and Seyfarth, R. M. 1983. Nonrandom dispersal in free-ranging vervet monkeys: Social and genetic consequences. *Am. Nat.* 122:392–412.

———. 1990. *How monkeys see the world.* Chicago: University of Chicago Press.

Cheney, D. L.; Seyfarth, R. M.; Andelman, S. J.; and Lee, P. C. 1986. Factors affecting reproductive success in vervet monkeys. In *Reproductive success,* ed. T. H. Clutton-Brock, 384–418. Cambridge: Cambridge University Press.

Cheney, D. L., and Wrangham, R. W. 1987. Predation. In *Primate societies,* ed. B. B. Smuts, D. L. Cheney, R. M. Seyfarth, R. W. Wrangham, and T. T. Struhsaker, 227–239. Chicago: University of Chicago Press.

Chepko-Sade, B. D., and Olivier, T. J. 1979. Coefficient of genetic relationship and the probability of intra-genealogical fission in *Macaca mulatta. Behav. Ecol. Sociobiol.* 5:263–278.

Chepko-Sade, B. D., and Sade, D. S. 1979. Patterns of group-splitting within matrilineal kinship groups. *Behav. Ecol. Sociobiol.* 5:67–86.

Chesser, R. K. 1991. Gene diversity and female philopatry. *Genetics* 127:437–447.

Chiarello, A. G., and Galetti, M. 1994. Conservation of the brown howler monkey in south-east Brazil. *Oryx* 28:37–42.

Chism, J.; Rowell, T.; and Olson, D. 1984. Life history patterns of female patas monkeys. In *Female primates: Studies by women primatologists,* ed. M. Small, 175–190. New York: Alan R. Liss.

Chivers, D. J. 1977. The lesser apes. In *Primate conservation,* ed. Prince Rainier III and G. H. Bourne, 539–598. New York: Academic Press.

———. 1984. Feeding and ranging in gibbons: A summary. In *The lesser apes,* ed. H. Preuschoft, D. J. Chivers, W. Y. Brockelman, and N. Creel, 267–284. Edinburgh: Edinburgh University Press.

———. 1994. Functional anatomy of the gastrointestinal tract. In *Colobine monkeys: Their ecology, behaviour and evolution,* ed. A. G. Davies and J. F. Oates, 205–228. Cambridge: Cambridge University Press.

Clark, A. B. 1978. Sex ratio and local resource competition in a prosimian primate. *Science* 201:163–165.

Clutton-Brock, T. H. 1989. Mammalian mating systems. *Proc. Roy. Soc. Lond.,* ser. B, 236:339–372.

Clutton-Brock, T. H., and Harvey, P. H. 1977. Species differences in feeding and ranging behaviour in primates. In *Primate ecology,* ed. T. H. Clutton-Brock, 557–584. London: Academic Press.

Cochrane, M. A., and Schulze, M. D. 1998. Forest fires in the Brazilian Amazon. *Cons. Biol.* 12:948–950.

Coe, M. J.; Cummings, D. H.; and Phillipson, J. 1976. Biomass and production of large African herbivores in relation to rainfall and primary production. *Oecologia* 22:341–354.

Coe, M. J., and Isaac, F. M. 1965. Pollination of the baobab (*Adansonia digitata* L.) by the lesser bushbaby (*Galago crassicaudatus* E. Geoffroy). *E. Afr. Wildl. J.* 3: 123–124.

Coelho, A. M., Jr.; Bramblett, C. A.; and Quick, L. B. 1977. Resource availability and population density in primates: A socio-bioenergetic analysis of diet and disease hypothesis. *Am. J. Phys. Anthropol.* 46:253–264.

Coelho, A. M., Jr.; Bramblett, C. A.; Quick, L. B.; and Bramblett, S. S. 1976. Resource availability and population density in primates: Socio-bioenergetic analysis of energy budgets of Guatemalan howler and spider monkeys. *Primates* 17:63–80.

Coimbra-Filho, A. F., and Mittermeier, R. A. 1977. Conservation of the Brazilian lion tamarins (*Leontopithecus rosalia*). In *Primate conservation*, ed. Prince Rainier III and G. H. Bourne, 59–95. New York: Academic Press.

Colchester, M. 1998. Who will garrison the fortress? A reply to Spinage. *Oryx* 32:245–248.

Cole, L. C. 1954. The population consequences of life history phenomena. *Quart. Rev. Biol.* 29:103–137.

Cole, F. R.; Reeder, D. M.; and Wilson, D. E. 1994. A synopsis of distribution patterns and the conservation of mammal species. *J. Mammal.* 75:266–276.

Colell, M.; Maté, C.; and Fa, J. E. 1994. Hunting among Moka Bubis in Bioko: Dynamics of faunal exploitation at the village level. *Biodiversity Cons.* 3:939–950.

Collias, N., and Southwick, C. 1952. A field study of population density and social organisation in howling monkeys. *Proc. Am. Philos. Soc.* 96:143–156.

Colvin, J. 1983. Influences of the social situation on male emigration. In *Primate social relationships*, ed. R. A. Hinde, 190–199. Oxford: Blackwell.

Colyn, M. M. 1988. Distribution of guenons in the Zaïre-Luabala-Lomani river system. In *A primate radiation: Evolutionary biology of the African guenons*, ed. A. Gautier-Hion, F. Bourlière, J.-P. Gautier, and J. Kingdon, 104–124. Cambridge: Cambridge University Press.

Connell, J. H., and Lowman, M. D. 1989. Low diversity tropical rainforests: Some possible mechanisms for their existence. *Am. Nat.* 134:88–119.

Conroy, G. C. 1990. *Primate evolution.* New York: Norton.

Conway, W. G. 1980. An overview of captive propagation. In *Conservation biology: An evolutionary-ecological perspective*, ed. M. E. Soulé and B. A. Wilcox, 199–208. Sunderland, Mass.: Sinauer.

Cook, R. R., and Hanski, I. 1995. On expected lifetimes of small-bodied and large-bodied species of birds on islands. *Am. Nat.* 145:307–315.

Corbet, G. B., and Hill, J. E. 1991. *A world list of mammalian species.* Oxford: Oxford University Press.

Cords, M. 1986. Interspecific and intraspecific variation in diet of two forest guenons, *Cercopithecus ascanius* and *C. mitis. J. Anim. Ecol.* 55:811–827.

———. 1987. Forest guenons and patas monkeys: Male-male competition in one-male groups. In *Primate societies*, ed. B. B. Smuts, D. L. Cheney, R. M. Seyfarth, R. W. Wrangham, and T. T. Struhsaker, 98–111. Chicago: University of Chicago Press.

———. 1990. Vigilance and mixed-species association of some East African forest monkeys. *Behav. Ecol. Sociobiol.* 26:297–300.

Cords, M., and Rowell, T. E. 1986. Group fission in blue monkeys of the Kakamega Forest, Kenya. *Folia Primatol.* 46:70–82.

Corlett, R. T., and Lucas, P. W. 1990. Alternative seed-handling strategies in primates: Seed-spitting by long-tailed macaques (*Macaca fascicularis*). *Oecologia* 82:166–171.

Corrêa, H. K. M., and Coutinho, P. E. G. 1997. Fatal attack of a pit viper, *Bothrops jararaca*, on an infant buffy tufted-ear marmoset (*Callithrix aurita*). *Primates* 38:215–216.

Cowlishaw, G. 1992. Song function in gibbons. *Behaviour* 121:131–153.

———. 1994. Vulnerability to predation in baboon populations. *Behaviour* 131: 293–304.

———. 1996. Sexual selection and information content in gibbon song bouts. *Ethology* 102:272–284.

———. 1997a. Trade-offs between foraging and predation risk in habitat use in a desert baboon population. *Anim. Behav.* 53:667–686.

———. 1997b. Refuge use and predation risk in a desert baboon population. *Anim. Behav.* 54:241–53.

———. 1998. The role of vigilance in the survival and reproductive strategies of a desert baboon population. *Behaviour* 135:431–452.

———. 1999. Predicting the decline of African primate diversity: An extinction debt from historical deforestation. *Cons. Biol.* 13:1183–93.

———. n d. A preliminary analysis of the metapopulation dynamics of the Tana River primates. Unpublished MS.

Cowlishaw, G., and Dunbar, R. I. M. 1991. Dominance rank and mating success in male primates. *Anim. Behav.* 41:1045–1056.

Cowlishaw, G., and Hacker, J. E. 1997. Distribution, diversity and latitude in African primates. *Am. Nat.* 150:505–512.

Crockett, C. M. 1998. Conservation biology of the genus *Alouatta. Int. J. Primatol.* 19:549–578.

Crockett, C. M., and Eisenberg, J. F. 1987. Howlers: Variations in group size and demography. In *Primate societies,* ed. B. B. Smuts, D. L. Cheney, R. M. Seyfarth, R. W. Wrangham, and T. T. Struhsaker, 54–68. Chicago: University of Chicago Press.

Crockett, C. M.; Kyes, R. C.; and Sajuthi, D. 1996. Modeling managed monkey populations: Sustainable harvest of longtailed macaques on a natural habitat island. *Am. J. Primatol.* 40:343–360.

Cronin, J. E., and Meikle, W. E. 1982. Hominid and gelada baboon evolution: Agreement between molecular and fossil time scales. *Int. J. Primatol.* 3:469–482.

Cronin, J. E., and Sarich, V. M. 1976. Molecular evidence for the dual origin of mangabeys among Old World monkeys. *Nature* (Lond.) 260:700–702.

Cunningham, A. A. 1996. Disease risks of wildlife translocations. *Cons. Biol.* 10: 349–353.

Damuth, J. 1981. Population density and body size in mammals. *Nature* 290:699–700.

———. 1991. Of size and abundance. *Nature* 351:268–269.

Dasilva, G. L. 1992. The western black-and-white colobus as a low-energy strategist: Activity budgets, energy expenditure and energy intake. *J. Anim. Ecol.* 61:79–91.

Datta, S. B. 1989. Demographic influences on dominance structure among female primates. In *Comparative socioecology,* ed. V. Standen and R. Foley, 265–284. Oxford: Blackwell.

Davies, A. G. 1987. Conservation of primates in the Gola Forest Reserves, Sierra Leone. *Primate Cons.* 8:151–153.

———. 1994. Colobine populations. In *Colobine monkeys: Their ecology, behaviour and evolution,* ed. A. G. Davies and J. F. Oates, 285–310. Cambridge: Cambridge University Press.

Davies, A. G.; Bennett, E. L.; and Waterman, P. G. 1988. Food selection by two south-east Asian colobine monkeys (*Presbytis rubicunda* and *Presbytis melalophos*) in relation to plant chemistry. *Biol. J. Linn. Soc.* 34:37–56.

Davies, J. G., and Cowlishaw, G. 1996. Interspecific competition between baboons and black kites following baboon-ungulate predation in the Namib Desert. *J. Arid Environ.* 34:247–249.

Davies, N. B. 1991. Mating systems. In *Behavioural ecology: An evolutionary approach*, ed. J. R. Krebs and N. B. Davies, 263–299. Oxford: Blackwell.

Defler, T. R., and Pintor, D. 1985. Censusing primates by transect in a forest of known primate density. *Int. J. Primatol.* 6:243–259.

Dei, G. J. S. 1989. Hunting and gathering in a Ghanaian rain forest community. *Ecology of Food and Nutrition* 22:225–243.

de Jong, G.; de Ruiter, J. R.; and Haring, R. 1994. Genetic structure of a population with social structure and migration. In *Conservation genetics*, ed. V. Loeschcke, J. Tomuk, and S. K. Jain, 147–164. Basel: Birkhauser.

Delson, E. 1985. Neogene African catarrhine primates: Climatic influence on evolutionary patterns. *S. Afr. J. Sci.* 81:273–274.

———. 1993. *Theropithecus* fossils from Africa and India and the taxonomy of the genus. In Theropithecus: *The rise and fall of a primate genus*, ed. N. Jablonski, 157–190. Cambridge: Cambridge University Press.

Delson, E., and Hoffstetter, R. 1993. *Theropithecus* from Ternifine, Algeria. In Theropithecus: *The rise and fall of a primate genus*, ed. N. Jablonski, 191–208. Cambridge: Cambridge University Press.

Demment, M. W. 1983. Feeding ecology and the evolution of body size of baboons. *Afr. J. Ecol.* 21:219–233.

de Ruiter, J. R., and Geffen, E. 1998. Relatedness of matrilines, dispersing males and social groups in long-tailed macaques (*Macaca fascicularis*). *Proc. Roy. Soc. Lond.*, ser. B, 265:79–87.

DeVore, I., and Hall, K. R. L. 1965. Baboon ecology. In *Primate behaviour*, ed. I. DeVore, 20–52. New York: Holt, Reinhart and Winston.

Dewar, R. E. 1984. Recent extinctions in Madagascar: The loss of subfossil fauna in Quartenary extinctions. In *Quartenary extinctions: A prehistoric revolution*, ed. P. S. Martin and R. G. Klein, 574–579. Tuscon: University of Arizona Press.

———. 1997. Were people responsible for the extinction of Madagascar's subfossils, and how will we ever know? In *Natural change and human impact in Madagascar*, ed. S. M. Goodman and B. D. Patterson, 364–377. Washington, D.C.: Smithsonian Institution Press.

Diamond, J. M. 1975. The island dilemma: Lessons of modern biogeographic studies for the design of natural preserves. *Biol. Cons.* 7:129–146.

———. 1984. "Normal" extinction of isolated populations. In *Extinctions*, ed. M. H. Nitecki, 191–246. Chicago: University of Chicago Press.

———. 1989. Overview of recent extinctions. In *Conservation biology for the twenty-first century*, ed. D. Western and M. Pearl, 37–41. New York: Oxford University Press.

Dias, P. C. 1996. Sources and sinks in population biology. *Trends Ecol. Evol.* 11:326–330.

Dietz, J. M.; Dietz, L. A.; and Nagagata, E. Y. 1994. The effective use of flagship species for conservation: The example of the lion tamarin. In *Creative conservation*, ed. P. J. S. Olney, G. M. Mace, and A. T. C. Feistner, 32–49. London: Chapman and Hall.

Di Fiore, A., and Randall, D. 1994. Evolution of social organisation: A reappraisal for primates by using phylogenetic methods. *Proc. Natl. Acad. Sci. USA* 91:9941–9945.

Disotell, T. R.; Honeycutt, R. L.; and Ruvulo, M. 1992. Mitochondrial DNA phylogeny of the Old World monkey tribe Papionini. *Mol. Biol. Evol.* 9:1–13.

Dittus, W. P. J. 1977. The social regulation of population density and age-sex distribution in the toque monkey. *Behaviour* 58:281–322.

———. 1985. The influence of cyclones on the dry evergreen forest of Sri Lanka. *Biotropica* 17:1–14.

———. 1986. Sex differences in fitness following a group take-over among toque monkeys: Testing models of social evolution. *Behav. Ecol. Sociobiol.* 19:257–266.

———. 1998. Birth sex ratios in toque macaques and other mammals: Integrating the effects of maternal condition and competition. *Behav. Ecol. Sociobiol.* 44:149–160.

Dobson, A. P., and Lyles, A. M. 1989. The population dynamics and conservation of primate populations. *Cons. Biol.* 3:362–380.

Dobson, F. S.; Smith, A. T.; and Yu, J. 1997. Static and temporal studies of rarity. *Cons. Biol.* 11:306–307.

Dobson, F. S., and Yu, J. 1993. Rarity in Neotropical forest mammals revisited. *Cons. Biol.* 7:586–591.

Dobson, F. S.; Yu, J.; and Smith, A. T. 1995. The importance of evaluating rarity. *Cons. Biol.* 9:1648–1651.

Doran, D. M., and McNeilage, A. 1998. Gorilla ecology and behaviour. *Evol. Anthropol.* 7:120–131.

Dracopoli, N. C.; Brett, F. L.; Turner, T. R.; and Jolly, C. J. 1983. Patterns of genetic variability in the serum proteins in the Kenyan vervet monkey (*Cercopithcus aethiops*). *Am. J. Phys. Anthropol.* 61:39–49.

Drickamer, L. C. 1974. A ten-year summary of reproductive data for free-ranging *Macaca mulatta*. *Folia Primatol.* 21:61–80.

Drucker, G. R. 1984. The feeding ecology of the Barbary macaque and cedar forest conservation in the Moroccan Moyen Atlas. In *The Barbary macaque: A case study in conservation*, ed. J. E. Fa, 135–164. New York: Plenum Press.

Duggleby, C. R. 1978. Rhesus of Cayo Santiago: Effective population size and migration rates. In *Recent advances in primatology*, vol. 1, *Behaviour*, ed. D. J. Chivers and J. Herbert, 189–191. London: Academic Press.

Dunbar, R. I. M. 1977. The gelada baboon: Status and conservation. In *Primate conservation*, ed. Prince Rainier III and G. H. Bourne, 363–383. New York: Academic Press.

———. 1978. Competition and niche separation in a high altitude herbivore community in Ethiopia. *E. Afr. Wildl. J.* 16:183–199.

———. 1979. Population demography, social organisation and mating strategies. In *Primate ecology and human origins*, ed. I. Bernstein and E. O. Smith, 65–88. New York: Garland.

———. 1980a. Causes and consequences of dominance in female gelada baboons. *Behav. Ecol. Sociobiol.* 7:253–265.

———. 1980b. Demographic and life history variables of a population of gelada baboons (*Theropithecus gelada*). *J. Anim. Ecol.* 49:485–506.

———. 1984. *Reproductive decisions: An economic analysis of gelada baboon social strategies.* Princeton: Princeton University Press.

———. 1987a. Habitat quality, population dynamics and group composition in colobus monkeys (*Colobus guereza*). *Int. J. Primatol.* 8:299–330.

———. 1987b. Demography and reproduction. In *Primate societies*, ed. B. B. Smuts, D. Cheney, R. Seyfarth, R. W. Wrangham, and T. T. Struhsaker, 240–249. Chicago: University of Chicago Press.

———. 1988. *Primate social systems.* London: Chapman and Hall.

———. 1990a. Environmental determinants of intraspecific variation in body weight in baboons (*Papio* spp.). *J. Zool., Lond.* 220:157–169.

———. 1990b. Environmental determinants of fecundity in klipspringer (*Oreotragus oreotragus*). *Afr. J. Ecol.* 28:307–313.

———. 1992a. Neocortex size as a constraint on group size in primates. *J. Human Evol.* 22:469–493.

———. 1992b. Time: A hidden constraint on the behavioural ecology of baboons. *Behav. Ecol. Sociobiol.* 31:35–49.

———. 1992c. A model of the gelada socioecological system. *Primates* 33:69–83.

———. 1992d. The behavioural ecology of the extinct papionids. *J. Human Evol.* 22:407–421.

———. 1993. Socioecology of the extinct theropiths: A modelling approach. In *Theropithecus: The rise and fall of a primate genus,* ed. N. G. Jablonski, 465–486. Cambridge: Cambridge University Press.

———. 1995a. The mating system of callitrichid primates. 1. Co-evolution of pair-bonding and twinning. *Anim. Behav.* 50:1057–1070.

———. 1995b. The mating system of callitrichid primates. II. The impact of helpers. *Anim. Behav.* 50:1071–1089.

———. 1996. Determinants of group size in primates: A general model. In *Evolution of social behaviour patterns in primates and man,* ed. G. Runciman, J. Maynard Smith, and R. Dunbar, 33–58. New York: Oxford University Press.

———. 1998a. The social brain hypothesis. *Evol. Anthropol.* 6:178–190.

———. 1998b. Impact of global warming on the distribution and survival of the gelada baboon: A modelling approach. *Global Change Biol.* 4:293–304.

———. 2000. Male mating strategies: A modelling approach. In *Primate males,* ed. P. M. Kappeler. Cambridge: Cambridge University Press.

Dunbar, R. I. M., and Bose, U. 1991. Adaptation to grass-eating in gelada baboons. *Primates* 32:1–7.

Dunbar, R. I. M., and Cowlishaw, G. 1992. Mating success in male primates: Dominance rank, sperm competition and alternative strategies. *Anim. Behav.* 44:1171–1173.

Dunbar, R. I. M., and Dunbar, E. P. 1974a. Ecological relations and niche separation between sympatric terrestrial primates in Ethiopia. *Folia Primatol.* 21:36–60.

———. 1974b. On hybridisation between *Theropithecus gelada* and *Papio anubis* in the wild. *J. Human Evol.* 3:187–192.

————. 1975. Guereza monkeys: Will they become extinct in Ethiopia? *Walia* 6:14–15.

Duncan, R. P. 1997. The role of competition and introduction effort in the success of passeriform birds introduced to New Zealand. *Am. Nat.* 149:903–915.

Durant, S. M., and Mace, G. M. 1994. Species differences and population structure in population viability analysis. In *Creative conservation: Interactive management of wild and captive animals,* ed. P. G. S. Olney, G. M. Mace, and A. T. C. Feistner, 67–91. London: Chapman and Hall.

Eames, J. C., and Robson, C. R. 1993. Threatened primates in southern Vietnam. *Oryx* 27:146–154.

Ebenhard, T. 1995. Conservation breeding as a tool for saving animal species from extinction. *Trends. Ecol. Evol.* 10:438–443.

Eck, G. G.; Jablonski, N. G.; and Leakey, M. 1989. *Les faunes Plio-Pléistocènes de la vallée de l'Omo (Éthiopie).* Vol. 3. *Cercopithecidae de la formation de Shungura.* Paris: CNRS.

Eeley, H. A. C., and Foley, R. A. 1999. Species richness, species range size and ecological specialization among African primates: Geographical patterns and conservation implications. *Biodiversity Cons.* 8:1033–1056.

Eisenberg, J. F.; O'Connell, M. A.; and August, P. V. 1979. Density, productivity and distribution of mammals in two Venezuelan habitats. In *Vertebrate ecology in the northern Neotropics,* ed. J. F. Eisenberg, 187–207. Washington D.C.: Smithsonian Institution Press.

Eisenberg, J. F., and Thorington, R. 1973. A preliminary analysis of a Neotropical mammal fauna. *Biotropica* 5:150–161.

Eley, R. M.; Strum, S. C.; Muchemi, G.; and Reid, G. D. F. 1989. Nutrition, body condition, activity patterns and parasitism of free-ranging troops of olive baboons (*Papio anubis*) in Kenya. *Am. J. Primatol.* 18:209–219.

Else, J. G. 1991. Nonhuman primates as pests. In *Primate responses to environmental change,* ed. H. Box, 155–165. London: Chapman and Hall.

Emmons, L. H. 1984. Geographic variation in densities and diversities of non-flying mammals in Amazonia. *Biotropica* 16:210–222.

————. 1987. Comparative feeding ecology of felids in a Neotropical forest. *Behav. Ecol. Sociobiol.* 20:271–283.

Emmons, L. H.; Gautier-Hion, A.; and Dubost, G. 1983. Community structure of the frugivorous-folivorous forest mammals of Gabon. *J. Zool., Lond.* 199:209–222.

Endler, J. A. 1982. Pleistocene forest refuges: Fact or fancy? In *Biological diversification in the tropics,* ed. G. T. Prance, 641–657. New York: Columbia University Press.

Environmental Investigation Agency. 1998. *The politics of extinction.* London: Emmerson Press.

Erwin, T. L. 1991. An evolutionary basis for conservation strategies. *Science* 253:750–752.

Estrada, A., and Coates-Estrada, R. 1985. A preliminary study of resource overlap between howling monkeys (*Alouatta palliata*) and other arboreal mammals in the tropical rain forest of Los Tuxtlas, Mexico. *Am. J. Primatol.* 9:27–37.

————. 1991. Howler monkeys (*Alouatta palliata*), dung beetles (Scarabaeidae) and seed dispersal: Ecological interactions in the tropical rain forest of Los Tuxtlas, Mexico. *J. Trop. Ecol.* 7:459–474.

———. 1996. Tropical rain forest fragmentation and wild populations of primates at Los Tuxtlas, Mexico. *Int. J. Primatol.* 17:759–783.

Estrada, A.; Coates-Estrada, R.; and Meritt, D. 1994. Non-flying mammals and landscape changes in the tropical rain forest region of Los Tuxtlas, Mexico. *Ecography* 17:229–241.

Estrada, A.; Coates-Estrada, R.; Meritt, D.; Montiel, S.; and Curiel, D. 1993. Patterns of frugivore species richness and abundance in forest islands and in agricultural habitats at Los Tuxtlas, Mexico. *Vegetatio* 107–108:245–257.

Eudey, A. A. 1987. *Action Plan for Asian primate conservation: 1987–1991.* Gland, Switz.: IUCN.

Fa, J. E. 1986. *Use of time and resources by provisioned troops of monkeys.* Basel: Karger.

———. 1987. Balancing the wild/captive equation—The case of the Barbary macaque (*Macaca sylvanus* L.). In *Primates: The road to self-sustaining populations,* ed. K. Benirschke, 197–212. New York: Springer.

———. 1994. Herbivore intake/habitat productivity correlations can help ascertain re-introduction potential for the Barbary macaque. *Biodiversity Cons.* 3:309–317.

———. 1999. Hunted animals in Bioko Island, West Africa: Sustainability and future. In *Hunting for sustainability in tropical forests,* ed. J. Robinson and E. Bennett, 165–95. New York: Columbia University Press. In press.

Fa, J. E.; Juste, J.; Del Val, J. P.; and Castroviejo, J. 1995. Impact of market hunting on mammal species in Equatorial Guinea. *Cons. Biol.* 9:1107–1115.

Fa, J. E., and Lind, R. 1996. Population management and viability of the Gibraltar Barbary macaques. In *Evolution and ecology of macaque societies,* ed. J. E. Fa and D. G. Lindburg, 235–262. Cambridge: Cambridge University Press.

Fa, J. E., and Purvis, A. 1997. Body size, diet and population density in Afrotropical forest mammals: A comparison with Neotropical species. *J. Anim. Ecol.* 66:98–112.

Fa, J. E.; Taub, D. M.; Ménard, N.; and Stewart, P. J. 1984. The distribution and current status of the Barbary macaque in North Africa. In *The Barbary macaque: A case study in conservation,* ed. J. E. Fa, 79–111. New York: Plenum Press.

Fa, J. E.; Juste, J. E. G.; and Castelo, R. n.d. Bushmeat markets on Bioko Island as "hunting barometers": A six-year comparison. *Cons. Biol.* In press.

Fairhead, J., and Leach, M. 1996. *Misreading the African landscape: Society and ecology in a forest-savanna mosaic.* Cambridge: Cambridge University Press.

———. 1998. *Reframing deforestation.* London: Routledge.

Fargey, P. J. 1992. Boabeng-Feima Monkey Sanctuary—An example of traditional conservation in Ghana. *Oryx* 26:151–156.

Faust, T.; Tilson, R.; and Seal, U.S. 1995. Using GIS to evaluate habitat risk to wild populations of Sumatran orangutans. In *The neglected ape,* ed. R. D. Nadler, B. F. M. Galdikas, L. K. Sheeran, and N. Rosen, 85–96. New York: Plenum Press.

Fay, J. M.; Carroll, R.; Peterhans, J. C. K.;and Harris, D. 1995. Leopard attack on and consumption of gorillas in the Central African Republic. *J. Human Evol.* 29:93–99.

Ferrari, S. F., and Diego, V. H. 1995. Habitat fragmentation and primate conservation in the Atlantic Forest of eastern Minas Gerais, Brazil. *Oryx* 29:192–196.

Ferrari, S. F., and Lopes Ferrari, M. A. 1989. A re-evaluation of the social organisation of the Callitrichidae, with reference to the ecological differences between genera. *Folia Primatol.* 52:132–147.

Ferrari, S. F., and Queiroz, H. L. 1994. Two new Brazilian primates discovered, endangered. *Oryx* 28:31–36.

Ferrari, S. F., and Strier, K. B. 1992. Exploitation of *Mabea fistulifera* nectar by marmosets (*Callithrix flaviceps*) and muriquis (*Brachyteles arachnoides*) in south-east Brazil. *J. Trop. Ecol.* 8:225–239.

Fimbel, C. 1994a. The relative use of abandoned farm clearings and old forest habitats by primates and a forest antelope at Tiwai, Sierra Leone, West Africa. *Biol. Cons.* 70:277–286.

———. 1994b. Ecological correlates of species success in modified habitats may be disturbance-and site-specific: The primates of Tiwai Island. *Cons. Biol.* 8:106–113.

Fimbel, C., and Fimbel, R. 1997. Rwanda: The role of local participation. *Cons. Biol.* 11:309–310.

Fitzgibbon, C. D.; Mogaka, H.; and Fanshawe, J. H. 1995. Subsistence hunting in Arabuko-Sokoke Forest, Kenya, and its effects on mammal populations. *Cons. Biol.* 9:1116–1126.

Fleagle, J. G. 1978. Locomotion, posture and habitat use of two sympatric leaf-monkeys in West Malaysia. In *Recent advances in primatology,* vol. 1, *Behaviour,* ed. D. J. Chivers and J. Herbert, 331–336. London: Academic Press.

———. 1999. *Primate adaptation and evolution.* 2d ed. New York: Academic Press.

Fleagle, J. G.; Janson, C.; and Reed, K. E. 1999. *Primate communities.* Cambridge: Cambridge University Press.

Fleagle, J. G., and Reed, K. E. 1996. Comparing primate communities: A multivariate approach. *J. Human Evol.* 30:489–510.

Flesness, N. R. 1987. Captive status and genetic considerations. In *Primates: The road to self-sustaining populations,* ed. K. Benirschke, 845–856. New York: Springer.

Foley, P. 1994. Predicting extinction times from environmental stochasticity and carrying capacity. *Cons. Biol.* 8:124–137.

Foley, R. A. 1993. African terrestrial primates: The comparative evolutionary biology of *Theropithecus* and the Hominidae. In Theropithecus: *The rise and fall of a primate genus,* ed. N. G. Jablonski, 245–270. Cambridge: Cambridge University Press.

———. 1994. Speciation, extinction and climatic change in hominid evolution. *J. Human Evol.* 26:275–289.

Foose, T. J.; Seal, U. S.; and Flesness, N. R. 1987. Captive propogation as a component of conservation strategies for endangered primates. In *Primate conservation in the tropical rain forest,* ed. C. W. Marsh and R. A. Mittermeier, 263–299. New York: Alan R. Liss.

Foster, R. B. 1980. Heterogeneity and disturbance in tropical vegetation. In *Conservation biology: An evolutionary-ecological perspective,* ed. M. E. Soulé and B. A. Wilcox, 75–92. Sunderland, Mass.: Sinauer.

Fox, B. J. 1987. Species assembly and the evolution of community structure. *Evol. Ecol.* 1:201–213.

Frankham, R., and Franklin, I. R. 1998. Response to Lynch and Lande. *Animal Cons.* 1:73.

Franklin, I. R. 1980. Evolutionary change in small populations. In *Conservation biology: An evolutionary-ecological perspective,* ed. M. E. Soulé and B. A. Wilcox, 135–149. Sunderland, Mass.: Sinauer.

Franklin, I. R., and Frankham, R. 1998. How large must populations be to retain evolutionary potential? *Animal Cons.* 1:69–70.

Freese, C. H.; Heltne, P. G.; Castro R. N.; and Whitesides, G. 1982. Patterns and determinants of monkey densities in Peru and Bolivia, with notes on distributions. *Int. J. Primatol.* 3:53–90.

Froelich, J. W., and Froelich, P. H. 1987. The status of Panama's endemic howling monkeys. *Primate Cons.* 8:58–62.

Gadsby, E. L. 1990. The status and distribution of the drill, *Mandrillus leucophaeus,* in Nigeria. Unpublished report of the Nigerian Government, WCI and WWF-US.

Gadsby, E. L.; Feistner, A. T. C.; and Jenkins, P. D. 1994. Coordinating conservation for the drill (*Mandrillus leucophaeus*): Endangered in forest and zoo. In *Creative conservation,* ed. P. J. S. Olney, G. M. Mace, and A. T. C. Feistner, 439–454. London: Chapman and Hall.

Gagneux, P.; Boesch, C.; and Woodruff, D. S. 1999. Female reproductive strategies, paternity and community structure in wild West African chimpanzees. *Anim. Behav.* 57:19–32.

Ganzhorn, J. U. 1987. A possible role of plantations for primate conservation in Madagascar. *Am. J. Primatol.* 12:205–215.

———. 1988. Food partitioning among Malagasy primates. *Oecologia* 75:436–450.

———. 1989. Niche separation of seven lemur species in the eastern rainforest of Madagascar. *Oecologia* 79:279–286.

———. 1992. Leaf chemistry and the biomass of folivorous primates in tropical forests: Tests of a hypothesis. *Oecologia* 91:540–547.

———. 1994. Lemurs as indicators for habitat change. In *Current primatology,* vol. 1, *Ecology and evolution,* ed. B. Thierry, J. R. Anderson, J. J. Roeder, and N. Herrenschmidt, 51–56. Strasbourg: Université Louis Pasteur.

———. 1995. Low-level forest disturbance effects on primary production, leaf chemistry, and lemur populations. *Ecology* 76:2084–2096.

———. 1997. Test of Fox's assembly rule for functional groups in lemur communities in Madagascar. *J. Zool., Lond.* 241:533–542.

Ganzhorn, J. U., and Abraham, J. P. 1991. Possible role of plantations for lemur conservation in Madagascar: Food for folivorous species. *Folia Primatol.* 56:171–176.

Ganzhorn, J. U., and Schmid, J. 1998. Different population dynamics of *Microcebus murinus* in primary and secondary deciduous dry forests of Madagascar. *Int. J. Primatol.* 19:785–796.

Garber, P. A. 1986. The ecology of seed dispersal in two species of callitrichid primates (*Saguinus mystax* and *Saguinus fuscicollis*). *Am. J. Primatol.* 10:155–170.

Gartlan, J. S. 1970. Preliminary notes on the behaviour and ecology of the drill, *Mandrillus leucophaeus* Ritgen 1824. In *Old World monkeys,* ed. J. H. Napier and P. Napier, 445–480. New York: Academic Press.

Gartlan, J. S., and Struhsaker, T. T. 1972. Polyspecific associations and niche separation of rain-forest anthropoids in Cameroon, West Africa. *J. Zool., Lond.* 168: 221–266.

Gaston, K. J. 1994. *Rarity.* London: Chapman and Hall.

Gaston, K. J., and Blackburn, T. M. 1995. Rarity and body size: Importance of generality. *Cons. Biol.* 10:1295–1298.

———. 1996a. Rarity and body size: Importance of generality. *Cons. Biol.* 10:1295–1298.

———. 1996b. Conservation implications of geographic range size-body size relationships. *Cons. Biol.* 10:638–646.

Gaston, K. J.; Blackburn, T. M.; and Lawton, J. H. 1997. Interspecific abundance-range size relationships: An appraisal of mechanisms. *J. Anim. Ecol.* 66:579–601.

Gates, D. M. 1993. *Climate change and its biological consequences.* Sunderland, Mass.: Sinauer.

Gautier-Hion, A. 1970. L'écologie du talapoin du Gabon. *Terre et Vie* 25:427–490.

———. 1983. Leaf consumption by monkeys in western and eastern Africa: A comparison. *Afr. J. Ecol.* 21:107–113.

———. 1984. La dissemination des graines par les cercopithecides forestiers africains. *Revue Ecol.* 39:159–165.

Gautier-Hion, A.; Duplantier, J.-M.; Quris, R.; Feer, F.; Sourd, C.; Decoux, J.-P.; Dubost, G.; Emmons, L.; Erard, C.; Hecketsweiler, P.; Roussilhon, C.; and Thiollay, J.-M. 1985. Fruit characters as a basis of fruit choice and seed dispersal in a tropical forest vertebrate community. *Oecologia* 65:324–337.

Gautier-Hion, A.; Gautier, J.-P.; and Maisels, F. 1993. Seed dispersal versus seed predation: An inter-site comparison of two related African monkeys. *Vegetatio* 107–108:234–244.

Gautier-Hion, A., and Michaloud, G. 1989. Are figs always keystone resources for tropical frugivorous vertebrates? A test in Gabon. *Ecology* 70:1826–1833.

Gautier-Hion, A.; Quris, R.; and Gautier, J.-P. 1983. Monospecific vs. polyspecific life: A comparative study of foraging and antipredatory tactics in a community of *Cercopithecus* monkeys. *Behav. Ecol. Sociobiol.* 12:325–335.

Ghiglieri, M. P. 1984. *The chimpanzees of Kibale Forest.* New York: Columbia University Press.

Giannecchini, J. 1993. Ecotourism: New partners, new relationships. *Cons. Biol.* 7:429–432.

Gibson, C. C., and Marks, S. A. 1995. Transforming rural hunters into conservationists: An assessment of community-based wildlife management programs in Africa. *World Development* 23:941–957.

Gilligan, D. M.; Woodworth, L. M.; Montgomery, M. E.; Briscoe, D. A.; and Frankham, R. 1997. Is mutation accumulation a threat to the survival of endangered populations? *Cons. Biol.* 11:1235–1241.

Ginsberg, J. R.; Mace, G. M.; and Albon, S. 1995. Local extinction in a small and declining population: Wild dogs in the Serengeti. *Proc. Roy. Soc. Lond.,* ser. B, 262:221–228.

Gittleman, J. L., and Purvis, A. 1998. Body size and species- richness in carnivores and primates. *Proc. Roy. Soc. Lond.,* ser. B, 265:113–119.

Glander, K. 1975. Habitat description and resource utilization: A preliminary report on mantled howling monkey ecology. In *Socioecology and psychology of primates,* ed. R. H. Tuttle, 37–57. The Hague: Mouton.

Glander, K. E.; Tapia R. J.; and Fachin, A. 1984. The impact of cropping on wild populations of *Saguinus mystax* and *Saguinus fuscicollis* in Peru. *Am. J. Primatol.* 7:89–97.

Glanz, W. E. 1990. Neotropical mammal densities: How unusual is the community on Barro Colorado Island, Panama? In *Four Neotropical forests,* ed. A. H. Gentry, 287–313. New Haven: Yale University Press.

———. 1991. Mammalian densities at protected versus hunted sites in central Panama. In *Neotropical wildlife use and conservation,* ed. J. G. Robinson and K. H. Redford, 163–173. Chicago: University of Chicago Press.

Godfrey, L. R.; Jungers, W. L.; Reed, K. E.; Simons, E. L.; and Chatrath, P. S. 1997. Inferences about past and present communities in Madagascar. In *Natural change and human impact in Madagascar,* ed. S. M. Goodman and B. D. Patterson, 218–256. Washington, D.C.: Smithsonian Institution Press.

Godoy, R.; Brokaw, N.; and Wilkie, D. 1995. The effect of income on the extraction of non-timber tropical forest products: Model, hypotheses, and preliminary findings from the Sumu Indians of Nicaragua. *Human Ecol.* 23:29–52.

Godoy, R.; Wilkie, D.; and Franks, J. 1997. The effects of markets on Neotropical deforestation: A comparative study of four Amerindian societies. *Current Anthropol.* 38:875–878.

Goldberg, T. L., and Wrangham, R. W. 1997. Genetic correlates of social behaviour in wild chimpanzees: Evidence from mitochondrial DNA. *Anim. Behav.* 45: 559–570.

Goldizen, A. W. 1987a. Tamarins and marmosets: Communal care of offspring. In *Primate societies,* ed. B. B. Smuts, D. Cheney, R. Seyfarth, R. Wrangham, and T. T. Struhsaker, 34–43. Chicago: University of Chicago Press.

———. 1987b. Facultative polyandry and the role of infant-carrying in wild saddleback tamarins (*Saguinus fuscicollis*). *Behav. Ecol. Sociobiol.* 20:89–109.

Goldizen, A. W.; Terborgh, J.; Cornejo, F.; Porras, D. T.; and Evans, R. 1988. Seasonal food shortage, weight loss, and the timing of births in saddle-back tamarins (*Saguinus fuscicollis*). *J. Anim. Ecol.* 57:893–901.

Goldstein, S. 1984. Ecology of rhesus monkeys, *Macaca mulatta,* in northern Pakistan. Ph.D. diss., Yale University.

Goodall, J. 1983. Population dynamics during a fifteen year period in one community of free-living chimpanzees in the Gombe National Park. *Z. Tierpsychol.* 61:1–60.

———. 1986. *The chimpanzees of Gombe: Patterns of behaviour.* Cambridge: Harvard University Press.

Goodman, D. 1987. The demography of chance extinction. In *Viable populations for conservation,* ed. M. E. Soulé, 11–34. Cambridge: Cambridge University Press.

Goodman, S. M.; O'Connor, S.; and Langrand, O. 1993. A review of predation on lemurs: Implications for the evolution of social behavior in small nocturnal primates. In *Lemur social systems and their ecological basis,* ed. P. M. Kappeler and J. U. Ganzhorn, 51–66. New York: Plenum Press.

Gould, L.; Sussman, R. W.; and Sauther, M. L. 1999. Natural disasters and primate populations: The effects of a two-year drought on a naturally occurring population of ring-tailed lemurs (*Lemur catta*) in southwestern Madagascar. *Int. J. Primatol.* 20:69–84.

Gray, A. P. 1972. *Mammalian hybrids: A checklist with bibliography.* 2d ed. Farnham, Eng.: Commonwealth Agricultural Bureau.

Green, G. M., and Sussman, R. W. 1990. Deforestation history of the eastern rain forests of Madagsascar from satellite images. *Science* 248:212–215.

Green, S., and Minowski, K. 1977. The lion-tailed macaque and its South Indian rain forest habitat. In *Primate conservation,* ed. Prince Rainier III and G. H. Bourne, 290–337. New York: Academic Press.

Grieser Johns, A. D., and Grieser Johns, B. G. 1995. Tropical forest primates and logging: Long-term coexistence? *Oryx* 29:205–211.

Grieser Johns, B. 1996. Responses of chimpanzees to habituation and tourism in the Kibale Forest, Uganda. *Biol. Cons.* 78:257–262.

———. 1997. *Timber production and biodiversity conservation in tropical rain forests.* Cambridge: Cambridge University Press.

Griffiths, B.; Scott, J. M.; Carpenter, J. W.; and Reed, C. 1989. Translocation as a species conservation tool: Status and strategy. *Science* 245:477–480.

Griffiths, M., and van Schaik, C. P. 1993. The impact of human traffic on the abundance and activity periods of Sumatran rain forest wildlife. *Cons. Biol.* 7:623–626.

Groombridge, B. 1992. *Global biodiversity: Status of the earth's living resources.* London: Chapman and Hall.

———. 1993. *1994 IUCN Red List of threatened animals.* Gland, Switz.: IUCN.

Groves, C. 1993. Order Primates. In *Mammal species of the world: A taxonomic and geographic reference,* ed. D. E. Wilson and D. M. Reeder, 243–277. Washington, D.C.: Smithsonian Institution Press.

Hacker, J. E.; Cowlishaw, G.; and Williams, P. H. 1998. Patterns of African primate diversity and their evaluation for the selection of conservation areas. *Biol. Cons.* 84:251–262.

Haffer, J. 1982. General aspects of the refuge theory. In *Biological diversification in the tropics,* ed. G. T. Prance. 6–24. New York: Columbia University Press.

Hall, J. S.; Saltonstall, K.; Inogwabini, B.-I.; and Omari, I. 1998. Distribution, abundance and conservation status of Grauer's gorilla. *Oryx* 32:122–130.

Hall, K. R. L. 1965. Behaviour and ecology of the wild patas monkey, *Erythrocebus patas,* in Uganda. *J. Zool., Lond.* 148:15–87.

Ham, R. 1998. Regional differences in hunting pressure on chimpanzees in the Republic of Guinea. Paper presented at the Bushmeat Hunting and African Primates Conference, Primate Society of Great Britain, Bristol.

Hamburg, S. P., and Cogbill, C. V. 1988. Historical decline of red spruce populations and climatic warming. *Nature* (Lond.) 331:428–431.

Hames, R. B. 1979. A comparison of the efficiencies of the shotgun and the bow in Neotropical forest hunting. *Human Ecol.* 7:219–252.

——— 1987. Game conservation or efficient hunting? In *The question of the commons,* ed. B. McCay and J. Acheson, 92–107. Tuscon: University of Arizona Press.

Hames, R. B., and Vickers, W. 1982. Optimal foraging theory as a model to explain variability in Amazonian hunters. *Am. Ethnol.* 9:358–378.

Hamilton, A. C. 1988. Guenon evolution and forest history. In *A primate radiation: Evolutionary biology of the African guenons,* ed. A. Gautier-Hion, F. Bourlière, J.-P. Gautier, and J. Kingdon, 13–34. Cambridge: Cambridge University Press.

Hamilton, W. J. 1982. Baboon sleeping site preferences and relationships to primate grouping patterns. *Am. J. Primatol.* 3:41–53.

———. 1986. Demographic consequences of a food and water shortage to desert chacma baboons, *Papio ursinus. Int. J. Primatol.* 6:451–462.

Hamilton, W. J.; Buskirk, R. E.; and Buskirk, W. H. 1976. Defence of space and resources by chacma (*Papio ursinus*) baboon troops in an African desert and swamp. *Ecology* 57:1264–1272.

Hamilton, W. J., and Tilson, R. L. 1982. Solitary male chacma baboons in a desert canyon. *Am. J. Primatol.* 2:149–158.

Hannah, A. C., and McGrew, W. C. 1991. Rehabilitation of captive chimpanzees. In *Primate responses to environmental change*, ed. H. O. Box, 167–186. London: Chapman and Hall.

Hanski, I. 1998. Metapopulation dynamics. *Nature* (Lond.) 396:41–49.

———. 1999. *Metapopulation ecology.* Oxford: Oxford University Press.

Happel, R. 1988. Seed-eating by West-African cercopithecines, with reference to the possible evolution of bilophodont molars. *Am. J. Primatol.* 75:303–327.

Happel, R. E.; Noss, J. F.; and Marsh, C. W. 1987. Distribution, abundance and endangerment of primates. In *Primate conservation in the tropical rain forest*, ed. C. W. Marsh and R. A. Mittermeier, 63–82. New York: Alan R. Liss.

Harcourt, A. H. 1980–1981. Can Uganda's gorillas survive?—A survey of the Bwindi Forest Reserve. *Biol. Cons.* 19:269–282.

———. 1986. Gorilla conservation: Anatomy of a campaign. In *Primates: The road to self-sustaining populations*, ed. K. Benirschke, 31–46. New York: Springer.

———. 1987. Dominance and fertility among female primates. *J. Zool., Lond.* 213:471–487.

———. 1989. Social influences on competitive ability: Alliances and their consequences. In *Comparative socioecology*, ed. V. Standed and R. Foley, 223–242. Oxford: Blackwell.

———. 1992. Coalitions and alliances: Are primates more complex than non-primates? In *Coalitions and alliances in humans and other animals*, ed. A. H. Harcourt and F. B. M. de Waal, 445–471. Oxford: Oxford University Press.

———. 1995. Population viability estimates: Theory and practice for a wild gorilla population. *Cons. Biol.* 9:134–142.

———. 1996. Is the gorilla a threatened species? How should we judge? *Biol. Cons.* 75:165–176.

———. 1998. Ecological indicators of risk for primates, as judged by species' susceptibility to logging. In *Behavioural ecology and conservation biology*, ed. T. M. Caro, 56–79. Oxford: Oxford University Press.

Harcourt, A. H., and Fossey, D. 1981. The Virunga gorillas: Decline of an "island" population. *Afr. J. Ecol.* 19:83–97.

Harcourt, A. H.; Fossey, D.; and Sabater-Pi, J. 1981. Demography of *Gorilla gorilla*. *J. Zool., Lond.* 195:215–233.

Harcourt, A. H.; Kineman, J.; Campbell, G.; Yamagiwa, J.; Redmond, I.; Aveling, C.; and Condiotti, M. 1983. Conservation and the Virunga gorilla population. *Afr. J. Ecol.* 21:139–142.

Harcourt, A. H., and Schwartz, M. W. n.d. Primate evolution: A comparative biology of Holocene extinction and survival on the Sunda Shelf. *Am J. Phys. Anthropol.* Submitted.

Harcourt, A. H.; Stewart, K. J.; and Inahoro, I. M. 1988. Nigeria's gorillas: A survey and recommendations. Unpublished Report.

———. 1989. Gorilla quest in Nigeria. *Oryx* 23:7–13.

Harcourt, C., and Thornback, J. 1990. *Lemurs of Madagascar and the Comoros: The IUCN Red Data Book.* Gland, Switz.: IUCN.

Hardin, G. 1968. The tragedy of the commons. *Science* 162:1243–1248.

Harrison, M. J. S. 1988. A new species of guenon (genus *Cercopithecus*) from Gabon. *J. Zool., Lond.* 215:561–575.

Harrison, R. L. 1992. Toward a theory of inter-refuge corridor design. *Biol. Cons.* 6:293–295.

Harrison, S., and Hastings, A. 1996. Genetic and evolutionary consequences of meta-population structure. *Trends Ecol. Evol.* 11:180–183.

Hart, T., and Hart, J. 1997. Zaire: New models for an emerging state. *Cons. Biol.* 11:308–309.

Hartwig, W. C. 1995. A giant New World monkey from the Pleistocene of Brazil. *J. Human Evol.* 28:189–195.

Harvey, P. H., and Krebs, J. R. 1990. Comparing brains. *Science* 249:140–146.

Harvey, P. H.; Martin, R. D.; and Clutton-Brock, T. H. 1987. Life histories in comparative perspective. In *Primate societies,* ed. B. B. Smuts, D. L. Cheney, R. M. Seyfarth, R. W. Wrangham, and T. T. Struhsaker, 181–196. Chicago: University of Chicago Press.

Harvey, P. H., and Pagel, M. D. 1991. *The comparative method in evolutionary biology.* Oxford: Oxford University Press.

Hashimoto, C. 1995. Population census of the chimpanzees in the Kalinzu Forest, Uganda: Comparison between methods with nest counts. *Primates* 36:477–488.

Hauser, M.; Cheney, D. L.; and Seyfarth, R. M. 1986. Group extinction and fusion in free-ranging vervet monkeys. *Am. J. Primatol.*

Hausfater, G. 1976. Predatory behaviour of yellow baboons. *Behaviour* 56:44–68.

Hausfater, G., and Hrdy, S. B., eds. 1984. *Infanticide: Comparative and evolutionary perspectives.* New York: Aldine.

Hawkes, K.; Hill, K.; and O'Connell, J. F. 1982. Why hunters gather: Optimal foraging and the Ache of eastern Paraguay. *Amer. Ethnol.* 9:379–398.

Heaney, L. R. 1986. Biogeography of mammals in SE Asia: Estimates of rates of colonization, extinction and speciation. *Biol. J. Linn. Soc.* 28:127–165.

Held, J. R., and Wolfle, T. L. 1994. Imports: Current trends and usage. *Int. J. Primatol.* 34:85–96.

Henzi, S. P., and Lucas, J. W. 1980. Observations on the inter-troop movement of adult vervet monkeys (*Cercopithecus aethiops*). *Folia Primatol.* 33:220–235.

Henzi, S. P.; Lycett, J. E.; and Weingrill, T. 1997. Cohort size and the allocation of social effort by female mountain baboons. *Anim. Behav.* 54:1235–1243.

Heymann, E. W. 1987. A field observation of predation on a moustached tamarin (*Saguinus mystax*) by an anaconda. *Int. J. Primatol.* 8:193–195.

———. 1998. Giant fossil New World primates: Arboreal or terrestrial? *J. Hum. Evol.* 34:99–101.

Hill, C. M. 1997. Crop-raiding by wild vertebrates: The farmer's perspective in an agricultural community in western Uganda. *Int. J. Pest Mgmt.* 43:77–84.

Hill, K., and Hawkes, K. 1983. Neotropical hunting among the Ache of eastern Paraguay. In *Adaptive responses of native Amazonians,* ed. R. B. Hames and W. T. Vickers, 139–188. New York: Academic Press.

Hill, K.; Padwe, J.; Bejyvagi, C.; Bepurangi, A.; Jakugi, F.; Tykuarangi, R.; and Tykuarangi, T. 1997. Impact of hunting on large vertebrates in the Mbaracayu Reserve, Paraguay. *Cons. Biol.* 11:1339–1353.

Hill, R. A., and Dunbar, R. I. M. 1998. An evaluation of the roles of predation rate and predation risk as selective pressures on primate grouping behaviour. *Behaviour* 135:1–20.

Hill, K., and Hurtado, A. M. 1996. *Ache life history: The ecology and demography of a foraging people.* New York: Aldine de Gruyter.

Hill, R. A., and Lee, P. C. 1998. Predation risk as an influence on group size in cercopithecoid primates: Implications for social structure. *J. Zool., Lond.* 245: 447–456.

Hill, R. A.; Lycett, J. E.; and Dunbar, R. I. M. n.d. Ecological and social determinants of birth intervals in baboons. *Behav. Ecol.* In press.

Hill, W. C. O. 1966. *The primates.* Vol. 6. *Cercopithecoidea.* Edinburgh: Edinburgh University Press.

Hladik, C. M. 1975. Ecology, diet, and social patterns in Old and New World primates. In *Socioecology and psychology of primates,* ed. R. H. Tuttle, 3–35. The Hague: Mouton.

Hladik, C. M., and Hladik, A. 1967. Observations sur le rôle des Primates dans la dissémination des végétaux de la forêt gabonaise. *Biol. Gabonica* 3:43–58.

Hofman, M. A. 1983. Energy metabolism, brain size and longevity in mammals. *Quart. Rev. Biol.* 58:495–512.

Horovitz, I., and MacPhee, R. D. E. 1999. The Quaternary Cuban platyrrhine *Paralouatta varonai* and the origin of Antillean monkeys. *J. Human Evol.* 36:33–68.

Horrocks, J. A., and Baulu, J. 1988. Effects of trapping on the vervet (*Cercopithecus aethiops sabaeus*) population in Barbados. *Am. J. Primatol.* 15:223–233.

Horwich, R. H. 1990. How to develop a community sanctuary—An experimental approach to the conservation of private lands. *Oryx* 24:95–102.

———. 1998. Effective solutions for howler conservation. *Int. J. Primatol.* 19:579–598.

Hough, J. L. 1988. Obstacles to effective management of conflicts between national parks and surrounding human communities in developing countries. *Environ. Cons.* 15:129–136.

Houghton, J. T.; Meira Filho, L. G.; Callander, B. A.; Harris, N.; Kettenberg, A.; and Maskell, K., eds. 1996. *Climate change 1995: The science of climate change.* Cambridge: Cambridge University Press.

Howe, H. F. 1980. Monkey dispersal and waste of a Neotropical fruit. *Ecology* 6:944–959.

Howe, H. F.; Schupp, E. W.; and Westley, L. C. 1985. Early consequences of seed dispersal for a Neotropical tree (*Virola surinamensis*). *Ecology* 66:781–791.

Huffman, M. A.; Gotoh, S.; Turner, L. A.; Hamai, M.; and Yoshida, K. 1997. Seasonal trends in intestinal nematode infection and medicinal plant use among chimpanzees in the Mahale Mountains, Tanzania. *Primates* 38:111–125.

Humphries, C. J., and Williams, P. H. 1994. Cladograms and trees in biodiversity. In *Models in phylogenetic reconstruction,* ed. R. W. Scotland, D. J. Siebert, and D. M. Williams, 335–352. Oxford: Clarendon Press.

Hutchinson, G. E. 1959. Homage to Santa Rosalia, or Why are there so many kinds of animals? *Am. Nat.* 93:137–159.

Idani, G. 1986. Seed dispersal by pygmy chimpanzees (*Pan paniscus*): A preliminary report. *Primates* 27:411–447.

IPCC. 1992. *Climate change: The IPCC 1990 and 1992 assessments.* New York: World Meteological Association and United Nations Environment Program.

Isbell, L. A. 1990. Sudden short-term increase in mortality of vervet monkeys (*Cercopithecus aethiops*) due to leopard predation in Amboseli National Park, Kenya. *Am. J. Primatol.* 21:41–52.

———. 1994. Predation on primates: Ecological patterns and evolutionary consequences. *Evol. Anthropol.* 3:61–71.

Isbell, L. A.; Cheney, D. L.; and Seyfarth, R. M. 1990. Costs and benefits of home range shifts among vervet monkeys (*Cercopithecus aethiops*) in Amboseli National Park, Kenya. *Behav. Ecol. Sociobiol.* 27:351–358.

———. 1991. Group fusions and minimum group sizes in vervet monkeys (*Cercopithecus aethiops*). *Am. J. Primatol.* 25:57–65.

IUCN. 1971. *United Nations list of national parks and equivalent reserves.* Brussels: Hayez.

———. 1988. *1988 Red List of threatened animals.* Gland, Switz.: IUCN.

———. 1990. *1990 Red List of threatened animals.* Gland, Switz.: IUCN.

IUCN, WWF, and UNEP. 1980. *World conservation strategy: Living resource conservation for sustainable development.* Gland, Switz.: IUCN, WWF and UNEP.

———. 1991. *Caring for the earth: A strategy for sustainable living.* Gland, Switz.: IUCN, WWF and UNEP.

Iwamoto, T. 1979. Feeding ecology. In *Ecological and sociological studies of gelada baboons,* ed. M. Kawai, 279–330. Basel: Karger.

Jablonski, D. 1991. Extinctions: A paleontological perspective. *Science* 253:754–757.

———. 1995. Extinctions in the fossil record. In *Extinction rates,* ed. J. H. Lawton and R. M. May, 25–44. Oxford: Oxford University Press.

Jablonski, N. G. 1993. The phylogeny of *Theropithecus.* In Theropithecus: *The rise and fall of a primate genus,* ed. N. G. Jablonski, 209–224. Cambridge: Cambridge University Press.

———. 1998. The response of catarrhine primates to Pleistocene environmental fluctuations in East Asia. *Primates* 39:29–37.

Jackson, G., and Gartlan, J. S. 1965. The flora and fauna of Lolui Island: A study of vegetation, men and monkeys. *J. Ecol.* 53:573–597.

Jackson, J. B. C. 1997. Reefs since Columbus. *Coral Reefs* 16:S23–S32.

James, R. A.; Leberg, P. L.; Quattro, J. M.; and Vrejenhoek, R. C. 1997. Genetic diversity in black howler monkeys (*Alouatta pigra*) from Belize. *Am. J. Phys. Anthropol.* 102:329–336.

Jans, L.; Poorter, L.; van Rompaey, R. S. A. R.; and Bongers, F. 1993. Gaps and forest zones in tropical moist forest in Ivory Coast. *Biotropica* 25:258–269.

Janis, C. 1976. The evolutionary strategy of the Equidae and the origins of rumen and caecal digestion. *Evolution* 30:757–774.

Janson, C. H., and Di Bitetti, M. S. 1997. Experimental analysis of food detection in capuchin monkeys: Effects of distance, travel speed, and resource size. *Behav. Ecol. Sociobiol.* 41:17–24.

Janson, C. H., and Emmons, L. H. 1990. Ecological structure of the nonflying mammal community at Cocha Cashu Biological Station, Manu National Park, Peru. In *Four Neotropical forests,* ed. A. H. Gentry, 314–338. New Haven: Yale University Press.

Janson, C. H.; Terborgh, J.; and Emmons, L. H. 1981. Non-flying mammals as pollinating agents in the Amazonian forest. *Biotropica* 13:1–6.

Janson, C. H., and van Schaik, C. P. 1988. Recognizing the many faces of primate food competition: Methods. *Behaviour* 105:165–186.

Janzen, D. 1974. Tropical blackwaters, animals, and mast fruiting by the Dipterocarpaceae. *Biotropica* 6:69–103.

Jernvall, J., and Wright, P. C. 1998. Diversity components of impending primate extinctions. *Proc. Natl. Acad. Sci. USA* 95:11279–11283.

Jiang, H.; Liu, Z.; Zhang, Y.; and Southwick, C. 1991. Population ecology of rhesus monkeys (*Macaca mulatta*) at Nanwan Nature Reserve, Hainan, China. *Am. J. Primatol.* 25:207–217.

Jiang, X. L.; Wang, Y. X.; Ma, S. L.; Sheeran, L.; and Wang, Q. 1994. Group size and composition of black-crested gibbon (*Hylobates concolor*). *Zool. Res.* 15:15–22.

Jiménez, J. A.; Hughes, K. A.; Alaks, G.; Graham, L.; and Lacy, R. C. 1994. An experimental study of inbreeding depression in a natural habitat. *Science* 266:271–273.

Joffe, T. 1997. Social pressures have selected for an extended juvenile period in primates. *J. Human Evol.* 32:593–605.

Joffe, T., and Dunbar, R. I. M. 1997. Visual and socio-cognitive information processing in primate brain evolution. *Proc. Roy. Soc. Lond.,* ser. B, 264:1303–1307.

Johns, A. D. 1985a. Behavioural responses of two Malaysian primates (*Hylobates lar* and *Presbytis melalophos*) to selective logging: Vocal behaviour, territoriality, and nonemigration. *Int. J. Primatol.* 6:423–433.

———. 1985b. Selective logging and wildlife conservation in tropical rain-forest: Problems and recommendations. *Biol. Cons.* 31:355–375.

———. 1985c. Differential detectability of primates between primary and selectively logged habitats and implications for population surveys. *Am. J. Primatol.* 8:31–36.

———. 1986. Effects of selective logging on the behavioural ecology of west Malaysian primates. *Ecology* 67:684–694.

———. 1988. Effects of "selective" timber extraction on rainforest structure and composition and some consequences for frugivores and folivores. *Biotropica* 20:31–37.

———. 1991. Forest disturbance and Amazonian primates. In *Primate responses to environmental change,* ed. H. O. Box, 115–135. London: Chapman and Hall.

———. 1992. Vertebrate responses to selective logging: Implications for the design of logging systems. *Phil. Trans. Roy. Soc. Lond.,* ser. B, 335:437–442.

Johns, A. D., and Skorupa, J. P. 1987. Responses of rainforest primates to habitat disturbance: A review. *Int. J. Primatol.* 8:157–191.

Johnson, C. N. 1988. Dispersal and the sex ratio at birth in primates. *Nature* (Lond.) 332:726–728.

Jolly, A. 1986. Lemur survival. In *Primates: The road to self-sustaining populations,* ed. K. Benirshcke, 71–98. New York: Springer.

Jolly, C. J. 1966. The classification and natural history of the *Theropithecus* (*Simopithecus*) (Andrews 1916) baboons of the African Plio-Pleistocene. *Bull. Brit. Mus. (Nat. Hist.), Geol.* 22:1–123.

———. 1993. Species, subspecies and baboon systematics. In *Species, species concepts and primate evolution,* ed. W. H. Kimbel and L. B. Martin, 67–108. New York: Plenum Press.

Jorgenson, J. P. 1995. Maya subsistence hunters in Quintana Roo, Mexico. *Oryx* 29:49–57.

Jori, F.; Mensah, G. A.; and Adjanohoun, E. 1995. Grasscutter production: An example of rational exploitation of wildlife. *Biodiversity Cons.* 4:257–265.

Julliot, C. 1996. Seed dispersal by red howling monkeys (*Alouatta seniculus*) in the tropical rain forest of French Guiana. *Int. J. Primatol.* 17:239–258.

———. 1997. Impact of seed dispersal by red howler monkeys *Alouatta seniculus* on the seedling population in the understorey of tropical rain forest. *J. Ecol.* 85:431–440.

Juste, J.; Fa, J. E.; Perez de Val, J.; and Castroviejo, J. 1995. Market dynamics of bushmeat species in equatorial Guinea. *J. Appl. Ecol.* 32:454–467.

Kaplan, H., and Hill, K. 1992. The evolutionary ecology of food acquisition. In *Evolutionary ecology and human behaviour,* ed. E. A. Smith and B. Winterhalder, 167–201. New York: Aldine de Gruyter.

Kappeler, P. M. 1996. Causes and consequences of life-history variation among strepsirhine primates. *Am. Nat.* 148:868–891.

———. 1997. Determinants of primate social organization: Comparative evidence and new insights from Malagasy lemurs. *Biol. Rev.* 72:111–151.

Kappeler, P. M., and Heymann, E. W. 1996. Nonconvergence in the evolution of primate life history and socio-ecology. *Biol. J. Linn. Soc.* 59:297–326.

Kavanagh, M. 1980. Invasion of the forest by an African savanna monkey: Behavioural adaptations. *Behaviour* 73:238–260.

Kavanagh, M.; Eudey, A. A.; and Mack, D. 1987. The effect of live trapping and trade on primate populations. In *Primate conservation in the tropical rain forest,* ed. C. W. Marsh and R. A. Mittermeier, 147–177. New York: Alan R. Liss.

Kay, R. F.; Madden, R. H.; van Schaik, C. P.; and Higdon, D. 1997. Primate species richness is determined by plant productivity: Implications for conservation. *Proc. Natl. Acad. Sci. USA* 94:13023–13027.

Kay, R. F., and Simons, E. L. 1980. The ecology of Oligocene African Anthropoidea. *Int. J. Primatol.* 1:21–37.

Kay, R. N. B., and Davies, G. A. 1994. Digestive physiology. In *Colobine monkeys: Their ecology, behaviour and evolution,* ed. G. A. Davies and J. F. Oates, 229–250. Cambridge: Cambridge University Press.

Kay, R. N. B.; Hoppe, P.; and Maloiy, G. M. O. 1976. Fermentative digestion of food in the colobus monkey, *Colobus polykomos. Experientia* 32:485–486.

Kenyatta, A. 1995. Ecological and social constraints on maternal investment strategies. Ph.D. diss., University of London.

Kershaw, M.; Mace, G. M.; and Williams, P. H. 1995. Threatened species, rarity, and diversity as alternative selection measures for protected areas: A test using Afrotropical antelopes. *Cons. Biol.* 9:324–334.

Kershaw, M.; Williams, P. H.; and Mace, G. M. 1994. Conservation of Afrotropical antelopes: Consequences and efficiency of using different site selection methods and diversity criteria. *Biodiversity Cons.* 3:354–372.

Kindvall, O.; Vessby, K.; Berggren, Å.; and Hartman, G. 1998. Individual mobility prevents an Allee effect in sparse populations of the bush cricket *Metrioptera roeseli:* An experimental study. *Oikos* 81:449–457.

King, F. A., and Lee, P. C. 1987. A brief survey of human attitudes to a pest species of primate—*Cercopithecus aethiops. Primate Cons.* 8:82–84.

King, S. 1994. Utilisation of wildlife in Bakossiland, West Cameroon: With particular reference to primates. *TRAFFIC Bull.* 14:63–73.

Kingdon, J. 1990. *Island Africa.* London: Collins.

———. 1997. *The Kingdon field guide to African mammals.* London: Academic Press.

Kinnaird, M. F. 1992. Competition for a forest palm: Use of *Phoenix reclinata* by human and nonhuman primates. *Cons. Biol.* 6:101–107.

Kinnaird, M. F., and O'Brien, T. G. 1991. Viable populations for an endangered forest primate, the Tana River crested mangabey (*Cercocebus galeritus galeritus*). *Cons. Biol.* 5:203–213.

———. 1996. Ecotourism in the Tangkoko DuaSudara Nature Reserve: Opening Pandora's box? *Oryx* 30:65–73.

———. 1998. Ecological effects of wildfire on lowland rainforest in Sumatra. *Cons. Biol.* 12:954–956.

Kinzey, W. G. 1982. Distribution of primates and forest refuges. In *Biological diversification in the tropics,* ed. G. T. Prance, 455–482. New York: Columbia University Press.

———. 1992. Dietary and dental adaptations in the Pitheciinae. *Am. J. Phys. Anthropol.* 88:499–514.

Kinzey, W. G., and Cunningham, E. P. 1994. Variability in platyrrhine social organization. *Am. J. Phys. Anthropol.* 34:185–198.

Kirkpatrick, R. C. 1995. The natural history and conservation of the snub-nosed monkeys (genus *Rhinopithecus*). *Biol. Cons.* 72:363–369.

Kleiber, M. 1969. *The fire of life.* New York: Wiley.

Kleiman, D. G.; Beck, B. B.; Dietz, J. M.; and Dietz, L. A. 1991. Costs of a reintroduction and criteria for success: Accounting and accountability in the Golden Lion Tamarin Conservation Program. *Symp. Zool. Soc. Lond.* 62:125–142.

Kleiman, D. G.; Beck, B. B.; Dietz, J. M.; Dietz, L. A.; Ballou, J. D.; and Coimbra-Filho, A. F. 1986. Conservation program for the golden lion tamarin captive research and management, ecological studies, educational strategies, and reintroduction. In *Primates: The road to self-sustaining populations,* ed. K. Benirschke, 959–980. New York: Springer.

Kleiman, D. G., and Mallinson, J. J. C. 1998. Recovery and management committees for lion tamarins: Partnerships in conservation planning and implementation. *Cons. Biol.* 12:27–38.

Kleiman, D. G.; Stanley Price, M. R.; and Beck, B. B. 1994. Criteria for reintroductions. In *Creative conservation,* ed. P. J. S. Olney, G. M. Mace, and A. T. C. Feistner, 287–303. London: Chapman and Hall.

Knott, C. D. 1998. Changes in orangutan caloric intake, energy balance, and ketones in response to fluctuating fruit availability. *Int. J. Primatol.* 19:1061–1071.

Koganezawa, M., and Imaki, H. 1999. The effects of food sources on Japanese monkey home range size and location, and population dynamics. *Primates* 40:177–185.

Kondo, M.; Kawamoto, Y.; Nozawa, K.; Matsubayashi, K.; Watanabe, T.; Griffiths, O.; and Stanley, M.-A. 1993. Population genetics of crab-eating macaques (*Macaca fascicularis*) on the island of Mauritius. *Am. J. Primatol.* 29:167–182.

Konstant, W. R., and Mittermeier, R. A. 1982. Introduction, reintroduction and translocation of Neotropical primates: Past experiences and future possibilities. *Int. Zoo. Yrbk.* 22:69–77.

Körn, H. 1994. Genetic, demographic, spatial, environmental and catastrophic effects on the survival probability of small populations of mammals. In *Minimum animal populations,* ed. H. Remmert, 33–49. Berlin: Springer.

Koyama, N. 1970. Changes in dominance rank and division of a wild Japanese monkey troop at Arishiyama. *Primates* 11:335–390.

Krebs, J. R., and McCleery, R. H. 1984. Optimization in behavioural ecology. In *Behavioural ecology: An evolutionary approach,* ed. J. R. Krebs and N. B. Davies, 91–121. Oxford: Blackwell.

Kuchikura, Y. 1988. Efficiency and focus of blowpipe hunting among Semaq Beri hunter-gatherers of peninsular Malaysia. *Human Ecol.* 16:271–305.

Kudo, H., and Mitani, M. 1985. New record of predatory behaviour by the mandrill in Cameroon. *Primates* 26:161–167.

Kumar, A. 1985. Patterns of extinction in India, Sri Lanka, and elsewhere in Southeast Asia: Implications for lion-tailed macaque wildlife management and the Indian conservation system. In *The lion-tailed macaque: Status and conservation,* ed. P. G. Heltne, 65–89. New York: Alan R. Liss.

Kummer, H. 1968. *Social organisation of hamadryas baboons.* Basel: Karger.

Kummer, H.; Banaja, A.; Abo-Khatwa, A.; and Ghandour, A. 1985. Differences in social behaviour between Ethiopian and Arabian hamadryas baboons. *Folia Primatol.* 45:1–8.

Lacy, R. C. 1992. The effects of inbreeding on isolated populations: Are minimum viable population sizes predictable? In *Conservation biology: The theory and practice of conservation, preservation and management,* ed. P. L. Fiedler and S. K. Jain, 277–296. London: Chapman and Hall.

———. 1993. VORTEX: A computer simulation model for population viability analysis. *Wildlife Res.* 20:45–65.

———. 1993–1994. What is population (and habitat) viability analysis? *Primate Cons.* 14–15:27–33.

Lahm, S. A. 1986. Diet and habitat preference of *Mandrillus sphinx* in Gabon: Implications of foraging strategy. *Am. J. Primatol.* 11:9–26.

Lahm, S. A.; Barnes, R. F. W.; Beardsley, K.; and Cervinka, P. 1998. A method for censusing the greater white-nosed monkey in northeastern Gabon using the population density gradient in relation to roads. *J. Trop. Ecol.* 14:629–643.

Lambert, J. E. 1998. Primate frugivory in Kibale National Park, Uganda, and its implications for human use of forest resources. *Afr. J. Ecol.* 36:234–240.

Lan, D., and Dunbar, R. I. M. n.d. Bird and mammal conservation in Gaolingongshan region and Jingdong county, Yunnan, China: Patterns of species richness and nature reserves. *Oryx.* Submitted.

Lande, R. 1988. Genetics and demography in biological conservation. *Science* 241:1455–1460.

———. 1993. Risks of population extinction from demographic and environmnetal stochasticity and random catastrophes. *Am. Nat* 142:911–927.

———. 1995. Mutation and conservation. *Cons. Biol.* 9:782–791.

Lande, R., and Barrowclough, G. F. 1987. Effective population size, genetic variation, and their use in population management. In *Viable populations for conservation,* ed. M. E. Soulé, 87–123. Cambridge: Cambridge University Press.

Lande, R.; Engen, S.; and Saether, B.-E. 1994. Optimal harvesting, economic discounting and extinction risk in fluctuating populations. *Nature* 372:88–90.

————. 1998. Extinction times in finite metapopulation models with stochastic local dynamics. *Oikos* 83:383–389.

Lande, R.; Saether, B.-E.; and Engen, S. 1997. Threshold harvesting for sustainability of fluctuating resources. *Ecology* 78:1341–1350.

Langholz, J. 1996. Economics, objectives and success of private nature reserves in sub-Saharan Africa and Latin America. *Cons. Biol.* 10:271–280.

Laurance, W. F. 1991. Ecological correlates of extinction proneness in Australian tropical rain forest mammals. *Cons. Biol.* 5:79–89.

Laurance, W. F.; Ferreira, L. V.; De Merona, J. M.; and Laurance, S. G. 1998. Rain forest fragmentation and the dynamics of Amazonian tree communities. *Ecology* 79:2032–2040.

Lawes, M. J. 1990. The distribution of the samango monkey (*Cercopithecus mitis erythrarchus* Peters, 1852 and *Cercopithecus mitis labiatus* I. Geoffroy, 1843) and forest history in southern Africa. *J. Biogeogr.* 17:669–680.

————. 1992. Estimates of population density and correlates of the status of the samango monkey *Cercopithecus mitis* in Natal, South Africa. *Biol. Cons.* 60: 197–210.

Lawler, S. H.; Sussman, R. W.; and Taylor, L. L. 1995. Mitochondrial DNA of the Mauritian macaques (*Macaca fascicularis*): An example of the founder effect. *Am. J. Phys. Anthropol.* 96:133–141.

Lawlor, T. E. 1986. Comparative biogeography of mammals on islands. *Biol. J. Linn. Soc.* 28:99–125.

Lawton, J. H. 1993. Range, population abundance and conservation. *Trends Ecol. Evol.* 8:409–413.

————. 1994. Population dynamic principles. *Phil. Trans. Roy. Soc. Lond.*, ser. B, 344:61–68.

————. 1995. Population dynamic principles. In *Extinction risk*, ed. J. H. Lawton and R. M. May, 147–163. Oxford: Oxford University Press.

Lawton, J. H.; Nee, S.; Letcher, A. J.; and Harvey, P. H. 1994. Animal distributions: Patterns and processes. In *Large-scale ecology and conservation biology*, ed. P. J. Edwards, R. M. May, and N. R. Webb, 41–58. Oxford: Blackwell.

Leader-Williams, N., and Albon, S. D. 1988. Allocation of resources for conservation. *Nature* 336:533–535.

Leader-Williams, N.; Albon, S. D.; and Berry, P. S. M. 1990. Illegal exploitation of black rhinoceros and elephant populations: Patterns of decline, law enforcement and patrol effort in Luangwa Valley, Zambia. *J. Appl. Ecol.* 27:1055–1087.

Leader-Williams, N., and Milner-Gulland, E. J. 1993. Policies for the enforcement of wildlife laws: The balance between detection and penalties in Luangwa Valley, Zambia. *Cons. Biol.* 7:611–617.

Leakey, M. 1993. Evolution of *Theropithecus* in the Turkana basin. In Theropithecus: *The rise and fall of a primate genus,* ed. N. Jablonski, 85–124. Cambridge: Cambridge University Press.

Lee, P. C., and Foley, R. 1993. Ecological energetics and extinction of giant gelada baboons. In Theropithecus: *The rise and fall of a primate genus,* ed. N. Jablonski, 487–498. Cambridge: Cambridge University Press.

Lee, P. C., and Hauser, M. D. 1998. Longterm consequences of changes in territory quality on feeding and reproductive strategies of vervet monkeys. *J. Anim. Ecol.* 67:347–358.

Lee, P. C.; Thornback, J.; and Bennett, E. L. 1988. *Threatened primates of Africa: The IUCN Red Data Book.* Gland, Switz.: IUCN.

Lee-Thorp, J. A.; van der Merwe, N. J.; and Brain, C. K. 1989. Isotopic evidence for dietary differences between two extinct baboon species from Swartkrans. *J. Human Evol.* 18:183–190.

Leigh, E. G., and Windsor, D. M. 1982. Forest production and regulation of primary consumers on Barro Colorado Island. In *The ecology of a tropical forest: Seasonal rhythms and longterm changes,* ed. E. G. Leigh, A. S. Rand, and D. M. Windsor, 111–122. Washington, D.C.: Smithsonian Institution Press.

Leighton, M., and Leighton, D. R. 1983. Vertebrate responses to fruiting seasonality within a Bornean rain forest. In *Tropical rain forest: Ecology and management,* ed. S. L. Sutton, T. C. Whitmore, and A. C. Chadwick, 181–196. Oxford: Blackwell.

Lernould, J.-C. 1988. Classification and geographical distribution of guenons: A review. In *A primate radiation: Evolutionary biology of the African guenons,* ed. A. Gautier-Hion, F. Bourlière, J.-P. Gautier, and J. Kingdon, 54–120. Cambridge: Cambridge University Press.

Leslie, P. H. 1945. The use of matrices in certain population mathematics. *Biometrika* 33:183–212.

Letcher, A. J., and Harvey, P. H. 1994. Variation in geographical range size among mammals of the Palearctic. *Am. Nat.* 144:30–42.

Lewis, D.; Kaweche, G. B.; and Mwenya, A. 1990. Wildlife conservation outside protected areas—Lessons from an experiment in Zambia. *Cons. Biol.* 4:171–180.

Lewis, D. M., and Phiri, A. 1998. Wildlife snaring—An indicator of community response to a community-based conservation project. *Oryx* 32:111–121.

Li Wenjun; Fuller, T. K.; and Sung, W. 1996. A survey of wildlife trade in Guangxi and Guangdong, China. *TRAFFIC Bulletin* 16:9–16.

Li Yiming and Li Dianmo. 1998. The dynamics of trade in live wildlife across the Guangxi border between China and Vietnam during 1993–1996 and its control strategies. *Biodiversity Cons.* 7:895–914.

Lieberman, D.; Hall, J. B.; Swaine, M. D.; and Lieberman, M. 1979. Seed dispersal by baboons in the Shai Hills, Ghana. *Ecology* 60:65–75.

Lindenfors, P., and Tullberg, B. S. 1998. Phylogenetic analyses of primate size evolution: The consequences of sexual selection. *Biol. J. Linn. Soc.* 64:413–447.

Lindenmeyer, D. B., and Nix, H. A. 1993. Ecological principles for the design of wildlife corridors. *Cons. Biol.* 7:627–630.

Lindstedt, S. L., and Boyce, M. S. 1985. Seasonality, fasting endurance, and body size in mammals. *Am. Nat.* 125:873–878.

Liu, Z.; Zhang, Y.; Haisheng, J.; and Southwick, C. 1989. Population structure of *Hylobates concolor* in Bawangalin Nature Reserve, Hainan, China. *Am. J. Primatol.* 19:247–254.

Lomolino, M. V. 1986. Mammalian community structure on islands: The importance of immigration, extinction and interactive effects. *Biol. J. Linn. Soc.* 28:1–21.

Lomolino, M. V., and Channell, R. 1995. Splendid isolation: Patterns of geographic range collapse in endangered mammals. *J. Mammal.* 76:335–347.

Long, Y. C.; Kirkpatrick, C. R.; Zhongtai; and Xiaolin. 1994. Report on the distribution, population and ecology of the Yunnan snub-nosed monkey (*Rhinopithecus bieti*). *Primates* 35:241–250.

López, G. S.; Orduña, F. G.; and Luna, E. R. 1988. The status of *Ateles geoffroyi* and *Alouatta palliata* in disturbed forest areas of Sierra de Santa Marta, Mexico. *Primate Cons.* 9:53–61.

Lovejoy, T. E.; Bierregaard, R. O., Jr.; Rylands, A. B.; Malcolm, J. R.; Quintela, C. E.; Harper, L. H.; Brown, K. S., Jr.; Powell, A. H.; Powell, G. V. N.; Schubart, H. O. R.; and Hays, M. B. 1986. Edge and other effects of isolation on Amazon forest fragments. In *Conservation biology: The science of scarcity and diversity,* ed. M. E. Soulé, 257–285. Sunderland, Mass.: Sinauer.

Lovejoy, T. E.; Rankin, J. M.; Bierregaard, R. O., Jr.; Brown, K. S., Jr.; Emmons, L. H.; and Van der Voort, M. E. 1984. Ecosystem decay of Amazon forest fragments. In *Extinctions,* ed. M. H. Nitecki, 295–325. Chicago: University of Chicago Press.

Lowen, C. B., and Dunbar, R. I. M. 1994. Territory size and defendability in primates. *Behav. Ecol. Sociobiol.* 35:347–354.

Loy, J. 1988. Effects of supplementary feeding on maturation and fertility in primate groups. In *Ecology and behavior of food-enhanced primate groups,* ed. J. E. Fa and C. H. Southwick, 153–166. New York: Alan R. Liss.

Luna, E. R.; Fa, J. E.; Orduna, F. G.; Lopez, G. S.; and Espinoza, D.C. 1987. Primate conservation in Mexico. *Primate Cons.* 8:114–118.

Lycett, J. E.; Henzi, S. P.; and Barrett, L. 1998. Maternal investment in mountain baboons and the hypothesis of reduced care. *Behav. Ecol. Sociobiol.* 42:49–56.

Lyles, A. M., and Dobson, A. P. 1988. Dynamics of provisioned and unprovisioned primate populations. In *Ecology and behavior of food-enhanced primate groups,* ed. J. E. Fa and C. H. Southwick, 167–198. New York: Alan R. Liss.

Lynch, M., and Lande, R. 1998. The critical effective size for a genetically secure population. *Animal Cons.* 1:70–72.

MacArthur, R. H., and Wilson, E. O. 1967. *The theory of island biogeography.* Princeton: Princeton University Press.

Mace, G. M. 1988. The genetic and demographic status of western lowland gorilla (*Gorilla g. gorilla*) in captivity. *J. Zool., Lond.* 216:629–654.

———. 1995a. Classification of threatened species and its role in conservation planning. In *Extinction rates,* ed. J. H. Lawton and R. M. May, 197–213. Oxford: Oxford University Press.

———. 1995b. An investigation in methods for categorizing the conservation status of species. In *Large-scale ecology and conservation biology,* ed. P. J. Edwards, R. M. May, and N. R. Webb, 293–312. Oxford: Blackwell.

Mace, G. M., and Balmford, A. 2000. Patterns and processes in contemporary mammalian extinction. In *Future priorities for the conservation of mammalian diversity,* ed. A. Entwhistle and N. Dunstone. Cambridge: Cambridge University Press.

Mace, G. M., and Lande, R. 1991. Assessing extinction threats: Toward a reevaluation of IUCN threatened species categories. *Cons. Biol.* 5:148–157.

Mace, G. M., and Stuart, S. 1994. Draft IUCN Red List categories, version 2. 2. *Species* 21–22:13–24.

Mack, D., and Mittermeier, R. A. 1984. *The international primate trade.* Washington, D.C.: WWF-US/TRAFFIC (USA).

Mackinnon, J. R. 1974. The ecology and behaviour of wild orangutans (*Pongo pygmaeus*). *Anim. Behav.* 22:3–74.

————. 1977. A comparative ecology of Asian apes. *Primates* 18:747–772.

Mackinnon, J. R., and Mackinnon, K. S. 1978. Comparative feeding ecology of six sympatric primates in West Malaysia. In *Recent advances in primatology,*vol. 1, *Behaviour,* ed. D. J. Chivers and J. Herbert, 305–321. London: Academic Press.

Mackinnon, K. 1998. Sustainable use as a conservation tool in the forests of South-East Asia. In *Conservation of biological resources,* ed. E. J. Milner-Gulland and R. Mace, 174–192. Oxford: Blackwell.

Mackinnon, K., and Mackinnon, J. 1991. Habitat protection and re-introduction programmes. *Symp. Zool. Soc. Lond.* 62:173–198.

Mahar, D., and Schneider, R. 1994. Incentives for tropical deforestation: Some examples from Latin America. In *The causes of tropical deforestation,* ed. K. Brown and D. W. Pearce, 159–171. London: UCL Press.

Maisels, F.; Gautier-Hion, A.; and Gautier, J.-P. 1994. Diets of two sympatric colobines in Zaire: More evidence on seed-eating in forests on poor soils. *Int. J. Primatol.* 15:681–701.

Mamede-Costa, A. C., and Gobbi, N. 1998. The black lion tamarin *Leontopithecus chrysopygus*—Its conservation and management. *Oryx* 32:295–300.

Maples, W. R.; Maples, M. K.; Greenhood, W. F.; and Walek, M. L. 1976. Adaptations of crop-raiding baboons in Kenya. *Am. J. Phys. Anthropol.* 45:309–315.

Marchesi, P.; Marchesi, N.; Fruth, B.; and Boesch, C. 1995. Census and distribution of chimpanzees in Côte D'Ivoire. *Primates* 36:591–607.

Margules, C. R.; Nicholls, A. O.; and Pressey, R. L. 1988. Selecting networks of reserves to maximize biological diversity. *Biol. Cons.* 43:63–76.

Marsh, C. W. 1986. A resurvey of the Tana River primates and their habitat. *Primate Cons.* 7:72–81.

————. 1988. Primates, forest conservation and development. *Primate Cons.* 9:150–156.

Marsh, C. W.; Johns, A. D.; and Ayres, J. M. 1987. Effects of habitat disturbance on rain forest primates. In *Primate conservation in the tropical rain forest,* ed. C. W. Marsh and R. A. Mittermeier, 83–107. New York: Alan R. Liss.

Marsh, C. W., and Wilson, W. L. 1981. *A survey of primates in peninsular Malaysian forests.* Kuala Lumpur: Universiti Kebangsaan Malaysia; Cambridge: Cambridge University Press.

Marshall, J., and Sugardjito, J. 1986. Gibbon systematics. In *Comparative primate biology,* vol. 1, *Systematics, evolution and anatomy,* ed. D. Swindler and J. Erwin, 137–185. New York: Alan R. Liss.

Martin, E. B., and Phipps, M. 1996. A review of the wild animal trade in Cambodia. *TRAFFIC Bulletin* 16:45–60.

Martin, G. H. G. 1983. Bushmeat in Nigeria as a natural resource with environmental implications. *Environ. Cons.* 10:125–132.

Martin, P. S., and Klein, R. G. 1984. *Quarternary extinctions.* Tuscon: University of Arizona Press.

Martin, R. B. 1999. The rule of law and African game, and social change and conservation misrepresentation—A reply to Spinage. *Oryx* 33:90–94.

Martin, R. D. 1981. Relative brain size and basal metabolic rate in terrestrial vertebrates. *Nature* (Lond.) 293:56–60.

————. 1990. *Primate origins and evolution.* London: Chapman and Hall.

————. 1993. Primate origins: Plugging the gaps. *Nature* 363:223–234.

Masters, J. C. 1993. Primates and paradigms: Problems with the identification of genetic species. In *Species, species concepts and primate evolution,* ed. W. H. Kimbel and L. B. Martin, 43–64. New York: Plenum Press.

Matsubayashi, K.; Gotoh, S.; and Suzuki, J. 1986. Changes in import of non-human primates after ratification of CITES (Washington Convention) in Japan. *Primates* 27:125–135.

Matzke G. E., and Nabane, N. 1996. Outcomes of a community controlled wildlife utilization program in a Zambezi Valley community. *Human Ecol.* 24:65–85.

May, R. M. 1971. Stability in model ecosystems. *Proc. Ecol. Soc. Aust.* 6:18–56.

———. 1988. Conservation and disease. *Cons. Biol.* 2:26–30.

———. 1994. The effects of spatial scale on ecological questions and answers. In *Large-scale ecology and conservation biology,* ed. P. J. Edwards, R. M. May, and N. R. Webb, 1–18. Oxford: Blackwell.

———. 1994. The economics of extinction. *Nature* 372:42–43.

May, R. M.; Lawton, J. H.; and Stork, N. E. 1995. Assessing extinction rates. In *Extinction rates,* ed. J. H. Lawton and R. M. May, 1–24. Oxford: Oxford University Press.

Maynard Smith, J. 1989. *Evolutionary genetics.* Oxford: Oxford University Press.

Mayr, E. 1942. *Systematics and the origin of species.* New York: Columbia University Press.

———. 1963. *Animal species and evolution.* Cambridge: Harvard University Press.

McCall, R. A. 1997. Implications of recent geological investigations of the Mozambique Channel for the mammalian colonization of Madagascar. *Proc. Roy. Soc. Lond.,* ser. B, 264:663–665.

McCarthy, M. A., and Lindenmayer, D. B. 1999. Incorporating metapopulation dynamics of greater gliders into reserve design in disturbed landscapes. *Ecology* 80:651–667.

McFarland Symington, M. 1988. Environmental determinants of population densities in *Ateles. Primate Cons.* 9:74–79.

McGraw, W. S. 1998. Three monkeys nearing extinction in the forest reserves of eastern Côte d'Ivoire. *Oryx* 32:233–236.

McKinney, M. L. 1997. Extinction vulnerability and selectivity: Combining ecological and paleonological views. *Ann. Rev. Ecol. Syst.* 28:495–516.

Medellin, R. A., and Equihua, M. 1998. Mammal species richness and habitat use in rainforest and abandoned agricultural fields in Chiapas, Mexico. *J. Appl. Ecol.* 35:13–23.

Medley, K. E. 1993. Primate conservation along the Tana River, Kenya: An examination of the forest habitat. *Cons. Biol.* 7:109–121.

Medway, Lord. 1976. Hunting pressure on orang-utans in Sarawak. *Oryx* 13:332–333.

Mehlman, P. T. 1984. Aspects of the ecology and conservation of the Barbary macaque in the fir forest habitat of the Moroccan Rif Mountains. In *The Barbary macaque: A case study in conservation,* ed. J. E. Fa, 165–199. New York: Plenum Press.

Meier, B., and Albignac, R. 1989. Hairy-eared dwarf lemur rediscovered (*Allocebus trichotis*). *Primate Cons.* 10:27–28.

Meikle, D. B., and Vessey, S. H. 1981. Nepotism among rhesus monkey brothers. *Nature* (Lond.) 294:160–161.

Melnick, D. J., and Hoelzer, G. A. 1992. Differences in male and female macaque dispersal lead to contrasting distributions of nuclear and mitochondrial DNA variation. *Int. J. Primatol.* 13:379–394.

———. 1994. The population genetic consequences of macaque social organisation and behaviour. In *Evolution and ecology of macaque societies,* ed. J. E. Fa and D. G. Lindburg, 413–443. Cambridge: Cambridge University Press.

Melnick, D. J.; Jolly, C. J.; and Kidd, K. K. 1986. The genetics of a wild population of rhesus monkeys (*Macaca mulatta*). 2. The Dunga Gali population in species-wide perspective. *Am. J. Phys. Anthropol.* 71:129–140.

Melnick, D. J., and Kidd, K. K. 1983. The genetic consequences of social group fission in a wild population of rhesus monkeys (*Macaca mulatta*). *Behav. Ecol. Sociobiol.* 12:229–236.

Melnick, D. J.; Pearl, M. C.; and Richard, A. F. 1984. Male migration and inbreeding avoidance in wild rhesus monkeys. *Am. J. Primatol.* 7:229–243.

Ménard, N., and Vallet, D. 1996. Demography and ecology of Barbary macaques (*Macaca sylvanus*) in two different habitats. In *Evolution and ecology of macaque societies,* ed. J. E. Fa and D. J. Lindburg, 106–131. Cambridge: Cambridge University Press.

Menon, S., and Poirier, F. E. 1996. Lion-tailed macaques (*Macaca silenus*) in a disturbed forest fragment: Activity patterns and time budget. *Int. J. Primatol.* 17:969–985.

Meyers, D. 1996. Update on the endangered sifaka of the north. *Lemur News* 2:13–14.

Milner-Gulland, E. J., and Leader-Williams, N. 1992. A model of the incentives for the illegal exploitation of black rhinos and elephants: Poaching pays in Luangwa Valley, Zambia. *J. Appl. Ecol.* 29:388–401.

Milner-Gulland, E. J., and Mace, R. 1998. *Conservation of biological resources.* Oxford: Blackwell.

Milton, K. 1982. Dietary quality and demographic regulation in a howler monkey population. In *The ecology of a tropical forest,* ed. E. G. Leigh, A. S. Rand, and D. M. Windsor, 273–289. Washington D.C.: Smithsonian Institution Press.

———. 1984. The role of food-processing factors in primate food choice. In *Adaptations for foraging in nonhuman primates,* ed. P. Rodman and J. Cant, 249–279. New York: Columbia University Press.

———. 1991. Comparative aspects of diet in Amazonian forest-dwellers. *Phil. Trans. Roy. Soc. Lond.,* ser. B, 334:253–263.

———. 1996. Effects of bot fly (*Alouattamyia baeri*) parasitism on a free-ranging howler monkey (*Alouatta palliata*) population in Panama. *J. Zool., Lond,* 239: 39–63.

Milton, K.; van Soest, P. J.; and Robertson, J. B. 1980. Digestive efficiencies of wild howler monkeys. *Physiol. Zool.* 53:402–409.

Mitani, J. C., and Rodman, P. 1979. Territoriality: The relation of ranging pattern and home range size to defendability, with an analysis of territoriality among primate species. *Behav. Ecol. Sociobiol.* 5:241–251.

Mitani, M. 1991. Niche overlap and polyspecific associations among symaptric cercopithecoids in the Campo Animal Reserve, southwestern Cameroon. *Primates* 32:137–151.

Mitchell, A. H., and Tilson, R. L. 1986. Restoring the balance: Traditional hunting and primate conservation in the Mentawai Islands, Indonesia. In *Primate ecology and conservation,* ed. J. G. Else and P. C. Lee. 249–260. Cambridge: Cambridge University Press.

Mittermeier, R. A. 1987a. Effects of hunting on rain forest primates. In *Primate conservation in the tropical rain forest,* ed. C. W. Marsh and R. A. Mittermeier, 109–146. New York: Alan R. Liss.

———. 1987b. Framework for primate conservation in the Neotropical region. In *Primate conservation in the tropical rain forest,* C. W. Marsh and R. A. Mittermeier, 305–320. New York: Alan R. Liss.

———. 1988. Primate diversity and the tropical forest. In *Biodiversity,* ed. E. O. Wilson, 145–154. Washington, D.C.: National Academy Press.

———. 1991. Hunting and its effect on wild primate populations in Suriname. In *Neotropical wildlife use and conservation,* ed. J. G. Robinson and K. H. Redford, 93–107. Chicago: University of Chicago Press.

Mittermeier, R. A.; Bailey, R. C.; Sponsel, L. E.; and Wolf, K. E. 1977. Primate ranching—Results of an experiment. *Oryx* 13:449–453.

Mittermeier, R. A., and Bowles, I. A. 1994. Reforming the approach of the Global Environment Facility to biodiversity conservation. *Oryx* 28:101–106.

Mittermeier, R. A., and Cheney, D. L. 1987. Conservation of primates and their habitats. In *Primate societies,* ed. B. B. Smuts, D. L. Cheney, R. M. Seyfarth, R. W. Wrangham, and T. T. Struhsaker, 477–490. Chicago: University of Chicago Press.

Mittermeier, R. A., and Coimbra-Filho, A. F. 1977. Primate conservation in Brazilian Amazonia. In *Primate conservation,* ed. Prince Rainier III of Monaco and G. H. Bourne, 117–167. New York: Academic Press.

Mittermeier, R. A.; Kinzey, W. G.; and Mast, R. B. 1989. Neotropical primate conservation. *J. Human Evol.* 18:597–610.

Mittermeier, R. A.; Konstant, W. R.; Nicoll, M. E.; and Langrand, O. 1992. *Lemurs of Madagascar: An action plan for their conservation, 1993–1999.* Gland, Switz.: IUCN.

Mittermeier, R. A.; Tattersall, I.; Konstant, W. R.; Meyer, D. M.; and Mast, R. B. 1994. *Lemurs of Madagascar.* Washington D.C.: Conservation International.

Mittermeier, R. A.; Valle, C. M. C.; Alves, M. C.; Santos, I. B.; Pinto, C. A. M.; Strier, K. B.; Young, A. L.; Veado, E. M.; Constable, I. D.; Paccagnella, S. G.; and Lemos de Sa, R. M. 1987. Current distribution of the muriqui in the Atlantic Forest region of eastern Brazil. *Primate Cons.* 8:143–149.

Mittermeier, R. A., and van Roosmalen, M. G. M. 1981. Preliminary observations on habitat utilization and diet in eight Surinam monkeys. *Folia Primatol.* 36:1–39.

Monkey fossils unearthed in Jamaica. 1997. *Neotrop. Primates* 5:115.

Mönkkönen, M., and Reunanen, P. 1999. On critical thresholds in landscape connectivity: A management perspective. *Oikos* 84:302–305.

Moore, J., and Ali, R. 1984. Are dispersal and inbreeding avoidance related? *Anim. Behav.* 32:94–112.

Morgan, G. S., and Woods, C. A. 1986. Extinction and the zoogeography of West Indian land mammals. *Biol. J. Linn. Soc.* 28:167–203.

Mori, A. 1979a. An experiment on the relation between the feeding speed and the caloric intake through leaf eating in Japanese monkeys. *Primates* 20:185–195.

————. 1979b. Analysis of population changes by measurement of body weight in the Koshima troop of Japanese monkeys. *Primates* 18:331–357.

Morin, P. A.; Moore, J. J.; Chakraborty, R.; Jin, L.; Goodall, J.; and Woodruff, D. 1994. Kin selection, social structure, gene flow and the evolution of chimpanzees. *Science* 265:1193–1201.

Moritz, C. 1995. Uses of molecular phylogenies for conservation. *Phil. Trans. Roy. Soc. Lond.*, ser. B, 349:113–118.

Muchaal, P. K., and Ngandjui, G. 1999. Impact of village hunting on wildlife populations in the Western Dja Reserve, Cameroon. *Cons. Biol.* 13:385–396.

Mugambi, K. G.; Butynski, T. M.; Suleman, M. A.; and Ottichilo, W. 1997. The vanishing De Brazza's monkey (*Cercopithecus neglectus* Schlegel) in Kenya. *Int. J. Primatol.* 18:995–1004.

Murcia, C. 1995. Edge effects in fragmented forests: Implications for conservation. *Trends Ecol. Evol.* 10:58–62.

Murdock, G. P. 1967. *Ethnographic atlas.* Pittsburgh: University of Pittsburgh Press.

Murphree, M. 1996. Approaches to community participation. Paper presented at the ODA African Wildlife Policy Consultation, England.

Myers, N. 1988. Threatened biotas: "Hot spots" in tropical forests. *Environmentalist* 8:187–208.

————. 1990. The biodiversity challenge: Expanded hot-spots analysis. *Environmentalist* 10:243–256.

————. 1994. Tropical deforestation: Rates and patterns. In *The causes of tropical deforestation,* ed. K. Brown and D. W. Pearce, 27–40. London: UCL Press.

Nagy, K. A., and Milton, K. 1979. Energy metabolism and food consumption by wild howler monkeys. *Ecology* 60:475–480.

Nash, L. 1976. Troop fission in free-ranging baboons in the Gombe Stream National Park, Tanzania. *Am. J. Phys. Anthropol.* 44:63–78.

Naughton-Treves, L. 1998. Predicting patterns of crop damage by wildlife around Kibale National Park. *Cons. Biol.* 12:156–168.

Naughton-Treves, L., and Sanderson, S. 1995. Property, politics and wildlife conservation. *World Development* 23:1265–1275.

Naughton-Treves, L.; Treves, A.; Chapman, C.; and Wrangham, R. 1998. Temporal patterns of crop-raiding by primates: Linking food availability in croplands and adjacent forest. *J. Appl. Ecol.* 35:596–606.

Nei, M. 1977. *F*-statistics and analysis of gene diversity in subdivided populations. *Ann. Hum. Genet.* 41:225–233.

Nepstad, D. C.; Veríssimo, A.; Alencar, A.; Nobre, C.; Lima, E.; Lefebvre, P.; Schlesinger, P.; Potter, C.; Moutinho, P.; Mendoza, E.; Cochrane, M.; and Brooks, V. 1999. Large-scale impoverishment of Amazonian forests by logging and fire. *Nature* 398:505–508.

Newmark, W. D. 1986. Species-area relationship and its determinants for mammals in western North American national parks. *Biol. J. Linn. Soc.* 28:83–98.

————. 1987. A land-bridge island perspective on mammalian extinctions in western North American parks. *Nature* 325:430–433.

Newmark, W. D.; Manyanza, D. N.; Gamassa, D. M.; and Sariko, H. I. 1994. The conflict between wildlife and local people living adjacent to protected areas in Tanzania: Human density as a predictor. *Cons. Biol.* 8:252–255.

Newton, P. N. 1989. Associations between langur monkeys (*Presbytis entellus*) and chital deer (*Axis axis*): Chance encounters or a mutualism? *Ethology* 83:89–120.

Nilsson, L. A.; Rabakonandrianina, E.; Pettersson, B.; and Grünmeier, R. 1993. Lemur pollination in the Malagasy rainforest liana *Strongylodon craveniae* (Leguminosae). *Evol. Trends Plants* 7:49–56.

Njiforti, H. L. 1996. Preferences and present demand for bushmeat in north Cameroon: Some implications for wildlife conservation. *Environ. Cons.* 23:149–155.

Noë, R., and Bshary, R. 1997. The formation of red colobus–diana monkey associations under predation pressure from chimpanzees. *Proc. Roy. Soc. Lond.*, ser. B, 264:253–259.

Norton-Griffiths, M. 1998. The economics of wildlife conservation policy in Kenya. In *Conservation of biological resources,* ed. E. J. Milner-Gulland and R. Mace, 279–293. Oxford: Blackwell.

Noss, A. J. 1997. Challenges to nature conservation with community development in central African forests. *Oryx* 31:180–188.

———. 1998. The impacts of BaAka net hunting on rainforest wildlife. *Biol. Cons.* 86:161–167.

Noss, R. F. 1987. Corridors in real landscapes: A reply to Simberloff and Cox. *Cons. Biol.* 1:159–164.

Nozawa, K. 1972. Population genetics of Japanese monkeys. I. Estimation of the effective troop size. *Primates* 13:389–393.

Nozawa, K.; Shotake, T.; Kawamoto, Y.; and Tanabe, Y. 1982. Population genetics of Japanese monkeys. 2. Blood protein polymorphisms and population structure. *Primates* 23:252–271.

Nozawa, K.; Shotake, T.; Minezawa, M.; Kawamoto, Y.; Hayasaka, K.; Kawamoto, S.; and Ito, S. 1991. Population genetics of Japanese monkeys. 3. Ancestry and differentiation of local populations. *Primates* 32:411–435.

Nunney, L., and Campbell, K. A. 1993. Assessing minimum viable population size: Demography meets population genetics. *Trends Ecol. Evol.* 8:234–239.

Oates, J. F. 1977. The guereza and man. In *Primate conservation,* ed. Prince Rainier III and G. H. Bourne, 420–467. New York: Academic Press.

———. 1981. Mapping the distribution of West African rain-forest monkeys: Issues, methods and preliminary results. *Ann. N.Y. Acad. Sci.* 76:53–64.

———. 1986a. *Action Plan for African primate conservation, 1986–90.* Stony Brook, N.Y.: IUCN/SSC Primate Specialist Group.

———. 1986b. African primate conservation: General needs and specific priorities. In *Primates: The road to self-sustaining populations,* ed. K. Benirschke, 21–29. New York: Springer.

———. 1987. Food distribution and foraging behaviour. In *Primate societies,* ed. B. B. Smuts, D. L. Cheney, R. M. Seyfarth, R. W. Wrangham, and T. T. Struhsaker, 169–209. Chicago: University of Chicago Press.

———. 1988. The distribution of *Cercopithecus* monkeys in West African forests. In *A primate radiation: Evolutionary biology of the African guenons,* ed. A. Gautier-Hion, F. Bourlière, J.-P. Gautier, and J. Kingdon, 79–103. Cambridge: Cambridge University Press.

———. 1994a. Africa's primates in 1992: Conservation issues and options. *Am. J. Primatol.* 34:61–71.

Oates, J. F. 1994b. The natural history of African colobines. In *Colobine monkeys: Their ecology, behaviour and evolution,* ed. A. G. Davies and J. F. Oates, 75–128. Cambridge: Cambridge University Press.

———. 1995. The dangers of conservation by rural development—A case-study from the forests of Nigeria. *Oryx* 29:115–122.

———. 1996a. *Status survey and Conservation Action Plan: African primates.* Gland, Switz.: IUCN/SSC Primate Specialist Group.

———. 1996b. Habitat alteration, hunting and the conservation of folivorous primates in African forests. *Aust. J. Ecol.* 21:1–9.

———. 1999. *Myth and reality in the rain forest: How conservation strategies are failing in West Africa.* Los Angeles: University of California Press.

Oates, J. F., and Davies, A. G. 1994. Conclusions: Past, present and future of the colobines. In *Colobine monkeys: Their ecology, behaviour and evolution,* ed. A. G. Davies and J. F. Oates, 347–358. Cambridge: Cambridge University Press.

Oates, J. F.; Davies, A. G.; and Delson, E. 1994. The diversity of living colobines. In *Colobine monkeys: Their ecology, behaviour and evolution,* ed. A. G. Davies and J. F. Oates, 45–73. Cambridge: Cambridge University Press.

Oates, J. F., and Whitesides, G. H. 1990. Association between olive colobus (*Procolobus verus*), diana guenons (*Cercopithecus diana*) and other forest monkeys in Sierra Leone. *Am. J. Primatol.* 21:129–146.

Oates, J. F.; Whitesides, G. H.; Davies, A. G.; Waterman, P. G.; Green, S. M.; Dasilva, G. L.; and Mole, S. 1990. Determinants of variation in tropical forest primate biomass: New evidence from West Africa. *Ecology* 71:328–343.

O'Brien, S. J.; Wildt, D. E.; Goldman, D.; Merril, C. R.; and Bush, M. 1983. The cheetah is depauperate in genetic variation. *Science* 227:459–462.

O'Brien, T. G., and Kinnaird, M. F. 1996. Changing populations of birds and mammals in north Sulawesi. *Oryx* 30:150–156.

O'Brien, T. G.; Kinnaird, M. F.; Dierenfeld, E. S.; Conklin-Brittain, N. L.; Wrangham, R. W.; and Silver, S. C. 1998. What's so special about figs? *Nature* (Lond.) 392:668.

O'Connor, S. 1996. Translocation and introduction of prosimans in Madagascar. *Lemur News* 2:11–13.

Ohsawa, H., and Dunbar, R. I. M. 1984. Variations in the demographic structure and dynamics of gelada baboon populations. *Behav. Ecol. Sociobiol.* 15:231–240.

O'Leary, H., and Fa, J. E. 1993. Effects of tourists on Barbary macaques at Gibralter. *Folia Primatol.* 61:77–91.

Olivier, T. J.; Ober, C.; Buettner-Janusch, J.; and Sade, D. S. 1981. Genetic differentiation among matrilines in social groups of rhesus monkeys. *Behav. Ecol. Sociobiol.* 8:279–285.

Oppenheimer, J. R., and Lang, G. E. 1969. *Cebus* monkeys: Effect on branching of *Gustavia* trees. *Science* 165:187–188.

Ostro, L. E. T.; Silver, S. C.; Koontz, F. W.; Young, T. P.; and Horwich, R. H. 1999. Ranging behaviour of translocated and established groups of black howler monkeys *Alouatta pigra* in Belize, Central America. *Biol. Cons.* 87:181–190.

Owen-Smith, N. 1988. *Megaherbivores.* Cambridge: Cambridge University Press.

Packer, C. R. 1979. Inter-troop transfer and inbreeding avoidance in *Papio anubis. Anim. Behav.* 27:1–36.

Pagel, M. D., and Harvey, P. H. 1989. How mammals produce large-brained off-spring. *Evolution* 42:948–957.

Pagel, M. D.; Harvey, P. H.; and Godfray, H. C. J. 1991. Species abundance, biomass and resource use distributions. *Am. Nat.* 138:836–850.

Pagel, M. D.; May, R. M.; and Collie, A. R. 1991. Ecological aspects of the geographical distribution and diversity of mammalian species. *Am. Nat.* 137:791–815.

Pagel, M. D., and Payne, R. J. H. 1996. How migration affects estimation of the extinction threshold. *Oikos* 76:323–329.

Palo, M. 1994. Population and deforestation. In *The causes of tropical deforestation,* ed. K. Brown and D. W. Pearce, 42–56. London: UCL Press.

Palombit, R. A. 1999. Infanticide and the evolution of pair bonds in nonhuman primates. *Evol. Anthropol.* 7:117–129.

Parker, G. A. 1974. Assessment strategy and the evolution of fighting behaviour. *J. Theoret. Biol.* 47:223–243.

Pastor-Nieto, R., and Williamson, D. 1998. The effect of rainfall seasonality on the geographic distribution of Neotropical primates. *Neotrop. Primates* 6:7–14.

Patterson, B. D., and Atmar, W. 1986. Nested subsets and the structure of insular mammalian faunas and archipelagos. *Biol. J. Linn. Soc.* 28:65–82.

Patterson, H. E. H. 1985. The recognition concept of species. In *Species and speciation,* ed. E. S. Vrba, 21–29. Pretoria: Transvaal Museum Monographs.

Payne, J. 1995. Links between vertebrates and the conservation of Southeast Asian rainforests. In *Ecology, conservation and management of Southeast Asian rainforests,* ed. R. B. Primack and T. E. Lovejoy, 54–65. New Haven: Yale University Press.

Pearce, J., and Amman, K. 1995. *Slaughter of the apes.* London: World Society for the Protection of Animals.

Pearce, D. W., and Turner, R. K. 1990. *Economics of natural resources and the environment.* Hemel Hampstead, Eng.: Harvester Wheatsheaf.

Peetz, A.; Norconck, M. A.; and Kinzey, W. G. 1992. Predation by jaguar on howler monkeys (*Alouatta seniculus*) in Venezuela. *Am. J. Primatol.* 28:223–228.

Peltonen, A., and Hanksi, I. 1991. Pattern of island occupancy explained by colonization and extinction rates in shrews. *Ecology* 72:1698–1708.

Peres, C. A. 1990. Effects of hunting on western Amazonian primate communities. *Biol. Cons.* 54:47–59.

———. 1991. Humboldt's woolly monkeys decimated by hunting in Amazonia. *Oryx* 25:89–95.

———. 1992. Prey-capture benefits in a mixed-species group of Amazonian tamarins, *Saguinus fuscicollis* and *S. mystax. Behav. Ecol. Sociobiol.* 31:339–347.

———. 1993a. Anti-predator benefits in a mixed-species group of Amazonian tamarins. *Folia Primatol.* 61:61–76.

———. 1993b. Structure and spatial organization of an Amazonian terra firme forest primate community. *J. Trop. Ecol.* 9:259–276.

———. 1994. Primate responses to phenological changes in an Amazonian terra firme forest. *Biotropica* 26:98–112.

———. 1996. Food patch structure and plant resource partitioning in interspecific associations of Amazonian tamarins. *Int. J. Primatol.* 17:695–723.

———. 1997a. Primate community structure at twenty western Amazonian flooded and unflooded forests. *J. Trop. Ecol.* 13:381–405.

————. 1997b. Effects of habitat quality and hunting pressure on arboreal folivore densities in Neotropical forests: A case study of howler monkeys. *Folia Primatol.* 68:199–222.

Peres, C. A., and Johns, A. D. 1991–1992. Patterns of primate mortality in a drowning forest: Lessons from the Turucuí dam, Brazilian Amazonia. *Primate Cons.* 12–13:7–10.

Peres, C. A.; Patton, J. L.; and da Silva, M. N. F. 1996. Riverine barriers and gene flow in Amazonian saddle-back tamarins. *Folia Primatol.* 67:113–124.

Peres, C. A., and Terborgh, J. W. 1995. Amazonian nature reserves: An analysis of the defensibility status of existing conservation units and design criteria for the future. *Cons. Biol.* 9:34–46.

Peters, R. H., and Raelson, J. V. 1984. Relations between individual size and mammalian population density. *Am. Nat.* 124:498–517.

Peters, R. L. 1980. *The ecological implications of body size.* Cambridge: Cambridge University Press.

————. 1988. The effect of global climatic change on natural communities. In *Biodiversity*, ed. E. O. Wilson, 450–461. Washington, D.C.: National Academy Press.

Petter, J. J., and Peyrièras, A. 1974. A study of population density and home ranges of *Indri indri* in Madagascar. In *Prosimian biology*, ed. R. D. Martin, G. A. Doyle, and A. C. Walker, 39–48. London: Duckworth.

Petter, J. J.; Schilling, A., and Pariente, G. F. 1975. Observations on the behaviour and ecology of *Phaner furcifer.* In *Lemur biology*, ed. I. Tattersall and R. W. Sussman, 209–218. New York: Plenum Press.

Phillips Conroy, J., and Jolly, C. J. 1988. Dental eruption schedules of wild and captive baboons. *Am. J. Primatol.* 15:17–29.

Pickford, M. 1993. Climate change, biogeography and *Theropithecus.* In *Theropithecus: The rise and fall of a primate genus*, ed. N. Jablonski, 227–244. Cambridge: Cambridge University Press.

Pimbert, M. P., and Pretty, J. N. 1995. *Parks, people and professionals: Putting "participation" into protected area management.* Geneva: United Nations Research Institute for Social Development.

Pimm, S. L., and Askins, R. A. 1995. Forest losses predict bird extinctions in eastern North America. *Proc. Natl. Acad. Sci. USA* 92:9343–9347.

Pimm, S. L.; Jones, H. L.; and Diamond, J. 1988. On the risk of extinction. *Am. Nat.* 132:757–785.

Pimm, S. L.; Moulton, M. P.; and Justice, L. J. 1995. Bird extinctions in the central Pacific. In *Extinction rates*, ed. J. H. Lawton and R. M. May, 75–87. Oxford: Oxford University Press.

Pimm, S. L.; Russell, G. J.; Gittleman, J. L.; and Brooks, T. M. 1995. The future of biodiversity. *Science* 269:347–350.

Plavcan, J. M., and van Schaik, C. P. 1992. Intrasexual selection and canine dimorphism in anthropoid primates. *Am. J. Phys. Anthropol.* 87:461–477.

————. 1997. Intrasexual competition and body weight dimorphism in anthropoid primates. *Am. J. Phys. Anthropol.* 103:37–68.

Plumptre, A. J. 1995. Modelling the impact of large herbivores on the food supply of mountain gorillas and implications for management. *Biol. Cons.* 75:147–155.

Plumptre, A. J.; Bizumuremyi, J.-B.; Uwimana, F.; and Ndaruhebeye, J.-D. 1997. The effects of the Rwandan civil war on poaching of ungulates in the Parc National des Volcans. *Oryx* 31:265–273.

Plumptre, A. J., and Harris, S. 1995. Estimating the biomass of large mammalian herbivores in a tropical montane forest: A method of faecal counting that avoids assuming a "steady state" system. *J. Appl. Ecol.* 32:111–120.

Plumptre, A. J., and Reynolds, V. 1994. The effects of selective logging on the primate populations in the Budongo Forest Reserve, Uganda. *J. Appl. Ecol.* 31:631–641.

Pope, T. R. 1992. The influence of dispersal patterns and mating system on genetic differentiation within and between populations of the red howler monkey (*Alouatta seniculus*). *Evolution* 46:1112–1128.

———. 1996. Socioecology, population fragmentation and patterns of genetic loss in endangered primates. In *Conservation genetics: Case histories from nature,* ed. J. C. Avise and J. L. Harwick, 119–159. London: Chapman and Hall.

———. 1998a. Effects of demographic change on group kin structure and gene dynamics of populations of red howling monkeys. *J. Mammal.* 79:692–712.

———. 1998b. Genetic variation in remnant populations of the woolly spider monkey (*Brachyteles arachnoides*). *Int. J. Primatol.* 19:95–109.

Pray, L. A.; Schwartz, J. M.; Goodnight, C. J.; and Stevens, L. 1994. Environmental dependency of inbreeding depression: Implications for conservation biology. *Cons. Biol.* 8:562–568.

Prendergast, J. R.; Quinn, R. M.; Lawton, J. H.; Eversham, B. C.; and Gibbons, D. W. 1993. Rare species, the coincidence of diversity hotspots and conservation strategies. *Nature* 365:335–337.

Pressey, R. L., and Nicholls, A. O. 1989. Efficiency in conservation evaluation: Scoring versus iterative approaches. *Biol. Cons.* 50:199–218.

Primate Specialist Group. 1996. IUCN/SSC Primate Specialist Group Triennial Report, 1994–1996. *Neotrop. Primates* 4:153–154.

Prins, H., and Reitsma, J. M. 1989. Mammalian biomass in an African equatorial rain forest. *J. Anim. Ecol* 58:851–861.

Proctor, J. 1995. Rainforests and their soils. In *Ecology, conservation and management of Southeast Asian rainforests,* ed. R. B. Primack and T. E. Lovejoy, 87–104. New Haven: Yale University Press.

Proctor, M.; Yeo, P.; and Lack, A. 1996. *The natural history of pollination.* London: HarperCollins.

Puertas, P., and Bodmer, R. E. 1993. Conservation of a high diversity primate assemblage. *Biodiversity Cons.* 2:586–593.

Purvis, A. 1995. A composite estimate of primate phylogeny. *Phil. Trans. Roy. Soc. Lond.,* ser. B, 348:405–421.

Purvis, A.; Gittleman, J. L.; Cowlishaw, G.; and Mace, G. M. n.d. Predicting extinction risk in declining species. Unpublished MS.

Purvis, A., and Harvey, P. H. 1995. Mammal life-history evolution: A comparative test of Charnov's model. *J. Zool., Lond.* 237:259–283.

Purvis, A.; Nee, S.; and Harvey, P. 1995. Macroevolutionary inferences from primate phylogeny. *Proc. Roy. Soc. Lond.,* ser. B, 260:329–333.

Pusey, A. E., and Packer, C. 1987. Dispersal and philopatry. In *Primate societies,* ed. B. B. Smuts, D. L. Cheney, R. M. Seyfarth, R. W. Wrangham, and T. T. Struhsaker, 250–266. Chicago: University of Chicago Press.

Rabinowitz, A., and Khaing, S. T. 1998. Status of selected mammal species in north Myanmar. *Oryx* 32:201–208.

Rabinowitz, D. 1981. Seven forms of rarity. In *The biological aspects of rare plant conservation,* ed. H. Synge, 205–217. New York: Wiley.

Ralls, K., and Ballou, J. 1982. Effects of inbreeding on infant mortality in captive primates. *Int. J. Primatol.* 3:491–505.

Ralls, K.; Ballou, J.; and Templeton, A. R. 1988. Estimates of lethal equivalents and the costs of inbreeding in mammals. *Cons. Biol.* 2:185–193.

Ramirez, M. 1984. Population recovery in the moustached tamarin (*Saguinus mystax*): Management strategies and mechanisms of recovery. *Am. J. Primatol.* 7:245–259.

Ranta, P.; Blom, T.; Niemelä, J.; Joensuu, E.; and Siitonen, M. 1998. The fragmented Atlantic rain forest of Brazil: Size, shape and distribution of forest fragments. *Biodiversity Cons.* 7:385–403.

Rao, P. S. S., and Inbaraj, S. G. 1980. Inbreeding effects on fetal growth and development. *J. Med. Genet.* 17:27–33.

Rapoport, E. H. 1982. *Areography: Geographical strategies of species.* Oxford: Pergamon.

Rasker, R.; Martin, M. V.; and Johnson, R. L. 1992. Economics: Theory versus practice in wildlife management. *Cons. Biol.* 6:338–349.

Rayner, J. M. V. 1985. Linear relations in biomechanics: The statistics of scaling functions. *J. Zool., Lond.* 206:415–439.

Redford, K. [H.], and Robinson, J. G. 1987. A game of choice: Patterns of Indian and colonist hunting in the Neotropics. *Am. Anthropol.* 89:650–667.

Redford, K. H., and Stearman, A. M. 1993. Forest-dwelling native Amazonians and the conservation of biodiversity: Interests in common or in collision? *Cons. Biol.* 7:248–255.

Reed, J. M. 1999. The role of behaviour in recent avian extinctions and endangerments. *Cons. Biol.* 13:232–241.

Reed, K. E., and Fleagle, J. G. 1995. Geographic and climatic control of primate diversity. *Proc. Natl. Acad. Sci. USA* 92:7874–7876.

Richard, A. F. 1985. *Primates in nature.* San Francisco: Freeman.

Richard, A. F.; Goldstein, S. J.; and Dewar, R. E. 1989. Weed macaques: The evolutionary implications of macaque feeding ecology. *Int. J. Primatol.* 10:569–594.

Ridley, M. 1986. The number of males in a primate troop. *Anim. Behav.* 34:1848–1858.

———. 1996. *Evolution.* 2d ed. Oxford: Blackwell.

Rijksen, H. D. 1978. *A field study on Sumatran orang utans* (Pongo pygmaeus abelii Lesson 1827): *Ecology, behaviour and conservation.* Wageningen: Veenman and Zonen.

Robinson, J. G. 1988. Group size in wedge-capped capuchin monkeys *Cebus olivaceus* and the reproductive success of males and females. *Behav. Ecol. Sociobiol.* 23:187–197.

———. 1993a. The limits to caring: Sustainable living and the loss of biodiversity. *Cons. Biol.* 7:20–28.

———. 1993b. "Believing what you know ain't so": Response to Holdgate and Munro. *Cons. Biol.* 7:941–942.

————. 1998. Limits to sustainable hunting in tropical forests. Paper presented at the Bushmeat Hunting and African Primates Conference, Primate Society of Great Britain, Bristol.

Robinson, J. G., and Bennett, E. 1999. *Hunting for sustainability in tropical forests.* New York: Columbia University Press.

Robinson, J. G., and Janson, C. H. 1987. Capuchins, squirrel monkeys and atelines: Socioecological convergence with old world primates. In *Primate societies,* ed. B. B. Smuts, D. L. Cheney, R. M. Seyfarth, R. W. Wrangham, and T. T. Struhsaker, 69–82. Chicago: University of Chicago Press.

Robinson, J. G., and Ramirez, C, J. 1982. Conservation biology of Neotropical primates. In *Mammalian biology in South America,* ed. M. A. Mares and H. H. Genoways, 329–344. Pittsburgh: University of Pittsburgh Press.

Robinson, J. G., and Redford, K. H. 1986. Body size, diet and population density of Neotropical forest mammals. *Am. Nat.* 128:665–680.

————. 1991. Sustainable harvest of Neotropical forest animals. In *Neotropical wildlife use and conservation,* ed. J. G. Robinson and K. H. Redford, 415–429. Chicago: University of Chicago Press.

————. 1994. Measuring the sustainability of hunting in tropical forests. *Oryx* 28:249–256.

Rodgers, W. A., and Homewood, K. M. 1982. Biological values and conservation prospects for the forests and primate populations of the Uzungwa Mountains, Tanzania. *Biol. Cons.* 24:285–304.

Rodman, P. S. 1973. Synecology of Bornean primates. 1. A test for interspecific interactions in spatial distribution of five species. *Am. J. Phys. Anthropol.* 38:655–660.

————. 1978. Diets, densities and distributions of Bornean primates. In *The ecology of arboreal folivores,* ed. G. Montgomery, 465–478. Washington, D.C.: Smithsonian Institution Press.

Rodman, P. S., and Mitani, J. C. 1987. Orangutans: Sexual dimorphism in a solitary species. In *Primate societies,* ed. B. B. Smuts, D. L. Cheney, R. M. Seyfarth, R. W. Wrangham, and T. T. Struhsaker, 146–154. Chicago: University of Chicago Press.

Rodríguez, J. P., and Rojas-Suárez, F. 1996. Guidelines for the design of conservation strategies for the animals of Venezuela. *Cons. Biol.* 10:1245–1252.

Rogers, C. M., and Caro, M. J. 1998. Song sparrows, top carnivores and nest predation: A test of the mesopredator release hypothesis. *Oecologia* 116:227–233.

Rogers, J., and Kidd, K. K. 1996. Nucleotide polymorphism, effective population size, and dispersal distances in the yellow baboons (*Papio hamadryas cynocephalus*) of Mikumi National Park, Tanzania. *Am. J. Primatol.* 38:157–168.

Ron, T.; Henzi, S. P.; and Motro, U. 1994. A new model of fission in primate troops. *Anim. Behav.* 47:223–226.

Rosenbaum, B.; O'Brien, T.; Kinnaird, M.; and Supriatna, J. 1998. Population densities of Sulawesi crested black macaque (*Macaca nigra*) on Bacan and Sulawesi, Indonesia: Effects of habitat disturbance and hunting. *Am. J. Primatol.* 44:89–106.

Rosenblum, L. L.; Supriatna, J.; Hasan, M. N.; and Melnick, D. J. 1997. High mitochondrial DNA diversity with little structure within and among leaf monkey populations (*Trachypithecus cristatus* and *Trachypithecus auratus*). *Int. J. Primatol.* 18:1005–1028.

Rosenzweig, M. L. 1995. *Species diversity in space and time.* Cambridge: Cambridge University Press.

Ross, C. 1988. The intrinsic rate of natural increase and reproductive effort in primates. *J. Zool., Lond.* 214:199–219.

————. 1992. Environmental correlates of the intrinsic rate of natural increase in primates. *Oecologia* 90:383–390.

Ross, C., and Srivastava, A. 1994. Factors influencing the population density of the Hanuman langur (*Presbytis entellus*) in Sariska Tiger Reserve. *Primates* 35:361–367.

Ross, C.; Srivastava, A.; and Pirta, R. S. 1993. Human influences on the population density of Hanuman langurs *Presbytis entellus* and rhesus macaques *Macaca mulatta* in Shimla, India. *Biol. Cons.* 65:159–163.

Rowe, N. 1996. *The pictorial guide to living primates.* New York: Pogonias Press.

Rowell, T. E., and Mitchell, B. J. 1991. Comparison of seed dispersal by guenons in Kenya and capuchins in Panama. *J. Trop. Ecol.* 7:269–274.

Rowell, T. E., and Richards, S. M. 1979. Reproductive strategies of some African monkeys. *J. Mammal.* 60:58–69.

Rudran, R. 1978. Socioecology of the blue monkeys (*Cercopithecus mitis stuhlmanni*) of the Kibale Forest, Uganda. *Smithsonian Contrib. Zool.* 249:1–88.

Ruggerio, A. 1994. Latitudinal correlates of the sizes of mammalian geographical ranges in South America. *J. Biogeog.* 21:545–559.

Russell, G. J.; Brooks, T. M.; McKinney, M. M.; and Anderson, C. G. 1998. Present and future taxonomic selectivity in bird and mammal extinctions. *Cons. Biol.* 12:1365–1376.

Ruvulo, M.; Pam, D.; Zehr, S.; Goldberg, T.; Disotell, T.; and von Dornum, M. 1994. Gene trees and hominoid phylogeny. *Proc. Natl. Acad. Sci. USA* 91:8900–8904.

Rylands, A. B. 1993–1994. Population viability analyses and the conservation of the lion tamarins, *Leontopithecus*, of south-east Brazil. *Primate Cons.* 14–15:34–42.

Rylands, A. B., and Bernardes, A. T. 1989. Two priority regions for primate conservation in the Brazilian Amazon. *Primate Cons.* 10:56–62.

Rylands, A. B.; Coimbra-Filho, A. F.; and Mittermeier, R. A. 1993. Systematics, geographic distribution, and some notes on the conservation status of the Callitrichidae. In *Marmosets and tamarins: Systematics, behaviour and ecology,* ed. A. B. Rylands, 11–77. Oxford: Oxford University Press.

Rylands, A. B., and Keuroghlian, A. 1988. Primate populations in continuous forest and forest fragments in central Amazonia. *Acta Amazonica* 18:291–307.

Rylands, A. B.; Mittermeier, R. A.; and Rodriguez-Luna, E. 1997. Conservation of Neotropical primates: Threatened species and an analysis of primate diversity by country and region. *Folia Primatol.* 68:134–160.

Ryman, N.; Baccus, R.; Reuterwell, C.; and Smith, M. H. 1981. Effective population size, generation interval, and potential loss of genetic variability in game species under different hunting regimes. *Oikos* 36:257–266.

Sabater-Pi, J. 1981. Exploitation of gorillas *Gorilla gorilla gorilla* Savage and Wyman 1847 in Rio Muni, Republic of Equatorial Guinea, West Africa. *Biol. Cons.* 19:131–140.

Sabater-Pi, J., and Groves, C. 1972. The importance of higher primates in the diet of the Fang of Rio Muni. *Man* 7:239–243.

Sacher, G. A., and Staffeldt, E. F. 1974. Relation of gestation time to brain weight for placental mammals: Implications for the theory of vertebrate growth. *Am. Nat.* 108:593–616.

Sade, D. S.; Cushing, K.; Cushing, P.; Dunaid, J.; Figueroa, A.; Kaplan, J.; Lauer, C.; Rhodes, D.; and Schneider, J. 1976. Population dynamics in relation to social structure on Cayo Santiago. *Ybk. Phys. Anthropol.* 20:253–262.

Salafsky, N. 1993. Mammalian use of a buffer zone agroforestry system bordering Gunung Palung National Park, West Kalimantan, Indonesia. *Cons. Biol.* 7: 928–933.

Samuels, A., and Altmann, J. 1991. Baboons of the Amboseli basin: Demographic stability and change. *Int. J. Primatol.* 12:1–19.

Sanchez-Azofeifa, G. A.; Quesada-Mateo, C.; Gonzalez-Quesada, P.; Dayanandan, S.; and Bawa, K. S. 1999. Protected areas and conservation of biodiversity in the tropics. *Cons. Biol.* 13:407–411.

Sauther, M. L. 1998. Interplay of phenology and reproduction in ring-tailed lemurs: Implications for ring-tailed lemur conservation. *Folia Primatol.* 69:309–320.

Sayer, J. A.; Harcourt, C. S.; and Collins, N. M. 1992. *The conservation atlas of tropical forests: Africa.* London: Macmillan.

Schaller, G. 1967. *The deer and the tiger.* Chicago: University of Chicago Press.

Scheffrahn, W.; de Ruiter, J. R.; and van Hooff, J. A. R. A. M. 1994. Genetic relatedness between populations of *Macaca fascicularis* on Sumatra. In *Evolution and ecology of macaque societies,* ed. J. E. Fa and D. G. Lindburg, 20–42. Cambridge: Cambridge University Press.

Scheffrahn, W.; Ménard, N.; Vallet, D.; and Gaci, B. 1993. Ecology, demography and population genetics of Barbary macaques in Algeria. *Primates* 34:381–394.

Scheffrahn, W.; Rabarivola, C.; and Rumpler, Y. 1998. Field studies of population genetics in *Eulemur:* A discussion of their potential importance in conservation. *Folia Primatol.* 69:147–151.

Schoener, T. W., and Spiller, D. A. 1992. Is extinction related to temporal variability in population size? An empirical answer for orb spiders. *Am. Nat.* 139:1176–1207.

Schupp, E. W. 1988. Seed and early seedling predation in the forest understorey and in treefall gaps. *Oikos* 51:71–78.

Schwarzkopf, L., and Rylands, A. B. 1989. Primate species richness in relation to habitat structure in Amazonian rainforest fragments. *Biol. Cons.* 48:1–12.

Seal, U. S.; Foose, T. J.; and Ellis, S. 1994. Conservation Assessment and Management Plans (CAMPs) and Global Captive Action Plans (GCAPs). In *Creative conservation,* ed. P. J. S. Olney, G. M. Mace, and A. T. C. Feistner, 312–325. London: Chapman and Hall.

Shaffer, M. L. 1981. Minimum population sizes for species conservation. *BioScience* 31:131–134.

Shipman, P.; Bosler, W.; and Davis, K. L. 1981. Butchering of giant geladas at an Acheulian site. *Current Anthropol.* 22:257–268.

Shotake, T., and Nozawa, K. 1984. Blood protein variations in baboons. 2. Genetic variability within and among herds of gelada baboons in central Ethiopian plateau. *J. Human Evol.* 13:265–274.

Shotake, T.; Nozawa, K.; and Tanabe, Y. 1977. Blood protein variations in baboons. 1. Gene exchange and genetic distance between *Papio anubis, Papio hamadryas* and their hybrid. *Jap. J. Gen.* 52:223–237.

Siegfried, W. R.; Benn, G. A.; and Gelderblom, C. M. 1998. Regional assessment and conservation implications of landscape characteristics of African national parks. *Biol. Cons.* 84:131–140.

Sigg, H.; Stolka, A.; Abegglen, J. J.; and Dasser, V. 1982. Life history of baboons: Physical development, infant mortality, reproductive parameters and family relationships. *Primates* 23:473–487.

Silk, J. B. 1983. Local resource competition and facultative adjustment of sex ratios in relation to competitive abilities. *Am. Nat.* 121:56–64.

———. 1988. Social mechanisms of population regulation in a captive group of bonnet macaques (*Macaca radiata*). *Am. J. Primatol.* 14:111–124.

Silkiluwasha, F. 1981. The distribution and conservation status of the Zanzibar red colobus. *Afr. J. Ecol.* 19:187–194.

Silva, M., and Downing, J. A. 1995. The allometric scaling of density and body mass: A nonlinear relationship for terrestrial mammmals. *Am. Nat.* 145:704–727.

Simberloff, D. 1992. Do species-area curves predict extinction in fragmented forest? In *Tropical deforestation and species extinction,* ed. T. C. Whitmore and J. A. Sayer, 75–89. London: Chapman and Hall.

Simberloff, D., and Abele, L. G. 1976. Island biogeography theory and conservation practice. *Science* 191:285–286.

Simberloff, D., and Cox, J. 1987. Consequences and costs of conservation corridors. *Cons. Biol.* 1:63–71.

Skole, D., and Tucker, C. 1993. Tropical deforestation and habitat fragmentation in the Amazon: Satellite data from 1978 to 1988. *Science* 260:1905–1910.

Skorupa, J. P. 1986. Responses of rainforest primates to selective logging in Kibale Forest, Uganda: A summary report. In *Primates: The road to self-sustaining populations,* ed. K. Benirshcke, 57–70. Berlin: Springer.

Slade, N. A.; Gomulkiewicz, R.; and Alexander, H. M. 1998. Alternatives to Robinson and Redford's method of assessing overharvest from incomplete demographic data. *Cons. Biol.* 12:148–155.

Smith, A. P.; Horning, N.; and Morre, D. 1997. Regional biodiversity planning and lemur conservation with GIS in western Madagascar. *Cons. Biol.* 11:498–512.

Smith, D. G. 1982. A comparison of the demographic structure and growth of freeranging and captive groups of rhesus monkeys (*Macaca mulatta*). *Primates* 23:24–30.

Smith, F. D. M.; May, R. M.; Pellew, R.; Johnson, T. H.; and Walter, K. S. 1993a. How much do we know about the current extinction rate? *Trends Ecol. Evol.* 8:375–378.

———. 1993b. Estimating extinction rates. *Nature* 364:494–496.

Smith, R. L., and Jungers, W. L. 1997. Body mass in comparative perspective. *J. Human Evol.* 32:523–559.

Smuts, B. B., and Nicholson, N. 1989. Reproduction in wild female olive baboons. *Am. J. Primatol.* 19:229–246.

Smuts, B. L.; Cheney, D. L.; Seyfarth, R. M.; Wrangham, R. W.; and Struhsaker, T. T., eds. 1987. *Primate societies.* Chicago: University of Chicago Press.

Snyder, N. F. R.; Derrickson, S. R.; Beissinger, S. R.; Wiley, J. W.; Smith, T. B.; Toone, W. D.; and Miller, B. 1996. Limitations of captive breeding in endangered species recovery. *Cons. Biol.* 10:338–348.

Sommer, V. 1987. Infanticide among free-ranging langurs (*Presbytis entellus*) at Jodhpur (Rajasthan/India): Recent observations and a reconsideration of hypotheses. *Primates* 28:163–197.

Sommer, V., and Rajpurohit, L. S. 1989. Male reproductive success in harem troops of Hanuman langurs (*Presbytis entellus*). *Int. J. Primatol.* 10:293–317.

Soulé, M. E. 1980. Thresholds for survival: Maintaining fitness and evolutionary potential. In *Conservation biology: An evolutionary-ecological perspective,* ed. M. E. Soulé and B. A. Wilcox, 151–169. Sunderland, Mass.: Sinauer.

———. 1991. Conservation: Tactics for a constant crisis. *Science* 253:744–750.

Soulé, M. E.; Wilcox, B. A.; and Holtby, C. 1979. Benign neglect: A model of faunal collapse in the game reserves of East Africa. *Biol. Cons.* 15:259–271.

Southwick, C. H., and Siddiqi, M. F. 1968. Population trends of rhesus monkeys in villages and towns of northern India, 1959–1965. *J. Anim. Ecol.* 37:199–204.

———. 1977. Population dynamics of rhesus monkeys in northern India. In *Primate conservation,* ed. Prince Rainier III and G. H. Bourne, 339–362. New York: Academic Press.

———. 1994. Population status of nonhuman primates in Asia, with emphasis on rhesus macaques in India. *Am. J. Primatol.* 34:51–59.

Southwick, C. H.; Siddiqi, M. F.; and Oppenheimer, J. R. 1983. Twenty-year changes in rhesus monkey populations in agricultural areas of northern India. *Ecology* 64:434–439.

Spinage, C. 1998. Social change and conservation misrepresentation in Africa. *Oryx* 32:265–276.

———. 1999. A reply to Colchester. *Oryx* 33:5–8.

Spinney, L. 1998. Monkey business. *New Scientist,* 2 May 1998, 18–19.

Srivastava, A., and Dunbar, R. I. M. 1996. The mating system of Hanuman langurs: A problem in optimal foraging. *Behav. Ecol. Sociobiol.* 39:219–226.

Stafford, B. J., and Ferreira, F. M. 1995. Predation attempts on callitrichids in the Atlantic coastal rain forest of Brazil. *Folia Primatol.* 65:229–233.

Stammbach, E. 1987. Desert, forest and montane baboons: Multi-level societies. In *Primate societies,* ed. B. B. Smuts, D. L. Cheney, R. M. Seyfarth, R. W. Wrangham, and T. T. Struhsaker, 112–120. Chicago: University of Chicago Press.

Stanford, C. B. 1995. The influence of chimpanzee predation on group size and antipredator behaviour in red colobus monkeys. *Anim. Behav.* 49:577–587.

Stanford, C. B.; Wallis, J.; Matama, H.; and Goodall, J. 1994. Patterns of predation by chimpanzees on red colobus monkeys in the Gombe National Park, 1982–1991. *Am. J. Phys. Anthropol.* 94:213–228.

Stanley Price, M. 1991. A review of mammal re-introductions, and the role of the Re-introduction Specialist Group of IUCN/SSC. In *Beyond captive breeding: Re-introducing endangered mammals to the wild,* ed. J. H. W. Gipps, 9–25. Symposia of the Zoological Society of London no. 62. Oxford: Clarendon Press.

Starin, E. D. 1989. Threats to the monkeys of the Gambia. *Oryx* 23:208–214.

Stelzner, J. K. 1988. Thermal effects on movement patterns of yellow baboons. *Primates* 29:91–105.

Sterck, E. H. M. 1997. Determinants of female dispersal in Thomas langurs. *Am. J. Primatol.* 42:179–198.

———. 1998. Female dispersal, social organization, and infanticide in langurs: Are they linked to human disturbance? *Am. J. Primatol.* 44:235–254.

———. 1999. Variation in langur social organization in relation to the socioecological model, human habitat alteration, and phylogenetic constraints. *Primates* 40: 199–213.

Sterling, E. J., and Rakotoarison, N. 1998. Rapid assessment of richness and density of primate species on the Masoala Peninsula, eastern Madagascar. *Folia Primatol.* 69 (suppl 1):109–116.

Stephens, D. W., and Krebs, J. R. 1986. *Foraging theory.* Princeton: Princeton University Press.

Stephenson, P. J., and Newby, J. E. 1997. Conservation of the Okapi Wildlife Reserve, Zaire. *Oryx* 31:49–58.

Stevens, G. C. 1989. The latitudinal gradient in geographic range: How so many species coexist in the tropics. *Am. Nat.* 133:240–246.

Stevenson, M.; Baker, A.; and Foose, T. J. 1991. Conservation assessment and management plan for primates. IUCN/SSC Captive Breeding Specialist Group and IUCN/SSC Primate Specialist Group report.

Stevenson, M.; Foose, T. J.; and Baker, A. 1992. Global Captive Action Plan for primates. IUCN/SSC Captive Breeding Specialist Group report.

Stoner, K. E. 1996. Prevalence and intensity of intestinal parasites in mantled howling monkeys (*Alouatta palliata*) in northeastern Costa Rica: Implications for conservation biology. *Cons. Biol.* 10:539–546.

Strier, K. B. 1991. Demography and conservation of an endangered primate, *Brachyteles arachnoides. Cons. Biol.* 5:214–218.

———. 1993–1994. Viability analyses of an isolated population of muriqui monkeys (*Brachyteles arachnoides*): Implications for primate conservation and demography. *Primate Cons.* 14–15:43–52.

———. 1997. Behavioural ecology and conservation biology of primates and other animals. *Adv. Study Behav.* 26:101–158.

Struhsaker, T. T. 1973. A recensus of vervet monkeys in the Masai-Amboseli Game Reserve, Kenya. *Ecology* 54:930–932.

———. 1976. A further decline in numbers of Amboseli vervet monkeys. *Biotropica* 8:211–214.

———. 1978. Food habits of five monkey species in the Kibale Forest, Uganda. In *Recent advances in primatology,* vol. 1. *Behaviour,* ed. D. J. Chivers and J. Herbet, 225–247. London: Academic Press.

———. 1981. Forest and primate conservation in East Africa. *Afr. J. Ecol.* 19:99–114.

———. 1997. *Ecology of an African rain forest.* Gainesville: University Press of Florida.

Struhsaker, T. T., and Leakey, M. 1990. Prey selectivity by crowned hawk-eagles on monkeys in the Kibale Forest, Uganda. *Behav. Ecol. Sociobiol.* 26:435–443.

Struhsaker, T. T., and Leland, L. 1980. Observations on two rare and endangered populations of red colobus monkeys in East Africa: *Colobus badius gordonorum* and *Colobus badius kirkii. Afr. J. Ecol.* 18:191–216.

Struhsaker, T. T., and Siex, K. S. 1996. The Zanzibar red colobus monkey *Procolobus kirkii:* Conservation status of an endangered island endemic. *Afr. Primates* 2:54–61.

———. 1998. Translocation and introduction of the Zanzibar red colobus: Success and failure with an endangered island endemic. *Oryx* 32:277–284.

Strum, S. C. 1981. Processes and products of change: Baboon predatory behaviour at Gilgil, Kenya. In *Omnivorous primates: Gathering and hunting in human evolution,* ed. R. S. Harding and G. Teleki, 255–302. New York: Columbia University Press.

Strum, S. C., and Southwick, C. H. 1986. Translocation of primates. In *Primates: The road to self-sustaining populations,* ed. K. Benirschke, 949–958. New York: Springer.

Strum, S. C., and Western, D. 1982. Variations in fecundity with age and environment in olive baboons (*Papio anubis*). *Am. J. Primatol.* 3:61–76.

Stuart, M. D., and Strier, K. B. 1995. Primates and parasites: A case for a multidisciplinary approach. *Int. J. Primatol.* 16:577–593.

Stuart, S. N. 1991. Re-introductions: To what extent are they needed? In *Beyond captive breeding: Re-introducing endangered mammals to the wild,* ed. J. H. W. Gipps, 27–37. Symposia of the Zoological Society of London, no. 62. Oxford: Clarendon Press.

Sugiyama, Y., and Ohsawa, H. 1982. Population dynamics of Japanese monkeys with special reference to the effect of artificial feeding. *Folia Primatol.* 39:238–263.

Sussman, R. W. 1978. Nectar-feeding by prosimians and its evolutionary and ecological implications. In *Recent advances in primatology,* vol. 3, *Evolution,* ed. D. J. Chivers and K. A. Joysey, 119–124. London: Academic Press.

Sussman, R. W.; Green, G. M.; and Sussman, L. K. 1994. Satellite imagery, human ecology, anthropology, and deforestation in Madagascar. *Human Ecol.* 22: 333–354.

Sussman, R. W., and Kinzey, W. G. 1984. The ecological role of the Callitrichidae: A review. *Am. J. Phys. Anthropol.* 64:419–449.

Sussman, R. W., and Phillips-Conroy, J. E. 1995. A survey of the distribution and density of the primates of Guyana. *Int. J. Primatol.* 16:761–791.

Sussman, R. W., and Raven, P. H. 1978. Pollination by lemurs and marsupials: An archaic coevolutionary system. *Science* 200:731–736.

Swanson, T. M. 1995. *The economics and ecology of biodiversity decline: The forces driving global change.* Cambridge: Cambridge University Press.

Swart, J., and Lawes, M. J. 1996. The effect of habitat patch connectivity on samango monkey (*Cercopithecus mitis*) metapopulation persistence. *Ecol. Modelling* 93: 57–74.

Swart, J.; Lawes, M. J.; and Perrin, M. R. 1993. A mathematical model to investigate the demographic viability of low-density samango monkey (*Cercopithecus mitis*) populations in Natal, South Africa. *Ecol. Modelling* 70:289–303.

Sweitzer, R. A.; Jenkins, S. H.; and Berger, J. 1997. Near-extinction of porcupines by mountain lions and consequences of ecosystem change in the Great Basin Desert. *Cons. Biol.* 11:1407–1417.

Taylor, B. 1995. The reliability of using population viability analysis for risk classification of species. *Cons. Biol.* 9:551–558.

Taylor, C. M., and Gotelli, N.J. 1994. The macroecology of *Cyprinella:* Correlates of phylogeny, body size and geographical range. *Am. Nat.* 144:549–569.

Teas, J.; Ritchie, T.; Taylor, H.; Siddiqi, M. F.; and Southwick, C. H. 1981. Natural regulation of rhesus monkey populations in Kathmandu, Nepal. *Folia Primatol.* 35:117–123.

Teleki, G. 1973. *The predatory behaviour of wild chimpanzees.* Lewisburg, Pa.: Bucknell University Press.

———. 1989. Population status of wild chimpanzees (*Pan troglodytes*) and threats to survival. In *Understanding chimpanzees,* ed. P. G. Heltne and L. A. Marquardt, 312–353. Cambridge: Harvard University Press.

Tenaza, R. R. 1976. Songs, choruses and countersinging of Kloss' gibbons (*Hylobates klossii*) in Siberut Island, Indonesia. *Z. Tierpsychol.* 40:37–52.

Tenaza, R. R., and Tilson, R. L. 1977. Evolution of long-distance alarm calls in Kloss's gibbon. *Nature* (Lond.) 268:233–235.

———. 1985. Human predation and Kloss's gibbon (*Hylobates klossii*) sleeping trees in Siberut Island, Indonesia. *Am. J. Primatol.* 8:299–308.

Terborgh, J. 1983. *Five New World primates: A study in comparative ecology.* Princeton: Princeton University Press.

———. 1986. Keystone plant resources in the tropical forest. In *Conservation biology: The science of scarcity and diversity,* ed. M. E. Soulé, 330–344. Sunderland, Mass.: Sinauer.

Terborgh, J., and van Schaik, C. P. 1987. Convergence vs. nonconvergence in primate communities. In *Organisation of communities: Past and present,* ed. J. H. R. Gee and P. S. Giller, 205–226. Oxford: Blackwell.

Terborgh, J., and Winter, B. 1980. Some causes of extinction. In *Conservation biology: An evolutionary:ecological approach,* ed. M. E. Soule and B. A. Wilcox, 119–133. Sunderland, Mass.: Sinauer.

Thirgood, J. V. 1984. The demise of the Barbary macaque habitat—Past and present forest cover of the Maghreb. In *The Barbary macaque: A case study in conservation,* ed. J. E. Fa, 19–69. New York: Plenum Press.

Thomas, C. D., and Lennon, J. J. 1999. Birds extend their ranges northwards. *Nature* 399:213.

Thomas, S. C. 1991. Population densities and patterns of habitat use among anthropoid primates of the Ituri Forest, Zaire. *Biotropica* 23:68–83.

Tilman, D.; Lehman, C. L.; and Yin, C. 1997. Habitat destruction, dispersal, and deterministic extinction in competitive communities. *Am. Nat.* 149:407–435.

Tilman, D.; May, R. M.; Lehman, C. L.; and Nowak, M. A. 1994. Habitat destruction and the extinction debt. *Nature* 371:65–66.

Tilson, R. L. 1977. Social organization of Simakobu monkeys (*Nasalis concolor*) in Siberut Island, Indonesia. *J. Mammal.* 58:202–212.

Tilson, R. L., and Tenaza, R. R. 1976. Monogamy and duetting in an Old World monkey. *Nature* (Lond.) 263:320–321.

Tingpalapong, M.; Watson, W. T.; Whitmire, R. E.; Chapple, F. E.; and Marshall, J. T., Jr. 1981. Reactions of captive gibbons to natural habitat and wild conspecifics after release. *Nat. Hist. Bull. Siam. Soc.* 29:31–40.

Tomiuk, J.; Bachmann, L.; Leipoldt, M.; Ganzhorn, J. U.; Ries, R.; Weis, M.; and Loeschckes, V. 1997. Genetic diversity of *Lepilemur mustelinus ruficaudatus,* a nocturnal lemur of Madagascar. *Cons. Biol.* 11:491–497.

Tracy, C. R., and George, T. L. 1992. On the determinants of extinction. *Am. Nat.* 139:102–122.

Treves, A. 1998. The influence of group size and neighbours on vigilance in two species of arboreal monkeys. *Behaviour* 135:453–482.

Treves, A., and Naughton-Treves, L. 1997. Case study of a chimpanzee recovered from poachers and temporarily released with wild conspecifics. *Primates* 38: 315–324.

Trivers, R. L., and Willard, D. E. 1973. Natural selection for parental ability to vary the sex ratio. *Science* 179:90–92.

Turner, R. K.; Pearce, D.; and Bateman, I. 1994. *Environmental economics: An elementary introduction.* Hemel Hempstead, Eng.: Harvester Wheatsheaf.

Turpie, J. K. 1995. Prioritizing South African estuaries for conservation: A practical example using waterbirds. *Biol. Cons.* 74:175–185.

Tutin, C. E. G. 1999. Fragmented living: Behavioural ecology of primates in a forest fragment in the Lopé Reserve, Gabon. *Primates* 40:249–265.

Tutin, C. E. G., and Fernandez, M. 1984. Nationwide census of gorilla (*Gorilla g. gorilla*) and chimpanzee (*Pan t. troglodytes*) populations in Gabon. *Am. J. Primatol.* 6:313–336.

Tutin, C. E. G.; Fernandez, M.; Rogers, M. E.; Williamson, E. A.; and McGrew, W. C. 1991. Foraging profiles of sympatric lowland gorillas and chimpanzees in the Lopé Reserve, Gabon. *Proc. Roy. Soc. Lond.,* ser. B, 334:179–186.

Tutin, C. E. G.; Ham, R. M.; White, L. J.; and Harrison, M. J. S. 1997. The primate community of the Lopé Reserve, Gabon: Diets, responses to fruit scarcity and effects on biomass. *Am. J. Primatol.* 42:1–24.

Tutin, C. E. G., and Oslisly, R. 1995. *Homo, Pan* and *Gorilla:* co-existence over 60,000 years at Lopé in central Gabon. *J. Hum. Evol.* 28:597–602.

Tutin, C. E. G., and White, L. J. T. 1998. Primate phenology and frugivory: Present, past and future patterns in the Lopé Reserve, Gabon. In *Dynamics of populations and communities in the tropics,* ed. D. M. Newbury, H. H. T. Prins, and N. Brown, 309–337. British Ecological Society Symposium 37. Oxford: Blackwell.

Tutin, C. E. G.; Williamson, E. A.; Rogers, M. E.; and Fernandez, M. 1991. A case study of a plant-animal relationship: *Cola lizae* and lowland gorillas in the Lopé Reserve, Gabon. *J. Trop. Ecol.* 7:181–199.

Uehara, S. 1997. Predation on mammals by chimpanzee (*Pan troglodytes*). *Primates* 38:193–214.

Vane-Wright, R. I.; Humphries, C. J.; and Williams, P. H. 1991. What to protect? Systematics and the agony of choice. *Biol. Cons.* 55:235–254.

van Hooff, J. A. R. A. M., and van Schaik, C. P. 1994. Male bonds: Affiliative relationships among nonhuman male primates. *Behaviour* 130:309–337.

van Schaik, C. P. 1983. Why are diurnal primates living in groups? *Behaviour* 87:120–144.

———. 1989. The ecology of female social relationships. In *Comparative socioecology,* ed. V. Standen and R. Foley, 195–218. Oxford: Blackwell.

van Schaik, C. P. 1999. The socioecology of fission-fusion sociality in orangutans. *Primates* 40:69–86.

van Schaik, C. P.; Assink, P. R.; and Salafsky, N. 1992. Territorial behaviour in southeast Asian langurs: Resource defence or mate defence? *Am. J. Primatol.* 26: 233–242.

van Schaik, C. P., and Dunbar, R. I. M. 1990. The evolution of monogamy in large primates: A new hypothesis and some critical tests. *Behaviour* 115:30–62.

van Schaik, C. P., and Hörstermann, M. 1994. Predation risk and the number of adult males in a primate group: A comparative test. *Behav. Ecol. Sociobiol.* 35:261–272.

van Schaik, C. P., and Hrdy, S. B. 1991. Intensity of local resource competition shapes the relationship between maternal rank and sex ratios at birth in cercopithecine primates. *Am. Nat.* 138:1555–1562.

van Schaik, C. P., and Kappeler, P. M. 1993. Lifehistory, activity period and lemur social systems. In *Lemur social systems and their ecological basis,* ed. P. M. Kappeler and J. U. Ganzhorn, 241–260. New York: Plenum Press.

———. 1996. The social systems of gregarious lemurs: Lack of convergence with anthropoids due to evolutionary disequilibrium? *Ethology* 102:915–941.

———. 1997. Infanticide risk and the evolution of male-female association in primates. *Proc. Roy. Soc. Lond.,* ser. B, 264:1687–1694.

van Schaik, C. P.; Terborgh, J. W.; and Wright, S. J. 1993. The phenology of tropical forests: Adaptive significance and consequences for primary consumers. *Ann. Rev. Ecol. Syst.* 24:353–377.

van Schaik, C. P., and van Noordwijk, M. 1985. The evolutionary effect of the absence of felids on the social organisation of the Simeulue monkey (*Macaca fascicularis fusca* Miller 1903). *Int. J. Primatol.* 6:180–200.

van Schaik, C. P., and van Noordwijk, M. A. 1988. Scramble and contest feeding competition among female long-tailed macaques (*Macaca fascicularis*). *Behaviour* 105:77–98.

———. 1989. The special role of male *Cebus* monkeys in predation avoidance and its effect on group composition. *Behav. Ecol. Sociobiol.* 24:265–276.

van Soest, P. J. 1982. *Nutritional ecology of the ruminant.* Ithaca: Cornell University Press.

Vàsàrhelyi, K., and Martin, R. D. 1994. Evolutionary biology, genetics and the management of endangered primate species. In *Creative conservation,* ed. P. J. S. Olney, G. M. Mace, and A. T. C. Feistner, 118–143. London: Chapman and Hall.

Veríssimo, A.; Júnior, C. S.; Stone, S.; and Uhl, C. 1998. Zoning of timber extraction in the Brazilian Amazon. *Cons. Biol.* 12:128–136.

Vickers, W. T. 1991. Hunting yields and game composition over ten years in an Amazon Indian territory. In *Neotropical wildlife use and conservation,* ed. J. G. Robinson and K. H. Redford, 53–81. Chicago: University of Chicago Press.

Vié, J.-C., and Richard-Hansen, C. 1997. Primate translocation in French Guiana: A preliminary report. *Neotropical Primates* 5:1–3.

Vogler, A. P., and Desalle, R. 1994. Diagnosing units of conservation management. *Cons. Biol.* 8:354–363.

Voysey, B. C.; McDonald, K. E.; Rogers, M. E.; Tutin, C. E. G.; and Parnell, R. J. 1999a. Gorillas and seed dispersal in the Lopé Reserve, Gabon. 1. Gorilla acquisition by trees. *J. Trop. Ecol.* 15:23–38.

———. 1999b. Gorillas and seed dispersal in the Lopé Reserve, Gabon. 2. Survival and growth of seedlings. *J. Trop. Ecol.* 15:39–60.

Wadley, R. L.; Colfer, C. J. P.; and Hood, I. G. 1997. Hunting primates and managing forests: The case of Iban Forest farmers in Indonesian Borneo. *Human Ecol.* 25:243–271.

Waser, P. M. 1977. Feeding, ranging and group size in the mangabey (*Cercocebus albigena*). In *Primate ecology,* ed. T. H. Clutton-Brock, 183–222. London: Academic Press.

————. 1982. Polyspecific associations: Do they occur by chance? *Anim. Behav.* 30:1–8.

————. 1987. Interactions among primate species. In *Primate societies,* ed. B. B. Smuts, D. L. Cheney, R. M. Seyfarth, R. W. Wrangham, and T. T. Struhsaker, 210–226. Chicago: University of Chicago Press.

Wasser, S. K., and Starling, A. K. 1988. Proximate and ultimate causes of reproductive suppression among female yellow baboons at Mikumi National Park, Tanzania. *Am. J. Primatol.* 16:97–121.

Watanabe, K. 1981. Variations in group composition and population density of the two sympatric Mentawaian leaf-monkeys. *Primates* 22:145–160.

Waterman, P. G. 1984. Food acquisition and processing as a function of plant chemistry. In *Food acquisition and processing in primates,* ed. D. Chivers, B. A. Wood, and A. Bilsborough, 177–211. New York: Plenum Press.

Waterman, P. G., and Kool, K. M. 1994. Colobine food selection and plant chemistry. In *Colobine monkeys: Their ecology, behaviour and evolution,* ed. G. L. Davies and J. F. Oates, 251–284. Cambridge: Cambridge University Press.

Waterman, P. G.; Ross, J. A. M.; Bennett, E. L.; and Davies, A. G. 1988. A comparison of the floristics and chemistry of the tree flora in two Malaysian rain forests and the influence of leaf chemistry on populations of colobine monkeys in the Old World. *Biol. J. Linn. Soc.* 34:1–32.

Watts, D. P. 1985. Relations between group size and composition and feeding competition in mountain gorilla groups. *Anim. Behav.* 33:72–85.

Weber, A. W. 1987. Socioecologic factors in the conservation of Afromontane forest reserves. In *Primate conservation in the tropical rain forest,* ed. C. W. Marsh and R. A. Mittermeier, 205–229. New York: Alan R. Liss.

Weber, A. W., and Vedder, A. L. 1983. Population dynamics of the Virunga gorillas: 1959–1978. *Biol. Cons.* 26:341–366.

Weisenseel, K.; Chapman, C. A.; and Chapman, L. J. 1993. Nocturnal primates of the Kibale Forest: Effects of selective logging on prosimian densities. *Primates* 34:445–450.

Western, D., and Gicholi, H. 1993. Segregation effects and the impoverishment of savanna parks: The case for ecosystem viability analysis. *Afr. J. Ecol.* 31:269–281.

Western, D., and Ssemakula, J. 1981. The future of the savanna ecosystems: Ecological islands or faunal enclaves? *Afr. J. Ecol.* 19:7–19.

Western, D., and van Praet, C. 1973. Cyclical changes in the habitat and climate of an East African ecosystem. *Nature* (Lond.) 241:104–106.

Wheatley, B. P.; Putra, D. K. H.; and Gonder, M. K. 1996. A comparison of wild and food-enhanced long-tailed macaques (*Macaca fascicularis*). In *Evolution and ecology of macaque societies,* ed. J. E. Fa and D. G. Lindburg, 182–206. Cambridge: Cambridge University Press.

White, L. J. T. 1994. Biomass of rain forest mammals in the Lopé Reserve, Gabon. *J. Anim. Ecol.* 63:499–512.

White, L. J. T.; Rogers, M. E.; Tutin, C. E. G.; Williamson, E. A.; and Fernandez, M. 1995. Herbaceous vegetation in different forest types in the Lopé Reserve, Gabon: Implications for keystone food availability. *Afr. J. Ecol.* 33:124–141.

White, L. J. T., and Tutin, C. E. G. n.d. Why chimpanzees and gorillas respond differently to logging: A cautionary tale from Gabon. In *African rain forest ecology and conservation*, ed. B. Weber, L. J. T. White, A. Vedder, and H. Simons Morland. New Haven: Yale University Press. In press.

Whiten, A.; Byrne, R. W.; Barton, R. A.; Waterman, P. G.; and Henzi, S. P. 1991. Dietary and foraging strategies of baboon. *Phil. Trans. Roy. Soc. Lond.*, ser. B, 334:187–197.

Whitesides, G. H.; Oates, J. F.; Green, S.; and Kluberdanz, R. P. 1988. Estimating primate densities from transects in a West African rain forest: A comparison of techniques. *J. Anim. Ecol.* 57:345–367.

Whitmore, T. C. 1984. *Tropical rain forests of the Far East.* Oxford: Clarendon Press.

Whitten, A. L. 1983. Diet and dominance among female vervet monkeys (*Cercopithecus aethiops*). *Am. J. Primatol.* 5:139–159.

Wilcox, B. A. 1980. Insular ecology and conservation. In *Conservation biology: An evolutionary-ecological perspective,* ed. M. E. Soulé and B. A. Wilcox, 95–117. Sunderland, Mass.: Sinauer.

Wilkie, D. S. 1989. Impact of roadside agriculture on subsistence hunting in the Ituri Forest of northeastern Zaire. *Am. J. Phys. Anthropol.* 78:485–494.

Wilkie, D. S.; Curran, B.; Tshombe, R.; and Morelli, G. A. 1998a. Modeling the sustainability of subsistence farming and hunting in the Ituri Forest of Zaire. *Cons. Biol.* 12:137–147.

———. 1998b. Managing bushmeat hunting in Okapi Wildlife Reserve, Democratic Republic of Congo. *Oryx* 32:131–144.

Wilkie, D. S., and Finn, J. T. 1988. A spatial model of land use and forest regeneration in the Ituri Forest of northeastern Zaire. *Ecol. Modelling* 41:307–323.

———. 1990. Slash-burn cultivation and mammal abundance in the Ituri Forest, Zaire. *Biotropica* 22:90–99.

Wilkie, D. S.; Sidle, J. G.; and Boundzanga, G. C. 1992. Mechanized hunting, market hunting, and a bank loan in the Congo. *Cons. Biol.* 6:570–580.

Williams, P. H. 1999. Key sites for conservation: Area-selection methods for biodiversity. In *Conservation in a changing world,* ed. G. M. Mace, A. Balmford, and J. R. Ginsberg, 211–250. Cambridge: Cambridge University Press.

Williams, P. H., and Gaston, K. J. 1996. Comparing character diversity among biotas. In *Biodiversity: A biology of numbers and difference,* ed. K. J. Gaston, 54–76. Oxford: Blackwell.

Williams, P. H.; Gaston, K. J.; and Humphries, C. J. 1994. Do conservationists and molecular biologists value differences between organisms in the same way? *Biodiversity Letters* 2:67–78.

Williams, P. H.; Gibbons, D.; Margules, C.; Rebelo, A.; Humphries, C.; and Pressey, R. 1996. A comparison of richness hotspots, rarity hotspots and complementary areas for conserving diversity using British birds. *Cons. Biol.* 10:155–174.

Williamson, D. 1997. Primate socioecology: Development of a conceptual model for the early hominids. Ph.D. thesis, University College London.

Williamson, D., and Dunbar, R. I. M. 1999. Energetics, time budgets and group size. In *Comparative primate socioecology,* ed. P. C. Lee, 320–328. Cambridge: Cambridge University Press.

Willis, J. C. 1922. *Age and area: A study in geographical distribution and origin of species.* Cambridge: Cambridge University Press.

Willner, L. 1989. Sexual dimorphism in primates. Ph.D. thesis, University of London.

Wilson, A. C., and Stanley Price, M. R. 1994. Reintroduction as a reason for captive breeding. In *Creative conservation*, ed. P. J. S. Olney, G. M. Mace, and A. T. C. Feistner, 243–264. London: Chapman and Hall.

Wilson, C. C., and Johns, A. D. 1982. Diversity and abundance of selected animal species in undisturbed forest, selectively logged forest and plantations in East Kalimantan, Indonesia. *Biol. Cons.* 24:205–218.

Wilson, E. O., and Bossert, W. H. 1977. *A primer of population biology.* Sunderland, Mass.: Sinauer.

Winkler, P.; Loch, H.; and Vogel, C. 1984. Life history of Hanuman langurs (*Presbytis entellus*): Reproductive parameters, infant mortality and troop development. *Folia Primatol.* 43:1–23.

Winterhalder, B., and Lu, F. 1997. A forager-resource population ecology model and implications for indigenous conservation. *Cons. Biol.* 11:1354–1364.

Wolf, C. M.; Garland, T., Jr.; and Griffith, B. 1998. Predictors of avian and mammalian translocation success: Reanalysis with phylogenetically independent contrasts. *Biol. Cons.* 86:243–255.

Wolf, C. M.; Griffith, B.; Reed, C.; and Temple, S. A. 1996. Avian and mammalian translocations: Update and reanalysis of 1987 survey data. *Cons. Biol.* 10:1142–1154.

Wolfheim, J. H. 1983. *Primates of the world: Distribution, abundance and conservation.* Seattle: Harwood.

Wong, G., and Carrillo, E. 1996. Squirrel monkey viewing and tourism in Costa Rica. In *Assessing the sustainability of uses of wild species*, ed. R. Prescott-Allen and C. Prescott-Allen, 37–39. Gland, Switz.: IUCN.

Wood, C. A., and Lovett, R. 1974. Rainfall, drought and the solar cycle. *Nature* (Lond.) 251:594–596.

Woodford, M. H., and Rossiter, P. B. 1994. Disease risks associated with wildlife translocation projects. In *Creative conservation*, ed. P. J. S. Olney, G. M. Mace, and A. T. C. Feistner, 178–200. London: Chapman and Hall.

World Resources Institute. 1994. *World resources, 1994–95: A guide to the global environment.* Oxford: Oxford University Press.

———. 1996. *World resources, 1996–97.* Oxford: Oxford University Press.

Wrangham, R. W. 1980. An ecological model of female-bonded primate groups. *Behaviour* 75:262–300.

———. 1987. Evolution of social structure. In *Primate societies*, ed. B. B. Smuts, D. L. Cheney, R. M. Seyfarth, R. W. Wrangham, and T. T. Struhsaker, 282–296. Chicago: University of Chicago Press.

Wrangham, R. W., and Bergmann Riss, E. 1990. Rates of predation on mammals by chimpanzees, 1972–1975. *Primates* 31:157–170.

Wrangham, R. W.; Chapman, C. A.; and Chapman, L. J. 1994. Seed dispersal by forest chimpanzees in Uganda. *J. Trop. Ecol.* 10:355–368.

Wrangham, R. W.; Conklin, N.; Chapman, C.; and Hunt, K. D. 1991. The significance of fibrous foods for Kibale Forest chimpanzees. *Phil. Trans. Roy. Soc. Lond.*, ser. B, 334:171–178.

Wrangham, R. W.; Conklin, N. L.; Etot, G.; Obua, J.; Hunt, K. D.; Hauser, M. D.; and Clark, A. P. 1993. The value of figs to chimpanzees. *Int. J. Primatol.* 14: 243–256.

Wright, P. C. 1989. The nocturnal primate niche in the New World. *J. Hum. Evol.* 18:635–658.

———. 1992. Primate ecology, rainforest conservation, and economic development: Building a national park in Madagascar. *Evol. Anthropol.* 1:29–33.

———. 1997. The future of biodiversity in Madagascar: A view from Ranomafana National Park. In *Natural change and human impact in Madagascar,* ed. S. M. Goodman and B. D. Patterson, 381–405. Washington, D.C.: Smithsonian Institution Press.

———. 1998. Impact of predation risk on the behaviour of *Propithecus diadema edwardsi* in the rain forest of Madagascar. *Behaviour* 135:483–512.

Wright, P. C.; Haring, D.; Simons, E. L.; and Anadu, P. 1987. Tarsiers: A conservation perspective. *Primate Cons.* 8:51–54.

Yeager, C. P. 1997. Orangutan rehabilitation in Tanjung Puting National Park, Indonesia. *Cons. Biol.* 11:802–805.

Yoder, A. D. 1997. Back to the future: A synthesis of strepsirhine systematics. *Evol. Anthropol.* 6:11–22.

Yoder, A. D.; Cartmill, M.; Ruvolo, M.; Smith, K.; and Vilgalys, R. 1996. Ancient single origin of Malagasy primates. *Proc. Natl. Acad. Sci. USA* 93:5122–5126.

Yost, J. A., and Kelly, P. M. 1983. Shotguns, blowguns, and spears: The analysis of technological efficiency. In *Adaptive responses of native Amazonians,* ed. R. B. Hames and W. T. Vickers, 189–224. New York: Academic Press.

Young, T. P. 1994. Natural die-offs of large mammals: Implications for conservation. *Cons. Biol.* 8:410–418.

Young, T. P., and Isbell, L. A. 1994. Minimum group size and other conservation lessons exemplified by a declining primate population. *Biol. Cons.* 68:129–134.

Zhang, Y.; Quan, G.; Lin, Y.; and Southwick, C. 1989. Extinction of rhesus monkeys (*Macaca mulatta*) in Xinglung, north China. *Int. J. Primatol.* 10:375–381.

Zhao, Q.-K. 1996. Etho-ecology of Tibetan macaques at Mount Emei, China. In *Evolution and ecology of macaque societies,* ed. J. E. Fa and D. G. Lindburg, 263–289. Cambridge: Cambridge University Press.

Zuberbühler, K.; Nöe, R.; and Seyfarth, R. M. 1997. Diana monkey long-distance calls: Messages for conspecifics and predators. *Anim. Behav.* 53:589–604.

INDEX

Species are indexed by Latin name (for a list of common and Latin names, see appendix 1). Entries for higher-taxon groupings (e.g., genus or family) do not incorporate lower-taxon entries within that grouping (e.g., species within genera or genera within families). Thus readers who wish to find all references to taxa in the subfamily Colobinae must seek entries under this subfamily heading, the constituent genera headings (such as *Presbytis, Procolobus*), and the constituent species headings (such as *Presbytis melalophos, Procolobus badius*). A similar arrangement is used for geographic regions: these are primarily indexed by continent and country but also include additional entries to subcontinental regions (e.g., Amazonia, Borneo, West Africa) and specific localities within countries (e.g., Kibale Forest in Uganda, Sulawesi in Indonesia). In each case the smaller-scale areas are not incorporated in the larger-scale entries, although subcontinental regions are cross-referenced at the continental entry and specific localities are cross-referenced at the country entry.

Abundance, 81, 105–106, 111; tables 5.2, 5.3; fig. 5.6
  body mass, 111–114, 277–280; figs. 5.7, 9.8, 9.9
  diagnosing population decline, 389–390, 393
  distribution, 114–117
  extinction risk classification, 295–298; tables 10.3, 10.4
  extinctions, global, 174, 179–181
  food availability, 106–111, 267; figs. 5.5, 9.6
  habitat fragmentation, 208, 211, 215, 233, 235–236, 388
  habitat modification, 216–217, 219–221, 226–228, 231–232, 237–240; table 8.7; figs. 8.7, 8.8, 8.10
  hunting, 246, 255–256, 267, 277–287, 388; figs. 9.6, 9.8, 9.9, 9.11
  niche, 111, 113–114, 178, 388
  population viability analysis, 311–312
  protected areas, 331
  rarity, 93–95, 111; table 5.1
Accessibility, 384–386
  conservation areas, 320, 342–343
  habitat disturbance, 193–194, 284, 325–326

  hunting, 194, 255–260, 267–268, 284, 325–326, 337; fig. 9.3
  rivers, 257, 342–343
  roads, 193–194, 214–215, 256–260, 284, 325–326, 395
Afghanistan, 9
Africa, 16–17, 20–22, 182–185; table 2.2; figs. 2.3, 2.5, 2.7. *See also* East Africa; North Africa; West Africa; *and under specific countries*
  community ecology, 63–67, 73, 109; table 4.2; figs. 4.1, 4.5, 4.6
  conservation tactics, 330, 360, 366; table 11.1
  distribution, 96–97, 101–102, 105; fig. 5.2
  extinction risk, 180, 299–301, 303; tables 10.5, 10.6, 10.9
  habitat disturbance, 193, 196, 205–206; tables 8.2, 8.4
  hunting, 242, 260, 265; tables 9.1, 9.2, 9.6
  priority setting, areas, 290–291, 313, 315, 317, 320; figs. 10.4, 10.5, 10.6, 10.7
  priority setting, species, 290–291; tables 10.1, 10.4, 10.5, 10.6, 10.9
  protected areas, 333, 337, 341; table 11.2

Agriculture, 192–196, 200, 203–204, 324,
    382, 387; tables 8.1, 8.3, 8.6; fig. 8.6. *See
    also* Crop raiding; Habitat modification
  effects on primates, 146, 218–222, 230,
    390–392, 395; table 8.7; fig. 8.8
  with hunting, 283–284
  interspecific vulnerability, 175, 229–232,
    238, 240, 285; table 8.11; fig. 8.12
Algeria, 154–155, 309; table 6.4
Allee effects, 142, 163, 171–173, 179, 184,
    189; fig. 7.4. *See also* Small-population
    processes
  density-dependent models, 311–312, 352
Allocebus trichotis, 160; table 12.1
Alouatta, 19, 38–39, 43, 89–90, 107, 110,
    171, 231; tables 2.1, 3.1
  hunting, 255, 267, 278, 348; figs. 9.6, 9.7
Alouatta belzebul, 228, table 6.4
Alouatta caraya, 251, table 5.1
Alouatta coibensis, 292
Alouatta fusca, 235, 285, 388
Alouatta palliata, 83, 85, 87–88, 90, 109
  habitat disturbance, 209–210, 219, 283,
    338; fig. 8.5
  hunting, 283
  population dynamics, 131, 134, 138–139,
    145; table 6.4; fig. 6.7
Alouatta pigra, 145, 344, 355, 362, 374, 395;
    tables 6.4, 8.7, 11.4, 11.10
Alouatta seniculus, 85, 87–88, 91; table 5.4;
    fig. 4.10
  habitat disturbance, 212, 235; tables 8.7,
    8.11, 8.12
  hunting, figs. 9.2, 11.4
  population dynamics, 131, 138, 145, 148–
    149, 154, 173; table 6.4; fig. 6.7
  translocation, 376; table 11.10
Alouattinae. See *Alouatta*
Altitude, 102, 108–109; fig. 5.4. *See also*
    Climate
  population biology, 124–125, 129, 143
  species richness, 62–67; table 4.2
Amazonia, 26, 76, 79, 109, 154; fig. 2.7
  conservation strategy, 320–321
  distribution, 99–100, 104–105; fig. 5.3
  habitat disturbance, 193, 200, 216, 228,
    326, 390
  habitat fragmentation, 201, 203
  hunting, 253–257, 265, 277–279, 325–326,
    348, 397; figs. 9.8, 9.9
  protected areas, 334, 340–343
  Waorani hunters, 252, 254–255

Amboseli (Kenya), 41, 49, 149, 151, 208; figs.
    3.5, 6.1
  community ecology, 63, 67; tables 4.2, 4.3
  environmental stochasticity, 134–135, 165,
    173, 181, 388
Americas, the, 16–22, 19–20, 25–26, 97;
    table 2.2. *See also* Amazonia; Caribbean;
    and under specific countries
  abundance, 109
  community structure, 63–65, 72–74; fig.
    4.5
  conservation strategy, 313, tables 10.1,
    10.4
  distribution, 102, 104; fig. 7.6
  extinction risk, 301; table 10.5
  habitat disturbance, 193, 196, 200, 326;
    table 8.2
  hunting, 242, 248, 256, 277, 386; tables
    9.1, 9.2, 9.3, 9.4, 9.6
  protected areas, 332–333, 335; table 11.2
  translocation projects, 373; table 11.10
Angola, tables 2.3, 8.3, 8.4, 11.3; fig. 11.3
Anthropoids, 9, 18–22, 29; fig. 2.1
Aotinae. See *Aotus*
Aotus, 11, 19, 35, 352; tables 2.1, 3.1; fig. 9.2
Aotus nancymae, 272; table 9.7
Archaeoindris, 185; table 7.3
Archaeolemur, 185; table 7.3
Asia, 16–17, 20–22, 63–65, 109, 148; table
    2.2; fig. 4.5. *See also* Borneo; Eurasia;
    Southeast Asia; *and under specific coun-
    tries*
  conservation strategy; tables 10.1, 10.4
  conservation tactics, 330, 359–361, 366;
    table 11.1
  extinction risk, 299–303; tables 10.5, 10.6,
    10.9
  habitat disturbance, 180, 193, 196, 200;
    table 8.2
  hunting, 180, 242, 247–248, 264–265;
    tables 9.1, 9.2, 9.6
  protected areas, table 11.2
  translocation projects, table 11.10
Ateles, 48–49, 89, 91, 107; table 5.4
  hunting, 278, 352; fig. 9.7
Ateles belzebuth, fig. 11.4
Ateles geoffroyi, 83, 87, 109, 366
  habitat disturbance, 219–220, 235, 283;
    table 8.7
  hunting, 283
  translocation success, 371
Ateles paniscus, 79, 85, 87–88, 91; fig. 4.10

habitat disturbance, 220; tables 8.7, 8.11, 8.12
hunting, 348; figs. 4.10, 11.4
Atelinae, 19, 188, 283; tables 2.1, 3.1
Atlantic Forest (Brazil), 162, 285
  conservation action, 310, 368, 395; fig. 10.3
  habitat disturbance, 193, 200–203, 205, 236–237, 285
Australasia, 16
Avahi laniger, 78; fig. 4.9

Babakotia, table 7.3
Barbados (Caribbean), 214, 272, 274
Barro Colorado Island (Panama), 83, 85, 91, 138, 163, 371; tables 4.3, 4.4; fig. 6.7
Belize, 344, 395; tables 6.4, 11.4
Bioko Island (Equatorial Guinea), 256–260, 351, 382, 397; tables 9.3, 9.4, 9.5
Biomass, 67–73; tables 4.3, 4.4
Birthrate, 28, 32–33, 49, 53, 83, 104, 113. *See also* Life history; Population dynamics
  extinction risk, 179–181
  food availability, 122–126, 130–134
  hunting, 255, 274, 281
  intraspecific determinants, 122–126; fig. 6.1
  population dynamics, 119–122, 137–138, 176; table 6.1
  population viability analysis, 305–312
  small-population processes, 163–165, 169–171, 178
  sustainable harvesting, 396–397
  tourism, 357
Body mass, 28–32; table 2.1; fig. 3.1. *See also* Energetics
  abundance, 108–109, 111–114, 277–280; figs. 5.7, 9.8, 9.9
  captive breeding, 364
  competition, interspecific, 76–77, 81, 83, 105
  distribution, 63, 100, 102–105, 154
  ecology, 30–31, 34–36, 73, 137, 178; figs. 3.2, 3.3
  extinction risk classification, 295
  extinctions, 175–181, 185, 188; table 7.3; fig. 7.8
  habitat disturbance, 226, 235, 237–238, 285
  hunting, effects on primates, 267, 269, 272, 277–281, 285, 287, 350, 388; figs. 9.7, 9.8, 9.9, 9.10, 11.5

hunting, hunter preferences, 242, 246, 252, 255, 259, 287, 348; table 9.5; fig. 9.2
  life history, 28, 31–32, 396–397
  population dynamics, 129, 131; table 6.2; fig. 6.4
  small-population processes, 176–179; fig. 7.6
  translocation success, 370
Bolivia, 19, 95, 101, 342; tables 2.3, 8.3, 11.3; fig. 11.3
Borneo, 60, 75, 95, 368. *See also* Brunei; Kalimantan; Sabah; Sarawak
  Iban hunters, 251, 254
Botswana, 99; table 4.2
Brachyteles. See *Brachyteles arachnoides*
Brachyteles arachnoides, 86, 145
  conservation tactics, 362, 367–368
  declining-population processes, 388, 392
  small-population processes, 164, 236–237, 310–311; table 6.4; fig. 10.3
Brazil, 18, 26, 59, 95, 154, 160; tables 2.3, 4.1, 6.4. *See also* Atlantic Forest; Ponta da Castanha
  community ecology, 35, 75, 79–80, 107
  conservation strategy, 320–321
  habitat disturbance, 63, 193, 196, 200–201; tables 8.1, 8.3
  hunting, 256
  protected areas, 340–343; table 11.3; fig. 11.3
  translocation, 366
Brunei, 95
Budongo Forest (Uganda), 212–213, 227–228, 239; tables 4.2, 8.10
Burma, 265
Bushmeat trade, 242, 247, 254, 257–260, 326, 382; tables 9.3, 9.4, 9.5; fig. 10.8. *See also* Hunting
  accessibility, 256–257, 284, 384–386
  conservation management, 351, 398–402
  sustainability, 260, 270, 284, 351–352, 396

Caipora bambuiorum, 188
Callicebinae. See *Callicebus*
Callicebus, 19, 26, 325–326, 388–389; tables 2.1, 5.4
Callicebus cupreus, table 8.11
Callicebus moloch, 101; table 8.7; fig. 9.2
Callicebus torquatus, tables 8.7, 8.11
Callimico goeldi, table 10.2
Callithrix, 9, 74, 280, 326

Callithrix emiliae, table 6.4

Callithrix flaviceps, 86, 388

Callithrix geoffroyi, table 6.4

Callithrix humeralifer, table 6.4

Callithrix jacchus, 95, 162; table 6.4

Callithrix penicillata, table 6.4

Callithrix pygmaea, 95; table 8.11

Callitrichidae, 19, 26, 45, 46, 99, 126; tables
   2.1, 5.2; fig. 2.1

   captive breeding, table 11.6

   crop raiding, table 8.5

   extinction risk, 303; tables 10.7, 10.8

   heterozygosity, 144–145, 153, 179, 236;
      table 6.4

   hunting, 242; tables 9.1, 9.6

   inbreeding, 171

Cameroon, 76, 79–80; tables 2.3, 4.4; fig. 4.8

   habitat disturbance, tables 8.1, 8.3, 8.4; fig.
      8.2

   hunting, 248, 251, 254–256, 258–259, 401;
      table 9.4

   protected areas, table 11.3; fig. 11.3

Canada, 261

Captive breeding, 359–361, 365, 399–400;
      tables 11.1, 11.6

   case studies, 364–365; table 11.7

   constraints, 362–364; fig. 11.9

   priority setting, 293, 359–361; table 11.6

   small-population processes, 169–170, 362–
      364; fig. 11.9

   translocation success, 370–373; table 11.9

Caribbean, 19, 189, 377, 383, 385. *See also*
      Barbados; Cayo Santiago (Puerto Rico)

Carrying capacity (*K*), 107–111. *See also*
      Food availability

   population dynamics, 121, 130–131; fig.
      6.5

   population viability analysis, 306, 309–
      312

   sustainable harvest, 269–270

   translocation, 367, 369, 374

Catarrhini, 9–11, 19–22, 48, 101, 121; table
      2.1

   community ecology, 72–73, 90–91

Catastrophes. *See* Environmental stochas-
      ticity

Cayo Santiago (Puerto Rico), 120, 123, 127–
      128, 147–149, 167; table 6.1; fig. 6.1

Cebidae, 26, 43, 46, 48, 76; tables 2.1, 5.2;
      figs. 2.1, 3.4

   captive breeding, 361; table 11.6

   crop raiding, table 8.5

extinction risk, 302; tables 10.7, 10.8

   hunting, 242, 349; tables 9.1, 9.6

Cebinae, 19; table 2.1

Cebus, 39–40, 110; table 5.4

   community ecology, 80–82, 89

   hunting, 255, 325–326; fig. 9.7

Cebus albifrons, 50, 74, 77, 79, 86, 230; table
      8.7

   hunting, 246

Cebus apella, 50, 74, 77, 79, 85–86; table
      10.2; fig. 4.10

   habitat disturbance, 208–209, 228, 388;
      tables 8.7, 8.12

   hunting, 246, 248, 251–252, 267–268, 388;
      tables 8.7, 8.12; figs. 9.2, 9.3

Cebus capucinus, 85, 87–88, 90, 356

Cebus nigrivittatus. See Cebus olivaceus

Cebus olivaceus, 45, 79, 85; fig. 4.10

Cebus xanthosternos, table 12.1

Central African Republic, 307, 342, 350, 386,
      401; tables 2.3, 8.3, 8.4, 11.3; fig. 11.3

Cercocebus, 15, 21, 25, 43

Cercocebus atys, table 8.7; figs. 8.8, 8.12

Cercocebus galeritus

   extinctions, local, 142, 161, 175, 178–179;
      figs. 6.9, 7.2

   habitat disturbance, 216, 218

   metapopulation, Tana River, 141–143, 161,
      175, 178–179, 211, 233, 308, 389; figs.
      6.8, 6.9, 7.2, 10.2

   population viability analysis, 308; fig. 10.2

Cercopithecidae, 149–150, 293–294; table
      2.1; figs. 2.1, 2.4

Cercopithecinae, 20–21, 89; tables 2.1, 5.2

   behavioral ecology, 20, 40, 46, 49, 149; fig.
      3.4

   captive breeding, 361; table 11.6

   extinction risk, 302–303; tables 10.7, 10.8,
      10.9

   habitat disturbance, 212, 230, 237; table
      8.5

   hunting, 242, 280, 285–287; tables 9.1, 9.6

Cercopithecoidea. *See* Cercopithecidae

Cercopithecus, 2, 13, 35–36, 43, 46, 99; table
      3.1

   community ecology, 72, 76, 80–81, 87–88;
      fig. 4.8

   live trapping, 264

   population biology, 124, 129, 136, 179

   speciation, 21, 25–26, 153

Cercopithecus aethiops, 13, 35–37, 41, 45,
      153; table 8.11; fig. 2.2

community ecology, 77–78, 89, 91
crop raiding, 212, 214
extinctions, local, 135, 165, 173, 181, 388,
    391; table 6.3
hunting, 262, 272, 276; fig. 9.4
population biology, 127, 134–136, 153–
    154; figs. 6.1, 6.3
translocation success, 377
Cercopithecus ascanius, 76, 85, 168; tables
    7.2, 8.11; fig. 4.8
    habitat disturbance, 212–213, 227, 231,
        239; table 8.7; figs. 8.6, 8.12
Cercopithecus campbelli, 221, 287; table 8.7;
    figs. 8.8, 8.12, 9.11
Cercopithecus cephus, 75, 80, 230, 233; table
    5.3; fig. 8.11
Cercopithecus diana, 36, 80–81, 212, 221,
    277; table 8.7; figs. 8.8, 8.12, 9.11
Cercopithecus erythrogaster, 344
Cercopithecus erythrotis, 76, 256, 259; table
    9.5; fig. 4.8
Cercopithecus lhoesti, 239
Cercopithecus mitis, 85, 107, 110, 137, 147;
    tables 5.4, 8.11, 10.2
    conservation tactics, 337–338, 367; fig.
        11.10
    crop raiding, 212–213, 243
    habitat disturbance, 217–218, 226–227,
        231–232, 239; table 8.7; figs. 8.7, 8.12
    hunting, 243, 267
    metapopulation, South Africa, 59, 99–100,
        312, 367, 394; figs. 4.2, 11.10
    metapopulation, Tana River, 141–143, 161,
        165–166, 175, 233; figs. 6.8, 6.9, 7.2
    population viability analysis, 171, 311–
        312
Cercopithecus mona, 76, 259; fig. 4.8
Cercopithecus neglectus, 259, 366
Cercopithecus nictitans, 75–76, 106, 233,
    256, 259–260; tables 5.3, 9.5, 10.2; figs.
    4.8, 8.11
Cercopithecus petaurista, 221, 287; table 8.7;
    figs. 8.8, 8.12, 9.11
Cercopithecus pogonias, 75–76, 90, 233,
    260, 337; tables 5.3, 8.7, 8.11, 9.5; fig.
    8.12
Cercopithecus preussi, 382; table 9.5
Cercopithecus solatus, 160
Cercopithecus wolfi, 90
Cheirogaleidae, 15; tables 2.1, 3.1; fig. 2.1
Cheirogaleinae, table 2.1
Cheirogaleus major, 78, 87; fig. 4.9

Cheirogaleus medius, 86
China, 1–2, 172, 181, 188; table 6.4. *See also*
    Hainan Island
    conservation tactics, 332, 355
    hunting, 181, 260, 262, 264–265; fig. 9.5
Chiropotes, 216, 228, 325–326
Chiropotes satanas, 79, 85; table 8.12; fig.
    4.10
CITES, 261–262; figs. 9.4, 9.5
Climate. *See also* Altitude; Climate change;
    Seasonality
    abundance, 107, 109, 267
    community structure, 73
    distribution, 99, 101–105; figs. 5.4, 7.6
    habitat disturbance, 226, 390
    population biology, 125–126, 129, 138–
        139, 176
    species richness, 62–67; table 4.2; figs. 4.5,
        4.6
    time budgets, 51–54; fig. 3.8
Climate change, historical, 18, 20
    extinctions, 22, 178, 182–189; fig. 7.7
    speciation, 22, 24–27, 182–183; fig. 2.7
Climate change, present
    effects on primates, 102, 206–208; figs.
        5.4, 8.4
    habitat disturbance, 194, 204
    protected areas, 338
Colobinae, 20, 26, 43, 46, 49, 89; table 2.1
    abundance, 107–108, 114; table 5.2; fig.
        5.5
    captive breeding, 361–362; table 11.6
    crop raiding, 212; table 8.5
    diet, 38–39, 41; fig. 3.4
    distribution, table 5.2
    extinction risk, 302–303; tables 10.7, 10.8,
        10.9
    habitat disturbance, 220–221, 237
    hunting, 242, 280; tables 9.1, 9.6
Colobus, 9, 26
Colobus angolensis, 214, 264–265, 285, 395;
    tables 8.7, 10.2
Colobus flandrini, 26–27
Colobus guereza, 137, 143; tables 8.7, 10.2
    crop raiding, 212
    habitat disturbance, 220, 226–227, 231,
        237, 239, 285; table 8.7; fig. 8.12
    hunting, 265–266, 285
Colobus polykomos, 265–266, 285; tables 8.7,
    10.2; fig. 9.11
Colobus satanas, 75, 233, 253, 285; tables
    5.3, 9.5; fig. 8.11

Colombia, 95, 261, 263, 333; tables 2.3, 8.3, 11.3; fig. 11.3
Colonization. *See also* Dispersal; Hetero-zygosity; Island biogeography theory; Metapopulations
  island species richness, 59–62
  regional, Americas, 19
  regional, Caribbean, 19–20, 61
  regional, Madagascar, 21–22, 61
Community-based conservation. *See* Sustain-able utilization
Community structure. *See also* Species richness
  biomass, 67–73; tables 4.3, 4.4
  extinctions, 179–180
  guilds, 69–70; fig. 4.7
  islands, 60–62
  keystone species, 73–75
Comoros Islands, 185
Competition, interspecific, 40, 75–84, 124; figs. 4.8, 4.9
  abundance, 110–111, 279–280; table 5.4
  competitive exclusion, 22, 35, 70–72, 75–76, 177
  distribution, 104–105
  habitat disturbance, 217, 236, 390
  introduced species, 162
  population viability, 306
  translocation success, 369
Competition, intraspecific, 42–45; fig. 3.6. *See also* Dispersal; Infanticide; Mating system; Philopatry; Social groups; Terri-toriality
  dominance rank, 43–45, 47, 120–121, 134, 139, 148, 151; figs. 3.6, 6.10
  female-female competition, 42–46, 50, 120–121, 125–126
  male-male competition, 29–30, 42–50, 120–121
  reproductive strategies, 42–50, 120–122
  translocation success, 370, 373
Conservation funding, 290, 343–347, 392–394, 402
  captive breeding, 364
  protected areas, 340–341, 343
  sustainable harvesting, 347, 351–352, 396–399
  tourism, 352–356; fig. 11.6
  translocation, 369, 376–377
  trust funds, 399
Conservation goals, 2, 289–291, 294–295, 313, 317, 330, 336–337

Conservation planning. *See* Conservation pri-ority setting; Conservation strategy design
Conservation priority setting, areas, 289–291, 313–321. *See also* IUCN Action Plans
  complementarity, 317; table 10.10; fig. 10.7
  currencies, 314–316; figs. 2.3, 5.2, 10.4, 10.5
  hot spots, 316–317; table 10.10; fig. 10.6
  units of conservation, 289–291, 313
Conservation priority setting, species, 289–293; table 10.1. *See also* Extinction risk; IUCN Action Plans; IUCN Conserva-tion Assessment and Management Plan; IUCN Global Captive Action Plan; IUCN Red List
  captive breeding, 359–361; table 11.6
  evolutionary uniqueness, 292–295, 314–315; tables 10.1, 10.2; fig. 10.5
  units of conservation, 289–292
Conservation problems, diagnosis of, 388–393
Conservation strategy design, 289–291, 393–402. *See also* IUCN Action Plans; IUCN Primate Specialist Group
  fragmented habitats, 394–396
  sustainable utilization, 396–399
Corridors, 209, 219
  fragmented habitats, 394–395, 402
  population viability analysis, 312, 367; fig. 11.10
  protected areas, 337–338; fig. 11.2
Costa Rica, 79, 85, 90, 165, 211, 231, 331, 355; table 6.4
Crop raiding, 211–215, 332, 384, 396; table 8.5; fig. 8.6. *See also* Pest control

Daubentonia madagascariensis, 294, 374, 377; tables 10.2, 11.10
Daubentoniidae, table 2.1; fig. 2.1
Declining-population processes, 161–163, 380. *See also* Habitat disturbance; Hunt-ing; Introduced species; Population dynamics; Secondary extinctions
  captive breeding, 359, 365
  conservation strategy, 316–317, 393–402
  conservation tactics, 344, 346
  diagnosis, 388–393
  population viability analysis, 306–307
  protected areas, 332
  reproductive value, 122
  translocation, 366–369, 374; table 11.8

Deforestation. *See* Habitat disturbance

Democratic Republic of Congo, 18, 90, 136; table 2.3. *See also* Ituri Forest; Virungas
  conservation tactics, 331, 342; table 11.3; fig. 11.3
  habitat loss, 284; tables 8.3, 8.4; fig. 8.2
  hunting, 254, 284; tables 9.3, 9.4

Demographic stochasticity, 163–164, 166, 168, 173–174, 178, 362; fig. 7.5. *See also* Small-population processes

Density dependence. *See* Population dynamics

Development. *See* Sustainable utilization

Diet, 30–31, 38–41, 43, 72; figs. 3.2, 3.3, 3.4. *See also* Niche
  abundance, 107–114, 278, 388; figs. 5.5, 5.8
  biomass, 70–73; table 4.4
  captive breeding, 362
  competition, interspecific, 75–84, 177
  crop raiding, 212, 276–277
  extinctions, global, 180–181
  extinctions, local, 142, 166, 178–179, 233–234
  flexibility, 40–42, 107, 140, 142, 166, 177, 230; fig. 3.5
  guild, 33–35, 69–73; tables 3.1, 4.4; fig. 4.7
  habitat disturbance, 175, 230, 233–235, 240; fig. 8.11
  hunting, 277, 388
  keystone resources, 74–75
  plant pollination, 86–87
  plant predation, 84–86; fig. 4.10
  plant seed dispersal, 87–91; fig. 4.10
  population dynamics, 129, 176
  small-population processes, 177–179
  translocation success, 370

Discounting, 323, 339, 352, 397, 400

Disease
  abundance, 106–107
  captive breeding, 364
  corridors, 337
  distribution, 105
  environmental stochasticity, 165
  habitat disturbance, 390
  population dynamics, 138–140, 145, 162; table 6.3
  population viability analysis, 309–310, 373
  tourism, 356, 358
  translocation success, 368, 370, 376

Dispersal, 49–50, 149, 274. *See also* Colonization; Competition, intraspecific; Metapopulations; Philopatry
  behavioral ecology, 120–121, 137
  colonization ability, 61–62, 97–105
  corridors, 337–338
  distribution, 98–100, 154; fig. 5.3
  habitat disturbance, 178, 208–210, 228, 236; fig. 8.5
  heterozygosity, 145, 147, 149–152; fig. 6.10
  hunting, 267, 270, 397
  population dynamics, 119
  population viability analysis, 307–308, 311–312
  secondary extinctions, 163
  speciation, 26–27
  translocation, 368

Distribution, 96–97; table 5.2; fig. 5.1
  abundance, 114–117
  climate, 63, 99, 102–104, 206–208, 388–389; figs. 5.4, 8.4
  declining-population processes, 206–208, 237, 285, 392; fig. 8.4
  dispersal ability, 98–100; fig. 5.3
  extinction risk classification, 295–298; tables 10.3, 10.4
  extinctions, global, 174, 179–181, 189, 285, 303, 383–384
  geographic barriers, 99–100; fig. 5.3
  interspecific interactions, 104–105
  niche specialization, 100–104; fig. 5.4
  range collapse, 115–116, 181, 392
  Rapoport's rule, 97, 102, 104–105
  rarity, 93–96; table 5.1
  translocation success, 370, 374, 400

East Africa, 41, 59, 266, 285, 335; fig. 4.3

Ecuador, 95

Education, 330–331, 353, 365, 369, 373, 399; tables 11.1, 11.4, 11.5, 11.7

Effective population size (N_e), 144, 147, 166–168, 273, 305, 308; table 7.2; fig. 7.3

Emigration. *See* Dispersal

Energetics
  abundance, 108–109
  energetic equivalence rule, 112–113; fig. 5.7
  fat reserves, 29–31, 63, 104, 124, 177, 256
  metabolic rate, 30–33, 51, 63, 104, 112–113, 177
  thermoregulation, 51, 63, 108–109, 124–126, 390

Environmental stochasticity, 130, 163–166, 168, 173–174, 177–178, 230; fig. 7.4. *See also* Carrying capacity; Disease; Fire;

Environmental stochasticity (*cont.*)
    Predation; Seasonality; Small-population
    processes
  drought, 134–135, 165
  earthquake, 165
  famine, 131–135, 162, 178
  hurricane, 131, 145, 165, 230
Eocene, 18–23
Equatorial Guinea, 333; tables 2.3, 8.3, 11.3;
    fig. 11.3. *See also* Bioko Island
  hunting, 259–260, 397; tables 9.3, 9.4
Erythrocebus patas, 78, 179
Ethiopia, 97, 102, 143; figs. 3.5, 5.4, 6.1. See
    also *Theropithecus gelada*
  community ecology, 62–63, 78, 82, 85;
    tables 4.2, 4.3
  hunting, 266, 273
Eulemur, 35, 48
Eulemur fulvus, 72, 78, 87; fig. 4.9
Eulemur macaco, 86, 236
Eulemur mongoz, 86
Eulemur rubriventer, 87
Eurasia, 16–22, 266. *See also* Mediterranean
Europe. *See* Eurasia
Evolution, 18–22; fig. 2.4. *See also* Speciation
Extinction events, 18–22, 158–160, 181–189,
    382–388; table 7.1; figs. 2.5, 7.7. *See also*
    Declining-population processes; Extinc-
    tion risk; Small-population processes
  interspecific vulnerability, global, 179–181,
    188–189, 302–305; tables 10.8, 10.9;
    figs. 7.8, 10.1
  interspecific vulnerability, local, 141–143,
    165–166, 178–179, 233–234; figs. 6.9,
    7.2
  island extinctions, 59, 61–62, 143, 181,
    189, 383
  island extinctions, Madagascar, 64–65, 72,
    159, 163, 180, 185–189, 242; table 7.3;
    fig. 7.8
Extinction processes. *See* Declining-
    population processes; Extinction pro-
    cesses, interspecific vulnerability to;
    Extinction risk; Small-population pro-
    cesses
Extinction processes, interspecific vulnerabil-
    ity to, 174–175, 393, 388–389
  agriculture, 175, 229–232, 238, 240, 285;
    table 8.11; fig. 8.12
  extinctions, global, 179–181, 188–189,
    302–305; tables 10.8, 10.9; figs. 7.8, 10.1

extinctions, local, 141–143, 165–166, 178–
    179, 233–234; figs. 6.9, 7.2
  forestry, 175, 229–232, 238–240, 285–287;
    table 8.11; figs. 8.12, 9.11
  habitat fragmentation, 229–237, 285;
    tables 8.11, 8.12; fig. 8.11
  hunting, 277–278, 280–287; figs. 9.10, 9.11
  small-population processes, 175–179; fig.
    7.6
Extinction risk, 295–299; tables 10.3, 10.4.
    See also IUCN Red List
  phylogenetic patterns, 302–305; tables
    10.8, 10.9; fig. 10.1
  spatial patterns, 299–301, 314, 316–317;
    tables 10.5, 10.6; figs. 10.4, 10.6
  temporal patterns, 301–302; table 10.7
  translocation success, 370
Extinction time lags, 61–62, 205–206, 382–
    383; table 8.4; fig. 4.4

F_{IS}. *See* Heterozygosity
F_{IT}. *See* Heterozygosity
F_{ST}. *See* Heterozygosity
Fat reserves. *See* Energetics
Figs (*Ficus*), 39, 74, 89, 216, 228
Fire, 131, 165, 230, 387
  anthropogenic, 187–188, 193, 209, 215–
    217, 307, 337, 391–392; fig. 8.7
Flagship species, 3, 369
Food availability. *See also* Carrying capacity;
    Diet; Provisioning
  competition, intraspecific, 43–46, 121,
    125–126; fig. 3.6
  density dependence, 130–131, 268; figs.
    6.5, 6.6, 6.7
  habitat disturbance, 208, 211–218, 221,
    223, 226, 228, 233–235, 237; fig. 8.11
  population dynamics, 130–135, 139–140,
    389, 391; table 6.2; figs. 6.5, 6.6, 6.7
Forestry, 193–196, 200, 203–204, 220, 344,
    387; tables 8.1, 8.3. *See also* Habitat
    modification
  conservation strategy, 320–321
  effects on primates, 63, 222–228, 230,
    390–392, 395; tables 8.8, 8.9; figs. 8.9,
    8.10
  with hunting, 284–287, 384–385; figs. 8.10,
    9.11
  interspecific vulnerability, 175, 229–232,
    238–240, 285–287; table 8.11; figs. 8.12,
    9.11

introduced species, 162
sustainable utilization, 346–347
Founder effects. *See* Heterozygosity
France, 262; fig. 9.5

Gabon, 35, 41, 83, 85, 90; tables 2.3, 4.3,
    11.9. *See also* Lopé
  abundance, 105–106, 110; table 5.3
  habitat disturbance, 284, 286–287, 307,
    342, 385; tables 8.3, 8.4; fig. 8.2
  hunting, 267, 284, 286–287, 385, 398
  protected areas, 342; table 11.3; fig. 11.3
Galagidae. *See* Galagonidae
Galago, 13, 46
Galagonidae, 9, 13, 15, 232, 294, 303; tables
    2.1, 3.1; fig. 2.1
Galago senegalensis, 101
Gambia, 217; table 11.9
Generation time (*T*), 122, 128, 273, 364
  effective population size, 167; table 7.2;
    fig. 7.3
  extinction risk, 301; table 10.3
  small-population processes, 145, 174, 178,
    273, 308, 362
Genetic drift. *See* Heterozygosity
Geographic range. *See* Distribution
Ghana, 91, 99, 160, 197, 253–255, 351; fig.
    8.2
Gibraltar (United Kingdom), 309–310, 357;
    figs. 6.1, 6.5
Gigantopithecus, 188
Gombe (Tanzania), 36, 41, 136, 138, 150–
    151, 358; fig. 3.5
Gorilla gorilla, 15; table 8.11
  abundance, 106, 110; table 5.3
  behavioral ecology, 28, 36, 40–41, 43, 48–
    49, 135, 185; table 3.1
  captive population, 164, 362, 400
  community ecology, 74–75, 82–83, 87–90
  extinction risk, 298
  habitat disturbance, 217, 233, 238, 283–
    285, 306–307
  hunting, 253, 267, 271, 280, 283–285, 323
  population biology, 135, 138–139, 179,
    358
  population management, 359; fig. 11.8
  population viability analysis, 306–307, 359;
    fig. 11.8
  protected areas, 339–340, 344, 400
  tourism, 353, 358, 399; table 11.5; fig. 11.6
Guatemala, 107

Guild. *See* Niche
Guinea, 254, 263
Guyana, 279; fig. 9.5

Habitat disturbance, 110, 162, 191; fig. 8.1.
    See also Accessibility; Agriculture;
    Declining-population processes; Fire;
    Forestry; Habitat fragmentation; Habi-
    tat loss; Habitat modification; Secondary
    forest
  conservation strategy, 320, 394–396, 402
  corridors, 337–338
  economic development, 325–327
  effects on primate populations, 390–392
  extinction risk, 295–298; tables 10.3, 10.4
  extinction risk, spatial variation, 299–301
  extinctions, global, 181, 187–188
  extinctions, local, 139, 174–175
  food availability, 208, 211–218, 221, 223,
    226, 228, 233–235, 237; fig. 8.11
  hunting, 279, 282–287, 384–385; figs. 8.10,
    9.11
  hydroelectric projects, 177, 193–194
  interspecific vulnerability, 229–232, 285–
    287; table 8.11
  introduced species, 162–163
  mining, 193–194
  predation, 218, 224, 226, 390–391
  proximate mechanisms, 192–194; table 8.1
  tourism, 353–355
  ultimate causation, 194–196, 384–386
Habitat fragmentation, 91, 200–203, 228; fig.
    8.3. *See also* Habitat disturbance; Meta-
    populations
  conservation management, 387–389,
    394–396
  corridors, 337–338
  edge effects, 203, 337–338, 387
  effects on primates, 63, 208–215, 230, 392;
    fig. 8.5
  extinctions, global, 183, 188
  extinctions, local, 139, 203
  heterozygosity, 146, 155
  interspecific vulnerability, 229–237, 285;
    tables 8.11, 8.12; fig. 8.11
  matrix, agricultural, 211–214; table 8.5; fig.
    8.6
  matrix, urban, 200, 214–215
  population viability analysis, 310–311; fig.
    10.3
  translocation, 365, 368

Habitat loss, 196–200; tables 8.1, 8.2, 8.3; fig. 8.2. *See also* Habitat disturbance
  effects on primates, 204–208; table 8.4; fig. 8.4
  extinction risk, 299–301
  extinctions, global, 187–188
  extinctions, local, 204–208
  geographic range collapse, 115–116, 181, 392
  population viability analysis, 307
  protected areas, 331–332
Habitat modification, 203–204, 230, 237–240, 283. *See also* Agriculture; Fire; Forestry; Habitat disturbance
  small-scale disturbances, 215–218; fig. 8.7
Hadropithecus, 185; table 7.3
Hainan Island (China), 164, 172, 341, 361, 380–381
Hapalemur, 35, 40; table 3.1
Hapalemur aureus, table 12.1
Hapalemur griseus, 78; table 5.1; fig. 4.9
Hapalemur simus, table 12.1
Haplorhini, 11
Heterozygosity, 144–147. *See also* Loss of heterozygosity
  fixation indices (F_{IS}, F_{IT}, F_{ST}), 146–147, 151–152; fig. 6.10
  founder effects, 144, 155–156, 174, 362; fig. 7.3
  genetic drift, 25, 144, 153–156
  group fission, 147–149
  inbreeding, 50, 144–147, 149, 153, 169–171
  mating system effects, 152–153, 179; figs. 6.10, 6.11
  migration effects, 149–151; fig. 6.10
  speciation, 153–156
  translocation, 367–368
Home range area, 45, 74, 78, 81, 114, 178; fig. 5.8. *See also* Territoriality
  corridors, 338
  extinctions, 142, 179–181, 233
  habitat fragmentation, 232–235; table 8.12
  habitat modification, 223–224, 227–228, 237–238, 240, 391; table 8.8
  small-population processes, 177–178
Hominidae, 15, 22, 158–159; tables 2.1, 7.1; fig. 7.7
Hominoidea, 10–11, 20, 28, 40, 167; figs. 2.4, 3.4
Homo erectus, 184–185, 242
Homozygosity. *See* Heterozygosity

Hunting, 162, 194, 323, 344; tables 9.1, 9.2, 9.3, 9.4. *See also* Accessibility; Bushmeat trade; Declining-population processes; Hunting, effects on primates; Hunting technology; Live trapping; Medicinal/ornamental trade; Pest control; Sustainable utilization
  body mass, hunting patterns, 242, 246, 252, 255, 259, 287, 348; table 9.5; fig. 9.2
  encounter rate (λ), 244–245, 248, 252, 254–255, 280–282, 287
  with habitat disturbance, 279, 282–287, 384–385; figs. 8.10, 9.11
  partial preferences, 245–247, 348
  prey algorithm, 245, 254, 281
  prey choice model, 244–245, 280–282, 348, 401; figs. 9.1, 9.10
  profitability (*e/h*), 244–246, 248, 252–255, 280–282, 285, 287, 348, 351; fig. 9.1
  seasonality, 251, 255–256, 259, 279
  taboos, 253–255, 280, 286, 385
  translocation success, 370, 374
Hunting, effects on primates. *See also* Hunting
  behavior, 267, 276–277
  body mass, 267, 269, 272, 277–281, 285, 350; figs. 9.7, 9.8, 9.9, 9.10, 11.5
  community structure, 277–280; figs. 9.8, 9.9
  extinction risk, 295–299; tables 10.3, 10.4
  extinctions, global, 181, 184–185, 187–188, 383–384
  extinctions, local, 174–175, 181, 206, 266–267, 269; 281–283; fig. 9.10
  with habitat disturbance, 279, 282–287, 384–385; figs. 8.10, 9.11
  interspecific vulnerability, 277–278, 280–287; figs. 9.10, 9.11
  population size, 266–272, 281–282; table 9.7; figs. 9.6, 9.7, 9.10
  population structure, 273–276, 348, 351; table 9.8; fig. 11.4
Hunting technology, 248–254, 280, 326, 382, 384–385; table 9.5; fig. 9.2. *See also* Hunting
  sustainability, 252–253, 271, 284, 348–350, 386, 397; fig. 11.5
Hybridization, 13, 208
Hylobates, 26, 45–46, 105, 167; tables 2.1, 3.1, 5.2, 5.4; fig. 2.1
  captive breeding, 361; table 11.6

crop raiding, 212; table 8.5
  extinction risk, 179, 301–303; tables 10.7,
    10.8
  extinctions, local, 188
  hunting, 260; tables 9.1, 9.6
Hylobates concolor, 164, 172, 341, 360–361
Hylobates hoolock, 264
Hylobates klossii, 276–277
Hylobates lar, 77, 108–110; tables 5.1, 5.4
  forestry, 223, 227, 240; tables 8.8, 8.9; figs.
    8.9, 8.10
  hunting, fig. 9.1
  translocation, 370
Hylobates moloch, table 12.1
Hylobates muelleri, 76, 95, 220, 230, 370;
    table 8.7
Hylobates syndactylus, 108–109; table 5.4
Hylobatidae. See Hylobates

Immigration. See Colonization; Dispersal
Inbreeding. See Heterozygosity; Loss of
    heterozygosity
India, 36, 148, 183, 366; table 6.4; fig. 6.1
  habitat disturbance, 214–215, 218–219,
    392, 394
  hunting, 261–262, 265, 323, 380
Indonesia, 18, 45, 155, 187; table 2.3. See
    also Java; Kalimantan; Mentawai
    Islands; Sulawesi; Sumatra
  conservation tactics, 342, 355; table 11.3;
    fig. 11.3
  habitat disturbance, 196, 200, 212, 386;
    tables 8.1, 8.3
  hunting, 261–264; table 9.4; fig. 9.5
Indriidae, 303; table 2.1; fig. 2.1
Indri indri, 78, 185, 238; fig. 4.9
Infanticide, 46, 48–50, 125. See also Competi-
    tion, intraspecific
  Allee effects, 171, 311–312
  secondary extinctions, 163
Interbirth interval. See Birthrate
Intrinsic rate of increase (r_m), 128–129; fig. 6.4
  hunting, 255, 268–270, 272, 280, 282
  population viability analysis, 305–306
  sustainable utilization, 352
Introduced species, 162–163, 337, 383, 398.
    See also Declining-population pro-
    cesses; Introductions (primates); Live-
    stock
Introductions (primates), 359, 374, 377; table
    11.10. See also Introduced species;
    Translocation

Island biogeography theory, 56–62; fig. 4.3
  habitat disturbance, 200, 204–206, 209–
    211; table 8.4; fig. 8.5
  protected areas, 334–337; fig. 11.2
  species-area curve, 56–59, 204–205, 335;
    table 4.1; figs. 4.1, 4.4
Islands. See also Island biogeography theory;
    Metapopulations
  continental (land-bridge) islands, 60–62,
    99, 205, 334–337; fig. 4.4
  extinction risk, 299–301, 303; table 10.6
  extinctions, 59, 61–62, 143, 181, 189, 383
  extinctions, Madagascar, 64–65, 72, 159,
    163, 180, 185–189, 242; table 7.3; fig.
    7.8
  heterozygosity, 155–156
  introduced species, 162
  montane, 208
  oceanic islands, 60–62, 99, 162, 383
  species richness, 56–62
  translocation projects, 377
Ituri Forest (Democratic Republic of
    Congo), 110, 340, 386
  habitat disturbance, 195, 221, 283–284;
    tables 8.6, 8.7
  hunting, 256, 259, 275, 283–284
IUCN Action Plans, 290–291, 330, table 11.1
  priority setting, areas, 313, 320
  priority setting, species, 292, 298, 359–
    360, 366; tables 10.1, 10.4, 10.6
IUCN Conservation Assessment and Man-
    agement Plan, 290–291, 298, 330; tables
    10.4, 10.6
IUCN Global Captive Action Plan, 291, 360–
    361, 363; table 11.6
IUCN Primate Specialist Group, 290–291,
    360
IUCN Red Data Books, 290, 298; table 10.4
IUCN Red List, 206, 290, 298–299
  biological correlates of categories, 179–181
  captive breeding, 361, 363
  categories, 295–299; tables 10.3, 12.1;
    appendix 1
  patterns of variation, 299–305; tables 10.5,
    10.7, 10.8, 10.9; figs. 10.1, 10.4
Ivory Coast, 160, 197, 201–202, 284, 386; fig.
    8.2. See also Taï Forest

Jamaica. See Caribbean
Japan, 9, 20, 146, 260–262, 355; tables 6.2,
    6.4; fig. 9.5
Java (Indonesia), 60, 242

K. See Carrying capacity

Kalimantan (Indonesia), 75, 95, 101, 230, 258; tables 4.3, 4.4

Karisoke. *See* Virungas

Kenya, 45, 83, 153–154, 324; table 4.2; figs. 3.5, 6.1. *See also* Amboseli; Tana River
  conservation tactics, 334, 366
  habitat disturbance, 214; fig. 8.2
  hunting, 243, 256, 266–267; table 9.4

Keystone species, 73–75, 230, 232, 387–388, 395
  primates as keystones, 86–91, 387–388, 396

Kibale Forest (Uganda)
  conservation, 356, 358, 399
  ecology, 35, 41, 63, 76, 84, 88, 396; tables 4.2, 4.5
  habitat disturbance, 212, 226–227, 232, 239–240, 387, 390–391; table 8.10; figs. 8.6, 8.12
  hunting, 256
  population biology, 107, 137, 167–168; table 7.2

Lagothrix, 255, 273, 348; fig. 9.7

Lagothrix flavicauda, table 12.1

Lagothrix lagotricha
  habitat disturbance, 216, 230, 283; table 8.7
  hunting, 267, 271, 273, 283, 325–326

Latin America. *See* Americas, the

Lemur catta, 124, 134–135, 162

Lemuridae, 303; table 2.1; fig. 2.1

Lemuriformes, 9, 15, 21–23; table 2.1. *See also* Madagascar
  abundance, 108–109, 111; table 5.2; fig. 5.5
  behavioral ecology, 29, 35–36, 38–40, 46, 49–50, 231; fig. 3.4
  captive breeding, 361; table 11.6
  community ecology, 62–65, 72–73, 78, 86–87; table 4.4; fig. 4.5
  crop raiding, table 8.5
  distribution, 96–97; table 5.2; fig. 5.2
  extinction risk, 301, 303; tables 10.5, 10.7, 10.8
  extinctions, 64–65, 72, 159, 163, 180, 185–189, 242; table 7.3; fig. 7.8
  habitat disturbance, 187–188, 226, 390, 392, 395; table 8.10
  hunting, 187–188, 242, 385; table 9.6
  population biology, 136; table 6.4

priority setting, species, 291, 293, 330, tables 10.1, 10.4

protected areas, 320, 344–346, 355–356, 395; table 11.4

secondary extinctions, 163

surveys, 393

sustainable utilization, 344–346, 355–356; table 11.4

translocations, 366, 373, 377; table 11.10

Leontopithecus, 236, 307–308, 366, 388–389

Leontopithecus caissara, 301; table 12.1

Leontopithecus chrysomelas, tables 6.4, 11.8

Leontopithecus chrysopygus, 395; tables 6.4, 11.8, 12.1

Leontopithecus rosalia, table 6.4
  captive breeding, 364–365, 400; table 11.7
  conservation, 115, 160, 307–308, 380, 400; table 12.1
  translocation, 365–366, 368–369, 371–373, 400; tables 11.7, 11.8

Lepilemur, table 3.1

Lepilemur mustelinus, 78; tables 6.4, 10.2; fig. 4.9

Leslie matrices. *See* Population dynamics

Liberia, 263, 371; table 11.9

Life history, 28–33. *See also* Birthrate; Mortality rate
  abundance, 113
  body mass, 28, 31–32, 396–397
  captive breeding, 362, 364
  extinctions, 179–181
  habitat fragmentation, 235
  hunting, 280–282
  life tables, 119–122, 126–129; figs. 6.3, 6.4
  population dynamics, 119, 391
  small-population processes, 175–178
  sustainable harvests, 269–270, 348; figs. 9.7, 11.4
  translocation success, 370

Life tables. *See* Population dynamics

Livestock, 139, 162, 396; table 11.5
  competition, interspecific, 82–83, 324, 369
  extinctions, global, 163, 188
  habitat disturbance, 162, 193, 196, 216–218, 324, 339, 392, 401; fig. 8.7
  hunting, 248, 398–399, 401
  secondary extinctions, 163, 188

Live trapping, 243, 260, 277, 280, 323, 365, 380; table 9.7. *See also* Hunting
  effects on population structure, 273–274; table 9.8

international trade, 243, 260–264; figs. 9.4, 9.5

sustainability, 264, 271–272, 285, 351–352

Lopé (Gabon), 72, 74–75, 204, 230; tables 4.3, 4.4

habitat disturbance, 233–234, 238–239; fig. 8.11

Lophocebus, 15, 21, 25, 43, 182

Lophocebus albigena, 75, 80, 85; tables 5.3, 5.4

habitat disturbance, 233, 235, 239; figs. 8.11, 8.12

Loridae. *See* Lorisidae

Lorisidae, 9; tables 2.1, 3.1; fig. 2.1

Lorisiformes, 9, 46; tables 2.1, 5.2

captive breeding, table 11.6

crop raiding, table 8.5

extinction risk, 301, 303; tables 10.7, 10.8, 10.9

hunting, 242; tables 9.1, 9.6

Loris tardigradus, 264

Loss of heterozygosity, 163, 166–171. *See also* Heterozygosity; Small-population processes

captive breeding, 362, 364

generation time, 178, 362; fig. 7.3

genetic drift, 144, 166, 168, 178, 236, 362

habitat disturbance, 236–237

hunting, 273, 275–276

inbreeding, 169–171, 174, 178, 236–237, 310, 362

mating system, 179, 275

minimum viable population (MVP), 166–169, 305, 308–309, 359, 361

mutational meltdown, 168–169

population viability analysis, 308–310

Los Tuxtlas (Mexico), 83, 90; table 4.3

habitat fragmentation, 201–202, 209–210, 338; fig. 8.5

Macaca, 13, 20, 43, 265, 355; table 3.1

habitat disturbance, 220, 237; table 8.7

heterozygosity, 150, 152–153, 154; fig. 6.10

weed macaques, 101, 215, 357–358

Macaca cyclopis, 264–265

Macaca fascicularis, 45, 48, 76, 88; tables 5.4, 8.11

competition, interspecific, 76–77

environmental stochasticity, 230

genetics, 150–151, 155; table 6.4

habitat disturbance, 215, 240; table 8.9; fig. 8.10

hunting, 262; figs. 9.1, 9.4

translocations, 377

Macaca fuscata, 16, 39, 99, 149, 357

population dynamics, 123, 146–148, 155, 171; table 6.4; fig. 6.2

Macaca mulatta, 85, 147–149

extinctions, 172, 181

genetics, 148, 155; table 6.4

habitat disturbance, 2, 214–215, 380–381

hunting, 2, 262, 264, 267, 273–274, 323, 380–381; table 9.8

population dynamics, 120, 123, 127–128, 148, 167; table 6.1; fig. 6.1

translocation project, 376–377; table 11.10

Macaca nemestrina, 76–77, 155, 223, 262, 356; figs. 9.1, 9.4

Macaca nigra, 162, 232, 284, 293, 390

tourism, 354, 356–357; fig. 11.7

Macaca pagensis, table 12.1

Macaca radiata, 45, 214–215; fig. 6.1

Macaca silenus, 201, 212, 219, 337, 366, 392, 394

Macaca sinica, 134, 155, 173, 215; table 6.4

Macaca sylvanus, 16, 183

captive breeding, 364

habitat disturbance, 201, 217, 394

population biology, 124, 357, 154–155; table 6.4; figs. 6.1, 6.5

population viability analysis, 309–310

translocation, 366, 369

tourism, 357

Macaca thibetana, 357

Madagascar, 16–18; tables 2.2, 2.3. *See also* Lemuriformes

habitat disturbance, 187–188, 194, 196–197; tables 8.2, 8.3; fig. 8.1

hunting, 247–248, 256, tables 9.1, 9.2

priority setting, areas, 313, 315, 317, 320, 330; figs. 10.4, 10.5, 10.6, 10.7

protected areas, 332–334, 342; tables 11.2, 11.3; figs. 11.1, 11.3

small-population processes, 165

Malawi, 214

Malaysia, 61–62, 70, 76–77, 110, 155; tables 4.3, 4.4; figs. 4.4, 9.1. *See also* Sabah; Sarawak; Tekam Forest

Semaq Beri hunters, 245, 248, 255; tables 9.3, 9.4; fig. 9.1

Mandrillus, 15, 21, 182

Mandrillus leucophaeus, 79, 382
  captive breeding, 293, 362, 364–365, 400
  hunting, 251, 260, 275, 382; table 9.5
Mandrillus sphinx, 41, 75, 79, 255, 232–233;
  table 5.3
Manu (Peru), 67, 72, 74, 80, 83–84, 86, 350;
  tables 4.3, 4.4; fig. 11.5
Mating system, 46–49; figs. 3.7, 4.7. *See also*
  Competition, intraspecific; Dispersal;
  Infanticide; Philopatry
  Allee effects, 142, 171–172; fig. 7.4
  genetics, 144–145, 147, 152–153, 167, 179;
    table 7.2; figs. 6.10, 6.11
  hunting, 274–276
  translocation success, 371
Mauritius, 155, 262, 377; fig. 9.5
Medicinal/ornamental trade, 242–243, 247,
  260, 264–266, 351–352; table 9.6. *See
  also* Hunting
  effects on primates, 277, 280, 382
Mediterranean, 183, 201
Megadiversity countries, 18; table 2.3
Megaladapidae, table 2.1
Megaladapis, 185; table 7.3
Mentawai Islands, 26, 276–277
  Siberut hunters, 248; tables 9.3, 9.4
Mesopropithecus, table 7.3
Metabolic rate. *See* Energetics
Metapopulations, 140–143, 205; figs. 6.8, 6.9
  captive breeding, 359
  conservation strategy, 316, 394–395
  population viability analysis, 312
  rescue effect, 59, 143, 189, 203, 205, 209,
    228, 394
  source-sink relationships, 100–101, 115,
    142–143, 316, 391–392, 395–396
  translocation, 365, 367–368, 377; fig. 11.10
Mexico, 219–221, 385; tables 8.6, 8.7. *See
  also* Los Tuxtlas
Microcebus, 32, 78, 167, 226
Microcebus murinus, 86, 136–137
Microcebus myoxinus, 28
Microcebus rufus, 87, 390
Migrant human populations, 193, 324–325,
  332, 385–386
Migration. *See* Dispersal
Minimum viable population (MVP). *See* Loss
  of heterozygosity
Miocene, 20, 27, 182
Miopithecus talapoin, 40, 45
Morocco, 155, 217, 309, 366

Mortality rate, 28, 53, 119–122, 134; table
  6.1. *See also* Life history; Population
  dynamics
  demographic stochasticity, 163–164
  disease, 138–139; table 6.3
  environmental stochasticity, 165
  food availability, 131–135; table 6.2
  habitat disturbance, 224, 236–237
  hunting, 263–264
  inbreeding, 169–171, 236–237
  population viability analysis, 305–312
  predation, 135–137; table 6.3
  rehabilitation/release project, 371
  translocation project, 374, 376; table 11.10
Mutational meltdown. *See* Loss of heterozygosity

N_e. *See* Effective population size
Namibia, 101, 134–135, 139; table 4.2; fig. 6.1
Nasalis, 26
Nasalis larvatus, 293, 356
National parks. *See* Protected areas
Neotropical primates. *See* Platyrrhini
Neotropics. *See* Americas, the
Nepal, 355; table 6.1
Netherlands, 261; fig. 9.5
Net reproductive rate (R_o), 126–127; fig. 6.3
New World. *See* Americas, the
Niche, 33–35, 69–70; table 3.1; fig. 4.7. *See
  also* Competition, interspecific; Diet
  abundance, 111–114, 116
  crop raiding, 212
  distribution, 100–104; fig. 5.4
  extinctions, 180–181
  habitat disturbance, 223–224, 230–240,
    285; table 8.8; fig. 8.11
  hunting, 272, 277, 285
  introduced species, 162
  secondary forest, 231–232; table 8.11
  small-population processes, 175, 178–179
  specialization, 93–95, 111–114, 231, 295;
    table 5.1
  translocation success, 374–376; table 11.10
Nigeria, 3, 99; table 2.3. *See also* Okomu
  Forest
  captive breeding, 364–365, 400
  habitat disturbance, 197, 324; tables 8.3,
    8.4; fig. 8.2
  hunting, 254, 258–260, 275; tables 9.3, 9.4;
    fig. 10.8
  protected areas, table 11.3; fig. 11.3

North Africa, 19, 183, 394
Nycticebus coucang, 264–265, 395

Okomu Forest (Nigeria), 325, 344, 346, 386
Old World. *See* Africa; Eurasia
Oligocene, 18–19
Okapi Wildlife Reserve. *See* Ituri Forest
Otolemur crassicaudatus, 86, 164; table 5.1

Pakistan, 9
Palaeopropithecus, 185; table 7.3
Palawan Islands (Philippines), 61–62; fig. 4.4
Pan, 15, 48–49, 51, 54
Panama, 165, 263, 366. *See also* Barro Colo-
    rado Island
Pan troglodytes, 36–37, 40–41, 50, 81, 136–
    137, 275
  abundance, 106, 110; tables 5.1, 5.3, 5.4
  captive breeding, fig. 11.9
  community ecology, 74–75, 85, 87–88, 90;
    tables 4.5, 5.4
  crop raiding, 212–213; fig. 8.6
  distribution, 392; table 5.1
  environmental stochasticity, 230
  habitat disturbance, 385, 391–392
  habitat fragmentation, 232–233, 235, 385,
    391; fig. 8.11
  habitat modification, 227, 238–239, 283–
    284; fig. 8.12
  heterozygosity, 150
  hunting, 253–254, 263–264, 267, 283–284,
    385, 392
  population biology, 138–139, 358; table
    6.3
  population declines, 138; table 6.3
  rehabilitation/release projects, 371–373;
    table 11.9
  tourism, 356, 358
  translocation, 371–373; table 11.9
Papio, 13, 15, 21, 129, 182–184
  Allee effects, 171
  behavioral ecology, 29, 36–37, 41–43; table
    3.1; fig. 3.5
  community ecology, 77–78, 80–81, 85
  crop raiding, 212–214, 332
  distribution, 102; fig. 5.4
  extinctions, global, 184
  group living, 50, 52–54; fig. 3.9
  hunting, 251
  systems model, 51–54, 102, 184; figs. 3.8,
    3.9, 5.4

Papio anubis, 45, 83, 87–88, 91, 147; fig. 3.5
  birthrate, 124–125, 164; fig. 6.1
  crop raiding, 212–213; fig. 8.6
  habitat disturbance, 83, 208
  hunting, 262; fig. 9.4
  small-population processes, 139, 170, 208
  translocation success, 374–377; table 11.10
Papio cynocephalus, 45, 99, 123; fig. 3.5
  crop raiding, 214, 243
  genetics, 150, 167, 170, 208
  habitat change, 134–135, 208, 388
  hunting, 243
Papio hamadryas, 9, 47–48, 77, 101, 123,
    164; figs. 3.5, 6.1
  Allee effects, 171–172
  crop raiding, 213–214
Papionidae, 22, 158–159, 182–185, 188–189;
    fig. 7.7
Papio papio, fig. 3.5
Papio ursinus, 99, 107, 148; figs. 3.5, 6.1
  environmental stochasticity, 134–135, 138–
    139, 358
Paraguay, 256, 267–268; fig. 9.3
  Ache hunters, 245, 251–252, 325; fig. 9.3
Paranthropus aethiopicus, 159
Parapapio, 182–183
People's Republic of Congo, 307; tables 2.3,
    8.3, 8.4, 11.3; figs. 8.2, 11.3
Peru, 18, 35, 95; tables 2.3. *See also* Manu
  habitat disturbance, 285; table 8.3
  hunting, 255, 261, 271–272, 276, 285, 352,
    397–398; tables 9.4, 9.7
  Machiguenga hunters, 252
  Piro hunters, 245–246, 252, 348–351; figs.
    9.2, 11.4, 11.5
  protected areas, table 11.3; fig. 11.3
Pest control (crop raiders), 242–243, 251,
    256, 388, 396; table 8.5. *See also* Crop
    raiding; Hunting
  effects on primates, 214, 243, 270, 277, 382
Phaerinae, table 2.1
Phaner furcifer, 86; table 5.1
Philippines, 61–62, 95, 155, 262, 264; figs.
    4.4, 9.5. *See also* Palawan Islands
Philopatry, 49–50, 121, 145, 149–151, 376.
    See also Competition, intraspecific; Dis-
    persal
Phylogeny, 4–5, 11–15, 73, 180–181, 237; fig.
    2.1
  conservation strategy, 293–295, 314–315;
    table 10.2; fig. 10.5

Pithecia albicans, 230; table 8.7
Pithecia pithecia, 79, 85, 235, 376; tables
  8.12, 11.10; fig. 4.10
Pitheciinae, 19, 39–40; table 2.1
Plant predation, 39–40, 84–86, 109; fig. 4.10.
  See also Diet
  abundance, 114, 116–117
  community ecology, 86, 90–91
  habitat disturbance, 209
  population variability, 176
Platyrrhini, 9, 35, 49; table 2.1. *See also*
  Americas, the
  abundance, 114, 116–117
  community ecology, 86, 90–91
  habitat disturbance, 209
  population variability, 176
Pleistocene, 18, 24–26, 59, 102, 182–183,
  188, 386; figs. 2.7, 4.3
Plesiadapids, 23, fig. 2.6
Pliocene, 182
Plio-Pleistocene, 9, 20, 183
Pollination, 86–87, 204, 387, 396
Polyspecific associations, 37, 80–82, 226, 272,
  391
Pongidae, 15, 20, 32, 149; table 2.1; fig. 2.1
  abundance, 111; table 5.2
  captive breeding, 361; table 11.6
  crop raiding, 212, table 8.5
  distribution, table 5.2
  extinction risk, 301–302; tables 10.7, 10.8
  hunting, 242, 285–286; tables 9.1, 9.6
  rehabilitation/release projects, 371–373;
    table 11.9
Pongo pygmaeus, 15, 31, 36, 46, 48, 76, 87
  captive breeding, 362
  corruption, 401
  distribution, 188, 242, 392
  environmental stochasticity, 230
  extinctions, local, 188, 242
  habitat disturbance, 238, 392
  hunting, 242, 245, 264, 392
  priority setting, species, table 10.2
  protected areas, 332
  rehabilitation, 353, 371–373
  tourism, 353, 356
  translocation, 366, 368, 373, 374–376;
    table 11.10
Ponta da Castanha (Brazil), 219–221, 228,
  231, 337; tables 8.6, 8.7, 8.10; figs. 9.8,
  9.9
Population density. *See* Abundance
Population dynamics, 115, 119. *See also*

Birthrate; Declining-population processes; Mortality rate; Provisioning;
  Small-population processes
  density dependence, 130–131, 171, 268,
    311–312; figs. 6.5, 6.6, 6.7
  disease, 134, 138–140, 389; table 6.3
  food availability, 130–135, 139–140, 389,
    391; table 6.2; figs. 6.5, 6.6, 6.7
  intrinsic rate of increase, 128–129; fig. 6.4
  Leslie matrices, 127, 171, 305–306; appen-
    dix 2
  life tables, 119–122, 126–129; figs. 6.3, 6.4
  logistic equation, 130–131, 269, 281, 305–
    306; fig. 6.5
  net reproductive rate, 126–127; fig. 6.3
  population pyramids, 126
  population variability, 175–176
  population viability analysis, 305–312
  predation risk, 134–137, 139–140, 389,
    391; table 6.3
Population genetics. *See* Heterozygosity;
  Loss of heterozygosity
Population size. *See* Abundance; Distribution; Population dynamics
Population viability analysis (PVA), 298, 305–
  312, 369; figs. 10.2, 10.3
  population and habitat viability analysis
    (PHVA), 307–308
  population management, 359, 367; figs.
    11.8, 11.10
Predation, 33–37, 45–46, 50–54, 73, 80–82,
  226. *See also* Hunting
  abundance, 106–107
  Allee effects, 171, 173, 183–184, 189
  captive breeding, 364
  distribution, 105
  habitat disturbance, 218, 224, 226,
    390–391
  introduced species, 162
  population dynamics, 135–137, 139–140,
    162, 165; table 6.3
  population viability analysis, 306
  translocation, 371
Presbytis, 26, 45, 149, 153
  habitat disturbance, 220; table 8.7
*Presbytis entellus. See Semnopithecus
    entellus*
Presbytis hosei, 110, 114
Presbytis melalophos, 70, 76–77, 108, 110
  forestry, 223, 226, 239; tables 8.8, 8.9; figs.
    8.9, 8.10
  hunting, table 9.1

Presbytis obscura. See *Trachypithecus obscurus*
Presbytis potenziana, 360
Presbytis rubicunda, 110, 212, 227
Procolobus badius, 50, 80–81, 84, 100; tables 5.4, 7.2
   effective population size, 168; table 7.2
   habitat disturbance, 214, 216–218, 226, 237, 239, 287; table 8.7; figs. 8.12, 9.11
   hunting, 287; table 9.5; fig. 9.11
   metapopulation, Tana River, 141–143, 161, 166, 178–179, 211, 233–234, 389, 394; figs. 6.8, 6.9, 7.2
   predation, 36–37, 136–137, 275
   tourism, 355, 357
   translocation, 374–376; table 11.10
Procolobus verus, 37, 80; fig. 9.11
Propithecus tattersalli, 385; table 12.1
Protected areas, 205, 313, 327, 331–332, 399–400; 402; table 11.1
   accessibility, 342–343; fig. 9.3
   climate change, 338
   corridors, 337–338, 395; fig. 11.2
   distribution, 333–334; tables 11.2, 11.3
   protection, active, 339–341, 353; table 11.5
   protection, passive, 320–321, 341–343; fig. 11.3
   spatial design, 307, 316, 320–321, 334–338, 341–343; figs. 11.1, 11.2, 11.3
   translocation success, 370–371
Provisioning, 215, 309, 351, 357, 382
   population dynamics, 123–124, 131–134, 148; table 6.2; fig. 6.2
   translocation, 370–371, 374–376; tables 11.9, 11.10
Property systems, 321–323, 392, 400
Prosimians, 9, 18–23, 28, 34, 46, 48; figs. 2.1, 2.6
   habitat disturbance, 212, 226
   hunting, 254, 264, 285
Pygathrix, 26

r_m. See Intrinsic rate of increase
R_o. See Net reproductive rate
Rapoport's rule. See Distribution
Rarity, 93–95, 116–117; table 5.1. *See also* Abundance; Distribution
   abundance, 105–106
   distribution, 96–97, 315–317; figs. 5.2, 10.6
   priority setting, species, 295; table 10.4
Rehabilitation, 353, 364–365, 371–373; table 11.9

Reintroduction, 359, 365–369, 371, 373–377, 400; tables 11.4, 11.7, 11.10. *See also* Translocation
   conservation strategy, 293, 308; table 11.1
Reproductive strategies. *See* Competition, intraspecific
Reproductive suppression, 45, 125, 144, 274
Rescue effect. *See* Metapopulations
Restocking, 359–360, 365–369, 373–377, 402; table 11.10; fig. 11.10. *See also* Translocation
Rhinopithecus, 1–2, 26, 40, 48
Rhinopithecus avunculus, 2; table 12.1
Rhinopithecus bieti, 2, 16, 360
Rhinopithecus brelichi, 16
Rhinopithecus roxellana, 16
Rivers. *See also* Accessibility
   abundance, 109, 267; fig. 9.6
   distribution, 99–100, 105; fig. 5.3
   gene flow, 151, 154
   population viability analysis, 308
Roads. *See* Accessibility
Rwanda, 307, 331, 344, 398. *See also* Virungas

Sabah (Malaysia), 95, 110, 227; table 8.10
Saguinus, table 5.4
   hunting, 276, 279–280, 352
Saguinus fuscicollis, 30, 35, 74, 77, 80–82, 86–87
   agriculture, table 8.7
   genetics, 154; table 6.4
   live trapping, 271–272; table 9.7
Saguinus imperator, 74, 77, 80–82, 86
Saguinus midas, 79, 85; table 6.4; fig. 4.10
   habitat fragmentation, 208–211; table 8.12
   hunting, 256
Saguinus mystax, 35–36, 77, 80, 87
   habitat modification, 230; table 8.7
   live trapping, 271–272, 276; table 9.7
Saguinus nigricollis, fig. 9.2
Saimiri, 230; table 8.7
Saimiri oerstedi, 29–30, 37, 124
   conservation management, 395
   environmental stochasticity, 165
   habitat disturbance, 211, 231, 355
   tourism, 355–356
Saimiri sciureus, 74, 77, 79, 85–86; fig. 4.10
   abundance, 110; table 5.4
   hunting, 262; figs. 9.2, 9.4
   polyspecific associations, 80–82

Sarawak (Malaysia), 95, 220–221, 284, 370; tables 8.6, 8.7
Saudi Arabia, 9, 101, 213–214
Seasonality, 41, 43, 46–47, 62–67, 75, 81–82. *See also* Climate
  abundance, 107, 110, 114, 267
  bottlenecks, 67, 72–73, 107, 230
  distribution, 102–104, 388–389
  habitat disturbance, 216, 221, 226, 232–233, 387; fig. 8.8
  hunting, 251, 255–256, 259, 279
  keystone resources, 74–75, 107, 232, 374
  population dynamics, 124–126, 129, 138–139; fig. 6.4
  population variability, 176–177, 389; fig. 7.6
  translocation success, 374
Secondary extinctions, 162–163, 188, 384, 387–388. *See also* Declining-population processes
Secondary forest, 203–204, 209, 220, 382
  corridors, 395, 402
  with hunting, 283–285, 392
  niche specialization, 210, 215, 231–232, 235, 388; table 8.11
Seed dispersal, 39–40, 87–91, 204, 387, 396; fig. 4.10
Seed predation. *See* Plant predation
Selective logging. *See* Forestry
Semnopithecus entellus, 2, 36, 49, 80, 153
  environmental stochasticity, 165
  habitat disturbance, 214–215, 218
  hunting, 265
  population biology, 124–125, 215; fig. 6.1
Senegal, 217; tables 4.2, 11.9; fig. 3.5
Sex ratios, 102–121, 164
  Allee effects, 171–172; fig. 7.4
  population viability analysis, 308, 310–311; fig. 10.3
Shifting cultivation. *See* Agriculture
Siberut. *See* Mentawai Islands
Sierra Leone, 263, 286–287; fig. 9.11. *See also* Tiwai
Simias, 26
Simias concolor, 248, 275–277, 360; table 5.1
Sink populations. *See* Metapopulations: source-sink relationships
Small-population processes, 161–162, 163–174, 188. *See also* Allee effects; Demographic stochasticity; Environmental stochasticity; Loss of heterozygosity; Population dynamics

captive populations, 169–170, 362–364; fig. 11.9
conservation strategy, 359, 394–397; fig. 11.8
extinction risk, 295–298; tables 10.3, 10.4
habitat disturbance, 204–205, 208
hunting, 266–267, 281, 397
interspecific vulnerability, 175–181
population viability analysis, 305–312; figs. 10.2, 10.3
rescue effect, 59, 143, 189, 203, 205, 209, 228, 394
translocation, 366–367, 376; tables 11.8, 11.10
Social groups, 37, 45–54, 125; figs. 3.7, 3.8, 3.9. *See also* Competition, intraspecific; Polyspecific associations
  Allee effects, 140, 171–173, 181, 183–184, 189
  climate change, 183–184, 189, figs. 5.4, 8.4
  dispersal, 99–100
  extinctions, global, 181, 189
  habitat fragmentation, 232–233, 235
  habitat modification, 217–218, 223, 226, 230, 238, 240, 390–391
  heterozygosity, 147–153, 179; figs. 6.10, 6.11
  hunting, 273–276, 287
  translocation, 376
Social relationships, 43, 49, 50–54, 147–151, 364
Social system. *See* Mating system
Soil chemistry, 67, 73, 109, 192, 230
Source populations. *See* Metapopulations: source-sink relationships
South Africa, 62–63, 107, 148; table 4.2; fig. 3.5
  metapopulation, 100, 110, 232, 311–312, 337–338, 367, 394; figs. 4.2, 11.10
South America. *See* Americas, the
Southeast Asia, 16, 26, 60–62, 73–74, 96, 155; fig. 4.4
  extinction processes, 180
  habitat disturbance, 193, 195, 205, 237, 390
Speciation, 21–27, 62, 105, 153–156, 182–183; figs. 2.2, 2.5, 2.6, 2.7
  conservation strategy, 294–295
Species definition
  primates, 8–11
  taxonomic unit, 11–15
Species endemism, 18; table 2.3

Species life span, 159, 303; fig. 7.1
Species richness, 16–18; tables 2.1, 2.2, 2.3;
    fig. 2.3. *See also* Island biogeography
    theory
    area effects, 56–59; table 4.1; fig. 4.1
    community structure, 69–73; tables 4.3,
        4.4; fig. 4.7
    environmental correlates, 62–67; table 4.2;
        figs. 4.5, 4.6
    on islands, 56–62; fig. 4.4
    isolation effects, 57–62; figs. 4.3, 4.4
    latitude, 62–67, 104–105; table 4.2; fig. 4.6
    priority setting, areas, 314–316; figs. 5.2,
        10.4, 10.5, 10.6
Sri Lanka, 134, 165; table 6.4
Strepsirhini, 11, 33
Sulawesi (Indonesia), 60, 232, 284, 354, 356–
    357; fig. 11.7
Sumatra (Indonesia), 60, 150–151, 155;
    tables 4.3, 4.4
    conservation, 332, 356, 368
    habitat disturbance, 230, 392
Sunda Shelf. *See* Southeast Asia
Suriname, 79, 88, 91; table 9.3; fig. 4.10
Surveys, 313, 330, 393, 399; table 11.1
Sustainable development, 325–327, 343–347;
    table 11.4; fig. 10.8
Sustainable utilization, 266–272, 281–282,
    387, 396–402; table 9.7. *See also* Hunt-
    ing; Migrant human populations;
    Tourism
    bushmeat trade, 257, 260, 270, 284, 351–
        352, 396
    conservation strategy, 396–402
    developing societies, 325–327, 343–347;
        table 11.4; fig. 10.8
    live trapping, 264, 271–272, 285, 351–352
    pest control, 243, 270
    subsistence hunting, 257, 347–351, 385;
        figs. 11.4, 11.5
    traditional societies, 323–325, 347–351;
        figs. 11.4, 11.5

Taï Forest (Ivory Coast), 26, 41, 81, 136,
    150
Tana River (Kenya), 100, 134, 394; fig. 6.1
    conservation, 338, 366, 389
    habitat disturbance, 193, 216, 218
    metapopulation, 100, 140–143, 165–166,
        175, 178–179, 211, 232–235; figs. 6.8,
        6.9, 7.2
    population viability analysis, 308; fig. 10.2

Tanzania, 36, 41, 99, 167; tables 4.2, 11.9;
    figs. 3.5. *See also* Gombe; Zanzibar
    declining-population processes, 217, 251,
        256; fig. 8.2
Tarsiidae. See *Tarsius*
Tarsiiformes. See *Tarsius*
Tarsius, 9, 11, 41, 96; tables 2.1, 3.1, 5.2, 8.5;
    fig. 2.1
    captive breeding, table 11.6
    extinction risk, tables 10.7, 10.8
    hunting, 242; tables 9.1, 9.6
Tarsius spectrum, 354, 357
Tarsius syrichta, 95, 362
Taxonomy, 11–15, 292–295; appendix 1
Tekam Forest (Malaysia), 193, 223–228, 231,
    284, 395; tables 8.8, 8.9, 8.10; figs. 8.9,
    8.10
Territoriality, 45, 50, 82, 149, 154, 179. *See
    also* Competition, intraspecific; Home
    range area
    environmental stochasticity, 165, 179
    habitat disturbance, 223, 236, 238–239
    hunting, 276, 280
    population viability, 312
    translocation success, 370–371, 374
Thailand, 261, 353, 370
Thermoregulation. *See* Energetics
Theropithecinae, 22, 105
    extinctions, 178, 182–185, 188–189
Theropithecus darti, 178
Theropithecus gelada, 21, 65, 178, 182–183
    abundance, 107
    Allee effects, 171–172, 189
    behavioral ecology, 40, 45–46, 48
    community ecology, 67–69, 77–78, 80–82,
        85
    distribution, 95, 102; fig. 5.4
    genetics, 13, 15, 147
    habitat disturbance, 206–208, 214; fig. 8.4
    hunting, 273, 275
    population biology, 124, 126, 129, 134,
        139, 143; fig. 6.1
Theropithecus oswaldi, 159, 178, 242
Tiwai (Sierra Leone), 204, 221–222, 238;
    tables 4.4, 8.6, 8.7; figs. 8.8, 8.12
Tourism, 343, 346, 352–356, 399–400; tables
    11.4, 11.5; fig. 11.6
    effects on primates, 356–358; fig. 11.7
    rehabilitation projects, 373
Trachypithecus, 45, 149, 153
Trachypithecus auratus, 150, 292
Trachypithecus cristatus, 292

Trachypithecus delacouri, table 12.1
Trachypithecus johnii, 264
Trachypithecus obscurus, 70, 76–77, 108,
    110, 114
    forestry, 226, 240; table 8.9; fig. 8.10
    hunting, 248; fig. 9.1
Trachypithecus phayrei, 264–265
Trachypithecus vetulus, 165; table 5.1
Translocation, 359, 399. *See also* Introduc-
    tions; Reintroduction; Restocking
    from captivity, 370–373; table 11.9
    criteria, 366–370; table 11.8; fig. 11.10
    declining-population processes, 366–369;
        table 11.8
    habitat fragmentation, 394
    hard/soft release, 370–371, 374–376;
        tables 11.9, 11.10
    population viability analysis, 312, 367; fig.
        11.10
    small-population processes, 366–367; table
        11.8
    from the wild, 373–377; table 11.10

Uganda, 78, 91, 135; tables 2.3, 4.2. *See also*
    Budongo Forest; Kibale Forest; Vir-
    ungas
    habitat disturbance, 197, 228, 284; tables
        8.3, 8.4; fig. 8.2
    protected areas, 340; table 11.3; fig. 11.3
    tourism, 353, 399
United Kingdom, 261–262; fig. 9.5. *See also*
    Gibraltar

United States of America, 260–264; fig. 9.5
Units of conservation. *See* Conservation prior-
    ity setting
Urbanization, 200, 211, 214–215

Varecia variegata, 48, 366; table 10.2
Venezuela, 138, 148–149, 154, 313; tables
    4.3, 6.4; figs. 6.7, 11.4
    Yanomamö hunters, 246, 251–252, 349–
        351; tables 9.3, 9.4
    Ye'kwana hunters, 246, 251–252, 349–351;
        tables 9.3, 9.4
Vietnam, 1–2, 260, 264
Virungas (Democratic Republic of Congo,
    Rwanda, Uganda), 82–83, 217, 307;
    table 4.3
    protected areas, 339–340, 353; table 11.5
    tourism, 353; table 11.5; fig. 11.6

West Africa, 41, 76, 79, 251, 255, 258
West Indies. *See* Caribbean

Xenothrix mcgregori, 383

Zambia, 339, 346
Zanzibar (Tanzania), 214, 216–217, 355, 376;
    table 11.10
Zimbabwe, 346
z-value. *See* Island biogeography theory:
    species-area curve